The Kowalevski Property

Sophie Kowalevski
1850–1891

Volume 32

CRM
PROCEEDINGS &
LECTURE NOTES

Centre de Recherches Mathématiques
Université de Montréal

The Kowalevski Property

Vadim B. Kuznetsov
Editor

The Centre de Recherches Mathématiques (CRM) of the Université de Montréal was created in 1968 to promote research in pure and applied mathematics and related disciplines. Among its activities are special theme years, summer schools, workshops, postdoctoral programs, and publishing. The CRM is supported by the Université de Montréal, the Province of Québec (FCAR), and the Natural Sciences and Engineering Research Council of Canada. It is affiliated with the Institut des Sciences Mathématiques (ISM) of Montréal, whose constituent members are Concordia University, McGill University, the Université de Montréal, the Université du Québec à Montréal, and the Ecole Polytechnique. The CRM may be reached on the Web at www.crm.umontreal.ca.

American Mathematical Society
Providence, Rhode Island USA

The production of this volume was supported in part by the Fonds pour la Formation de Chercheurs et l'Aide à la Recherche (Fonds FCAR) and the Natural Sciences and Engineering Research Council of Canada (NSERC).

2000 *Mathematics Subject Classification.* Primary 14H70, 17B80, 37J35, 35Q58, 37K10, 70E40, 70H06, 81R12, 81U15, 37K15.

Library of Congress Cataloging-in-Publication Data

The Kowalevski property / Vadim B. Kuznetsov, editor.
p. cm. — (CRM proceedings & lecture notes ; v. 32)
In English, with a reprint of Kowalevski's paper in the original French.
Includes bibliographical references.
ISBN 0-8218-2885-1 (alk. paper)
1. Differentiable dynamical systems. 2. Geometry, Algebraic. 3. Differential equations, Partial. 4. Kovalevskaia, S. V. (Sof'ia Vasil'evna), 1850–1891. I. Kuznetsov, Vadim B., 1963– II. Series.

QA614.8 .K695 2002
514′.74—dc21 2002066671

♾ The paper used in this book is acid-free and falls within the guidelines
established to ensure permanence and durability.
This volume was submitted to the American Mathematical Society
in camera ready form by the Centre de Recherches Mathématiques.
Visit the AMS home page at http://www.ams.org/

10 9 8 7 6 5 4 3 2 1 07 06 05 04 03 02

Contents

Preface

A famous mathematician with a Russian heart, a nihilist and romantic, a passionate litterateur, a breaker of many rules and a dedicated fighter for her ideals, Sophie Kowalevski is a legendary figure in the history of science.

Born in 1850, she grew up on the family estate in central Russia where she was educated by private teachers. After the marriage she travelled widely in Europe visiting many centres of scientific excellence. In 1874 she earned her doctoral degree in mathematics in spite of the fact that in almost all Universities of the time, women were even forbidden to attend lectures. Sophie Kowalevski led a turbulent life switching between mathematical research, writing novels, business ventures and helping her friends to gain an education.

In 1889 Sophie Kowalevski wrote a fundamental paper published in *Acta Mathematica* [Acta Math. **12** (1889), 177–232] for which she was awarded a prestigious Bordin Prize. This ground breaking paper opened up whole new area of mathematical research. In the paper a new technique, now known as the *Kowalevski property*, was introduced and ingeniously used to solve the first integrable system of enormous complexity, the famous *Kowalevski top*. This result was a beautiful demonstration of the triumph of the classical theory of complex analysis developed by Cauchy, Abel, Jacobi, Riemann and Weierstrass in the second half of the 19th century. This is the reason for choosing the title of the present book. Because of the importance of Sophie Kowalevski's 1889 paper and as a tribute to her achievements and because of a large body of mathematical research generated by it, we decided to reproduce the original paper at the end of this volume[1].

Roger L. Cooke kindly agreed to write two articles for this volume about Sophie Kowalevski's contributions to mathematics, *Kovalevskaya's mathematical work*, and her interesting life story, *The life of S. V. Kovalevskaya* [2].

In 2000 I organized a Workshop on *Mathematical Methods of Regular Dynamics* dedicated to the 150th anniversary of Sophie Kowalevski's birth. It was supported by the London Mathematical Society and was held in the University of Leeds during April 12–15, 2000. Workshop's web-page is still available at the address: http://maths.leeds.ac.uk/~vadim/work.html. The meeting was well-attended with more than 70 participants from many countries. A book of Workshop's Proceedings featuring 39 research papers was published as a separate volume of the *Journal of*

[1]We thank the Director of Institut Mittag-Leffler for granting their permission to include a reprint of the Kowalevski paper into this volume.

[2]In the present volume we tried to adopt everywhere, if it was possible, the name Sophie Kowalevski, as the one she was known through her work related to mathematics. By this name she signed her Acta Mathematica papers and was known as an Editorial Board member of this journal. We want to notice that there are other variants of her name used in the literature about her.

Physics A: Math&Gen.[3] edited by Frank Nijhoff and myself. After the meeting I was inspired to edit a book, composed of survey articles based on Workshop's talks together with a few invited contributions. The result of this idea is the present volume of 15 surveys in various subjects, all being within the integrability area pioneered by Sophie Kowalevski.

The survey by Benjamin Enriquez, *The KZB connection: parametrizations, flat sections and q-deformation*, exposes the reader to the theory of the Knizhnik–Zamolodchikov–Bernard connection and its q-deformation and explains recent results obtained by the author.

The work by Lubomir Gavrilov, *Jacobians of singularized spectral curves and completely integrable systems*, reviews important developments in the subject of generalized Jacobians.

The contribution by Luc Haine, *The q-hypergeometric equation, Askey–Wilson type solitons and rational curves with singularities*, surveys recent work describing families of functions which are eigenfunctions of a q-difference operator of arbitrary order.

The survey by Kazuhiro Hikami, *Quantum discrete soliton equations*, is dedicated to the study of integrable structures of discrete soliton equations, in particular the discrete affine Toda field theory and the N-reduced KP equation.

The work by Emil I. Horozov, *Dual algebras of differential operators*, introduces the reader to the subject of bispectral operators. An algebraic idea helps to significantly simplify both construction and classification of such operators. The author discusses important recent results in the subject obtained by him and other researchers.

In the joint paper, *A link between two fundamental contributions of Kowalevski*, Ji-Shan Hu and Min Yan demonstrate an intimate relation between two important results obtained by Sophie Kowalevski in her Doctoral Dissertation of 1874 and in her famous Prize Bordin paper published in *Acta Mathematica* in 1889. The key idea is to convert the Laurent series in the Kowalevski test[4] to power series by means of a meromorphic change of dependent variables.

The work by Andrei A. Kapaev, *Monodromy deformation approach to the scaling limit of the Painlevé first equation*, reviews an extensive list of publications on the subject and applies the isomonodromy deformation method to the degenerate solution of the Painlevé first equation.

In my own contribution, *Kowalevski top revisited*, I collect together all known results about separation of variables for the Kowaleski top and, based on Kowalevski's separation, construct a new 2×2 Lax pair which respects the Poisson structure of the model and gives a homomorphism of the top into the B_2-type Heisenberg magnet. This result provides an algebraic 'explanation' or interpretation of what Sophie Kowalevski achieved in her famous paper. Namely, one can say that her integration of the top amounts to the found homomorphism between these two integrable systems.

The survey by Dimitri Markushevich, *Some algebro-geometric integrable systems versus classical ones*, examines Mukai's integrable system and its generalisation by Beauville and identify several classical integrable systems, like Kowalevski

[3] J. Phys. A **34**, no. 11 (2001), 2071–2523

[4] a system is said to possess the Kowalevski property when it passed the Kowalevski test (sometimes called Painlevé or Kowalevski–Painlevé test/property)

top, Jacobi problem and 3-particle periodic Toda lattice, as the degenerate cases related to singular $K3$ surfaces. New integrable systems that are Lagrangian families of Prym varieties are constructed. The restricted Kowalevski top is studied in detail from the point of view of different families of small genus curves.

The work by Marta Mazzocco, *Painlevé sixth equation as isomonodromic deformations equation of an irregular system*, is dedicated to a thorough study of the PVI equation by constructing a map between an irregular system of first order differential equations and a special Fuchsian system.

In the work *Euler characteristics of theta divisors of Jacobians for spectral curves* Atsushi Nakayashiki and Fedor A. Smirnov compute the Euler characteristic of the affine part of the Jacobian, and hence of the theta divisor itself. Besided being fundamental in algebraic geometry, Jacobians are also known to be very useful in number theory and in the theory of integrable systems. The work is accessible to people who are not specialists in this field.

The review by Emma Previato, *Reduction theory, elliptic solitons and integrable systems*, explains the connection between "two of the thesis topics of Sophie Kowalevski, namely reduction theory for Abelian functions and integrability of a dynamical system". Apart from giving a stimulating guided tour into the classical and current results of this huge area of modern research into integrability, the author presents many open questions.

The survey by Takeshi Sasaki and Masaaki Yoshida, *Schwarzian derivatives and uniformization*, discusses various generalizations of the Scharzian derivative and associated geometry.

In the work *Elliptic solitons and Heun's equation* Alexander O. Smirnov gives a review of finite-gap solutions of the Heun equations, from Darboux to the present time.

Finally, in the work *Generalized Kowalevski top: new integrable cases on* e(3) *and* so(4), Vladimir V. Sokolov finds an integrable deformation of the Kowalevski top with quadratic terms mixing angular momenta and gravitation field directional cosines.

The present book is a slice of the modern integrability pie and I am sure that these wonderful surveys will give the reader a good taste of this diverse and ever-expanding subject.

Finally, as an organiser of the Kowalevski Workshop 2000 I want to thank all participants who made the meeting a success. I am grateful to all the referees for all their un-paid, but enormously valuable job and, especially, to Brian D. Sleeman for suggested corrections to this Preface. I thank all the staff of the *Centre de Recherches Mathématiques*, Université de Montréal, who in any way did anything in connection with the production of this volume for their help and support which is hugely appreciated.

Vadim B. Kuznetsov
University of Leeds
20.02.2002

List of Participants
Kowalevski Workshop (MMRD)
University of Leeds
April 12–15, 2000

S. Abenda (Bologna)
E. D. Belokolos (Kiev)
Yu. Berest (New York)
A. Bobenko (Berlin)
H. Braden (Edinburgh)
R. Caseiro (Coimbra)
O. Chalykh (Loughborough)
E. Corrigan (York)
S. Derkachov (St.Petersburg and Leipzig)
H. Dullin (Loughborough)
J. C. Eilbeck (Heriot-Watt)
V. Enol'skii (Kiev and Heriot-Watt)
B. Enriquez (Paris)
G. Falqui (Trieste)
Yu. Fedorov (Moscow)
M. Feigin (Loughborough)
E. Ferapontov (Loughborough)
A. Fordy (Leeds)
B. Gaffet (Saclay)
A. Grunbaum (Berkeley)
L. Haine (Louvain-la-Neuve)
R. Heredero (Madrid)
K. Hikami (Tokyo)
A. Hone (Adelaide)
J. Hurtubise (Montréal)
A. Its (Indianapolis)
A. Kapaev (St.Petersburg)
A. Kitaev (St.Petersburg and Adelaide)
N. Kitanine (St.Petersburg and York)
I. V. Komarov (St.Petersburg)
Y. Kosmann-Schwarzbach (Paris)
V. Kuznetsov (Leeds)
D. Lebedev (Moscow)
S. Lombardo (Leeds and Rome)
F. Magri (Milano)
J.-M. Maillet (Lyon)
D. Markushevich (Lille)
I. Marshall (Lausanne)
V. Matveev (Chelyabinsk and Warwick)
M. Mazzocco (Oxford)
O. McCarthy (Hull)
A. Mikhailov (Leeds)
P. van Moerbeke (Louvain-la-Neuve)
F. W. Nijhoff (Leeds)
D. F. Parker (Edinburgh)
H. Pfeiffer (Cambridge)
L. A. Piovan (Baha Blanca)
E. Previato (Boston)
S. Puttock (Leeds)
S. Rauch-Wojciechowski (Linkping)
N. Reshetikhin (Berkeley)
P. Richter (Bremen)
S. Saito (Tokyo)
M. Salerno (Salerno)
M. Semenov-Tian-Shansky (Dijon)
E. Sklyanin (St.Petersburg)

V. Sokolov (Moscow)

Yu. Suris (Berlin)

T. Takebe (Tokyo)

V. Terras (Lyon)

P. Vanhaecke (Poitiers)

C. Verhoeven (Brussels)

A. Veselov (Loughborough)

F. Wagner (Cambridge)

K. Wakatsuki (Tokyo)

A. Walker (Leeds)

M. Yan (Hong Kong)

M. Yoshida (Kyushu)

Centre de Recherches Mathématiques
CRM Proceedings and Lecture Notes
Volume **32**, 2002

The Life of S. V. Kovalevskaya

Roger L. Cooke

ABSTRACT. We sketch a selection of the more dramatic points in the life of S. V. Kovalevskaya, some of which have not been previously published in detail because they are peripheral to the social and intellectual issues that her life represents. They are imbedded here among the more standard events in her biography to add a sense of her personality to the historical role that she played as a pioneering woman mathematician.

Introduction

The following informal essay highlights some intriguing moments in the life of the distinguished Russian mathematician Sof′ya Vasil′evna Kovalevskaya (1850–1891). It is by no means intended to replace the scholarly work that is the basis for a good deal of the following information, and if it leads the curious reader to the excellent 1983 biography *A Convergence of Lives* by Ann Hibner Koblitz, its purpose will have been more than fulfilled. I am including some trivial gossip just because it illustrates some of the more colorful aspects of her personality; for Kovalevskaya was a very strong and determined person, though subject to fits of depression. When I wrote my analysis of her mathematical work in 1984, I thought I had finished writing about her, having said all that I could. I was wrong; the more I reflect on her life and consider the magnitude of her achievements set against the weight of the obstacles she had to overcome, the more I admire her. For me she has taken on a heroic stature achieved by very few other people in history. To venture, as she did, into academia, a world almost no woman had yet explored, and to be consequently the object of curious scrutiny, while a doubting society looked on, half-expecting her to fail, took tremendous courage and determination. To achieve, as she did, at least two major results of lasting value to scholarship, is evidence of a considerable talent, developed through iron discipline. In this essay I hope to pay due respect to that courage, determination, talent, and discipline.

This biography consists of four parts: Childhood (1850–1868), University Years (1869–1874), "Wilderness" Years (1875–1883), and Professional Years (1883–1891).

2000 *Mathematics Subject Classification.* Primary: 01A70; Secondary: 01A55.

The author is grateful to an unknown referee for a number of corrections and helpful suggestions.

This is the final form of the paper.

1. Childhood (1850–1868)

Sof′ya Vasil′evna Kryukovskáya was born January 15, 1850 (January 3 on the Julian Calendar used in Russia at the time). The family was in Moscow because her father Vasilii Vasil′evich Kryukovskói (1800–1874), an artillery officer, was stationed there. They were minor gentry, not at the time listed in the Russian equivalent of Burke's *Peerage*, despite having petitioned for such recognition for many years. Kovalevskaya's mother, Elizaveta Fyodorovna Shubert (1820–1879), was the great granddaughter of Johann Ernst Schubert, a Lutheran theologian driven from Braunschweig to Pomerania by the Seven Years' War (1757–1763). His sixth and last son, Theodor, Elizaveta's grandfather, emigrated to Russia, becoming a surveyor in what is now Estonia. In Russia he followed the practice of russifying his name and using his father's name as his middle name (patronymic). Hence he became Fyodor Ivanovich (Theodor, son of Johann) Shubert. (The German Sch is represented by the single letter Ш in Russian, transliterated into English as Sh.) Fyodor Ivanovich Shubert became a competent mathematician and astronomer, and was eventually elected to the Petersburg Academy of Sciences. He corresponded with some of the great German scientists of his time, including C. F. Gauss and F. W. Bessel, and with the French astronomer/mathematician Pierre-Simon Laplace. His son Fyodor Fyodorovich, Elizaveta's father, was a distinguished geodesist, whose papers on map-making were read with interest by mathematicians, including one who was to play an important role in the life of our subject, Karl Weierstrass. In 1861, nine years before he ever met or heard of Sof′ya Vasil′evna Kovalevskaya, Weierstrass wrote a paper on geodesics on general ellipsoids, in which he cited findings of "the Russian general Schubert" on the shape of the earth. Elizaveta Fyodorovna Shubert was a talented woman, an excellent pianist, who responded to the confinement imposed on Russian women by organizing a home theater. Having nannies to look after her children, she took very little interest in their upbringing.

Vasilii Vasil′evich Kryukovskói likewise entrusted the upbringing of his children to the servants, and in addition, neglected his wife. Having served in the army from the age of 17, he had risen to the rank of artillery colonel by 1848. After becoming commander of the Moscow Artillery Garrison, he spent a great deal of time gambling at the English Club. Following in the tradition of Tolstoy, he had to sell some of his estates to pay off his gambling debts. He did, however, show a desire to become a proper country gentleman. To that end, he sought recognition as a member of the nobility. Although the exact details are murky, it appears the family was able to parlay its surname into a genealogical connection with the family Krukóvskii and thereby trace a descent from a Hungarian king Matthew Corvus ("Raven"). In 1858 their claim was recognized, and they were granted the right to use the name Korvin-Krukóvskii and a family crest with a raven. By this time Vasilii Vasil′evich had risen to the rank of lieutenant-general, and, because the emancipation of the serfs was imminent, decided to retire to his estate, named Palibino, in western Russia. There he became Marshal of the Nobility (Предводитель Дворянства) in 1863.

Vasilii Vasil′evich Kryukovskói and Elizaveta Fyodorovna Shubert were married in 1843. Within a year of their marriage their first daughter Anna was born. There was apparently another child in the 1840s, who did not survive. Then, in 1850, Sof′ya was born, followed by the only surviving male child, Fyodor, in 1855. Thus the first eight years of our subject's childhood was spent in Kaluga, south of

Moscow, with a much older sister, joined when she was five by a younger brother. The main adult in her life was her nanny. When their father retired in 1858, he discovered to his horror that his daughters were very ignorant. He replaced the nanny with an English governess, Margaret Smith, and hired a Polish tutor named Joseph Malevich.

The move to Palibino coincided with the beginning of the intellectual awakening of the eight-year-old Sof′ya. She later recalled that the walls in one room of the house had, for lack of proper wallpaper, been covered with the notes from a mathematics course her father had taken. (As he had taken calculus from the great mathematician Ostrogradskii, co-discoverer with Gauss of the divergence theorem, these haphazardly arranged notes may well have been from that course.) This room fascinated the 8-year-old. She also enjoyed listening to her mother's brother tell her about the intricacies of infusoria and algae. Her father's brother told her, as she recalled, "about the quadrature of the circle, about the asymptotes toward which a curve continually approaches but never reaches, and many other things of that sort, whose meaning I of course could not yet understand, but which influenced my imagination, inculcating in me a reverence for mathematics as the highest and most mysterious science, which opens to its initiates a new and wondrous world that is inaccessible to ordinary mortals." This material is of course beyond the ken of a preadolescent, but it certainly piqued her interest, else she would not have remembered it all into adulthood. Her memories of this life were elegantly written up in an autobiographical book called *Memories of Childhood*, translated into English by B. Stillman and published as *A Russian Childhood* (Springer-Verlag, 1978).

Here is a description of her by her first teacher, Joseph Malevich [**9**]:

At my first meeting with my gifted pupil in October 1858 *I saw an eight-year-old girl rather strongly built, of a pleasant and attractive appearance, whose brown eyes shone with a receptive intelligence and a heartfelt kindness. From the very beginning of her studies she exhibited a rare attentiveness and rapid assimilation of what was taught, a completely methodical approach, so to speak, a precise fulfillment of what was required and consistently good understanding of the lessons... Three or four years of entirely successful lessons passed without the occurrence of anything notable. But when our study of geometry reached the ratio of the circumference of a circle to its diameter... my pupil, when explaining this topic in the next lesson, astonished me by arriving at the same result in a completely different way using her own reasoning.* [*Malevich adds that when he pointed out that her approach was rather roundabout, she grew embarrassed and burst into tears.*]

Three other influential people can be singled out as having recognized Sof′ya's talent and encouraged the young scholar to pursue her education. The first of these was a neighbor, the physicist N. P. Tyrtov, who brought her father a textbook of optics some time during the early 1860s. In 1864 he was astonished to learn that the 14-year-old Sof′ya had been reading the book. Not knowing trigonometry, she had been obliged to do her best to make sense of the sine function. Her guess was that the sine is proportional to the chord in a circle subtended by a central angle. In fact it is proportional to the chord subtended by an inscribed angle, but for small angles the difference is negligible. Tyrtov called the attention of Krukovskii to his daughter's extraordinary abilities. Krukovskii thereby became the second beneficent influence by agreeing to find a more advanced tutor for her. The tutor eventually found, A. N. Strannolyubskii (1839–1903), was the third major influence.

Strannolyubskii, an excellent teacher, and Sof′ya, an excellent pupil, were ideally matched. In three years of study during the family's winter stays in St. Petersburg, they progressed through calculus and perhaps beyond.

Although these notes have been written so far with an emphasis on Sof′ya's education, we cannot ignore the most important influence of all, her older sister Anyuta, whom she called "my spiritual mother." Anyuta was not scientifically inclined. Her interests were literary and romantic. Her personality was much more flamboyant and outgoing than that of her morose younger sister, leading the nanny to encourage little Sof′ya in the belief that she was not loved as much as she deserved. Anyuta wrote short stories and communicated her enthusiasm for radical causes to her younger sister. A beautiful and talented 20-year-old, she sent some of her short stories to Dostoevsky in 1863, one of which he printed in his journal *Epoch.* Anyuta's father was outraged that Anyuta had had the effrontery to pursue a career without his consent. Nevertheless, Dostoevsky visited Anyuta, and even proposed marriage to her. Sof′ya, only 14 at the time, developed a crush on Dostoevsky, and was appalled that Anyuta refused him. Her adolescent pride was also wounded when she found Dostoevsky wooing Anyuta while she, Sof′ya was attempting to get his attention by playing his favorite music on the piano (Beethoven's *Sonate Pathétique*).

During the Polish uprising of 1863, both sisters sided with the Polish rebels, against the government of the Tsar. In sympathy Sof′ya began to take lessons in Polish from Malevich. The rebellion had been brutally crushed, and the rebels who were not killed were either in exile in Siberia or had fled to the West. Among them was a young man named Bujnicki, who had shown some interest in her and had written in the personal album that she kept (as did all young women at the time) the following lines, in Polish of course:

My child, if I never see thee again, I shall forever keep the bright memory of thee. How happy I had been, had it been granted me to see the flowering of thy bud, which is even now about to bloom. But fate will not grant me that happiness, and I can only salute thy beauty in farewell.

Sof′ya later wrote of these lines, "What could these verses mean? I was happy and proud that he had dedicated them to me, but at the same time my heart was oppressed with a sense of foreboding." Bujnicki disappeared a few days later.

It was in this frame of mind that Sof′ya approached a reception at the family estate in April 1865, to be attended (against the family's wishes) by one Colonel Yakovlev, who had helped to crush the rebellion. She began to fantasize, "Tomorrow, as soon as he sits down to table, I'll take a large knife and plunge it into his heart, crying 'That's for Poland!' Then they'll seize me, of course, clap me in irons, and send me to Siberia, where I'll meet Pan Bujnicki." Needless to say, the affair turned out differently. Yakovlev tried to show off his artistic talent and asked for an album in which to draw a picture. Sof′ya was ordered to produce her album, the one containing the verses quoted above. Reluctantly she did so; and as soon as Yakovlev had drawn his picture, she ripped it out, tore it into pieces, and stepped on it. She was duly punished for this impertinence, and the family explained to Yakovlev that she was jealous of his ability.

Childhood does come to an end. To continue her education and get that important documentation of achievement, the university degree, was a problem for young women of her time, one that, in retrospect seems almost unsolvable. Absolutely every instant of socialization of young girls at the time was directed toward making them wives and mothers. If a girl somehow overcame her own natural inclination to fit into her society and persisted in wanting an education, she most likely ran into such formidable opposition from her family that only the most heroic would continue the struggle. If by some happy circumstance her family could be persuaded to consent, the larger society itself made it essentially impossible to get any recognized degree by closing all the universities to women, confining them to what were called in Russia the Women's Higher Courses. These were, as one might expect, rather weak in science. How was the eighteen-year-old Sof′ya to continue the education she had begun so auspiciously?

It seems likely that Sof′ya hoped to persuade her father to support her enrollment in some institution, perhaps even allowing her to travel abroad, if necessary. Although he was very old and of military and conventional background, she was his favorite child and she knew that she had considerable influence over him. However, her sister Anyuta was also to be considered, in her early 20s by 1868, and the two had girl friends who also wanted to pursue an education. One way of getting out was to find a politically conscious young man willing to enter into a formal marriage, while observing all the Victorian proprieties of behavior between people who were not actually married. In other words, the young man had to be a rebel willing to flout the conventions of society, yet enough of a traditional gentleman to treat a lady with the proper respect. Having a husband of this sort brought a measure of freedom and independence. It was, however, a drastic solution to the problem of women's emancipation, ironically well-described in the words of the English marriage ceremony: an estate not to be entered into lightly. That it was resorted to as often as it was shows the difficulties in the way of women's education at the time.

A young man more or less fitting the qualifications described above was Vladimir Onufrevich Kovalevskii (1842–1883), a young radical who was making his living as a translator and publisher, and who had an interest in biology. It may have been, as he stated, his desire to serve science that caused him to revoke his agreement to "marry" Anyuta and promise to "marry" Sof′ya, instead. That is, he may have perceived that Sof′ya was serious about science, but Anyuta wasn't. He had begun a correspondence with Darwin in 1867, in connection with his Russian translation of the second volume of Darwin's work on domesticated animals. (Sof′ya helped to edit this translation.) Together they began to dream of an ascetic life devoted to the cause of science and the reform of society. This may be the appropriate place to mention the movement in Russia at the time known as nihilism. The nihilists, as depicted by Turgenev in the character Bazarov of his novel *Fathers and Sons*, were people who thought the current state of society was radically hopeless, and that it was necessary to start over *ex nihilo*, on a foundation of science and humanitarian values. By the 1860s this term was being used to describe the radical youth of the time. Kovalevskaya considered herself a nihilist, as was her older sister.

Sof′ya and Vladimir were married on 15/27 September 1868. They immediately set out for Petersburg, where both attended lectures on anatomy and physiology. (In Sof′ya's case, even though anyone could legally attend lectures, she entered in the company of her husband, her uncle, and another trustworthy man, so as to avoid attracting attention.) One brilliant opportunity was lost to her in St. Petersburg. She made the acquaintance of the distinguished mathematician Pafnutii L'vovich Chebyshev (1821–1894). Chebyshev had grown wealthy by speculating in real estate, and he was by no means a reclusive academic. He held weekly open houses to which anyone could come and consult with him on mathematical problems. If only the university had been open to women, Sof′ya would undoubtedly have become his student. She became acquainted with him at this time, but as women could not take examinations or be granted degrees, there seemed to be no future in this direction. Nevertheless, this acquaintance was later to be of mutual benefit, as Chebyshev supported Sof′ya's eventual appointment as a corresponding member of the Petersburg Academy of Sciences, and she in turn helped him to put his students in touch with Western journals and mathematicians.

2. University Years (1869–1874)

Sof′ya, Anyuta, and Vladimir left Petersburg for Vienna on 16/28 April 1869. The high cost of living and the search for a mathematician with whom Sof′ya could work, however, impelled them to move on to Heidelberg, where their requests for admission became something of a cause célèbre. In the end it was left to individual professors to grant women the right to attend their classes if they wished. Sof′ya was able to study with some very well-known professors: Helmholtz, Kirchhoff, Bunsen, Leo Königsberger, and Paul Du Bois–Reymond. Her efforts on behalf of another young woman, Yuliya Lermontova, with whom she became good friends, led to an amusing incident with Bunsen. The latter had sworn never to allow women into his laboratory, especially not Russian women. He not only didn't want to let Lermontova study with him, he didn't even want to hear about her. Sof′ya, however, asked him so tenderly that he changed his mind. Later he wrote to Sof′ya's professor Weierstrass that Sof′ya was a "dangerous woman." Well, if she did sometimes take advantage of the fact that men found her attractive, who can blame her? This small asset was hardly sufficient to counterbalance the overwhelming weight of society against a woman's career in her day. In her work she was always earnest, and never wanted anything she hadn't honestly earned by her talent and effort. Indeed she often underestimated the profundity of her own work. From Heidelberg Sof′ya and Vladimir frequently traveled to different parts of Europe, including England, where they visited Darwin and Thomas Huxley, and were once guests of the writer George Eliot. Eliot described Sof′ya as, "a pretty creature, with charming modest voice and speech, who is studying mathematics (by allowance, through the aid of Kirchhoff at Heidelberg)...." Kovalevskaya was herself a writer of considerable talent and later wrote an essay on her acquaintance with George Eliot.

Anyuta soon left Heidelberg for Paris. To keep her parents believing that she was still properly chaperoned by her married younger sister, she sent letters to her parents via Sof′ya, who repackaged them so as to make it appear that they came from Heidelberg. (Eventually their father learned of the deception and terminated Anyuta's allowance.) In the end, Heidelberg also proved unsuitable as a place to obtain a degree. It was clear that Sof′ya would need the support of some very influential people to achieve that end. Such a person was Karl Weierstrass (1815–1897), one of the giants of nineteenth-century mathematics, a professor at the University of Berlin, and by good fortune the teacher of two of her teachers at Heidelberg. As it must have seemed to her in 1870, even if she couldn't obtain a university degree, the mere fact of having been a student of Weierstrass would establish her credentials in a very impressive way. It should be said that Vladimir had involved himself in her life more than she may have wanted or expected from a fictitious husband, and certainly more than Anyuta wanted. However, had he not appeared on the scene to vouch for her marital state, the university authorities might not have allowed her to attend classes at all. On the occasions when he visited Heidelberg, the two frequently quarreled. Therefore, when Kovalevskaya (as I shall henceforth refer to Sof′ya) set out for Berlin in the late summer of 1870, she was probably feeling a mixture of hope and relief. Anyuta, meanwhile, was still in Paris, where she had entered into a liaison with a French radical named Victor Jaclard.

World events sometimes break into the lives of ordinary people with momentous consequences. In 1870 the major event in Europe was the Franco-Prussian War, which began in August. In mathematics this war between two of the leading centers of mathematical excellence had perhaps less effect than one might expect, although it did force Felix Klein to return from Paris earlier than he had planned. Its first consequence for Kovalevskaya was that, when she knocked on Weierstrass' door, the 55-year-old confirmed bachelor, an opponent of admitting women to the universities, was favorably inclined to her for at least two reasons. First, her visit had been preceded by good letters of recommendation from her Heidelberg professors. Second, his auditory of 50 students the previous year had shrunk to 20 because of the war. Still, since she couldn't attend classes legally, he proposed private lessons. (I am trying hard not to make that proposal sound sinister. There is not the slightest reason to believe their relationship was anything but that of a friendly professor and student, despite some outrageous insinuations to the contrary in the book by E. T. Bell, in which Kovalevskaya, the only woman in a book appropriately named *Men of Mathematics*, is trivialized and depicted as a seductive playmate.)

Kovalevskaya proved to be very well prepared and was able to solve the test problems Weierstrass proposed to her. She began a correspondence with Weierstrass at this time, of which about 150 letters from Weierstrass to her have been preserved, but only a draft of one of her letters to Weierstrass. (Weierstrass burned her letters when he learned of her death in 1891. The draft of the letter she wrote to him on her arrival in Stockholm in 1883 was only recently uncovered by Professor Reinhard Bölling.)

Like all her earlier tutorials, this one appears to have gone very well and fallen into a pleasant routine; but that routine underwent a hiatus in April of 1871 when the Franco-Prussian War irrupted into Kovalevskaya's life for the second time. The Prussians had essentially won the war, having surrounded Paris, but the National Guard defending Paris refused to recognize any peace with the Prussians. The defenders of Paris included Anyuta's lover Victor Jaclard, and of course Anyuta herself. Sof′ya and Vladimir managed to get through the German lines into Paris. They lived in the famous Paris Commune for most of its brief existence, Sof′ya serving as a nurse and Vladimir visiting the Paris museums to inspect the fossils. (He received the degree of doctor of philosophy from the University of Jena in 1872.) When they finally left Paris, it appeared that the Commune would hold out for a while, yet they had hardly departed when the news came that the Commune had fallen. Jaclard had been arrested, and was probably facing the death penalty. At this point it was necessary to appeal to their father. He came to Paris, arranged to have Jaclard spirited away from the line of convicts who were being marched to show trials, and saw to it that Victor and Anyuta were legally married in Switzerland.

Back in Berlin, Kovalevskaya began to suffer from overwork. Although Weierstrass and his two unmarried sisters, with whom he shared a house, were very cordial to Kovalevskaya, they were somewhat puzzled by her mysterious relationship with her husband. Kovalevskaya did not attempt to explain it, for the reason best expressed by another Russian woman who went abroad around this time, Elizaveta Litvinova: "You don't have to be a genius to understand that, but you do have to be a Russian." However, by October of 1872, quarrels with Vladimir had made Sof′ya so miserable that she made a full explanation to Weierstrass. Full of sympathy for her situation, he cast about for ways to help her. The result was the following letter of 26 October 1872:

...I have been very preoccupied with you this night, and indeed, how could it have been otherwise? My thoughts have run in many different directions, but have always come back to the same point, which I must discuss with you today. Do not fear that I will mention things that we have agreed not to talk about, at least not yet. What I have to say to you is much more concerned with your scientific ambitions. However, I am not sure whether, given the endearing modesty with which you judge what you are now able to achieve, you will be inclined to agree with my plan... ([**1**, p. 58].)

Obviously Weierstrass intended to propose that Kovalevskaya write a dissertation. From this point on in his letters to her, he always uses the informal Du-form of address (as she does with him, in the draft discovered by Prof. Bölling). Over the next 18 months Kovalevskaya, guided by Weierstrass, wrote a number of mathematical works, three of which Weierstrass cited in requesting a degree for her from the more liberal University of Göttingen. The emotional bond between them, though always tinged on Kovalevskaya's side by the respect due from a young scholar to a teacher 35 years her senior, became ever closer. It was poignantly expressed by Weierstrass in a letter of 25 April 1873:

...That you, so long as you can learn something from me and wish to do so, will remain my student in the best sense of the word, I wouldn't even bother to mention if not for one place in your letter. You think that the student could be a burden (such is the ugly word you use), even if the friend could not. If I were malicious, I could interpret the fact that you have no such fear for the friend in a sense that you might vigorously protest, but in all seriousness, my dear, beloved Sonia, be assured that I shall never forget that it is the gratitude of my student to which I owe the possession of not my best but my only true friend... ([**1**, p. 87])

Later that year, on 20 August, he wrote to her:

I hope that by now you have escaped from the Zürich atmosphere and are breathing the pure mountain air. During my stay here [in Sassnitz] I have often thought of you and pictured to myself how wonderful it would be if I could spend a few weeks with you, my soulmate [Herzensfreundin], in such a magnificent natural setting. How wonderful it would be for us here—you with your imaginative soul, I stimulated and refreshed by your enthusiasm—to dream and contemplate the many problems that remain for us to solve, on finite and infinite spaces, the stability of the solar system, and all the other great problems of the mathematics and physics of the future. However, I learned long ago to be reconciled to the fact that not every beautiful dream comes true ([**1**, p. 90].)

It is impossible to doubt that Weierstrass was in love with Kovalevskaya, in a way that only a middle-aged bachelor can be, when an emotional bond that he has long ago renounced and believed impossible suddenly becomes a glorious reality. It is equally impossible to imagine the reticent and proper Weierstrass, a sincere Catholic, taking advantage of Kovalevskaya's gratitude in order to obtain sexual favors, if for no other reason than the realization on his part that such a step would destroy the whole relationship. (I say this despite the fact that Prof. Biermann's acute detective work has recently shown that Weierstrass may have fathered a child by the widow of his good friend and mentor Borchardt.)

To return to less exciting matters, Weierstrass, as mentioned, persuaded the University of Göttingen to award the doctorate to Kovalevskaya in 1874, in absentia, and without the usual oral defense. (He wrote that neither her German nor her nerves were strong enough to display her true talents in such a setting.) She returned to Russia in September 1874, to be joined there by Yuliya Lermontova and Vladimir in a well-deserved celebration of three doctoral degrees earned.

3. "Wilderness" Years (1875–1883)

The barriers to a woman's career in the nineteenth century were multitudinous. Like a course of high hurdles, each barrier surmounted with great expenditure of energy was soon followed by another requiring yet more prodigious effort. Having obtained a degree against all odds, Kovalevskaya now found no employment. Vladimir, who could have found academic employment, had the indiscretion to criticize the work of one of his examiners when taking the examination for certification to teach. As a result, he was not certified. Kovalevskaya's father had died in 1874, leaving her a small legacy, which the couple invested in real estate, hoping by following the example of Chebyshev mentioned above, to become independently wealthy and free to pursue their scholarly interests. Unfortunately academic brilliance is not necessarily correlated with financial shrewdness. Although they adopted an opulent lifestyle consistent with their anticipated revenues, their enterprise ultimately ended in bankruptcy. Meanwhile they took the momentous decision to end the charade of their fictitious marriage by consummating it. The unsurprising result of this decision was the birth of a daughter, Sof′ya Vladimirovna Kovalevskaya, in October of 1878, affectionately referred to as "Fufa." Difficult as it may be to remember at the end of the twentieth century, a century ago there were many people who believed women were so frail that they couldn't possibly withstand the rigors of university life. This belief was to some degree absorbed even by women who had obtained an education. At the time of her daughter's birth, Kovalevskaya had been away from academic work for three years, and she actually took pride in that fact, remarking to Elizaveta Litvinova, "Thank heaven I did not completely waste my strength studying mathematics; now at least my little girl will inherit some intellectual ability." The pathos of that statement, I believe, matches anything one can find in literature.

Fufa's first year of life was not auspicious for the family, as Kovalevskaya's mother died in February 1879, and this heavy blow was soon followed by the bankruptcy just mentioned. Such stresses are notoriously bad for unstable marriages, and relations between Sof′ya and Vladimir were never again close. Vladimir, despite his generous liberal heart, could not overcome the feeling that he should be the protector of the family. He fell into a severe depression, and began to pick quarrels with Sof′ya. By 1879, Kovalevskaya began hesitantly to try to resume the mathematical career she had abandoned three years earlier. Even before the birth of Fufa she had renewed her correspondence with Weierstrass. At the end of 1879 she went to a scientific meeting in St. Petersburg, where she gave a talk on one of the three papers she had submitted as a dissertation. (Only one of the three papers had been published; the paper she spoke on at this meeting was published in the mid 1880s and was later highly praised by the great French mathematician Henri Poincaré.) This meeting was attended by another of Weierstrass' students, the Swedish mathematician Gösta Mittag–Leffler (1846–1927), then at the University of Helsingfors (Helsinki). He had met her in St. Petersburg on 10 February 1876. He described the earlier meeting in a 1923 quotation from his own diary (translated into French):

What interested me most in St. Petersburg was making the acquaintance of Madame Kovalevsky. I spent several hours with her today. She is a delightful woman. She is beautiful and when she talks her face lights up with an expression of feminine good will and superior intelligence that one cannot help being dazzled by. Her manners are simple and natural with no trace of pedantry or affectation. In short, a "high-class woman" in all respects. As a scholar she is distinguished by a rare clarity and a precision of expression, as well as an extraordinarily quick perception. It is also easy to see the degree of profundity to which she has pursued her studies, and I understand perfectly why Weierstrass regards her as the most talented of his students ([**6**, p. 196]).

From that time on Mittag–Leffler was determined to find an academic position for Kovalevskaya. Wisely, both Kovalevskaya and Mittag–Leffler allowed themselves to be guided by Weierstrass, who showed considerable shrewdness in regard to academic politics. Resented as a Swede among Finns, Mittag–Leffler was himself eventually forced out of Helsingfors. He returned to his native Stockholm and became one of the founders of the institution that is now the University of Stockholm, an innovative institution, contrasting with the more traditional universities of Uppsala and Lund. There he would eventually be able to arrange a position for her. Kovalevskaya, in turn, eagerly accepted Weierstrass' invitation to apply a mathematical technique he had developed to solve a problem in mathematical physics. Conditions for working on the problem were far from ideal, however. Her marriage was strained. She was in poverty, having had to take a loan of 65,000 rubles, guaranteed by her brother; and she had a baby daughter to look after (although she received so much help in that regard from her friend Yuliya Lermontova, Fufa's godmother, that she was sometimes accused of exploiting Yuliya's good nature.)

In October 1880 she went back to Berlin to consult with Weierstrass, leaving Fufa with Yuliya, for a two-month stay. Meanwhile Vladimir had decided to try his hand at making money again, with results that were even more disastrous than the real estate speculation. This time he would find himself set up by his unscrupulous business partners as the legally responsible person in a stock scandal, facing prosecution in addition to bankruptcy. However, this situation took several years to develop fully. In April 1881 Kovalevskaya returned to Berlin with Fufa and (despite her poverty!) a nursemaid to free her from the necessity of looking after her 2-year-old daughter. At this time Mittag–Leffler, still in Helsingfors, was exploring the possibility of a position there for Kovalevskaya. Once again, however, world events intervened, this time against Kovalevskaya's interests. The University of Helsingfors, though nominally under control of the Russian empire, enjoyed a great deal of autonomy. The assassination of the liberal Tsar Alexander II in March of

1881 was sure to provoke a crackdown on radicals, and Kovalevskaya was of course tainted in the eyes of the government by her nihilist associations. The University declined to take a chance on Kovalevskaya.

Kovalevskaya's life remained unsettled throughout 1881 and 1882. Vladimir's position became more and more precarious. In March of 1882 Kovalevskaya sent Fufa away from Paris, where she had moved in the fall of 1881, back to Russia with her nursemaid. For the next four years she would see Fufa only on occasional visits. After Fufa's departure she continued to work on the problem Weierstrass had set for her, but she also became involved in some radical political circles in Paris, making the acquaintance of the Polish revolutionary Maria Jankowska (recently released from the prison at Poznan), the German Social Democrat Georg Volmar, and a young Russian expatriate, a student of mathematics named Joseph Perott. An incident of great dramatic impact, though not a major event in Kovalevskaya's life, took place at this time. Considering it was such a trivial incident, it is surprisingly well documented, and I shall tell the story in documents, in keeping with the principle mentioned above of including interesting gossip that illustrates some aspect of Kovalevskaya's character.

The story begins with a reminiscence by Maria Jankowska [**10**].

Once, when she was visiting me I noticed a cloud on her brow. She was distracted and her nervousness was increasing. Our conversation, despite efforts from both sides, was constantly breaking off. Finally, unable to contain herself any longer, she burst out: "Excuse me, but I must involve you in a matter that has me completely absorbed at the moment. I would like to ask you to give me some advice as to how to proceed in this situation. One of my friends, a mathematician, has a 16-year-old sister who also wants to devote herself to science and study mathematics. Her parents won't hear of it. This is all taking place in Russia, where women are forbidden access to the universities, so that the girl can't even dream of achieving her goal. One might consider a fictitious marriage," she added, and a wistful smile passed across her face, "but it's not easy to find a man willing to sacrifice his own freedom just to make the path to science and higher education easier for a young girl. That way out isn't available in this case. I have advised my friend to get the girl out secretly. But so that she couldn't be stopped on the road through a telegram from her parents I loaned her my passport. My friend was supposed to telegraph me as soon as they reached Prussia. By my calculations they should have been there by now, and I'm very worried that I haven't had any news." In this whole affair Sof′ya's behavior is very typical. She had formed a plan in complete accord with her nature, impetuous and romantic.

"Have you considered how many unpleasant consequences your friend's impetuous action might involve? If the young lady is detained at the border, she will certainly be turned over to her parents. And if it becomes necessary for you to go to Russia, you will find yourself in trouble, not to mention that you might be accused of abetting the abduction of an underage girl..."

"All that may happen," the worried Sof′ya interrupted, "but I couldn't do otherwise. Just consider, how can one refuse to help a person who is thirsting for knowledge and unable to reach its source? After all, so many obstacles are piled up in a woman's way whenever she doesn't want to follow the well-trodden path to marriage! I myself have encountered many of them, and so I consider it my duty to remove as many as possible from the paths of others. The girl I'm talking about

possesses, according to her brother, a remarkable talent for the exact sciences. Who can say whether a great scholar will emerge from this? I shall be proud if I am able to advance a young talent in the arena of science !"

The story continues with an undated newspaper clipping in Swedish (from the archives of the Institute Mittag–Leffler, henceforth referred to as IML).

Suspected nihilists. ... *Two travelers from the Kuopio district, one a young man, the other an even younger, barely full-grown woman, were detained by the Uleåborg police from the* 8*th to the* 11*th of October. The man had a passport for Mr. Josef Perott, and the girl had her school report card from St. Petersburg. But the police thought they saw signs of nihilism nevertheless, if only because the girl was very beautiful with dark hair and rosy cheeks. After an exchange of telegrams it was clear that a mistake had been made and the detainees were allowed to continue their interrupted journey... Yet [it appears] that the travelers were, after all, fugitives of a kind, a pair of unlucky lovers, whose parents had refused consent to their marriage. As a result, they were forced to spend three days in the Uleåborg "guest house".*

Those who lived through the late 1960s will recall how easy it seemed to identify young anarchistic rebels at the time, merely from the length of their hair. Again, it bears repeating that those inside the nihilist movement thought of themselves as reforming society, but the general public regarded them as violent anarchists. It appears from this story that the Prussian route was not the one chosen for escape. The Swedish route they chose was fortuitous (or perhaps advised by Kovalevskaya), since it brought them the assistance of Mittag–Leffer. The next relevant documents are a telegram and letter from Kovalevskaya to Mittag–Leffer, dated October 18, 1882. First the telegram (IML):

Please do me the great favor of providing 100 *francs to M. Perott on board the steamship* Norra Sverige. *Letter and money follow...* [IML] .

Next the letter (IML):

I have taken the great liberty of sending you a telegram requesting the loan of 100 *francs to one of my friends upon his arrival in Stockholm. The case was an extreme emergency, and I thought you would excuse me for appealing to your friendly feelings in this way. M. Perott, who is also a mathematician, was forced to leave Russia suddenly for very important reasons, which are, however, not of a political nature. I have received a telegram from him saying that he will arrive in Stockholm on the steamship "Norra Sverige" and asking me to send him money by telegraph. I could not do that, since the telegraph here does not accept money for Sweden, and no banker would agree to send such a small sum by telegraph. I had no choice but to resort to your aid, dear sir. I very much fear that you will think me impertinent and that this request will cause you some trouble.* [IML]

Kovalevskaya need not have worried. Perott and Mittag–Leffler got along famously, and for the next few years the latter occasionally reminded Kovalevskaya to seek out papers by Perott that he could publish in his journal *Acta Mathematica*. A telegram from Perott to Kovalevskaya dated 28 October 1882, says simply, "Arrested twice, released twice. All going well now." [IML]

To close out the story, we return to Maria Jankowska [**10**]:

A week later Sof'ya appeared again at my door, but this time accompanied by a beautiful young girl with black curls and laughing eyes, full of life. She was in ecstasy at everything she had found in Paris and absolutely worshipped her new friend and mentor. To be sure, she would mention her love for knowledge and her intention of attending lectures at the Sorbonne, but her beauty and her laughing eyes told me clearly that Paris appealed to her more as a city full of diversions than as the cynosure of world enlightenment and the treasure trove of knowledge. Sof′ya was sharing an apartment with her and had already begun to teach her mathematics. These lessons did not last long: Mathematics turned out to be too difficult a subject, too dry for a girl not trained in mental labor. The young Zoya, beautiful and aware of her beauty, preferred having fun, getting dressed up, making the rounds of the Paris shops, and chatting with young professors and students, of whom there were many around Sof′ya Kovalevskaya. Sof′ya once confided to me that she was a little disappointed, and that science in general would not gain much from her young friend, and mathematics even less. As she said this, she gave a deep sigh and adopted a comically concerned expression. She was extremely pleasant whenever she returned to reality from her golden daydreams; she seemed to be laughing at herself, confessing that she didn't understand either the world or people. Nevertheless, she maintained cordial relations with Zoya, for which the girl loved her sincerely. Circumstances were soon to provide the young girl with an occasion to demonstrate her gratitude and devotion to her brother's friend.

We shall take up those circumstances shortly. However, at this point we should note a few points on which Jankowska was slightly misinformed. First, the girl was Zinaida (Zina), not Zoya, a common trick of memory, to which we are all subject (or perhaps she really did have the nickname Zoya). Second, Perott was Russian on his mother's side; only his father was Polish. Jankowska describes his relationship to Kovalevskaya as she had heard it from Kovalevskaya herself, saying that Kovalevskaya was fascinated with Poland and Poles and had been enchanted by Perott's ability to convey Polish culture and literature to her. As soon as she stopped looking at him as a representative of Polish culture, however, and saw him merely as a man, she lost all interest in him. She also says, by the way, that Kovalevskaya confessed to being unable to feel the excitement other women expressed when they were courted by handsome young men. Without speculating on what that may mean in the context of the late twentieth century, it should be noted that Kovalevskaya certainly tried to find out what it was about men that interested other women, and she became deeply infatuated with at least one man a few years later. In fact, she expressed herself in an entirely different way when talking to Mittag–Leffler, saying that she was annoyed by the fact that so many men wanted to marry her, when she would much prefer to be a mistress. Mittag–Leffler commented in his diary,

Unfortunately for her she is not such that a man would desire to get her as his mistress whereas many would want her for a wife. Men do require from a mistress primarily physical attraction, which she lacks, but in a wife they can well accept the absence of these if compensated by other qualities, which she has in such eminent degree ([**6**, p. 201]).

This quotation seems to contradict Mittag–Leffler's first impression of Kovalevskaya's attractions, quoted above. However, the apparent discrepancy may be explained by the passage of time and his increased familiarity with her.

Jankowska's description of Zinaida as a frivolous party-goer seems to be borne out by at least one letter that the latter wrote to Kovalevskaya (in Russian) from Port-Navalo, France in 1885, full of utterly trivial gossip about her cousin Rudolph Becker from Kaunas who posed as her brother in order to visit her. She confirms that she and her actual brother posed as newlyweds during their journey from St. Petersburg. I have found no trace of her after 1885.

Finally, let us finish with this rather pompous fellow Joseph Perott. He left for Port-Navalo, on the south coast of Brittany, almost immediately, leaving Zinaida in Kovalevskaya's care. He kept up a correspondence with Kovalevskaya and his sister for a brief time. The last letter from him that Kovalevskaya kept was an 1889 request for a set of mathematical notes. In this letter he expresses his extreme reluctance to write to her and refers to some unpleasantness between them, as he finishes his request by saying, "**One word** on any subject different from that which forms the object of this letter will be for me a sufficient motive to refuse to accept what I am asking, even if you should grant it." At the same time that he wrote this letter, he departed for America, where he studied briefly at Johns Hopkins University, then became a lecturer in mathematics at Clark University in Worcester, Massachusetts, from its third year of existence until his retirement in 1921. He died in 1924 and is buried in Worcester. The late R. C. Archibald, secretary of the Mathematical Association of America was once led to ask Mittag–Leffler if it was true, as Perott claimed, that he had had an intimate acquaintance with Kovalevskaya. As Archibald remarked, "I have met the man and can scarcely credit that even when he was much younger he could be at all attractive to such a brilliant woman" ([**5**, p. 166]).

The preceding "soap-opera" story is an example of the capriciousness of historical documents. The documents that survive do so in a seemingly random manner. It is both frustrating and ironic that so many documents related to this trivial incident were preserved, when so many other things one would like to know about the life and work of Kovalevskaya cannot be found.

To resume Kovalevskaya's life from early 1883, the Furies finally cornered poor Vladimir. Faced with prosecution in the stock scandal mentioned above, on the night of 27 April 1883 he partly inhaled, partly drank a bottle of chloroform, leaving behind a note imploring Sof′ya's forgiveness for the financial and emotional mess he had made of their lives, protesting his innocence of the charges he knew he would be facing, and asking Anyuta to look after Sof′ya and Fufa. Sof′ya, in Paris, was devastated by the news of his death. According to Jankowska, for five days she sat alone in her room with the curtains drawn, weeping and refusing all food. A doctor was called; she refused to see him. Finally, after she fell into a coma, the doctor opened her mouth and forced some liquid food into her stomach. Zinaida attended her faithfully, never leaving her side for a minute. On the sixth day, Zinaida greeted Jankowska with the news that Kovalevskaya was saved, saying, "This morning she woke up and, without saying a word, sat up in bed and began to trace some symbols on the blanket. Then she asked me for pencil and paper. She is engrossed in some mathematical problem."

The mathematical problem was most likely the one Weierstrass had wanted her to solve to re-establish her mathematical credentials preparatory to taking up a position in Stockholm. All summer she worked on that problem, getting no assistance from Weierstrass, a notorious valetudinarian, who didn't feel well enough to write to her. She finished the problem in the autumn[1] and sailed for Stockholm at

[1]This paper contained a mistake, discovered a few weeks after her death by the Italian mathematician Vito Volterra.

the end of November, ready to go (boldly, but with great diffidence) where (almost) no woman had gone before.

4. Professional Years (1883–1891)

Having devoted considerable space to the early part of Kovalevskaya's life, I find it is possible to describe her entire professional career—the last eight years of her too-short life—in many fewer words. She came to Stockholm to lecture at the newly opened university there. As a fallback position, to avoid compromising the cause of women's careers in case her lectures were not successful, Mittag–Leffler had caused it to be believed that these lectures were a one-time special occasion, the opportunity to hear the words of a distinguished woman scholar. His caution is understandable, but fortunately was not necessary. Her lectures were well received. After she lectured for one semester in German, her Swedish was adequate for future semesters. She immediately began collaborating with Mittag–Leffler as an editor of *Acta Mathematica*, her Russian contacts being particularly valuable in procuring good papers to establish the reputation of this new journal. She soon brought Fufa to live with her in Stockholm and lived the life of a single mother with a career that may have been not very different from what such a life would be today. Looked at as a whole, her career cut an amazingly wide swath through the economic and social structure of her times. It is as if a pioneer went into an uncharted wilderness and built there a comfortable home with plastered walls, wallpaper, and modern appliances.

Nevertheless, she was continually forced to confront new gender-related issues. The first was unabashed prejudice. Kovalevskaya, as mentioned, had literary talent and collaborated on the writing of plays with Mittag–Leffler's sister Anne-Charlotte. Her presence in Stockholm was an irritant to the famous playwright and misogynist August Strindberg (1849–1912). In December of 1884 she wrote to Mittag–Leffler: *As a Christmas present I have received from your sister an article by Strindberg in which he proves as clearly as two times two equals four how monstrous a phenomenon is a woman professor of mathematics, how pernicious, useless, and unpleasant. I think he is essentially correct; the only thing I object to is his assertion that there are many male mathematicians in Sweden who are better than I and that I have been invited here only out of a sense of chivalry* ([**8**, p. 77]).

The second gender issue she faced was the perennial family/career conflict that women are still dealing with today. On 19 August 1886 Mittag–Leffler wrote the following to Kovalevskaya, who had gone to Russia to be with her gravely ill sister Anyuta. (Anyuta died the following year.)

You do me a great injustice if you think that only egotism leads me to fear the consequences if you should not return by the beginning of the semester. That is not at all the case, and I deeply share the feeling that leads you to remain with your sick sister. But in Sweden, as in the majority of other countries, leave of absence is never granted for personal reasons. A man may not request, and would never be granted, leave to care for a sick wife, child, or other relative. He can obtain a leave for his own medical treatment, but in that case it is granted because he is unable to work... If you request leave to care for your sister, it will provoke a storm, and will become an issue in the women's question... Be that as it may, I will do whatever I can in your interest ([**8**, p. 136]).

Kovalevskaya returned to Sweden. The unfairness of the gender comparison reported by Mittag–Leffler, and his seeming unawareness of its injustice, strikes the modern reader as both sad and curious. Did he really not understand that the male professors he spoke of were expected to have female relatives to bear the burden of caring for the sick, a convenience not enjoyed by Kovalevskaya? In fairness to Mittag–Leffler, however, it must be said that he was only thinking of the reaction of a hostile society. He remained Kovalevskaya's loyal friend and protector even in the face of what he regarded as callous ingratitude and betrayal on her part. She seemed to count for nothing the extraordinary efforts he had made in order to procure her a position and keep it in the face of powerful opposition. She was continually trying to find a more advantageous position for herself elsewhere, and would have left Stockholm at a moment's notice, leaving him to face the consequences, if only she could have arranged it.

Naturally one wonders if she was paid less than her male colleagues in comparable positions. The answer, unsurprisingly, seems to be, "Yes." According to excerpts from Mittag–Leffler's diary, recently published by L. Hörmander, Kovalevskaya had originally been willing to work in Stockholm without salary, and then declared that 2,000 kronor would suffice. Mittag–Leffler generously arranged for her to receive 4,000 kronor. In 1887 he arranged for her to teach an extra course in mechanics for a supplement of 3,000 kronor. After that she found she couldn't live on less. In order to get her 6,000 kronor for 1888 Mittag–Leffler paid 1,000 kronor out of his own pocket. Still she complained that it wasn't enough. According to Hörmander, Strindberg had written in his play *Getting Married*:

When the University of Stockholm divided the salary of a male mathematics professor to give half of it to a woman, it was a crime—against justice. And the men rejoiced ([**6**, p. 195]).

The editors of the 1982 edition of Strindberg's works comment that, "the salary was in no way divided, but Sonja Kovalevsky was paid far less than her male colleague."

Tragic though Vladimir's death had been, it was propitious for Sof′ya's career. She had the respectability of a married woman with a legitimate child, without the burden of a husband whose companion and support she would have been expected to be. She was as free as a woman could be at the time. One can therefore only conclude that it was a sincere emotional attachment that she developed to one Maksim Kovalevsky (1851–1916). It is not entirely a coincidence that her second great love in life had the same surname as the first. The Swedish post office mixed up their mail when he came to Stockholm in 1885 to deliver a series of lectures. He was a radical sociologist and legal scholar, and he spent a good deal of time outside Russia because of his radical views.

His entrance into her life came at an inconvenient time, but perhaps she was pleased to have the distraction from the concentrated effort she was expending at the time. She was writing a play jointly with Anne-Charlotte Leffler and, more importantly, was developing a mathematical discovery that she had made and was planning to submit to a prize competition at the Paris Academy in 1888. At Mittag–Leffler's request, Kovalevsky went away briefly to allow her to complete the work.

She apparently made this discovery in 1886 and communicated it to several prominent French mathematicians in June of that year. The acknowledged leader of French mathematics at the time was Charles Hermite (1822–1902), who routinely arranged competitions to obtain recognition for his protégés. He did this both for his son-in-law Émile Picard and for the greatest French mathematician of the time, Henri Poincaré. As Kovalevskaya wrote to Mittag–Leffler on 26 June 1886:

Yesterday Hermite, Bertrand, Camille Jordan, and Darboux, who are all members of the committee [to choose the topic for the prize], discussed the project with me... and they seemed to find it so interesting that they believed it has a good chance of winning. The only disadvantage is that in that case I shall have to postpone publication until 1888. *I shall have to submit it to the Academy, and the decision will be made in* 1888. *Only in that case it is impossible for me to communicate my work at Christiania [Oslo] this year. That would be much too risky...* ([**8**, p. 133]).

As Darboux was a member of the jury for this competition, one might speculate that the competition itself was not entirely as objective and anonymous as the rules required. However, that impression would be misleading. This competition was exactly like many others. As Hermite described it in a letter to Thomas Jan Stieltjes a few years later, when he was proposing to render Stieltjes the same service,

The problem proposed for the prize will be the one that you yourself indicate. You will choose among the problems you are currently studying, the one that suits you. You may trust me, that I am not going beyond my authority in asking you. I have the complete assurance that it will be accepted immediately and with dispatch by all the members of the committee charged with choosing a prize topic. Moreover, we are only following traditional usage, so that there is no question of any special favor being done for you.

The prize awarded to Mme de Kowalewski and the prize won by M. Picard were given for problems that everyone knew had been investigated by M. Picard and Mme de Kowalewski, whose work it was desired to reward with distinction ([**15**, p. 81]).

The competition was, as everyone expected, won by Kovalevskaya, and Hermite arranged for an increase in the value of the prize (a gold medal) from 3,000 to 5,000 francs. The strain of preparing a long and difficult manuscript for the competition and revising it for publication were so great that she experienced a terrible depression immediately after the award ceremony, one that took her some time to recover from. The ceremony itself miffed Maksim Kovalevsky, who was not used to sitting on the sidelines and cheering while others collected honors. Relations between the two of them were strained from that time on.

The competition was the certification that Mittag–Leffler needed to get Kovalevskaya tenure at Stockholm University. However, despite her gratitude to Sweden and Mittag–Leffler, she had already begun to dream of a more exciting place to live. She wanted to present her prize-winning paper as a dissertation to a French university so that she could become a naturalized French citizen and procure a position there. How much her desire to live with Maksim Kovalevsky influenced this project is impossible to establish. Certainly she had begun to express her boredom with Stockholm to her Swedish friends in the most tactless manner. She confided her plan in a letter to Weierstrass, who was appalled. He told her in no uncertain terms that it was an unforgivable insult to the University of Göttingen to seek a second doctorate in mathematics from another university and mentioned that if she did so, she might well find her degree revoked. She eventually abandoned the plan.

Compared with the nearly continuous round of adventures she had lived through up to 1888 the rest of her life was uneventful. After the astounding achievement of becoming a full member of the mathematical community of her time, respected for her mathematical achievements and in no way patronized by male mathematicians as being "good—for a woman," she was undoubtedly rather tired. She produced no more original papers, but in 1889 presented an old one dating to her unversity days. It had never been published and was not used in her dissertation, although Weierstrass had alluded to it in his letter to the University of Göttingen requesting a degree for her. She kept trying to make the relationship with Maksim Kovalevsky work out. (For her "working out" would have meant living together as lovers. According to Mittag–Leffler, she was adamant that she would not marry Maksim, fearing that if she did, he would begin to take her for granted and look for a mistress.) It never did. After they traveled to Genoa for the holidays at the end of 1890, they parted for the last time. On the way back to Stockholm, she took a route that avoided Copenhagen because of a current epidemic. Having no Danish money to tip a porter, she had to carry her own bags in a driving rain. She returned to Stockholm with a bad cold, which turned into pneumonia. She died on 10 February 1891 and is buried at the Norra Begravningsplats in Stockholm.

Weierstrass was so overcome with grief at the news of her death that his friends feared for his own life. Fufa was briefly cared for by the family of the Swedish astronomer Hugo Gyldén, who wished to adopt her. In the end she was adopted by Yuliya Lermontova, returned to Russia, and served with the Red Cross during World War I. She became a physician, never married, and had no children. She lived to help with the celebration of the centenary of her mother's birth in 1950, and died the following year.

References

1. Reinhard Bölling, ed., *Briefwechsel. Karl Weierstraß-Sofia Kowalewskaja*, Akademie Verlag, Berlin.
2. ———, *Deine Sonia: A reading from a burned letter*, Math. Intelligencer **14** (1992), no. 3, 24–30.
3. ———, *Zum ersten Mal: Blick in einen Brief Kowalewskajas an Weierstraß*, Historia Math. **20** (1993), 126–150.
4. Roger Cooke, *The mathematics of Sonya Kovalevskaya*, Springer-Verlag, New York, 1984.
5. Ann Hibner Koblitz, *A convergence of lives. Sof′ya Kovalevskaia: Scientist, writer, revolutionary*, Birkhäuser, Boston, 1983.
6. Lars Hörmander, *The first woman professor and her male colleague* (P. Hilton, F. Hirzebruch, and R. Remmert, eds.), Miscellanea Mathematica, Springer-Verlag, Berlin, 1991.
7. P. Ya. Kochina, *Sof′ya Vasil′evna Kovalevskaya* (Софья Васильевна Ковалевская), Nauka, Moscow, 1981.
8. ———, *Correspondence of S. V. Kovalevskaya with G. Mittag–Leffler* (Переписка С. В. Ковалевской Г. Миттаг–Леффлера), Nauka, Moscow, 1984.
9. Joseph Ignat′evich Malevich, *S. V. Kovalevskaya, doctor of philosophy and professor of mathematics, in the reminiscences of her chronologically first teacher* (С. В. Ковалевская, доктор философии и профессор математики, в воспоминаниях ее первого по времени учителя), Russian Antiquity **12** (1890), 615–654.
10. Maria Mendelssohn, *Reminiscences of Sof′ya Kovalevskaya* (Воспоминания о Софе Ковалевской), Modern World **2** (1912), 134–176.
11. Beatrice Stillman, *Sof′ya Kovalevskaya, a Russian childhood*, Springer-Verlag, New York, 1978.
12. Wilderich Tuschmann and Peter Hawig, *Sofia Kowalewskaja, ein Leben für Mathematik und Emanzipation*, Birkhäuser, Basel, 1993.

13. *In memoriam S. V. Kovalevskaya* (Памяти С. В. Ковалевской. Сборник Статей), USSR Academy of Sciences, Moscow, 1951.
14. *The legacy of Sonya Kovalevskaya*, Proc. of a Symposium held at Radcliffe College (Cambridge, 1985) (Linda Keen,ed.), Contemporary Mathematics, vol. 64, Amer. Math. Soc., Providence, 1987.
15. *Partie inédite de la correspondance d'Hermite avec Stieltjes*, Proc. of the Seminar on the History of Mathematics, vol. 4, Inst. Henri Poincaré, Paris, 1983, pp. 75–87.

DEPARTMENT OF MATHEMATICS, UNIVERSITY OF VERMONT,BURLINGTON, VT 05401, USA.
E-mail address: cooke@emba.uvm.edu
URL: http://www.emba.uvm.edu/~cooke

Centre de Recherches Mathématiques
CRM Proceedings and Lecture Notes
Volume **32**, 2002

Kovalevskaya's Mathematical Work

Roger L. Cooke

ABSTRACT. We sketch the two major mathematical achievements of S. V. Kovalevskaya (1850–1891), placing them in the historical context of their time, and survey her four minor works.

Introduction

The present essay is an analysis of the mathematical achievements of Sof′ya Vasil′evna Kovalevskaya (1850–1891); it is divided into five parts. The first part reviews the seventeenth- and eighteenth-century mathematics that she inherited from her predecessors. The second part describes her education and the mathematical areas in which she specialized. Parts three and four are devoted to analysis of her two greatest results; and the final section contains a brief look at her minor papers and gives an overall assessment of her place in the history of mathematics. Although this essay makes some mathematical demands on the reader, at least to the extent of assuming that partial differential equations are a familiar concept, I have tried to keep the descriptions as elementary as possible.

1. From Calculus to Modern Algebra and Analysis

Although the problems that motivated them come from geometry and physics, the basic techniques of calculus—differentiation, integration, and infinite series—are essentially algebraic subjects. In fact the equation "algebra + infinitesimals = calculus" is a fair approximation to the truth. Infinitesimal methods originally arose in geometry in problems of constructing tangents to curves or finding the lengths of curved lines and the areas enclosed by such lines, and the areas of curved surfaces and the volumes enclosed by such surfaces. These methods could not be systematized, however, until they were suitably formulated using the symbolic methods of algebra. Once this labor had been performed, a powerful mathematical technique arose through the application of the fundamental theorem of calculus to unite problems that originally seemed to have no connection. The laws of nature turned out to be differential equations. Differential calculus became important as

2000 *Mathematics Subject Classification.* Primary: 01A70; Secondary: 01A55.

The author is grateful to the editor and to an unknown referee for a number of corrections and helpful suggestions.

This is the final form of the paper.

the way of deriving such equations, while integral calculus and infinite series were the way to solve them. In this way many deep problems of mechanics, heat, and ultimately electricity, magnetism, and light, all became amenable to study through models that involved differential equations.

1.1. Algebra, differentiation, and integration. The full understanding of the nature of these algebraic problems, however, required a great deal of basic research into the nature of numbers and the fundamental problem of algebra: finding the roots of a polynomial. It required, for example, complex numbers to reveal hitherto hidden connections between the exponential and trigonometric functions in the formula $e^{ix} = \cos(x) + i\sin(x)$ and to explain why the formula $1/(1+x^2) = 1 - x^2 + x^4 - x^6 + \cdots$ is valid only when $-1 < x < 1$. A very important discovery was that the operation of integration leads out of the domain of algebraic functions, while differentiation does not. For example, the integral of the function $f(t) = 1/t$ is $\ln(t)$. The effort to explain these mysteries led mathematicians far afield and generated a wide variety of techniques that seemingly had little to do with one another, despite having a common source in the problems of solving differential equations.

1.2. Obstacles to the solution of differential equations. The solution of a differential equation could be stymied in two different ways. First, it might not be possible to manipulate the equation in such a way that the solution could be exhibited as an integral. Second, when the solution was exhibited in this way (reduced to quadratures, as the mathematicians of the time said), the integrals might not be expressible in terms of known functions. An example of the first phenomenon occurs in the case of Bessel's equation

$$x^2y''(x) + xy'(x) + (p^2 - x^2)y(x) = 0,$$

which arises when the variables are separated in Laplace's equation in cylindrical coordinates. The second phenomenon occurs even more frequently, for example in the equation of pendulum motion

$$y''(t) + \omega^2 \sin\bigl(y(t)\bigr) = 0,$$

which can be reduced to quadratures in terms of y or the variable $u = \cos y$, as

$$dt = \frac{dy}{\sqrt{C - 2\omega^2 \cos y}}, \quad \text{or} \quad dt = -\frac{du}{\sqrt{(C - 2\omega^2 u)(1 - u^2)}}.$$

These last integrals turn out to be inexpressible in terms of the elementary algebraic, trigonometric, and exponential functions.

1.3. Power-series solutions. In the first case the pioneers of calculus did not hesitate to use infinite series, regarding a function as known if its Taylor coefficients could be generated recursively; later they added trigonometric (Fourier) series to their repertoire of techniques. In the second case an integral that could not be evaluated was given a name, thereby becoming a new function whose relations to known functions could be investigated. Thus, for example, integrals involving the square root of a cubic or quartic polynomial were classified and generically referred to as *elliptic functions* because of their connection with the rectification of the ellipse.

1.4. Integrals of algebraic functions. The systematic study of algebraic integrals proceeded in parallel with the development of algebra itself. Both subjects were advanced spectacularly in the work of the short-lived Norwegian prodigy Niels Henrik Abel (1802–1829). In his youth Abel had the ambition of settling these problems once and for all by finding a general formula for the roots of any polynomial and a general form for the integral of any algebraic function. Eventually he realized that such formulas could not be obtained and produced a proof that the general quintic equation could not be solved by a finite number of algebraic operations. As for integrals, he brought a great deal of order into this area and produced what is arguably the most profound theorem of the early nineteenth century, the theorem now known by his name. He considered two variables x and y constrained to satisfy a polynomial equation $p(x, y) = 0$. A general algebraic integral, now known as an *Abelian integral*, is an integral of the form

$$\int R(x, y)\, dx$$

where $R(x, y)$ is a rational function of the two variables. For example, the elliptic integral

$$\int \frac{1}{\sqrt{1 - x^4}}\, dx$$

is of this form with $p(x, y) = x^4 + y^2 - 1$ and $R(x, y) = 1/y$.

Abel's great result was that for any given polynomial $p(x, y)$ and rational function $R(x, y)$ there is a number g such that the sum of any given number of indefinite integrals can be expressed as a sum of g integrals whose limits of integration are algebraic functions of the limits in the given integrals.

1.5. Jacobi and theta functions. Abel died without being able to explore fully the ramifications of this theorem, but his work was ably continued by C. G. J. Jacobi (1804–1851). (In fact it was Jacobi who proposed the name *Abel's theorem.*) In a series of papers devoted to elliptic functions and extending to more general algebraic integrals, Jacobi showed that these algebraic integrals could not be understood fully without the use of complex variables, and that their inverse functions were always functions of n complex variables having $2n$ independent periods. In the case of elliptic functions, for which Abel's number g equals 1, Jacobi showed that the inverse functions could be elegantly expressed as quotients of functions that later came to be known as *theta functions.* The two theta functions that he introduced were

$$\Theta\left(\frac{2Kx}{\pi}\right) = 1 - 2q\cos 2x + 2q^4\cos 4x - 2q^9\cos 6x + \cdots,$$

$$H\left(\frac{2Kx}{\pi}\right) = 2\sqrt[4]{q}\sin x - 2\sqrt[4]{q^9}\sin 3x + 2\sqrt[4]{q^{25}}\sin 5x - 2\sqrt[4]{q^{49}}\sin 7x + \cdots.$$

The constant K here is known as the complete elliptic integral of first kind. It depends on a parameter k via the formula

$$K = \int_0^1 \frac{1}{\sqrt{(1 - t^2)(1 - k^2t^2)}}\, dt.$$

In 1849 Jacobi was able to use these theta functions to give a very elegant expression of the functions that describe the motion of a rigid body free of external torque, a case of motion that had been studied previously by Leonhard Euler (1707–1783).

1.6. Mathematical complications in the two techniques. The two techniques—reduction to quadratures and power-series expansions—turned out to share a common domain, namely the complex numbers. We noted above that algebraic functions really cannot be understood on any other basis. Now the theory of functions of a complex variable was constructed by Kovalevskaya's adviser Karl Weierstrass (1815–1897) strictly on the basis of power series. Weierstrass had been anticipated in this work by Augustin-Louis Cauchy (1789–1856), who had based this theory on the use of complex integrals, but had also shown that the functions involved could always be represented by power series. Weierstrass' contemporary Bernhard Riemann (1826–1866) based his development of the subject on geometric transformations, but he too employed power series. The theory of analytic functions of a complex variable was the vantage point from which Weierstrass and his students studied both algebraic functions and differential equations. One peculiar consequence of this approach was that the fundamental physical variable, time, which is intuitively a linearly ordered set, must be represented as a complex variable, which is *not* linearly ordered in any natural way.

1.7. General algebraic functions. Jacobi was the first to note that if $\Pi(x)$ is an elliptic integral, Abel's theorem can be stated as the equation

$$\Pi(x) + \Pi(y) = \Pi(a)$$

where a is an algebraic function of x and y. At the same time he noted that for a more complicated integral, say where

$$\Pi(x) = \int_0^x \frac{1}{\sqrt{1 - t - t^5}}\, dt,$$

Abel's theorem seems to lead to indeterminacy, since it provides only one equation

$$\Pi(x) + \Pi(y) + \Pi(z) = \Pi(a) + \Pi(b),$$

in which a and b are two algebraic functions of x, y, and z. In 1832 he proposed that a second integral be considered, namely

$$\Phi(x) = \int_0^x \frac{t}{\sqrt{1 - t - t^5}}\, dt,$$

and posed the problem of finding x and y in terms of u and v from the equations

$$\Pi(x) + \Pi(y) = u, \quad \Phi(x) + \Phi(y) = v.$$

This problem, in general form, became known as the *Jacobi inversion problem.* A complete solution of it required 25 years of work, the final step being taken almost simultaneously in 1856 by Riemann and Weierstrass, using two very different approaches.

1.8. Weierstrass' development of algebraic function theory. Weierstrass was so intrigued by the fact that Riemann and he had both solved the same problem from different perspectives that he withdrew his paper in order to investigate the relation between them. By the mid-1860s he had organized the subject to his own satisfaction, in a large set of notes published as vol. 4 of his collected works. The secret of solving the Jacobi inversion problem turned out to be multivariable theta functions, whose theory was developed during the 1840s and 1850s.

Bearing in mind that elliptic integrals were originally encountered in solving simple problems of mechanics, such as that of pendulum motion, one would naturally inquire about the practical application, if any, of the more complicated algebraic integrals or the theta functions of two variables used to express them.

1.9. Power series solutions. The power series technique also required mathematical investigation. The simple problem was the following: A differential equation can be used to generate the coefficients of a power series recursively. To take, for example, the Bessel equation of order 0

$$x^2y''(x) + xy'(x) - x^2y(x) = 0$$

with the initial conditions $y(0) = 1$, $y'(0) = 0$, and assuming a power-series expansion

$$y(x) = \sum_{n=0}^{\infty} a_n x^n,$$

one finds the relations $a_0 = 1$, $a_1 = 0$ and

$$n(n-1)a_n + na_n + a_{n-2} = 0,$$

that is

$$a_n = -\frac{a_{n-2}}{n^2}$$

from which it is easy to prove recursively that

$$a_{2n-1} = 0, \quad a_{2n} = (-1)^n \frac{1}{2^{2n}(n!)^2},$$

and hence that the solution of this equation, usually denoted $J_0(x)$, is

$$J_0(x) = \sum_{n=0}^{\infty} (-1)^n \frac{(x/2)^{2n}}{(n!)^2}.$$

It is not difficult to verify that in this particular case the series converges for all values of x (real and complex), and therefore provides a genuine solution of the initial-value problem. The difficulty mathematicians see in the technique, however, is that it tells only what the coefficients of a series *must* be if the series is to represent a function satisfying the differential equation. It is necessary at the very least to verify that the recursively generated series does converge. It would be desirable to prove a general theorem to the effect that a given differential equation has an analytic solution. For Weierstrass this was particularly important, since his definition of an analytic function involved a convergent power series. If such a theorem could be proved, then a differential equation itself could be used as the definition of an analytic function. Such was the theme of a paper Weierstrass wrote in 1841 (but did not publish until 1894).

These two topics, Abelian integrals and power-series solutions of differential equations, were the areas in which Kovalevskaya most distinguished herself. As we have seen, they have a common root in the basic problem of solving differential equations and a common domain in the analytic functions of a complex variable. A third common element is that both areas were greatly advanced by Jacobi, whose influence on both Kovalevskaya and Weierstrass is especially noticeable in the analytic style in which all three wrote.

TABLE 1

Winter 1870–1871	Elliptic Functions
Summer 1871	Recent Synthetic Geometry
Summer 1871	Selected Problems of Geometry and Mechanics Solvable by Elliptic Functions
Winter 1871–1872	Abelian Functions
Summer 1872	Introduction to Analytic Functions
Summer 1872	Calculus of Variations
Winter 1872–1873	Elliptic Functions
Summer 1873	Elements of Recent Synthetic Geometry
Summer 1873	Selected Problems of Geometry and Mechanics Solvable by Elliptic Functions
Winter 1873–1874	Abelian Functions

2. Kovalevskaya's Education

Kovalevskaya studied with Weierstrass for four years, from 1870 to 1874. Although she was not allowed to attend his lectures, he received her in his home on Sundays and gave her lessons there. Some idea of the contents of this private instruction can be gained from the list of courses he was giving at the University of Berlin at this time and his letters to her. The courses listed in his collected works are shown in Table 1.:

It seems clear from Weierstrass' letters to Kovalevskaya that he taught her essentially the same things he taught his other students. He was teaching Abelian functions in January 1872, and on 4 January he sent her a letter cancelling their appointment for the next day and enclosing some notes on the topic of their next appointment. These notes are concerned with hyperelliptic functions as a special case of Abelian functions, and Weierstrass says explicitly that he needs these notes for his next lecture. Letters from November 1872 contain discussions of problems in the calculus of variations, consistent with the offering of that course in the summer semester. However, several other letters from the same time discuss theta functions. A year later, when Weierstrass was teaching geometry and its applications, the letters discuss minimal surfaces.

2.1. Kovalevskaya's dissertations. Weierstrass mentioned three papers of Kovalevskaya when he wrote to the University of Göttingen to request that she be granted the doctoral degree. By far the most important of these was her paper on analytic solutions of partial differential equations, which contains the result now known as the Cauchy–Kovalevskaya theorem. Almost any university, however, would have awarded the degree on the basis of a second paper that she wrote, on Abelian integrals that degenerate to elliptic functions. For good measure Weierstrass included a third paper, an analysis of the shape of the cross-section of a ring around a planet, carried out on the assumption that a thin layer of liquid on the surface of the ring would be in equilibrium. The only paper she published at the time was the first of these three. Judging from the correspondence, one would conclude that this paper must have been written in a very short time. The dissertation was completed in the summer of 1874, yet no letter mentions this topic until 6 May of that year. The topic then occurs frequently in letters through 4 July, on which date Weierstrass sent her a long letter containing some formulas that actually occur in her dissertation. His letter to Lazarus Fuchs requesting the degree was written on almost the same day.

2.2. Kovalevskaya's specialty. Even 125 years ago the subject of algebraic functions was sufficiently complicated that a thorough mastery of it required several years of study. By the time she had finished studying with Weierstrass Kovalevskaya was one of the world's experts in this area. In addition, she had a thorough grounding in the more analytic parts of geometry and differential equations. These were the subjects she was prepared to teach when, eight years later, she took up her position at Stockholm University.

3. Partial Differential Equations

As mentioned above, the first person to ask whether a differential equation automatically generates a convergent power-series solution via the so-called method of undetermined coefficients was Cauchy, who first considered this problem in his lectures at the École Polytechnique in 1823, then again in a series of memoirs in 1835. In 1842 he took a third approach to the problem and established the existence of an analytic solution to a partial differential equation having analytic coefficients using a method that he called the *calcul des limites*, but which is commonly called the *method of majorants* in English. The idea behind the method is that, when a power series $\sum a_n x^n$ converges for any nonzero value of x, say $x = x_0$, it must converge for all values of x such that $|x| < |x_0|$ by comparison with a geometric series. That is, if

$$\sum a_n x_0^n$$

converges, then its terms must tend to zero. As a consequence there is a constant M such that $|a_n x_0^n| \le M$ for all n. But then $|a_n x^n| \le M \big|x/x_0\big|^n$, so that for $|x| < |x_0|$, the terms of the original series are majorized by those of the geometric series

$$\sum M\left(\frac{x}{x_0}\right)^n = \frac{M}{1 - x/x_0}.$$

This observation, that the Cauchy kernel $1/1 - z$ has a Taylor series with positive coefficients that majorizes the terms of any convergent power series, provided Cauchy, and almost simultaneously Weierstrass, with a technique for proving that the method of undetermined coefficients must produce a convergent series. The idea is to exploit the positivity of the coefficients of this series, replace all the analytic data in the equation with the Cauchy kernel, and thereby generate a power series that majorizes, term by term, the series generated by the original equation. Since the equation with the Cauchy kernel as data can be solved explicitly and shown to have an analytic solution, it follows that this series converges and hence that the original equation has an analytic solution.

Except for technical details (which are considerable) the process just described is the essence of the Cauchy–Kovalevskaya theorem.

3.1. Cauchy's 1842 papers. Cauchy considered a quasi-linear partial differential equation

$$D_t\varpi = AD_x\varpi + BD_y\varpi + \cdots + K$$

where the coefficients $A, B, \ldots, K$ are functions of both the unknown ϖ and the independent variables t, x, y,.... For the case when the coefficients are $A = a/xyz\cdots t$, $B = b/xyz\cdots t$, ..., $K = k/xyz\cdots t$, Cauchy solved this differential equation with the initial conditions

$$\varpi = \omega, \quad D_x\varpi = 0, \quad D_y\varpi = 0, \ldots,$$

at the point $t = \tau$, giving the solution as the integral

$$\log\left(\frac{t}{\tau}\right) = \int_0^{\varpi-\omega} (\varpi - \theta)\left(x + a\frac{\theta}{k}\right)\left(y + b\frac{\theta}{k}\right)\cdots\frac{1}{k}\,d\theta.$$

Since, as remarked above, the terms of the Taylor series of this analytic function majorize those that result from any given analytic coefficients $A, \ldots, K$, it follows that the initial-value (Cauchy) problem for this quasi-linear equation must have an analytic solution.

3.2. Jacobi's canonical and normal form. A property of differential equations that makes their solution more feasible was discovered by Jacobi,who defined an equation to be in "normal form" if it could be solved for the highest order derivative with respect to some variable (and apparently highest order meant the pure derivative of order n with respect to some variable when the equation is of order n).[1] He did not make any connection between normal form and the method of undetermined coefficients. That connection was an important part of Kovalevskaya's achievement.

He called the same concept "canonical form" in a later paper, published in vol. 64 of the *Journal*, saying that "a system of ordinary differential equations is *noncanonical* if the equations contain the highest-order derivatives of the dependent variables in such a way that they cannot be determined from the equations." (That is, the equations cannot be solved for the highest-order derivatives.) In the papers just mentioned he gave a combinatorial procedure for reducing equations to canonical or normal form.

3.3. Kovalevskaya's theorems. Kovalevskaya considered a very general ordinary differential equation

$$G\left(x, y, \frac{dy}{dx}, \ldots, \frac{d^ny}{dx^n}\right) = 0$$

---

[1]In the same paper in which he introduced normal form Jacobi developed his "last multiplier method," which turns out to have an important connection with Kovalevskaya's other major papers.

where G is a polynomial. Using Jacobi's techniques for bringing the equation into canonical form, she reduced this equation to a first-order system

$$D_{n+2}G(x, y_0, y_1, \ldots, y_n)\frac{dy_1}{dx} - G_1(x, y_0, y_1, \ldots, y_n) = 0$$

$$\vdots$$

$$D_{n+2}G_n(x, y_0, y_1, \ldots, y_n)\frac{dy_n}{dx} - G_n(x, y_0, y_1, \ldots, y - n) = 0$$

where $G_1, \ldots, G_n$ are polynomials. The method of majorants can then be applied to show that the power series generated assuming any initial values for $y_0(0), \ldots, y_n(0)$ actually has a positive radius of convergence.

Kovalevskaya's most general theorem concerned a partial differential equation

$$G\left(x, x_1, \ldots, x_r, \varphi, \ldots, \frac{\partial^{\alpha+\alpha_1+\cdots+\alpha_r}\varphi}{\partial x^{\alpha}\partial x_1^{\alpha_1}\cdots\partial x_r^{\alpha_r}}, \ldots\right) = 0$$

where G is an irreducible polynomial in its variables and the maximum order of the derivatives is n, that is, $\alpha + \alpha_1 + \cdots + \alpha_r \leq n$. Kovalevskaya described normal form as follows:

I shall now consider the case, which I shall call normal, *in which at least one of the derivatives* $(\partial^n\varphi)/(\partial x^n), (\partial^n\varphi)/(\partial x_1^n), \ldots, (\partial^n\varphi)/(\partial x_r^n)$, *say* $(\partial^n\varphi)/(\partial x^n)$, *actually occurs in* G.

The vague phrase "actually occurs" seems to mean that the equation can be *solved* for that derivative.

Under this assumption she was able to show that it was possible to find a function $\varphi(x, x_1, \ldots, x_n)$ that is analytic in a neighborhood of the point $(a, a_1, \ldots, a_n)$ and satisfies the equation and the initial conditions

$$\varphi(a, x_1, \ldots, x_r) = f_0(x_1, \ldots, x_r),$$

$$\frac{\partial\varphi}{\partial x}(a, x_1, \ldots, x_r) = f_1(x_1, \ldots, x_r),$$

$$\vdots$$

$$\frac{\partial^{n-1}\varphi}{\partial x^{n-1}}(a, x_1, \ldots, x_r) = f_{n-1}(x_1, \ldots, x_r),$$

provided the functions $f_0, \ldots, f_{n-1}$ are analytic on a neighborhood of $(a_1, \ldots, a_{n-1})$.

This result is now known as the *Cauchy–Kovalevskaya* theorem, and it remains to this day one of the fundamental results in this area. As with any mathematical result, it depends on clever and efficient use of results and techniques derived by earlier mathematicians. Kovalevskaya's main contribution, as I stated above, was to recognize the importance of normal form as a sufficient condition for the functioning of the method of majorants and hence for the validity of the method of undetermined coefficients.

In this paper Kovalevskaya unified a number of ideas of earlier mathematicians and brought clarity and systematization to an area that had previously lacked coherence. She provided a counterexample, showing that a problem might fail to have an analytic solution in the absence of normal form. She considered the one-dimensional heat equation

$$\frac{\partial u}{\partial t} = \frac{\partial^2 u}{\partial x^2},$$

with the *spatial* initial condition $u(x,0) = 1/1 - x$. For this problem the method of undetermined coefficients generates the power-series

$$u(x,t) = \sum_{\nu=0}^{\infty} \frac{(2\nu)!}{\nu!} \frac{t^\nu}{(1-x)^{2\nu+1}},$$

which diverges for all $t \neq 0$.

A physicist might object that Kovalevskaya's initial temperature distribution becomes infinite at the point $x = 1$. However, the same result would be achieved with the initial condition $u(x,0) = 1/1 + x^2$, which is physically realizable. The series generated by the method would diverge for all values of t except $t = 0$.

By contrast, since the equation is of order 2, Kovalevskaya's theorem guarantees that any prescribed temperature and temperature gradients at the point $x = 0$ that are analytic functions of time will yield an analytic temperature $u(x,t)$. The crucial property needed to get analytic solutions is being able to solve the differential equation for a pure derivative *of the same order as the equation.*

Weierstrass was astounded by this counterexample. He stated in his letter recommending the degree for Kovalevskaya that he had believed that the equations of mathematical physics would necessarily have analytic solutions. Apparently he was guided by physical intuition to a greater degree than the prevailing view of him would suggest. Analytic functions have a deterministic property that harmonizes very well with the possibility of physical prediction, as Lagrange had pointed out. (Knowing the values of a physical system over any finite interval of time makes it possible to predict the future state of the system for all time, provided the functions describing it are analytic.) Kovalevskaya had shown the presence of a previously unsuspected subtlety in this determinism.

4. The Rotation of a Rigid Body

Kovalevskaya's papers on the rotation of a rigid body about a fixed point represent an attempt to penetrate an unexplored region of mathematical physics using the resources of analytic function theory. The problem she was attempting to solve is a system of six ordinary differential equations. Her approach was a combination of the two approaches—power-series and closed-form—mentioned above. The way in which the two approaches were combined is interesting in itself. She first asked whether the equations could have meromorphic solutions, that is, solutions expressible as power series in time (including possibly a finite number of negative powers of time). That portion of the problem is the power-series approach. Two such cases were already known and had been studied in detail by Euler and Lagrange. Having determined the third and only new case in which such solutions exist, she set out to find the solution in closed form, thereby making use of the very latest results of mathematical research in the form of theta functions of two variables.

4.1. The differential equations of motion. Physicists know very well that the particular differential equations that describe a physical system are not obvious consequences of the general laws of physics such as conservation of energy or Newton's second law of motion. Every mathematical model results from a conscious decision to include certain properties and neglect others. For example, in the famous description of a vibrating string held fast at two points, one can represent the displacement of the string above the point with coordinate x at time t as a function

$u(x, t)$ and derive the equation

$$\frac{\partial^2 u}{\partial t^2} = c^2 \frac{\partial^2 u}{\partial x^2}.$$

This model assumes that the tension in the string is the same at all times and all points, that an individual point on the string moves only vertically, not horizontally, that all powers of the slope $\partial u/\partial x$ higher than the first can be ignored, and so forth. It requires considerable insight and a bit of luck to come up with a judicious selection of properties yielding a model that provides accurate predictions. With the passage of time, good models get reformulated and refined, as Newton's basic outline of the laws of motion was refined to include the concept of energy and given a more elegant mathematical formulation in Lagrange's exposition of mechanics, which was then further refined in the presentation of Hamilton.

So it was with the equations of motion of a rotating rigid body. It was Euler who first discovered that every solid body has a set of three mutually perpendicular axes such that the products of inertia of the body become zero in these axes. That is, if the axes are labeled x, y, and z, the body occupies a region $\mathcal{B}$, and the mass distribution of the body is dM, then

$$\int_{\mathcal{B}} yz\, dM = \int_{\mathcal{B}} zx\, dM = \int_{\mathcal{B}} xy\, dM = 0.$$

The motion of the body is most easily described by giving the nine direction cosines of these canonical body axes at any time t in terms of a set of mutually perpendicular axes fixed in space. (In modern terms, this amounts to giving the rotation matrix $R(t)$ that changes spatial coordinates into body coordinates.) To solve this problem for the case when no external torque acts on the body, Euler derived a set of nine differential equations, only the first three of which became a permanent addition to the literature on this problem. These three equations were

$$dx + \frac{c^2 - b^2}{a^2} yz\, dt = 0, \quad dy + \frac{a^2 - c^2}{b^2} zx\, dt = 0, \quad dz + \frac{b^2 - a^2}{c^2} xy\, dt = 0.$$

Here the quantities a^2, b^2 and c^2 are the moments of inertia of the body about its principal axes, that is,

$$a^2 = \int_{\mathcal{B}} (y^2 + z^2)\, dM, \quad b^2 = \int_{\mathcal{B}} (z^2 + x^2)\, dM, \quad c^2 = \int_{\mathcal{B}} (x^2 + y^2)\, dM.$$

The variables x, y, and z are the coordinates of a point on the body. Euler discussed in detail the motion of a body upon which no external torque is acting and showed that the direction cosines needed to describe the motion in this case can be expressed as elliptic integrals, reducing to elementary functions if two of the three principal moments of inertia are equal.

The fact that elliptic integrals arise even in the simplest case of the motion of a rigid body showed that the mathematics involved in describing such motion in general was likely to be very complicated. As Euler wrote,

... when difficulties are... encountered because of the large number of variables, it is no longer in Mechanics that one should seek the means to overcome them, since it seems that the nature of such a motion is not susceptible of any simpler calculation. Everything then devolves onto the calculator, who must render the necessary support in analysis so as to solve the equations that determine the motion; but there is no

doubt that an infinite number of cases exist that are absolutely unsolvable due to the limitations of analysis....

4.2. The Lagrange case. The next advance in the subject came with Lagrange's reformulation of Newtonian mechanics in terms of variational principles. Lagrange was able to study the general case of motion of a body having equal moments of inertia about two of its principal axes and its center of gravity on the third principal axis provided the external torque acting on the body results purely from gravitational attraction. To this day Lagrange's case is the one most frequently seen in practice. It was referred to in the nineteenth century as the case of a "heavy" body. In this case also, the solutions turned out to be expressible as elliptic functions.

4.3. The work of Jacobi. Once again the work of Jacobi appears as one of the key elements in Kovalevskaya's work. Although his influence does not show directly in the papers she wrote, it was crucial in setting the problem she solved in two different ways. First, his work on methods of solving equations in closed form had led him to a generalization of the technique that applies to exact differential equations. In general solving a system of $n-1$ ordinary differential equations of the form

$$\frac{dx_1}{X_1} = \frac{dx_2}{X_2} = \cdots = \frac{dx_n}{X_n},$$

requires finding $n-1$ independent "integrals," that is, functions $\varphi(x_1, \ldots, x_n)$ that are constant when the variables $x_1, \ldots, x_n$ satisfy the differential equation. For a system in which the divergence of the vector $(X_1, \ldots, X_n)$ vanishes, that is,

$$\frac{\partial X_1}{\partial x_1} + \cdots + \frac{\partial X_n}{\partial x_n} = 0,$$

and $n-2$ independent integrals have been found, Jacobi showed how to form the last independent integral by integrating an exact differential equation in two variables. This method became known as his "last-multiplier" method. The significance of this result for the problem of rotation of a rigid body will appear below.

Second, since Jacobi had shown that elliptic functions can be most elegantly expressed as quotients of theta functions of one variable, it followed implicitly that the solution of the problem of a rotating rigid body in the cases of Euler and Lagrange must be expressible in terms of these functions. Jacobi made this implicit conclusion explicit in an 1849 paper in which he wrote out the direction cosines of the body axes in terms of space axes using the two theta functions θ and H exhibited above. For example, the entry in the first row and first column of the rotation matrix $R(t)$ mentioned above is

$$\frac{\theta(K)[H(u+ia) + H(u-ia)]}{2H(K-ia)\theta(u)}$$

where u is a linear function of time ($u = nt + \tau$).

The importance of Jacobi's work was shown by the problem posed as the topic for the Leibniz Prize at the Prussian Academy of Sciences in 1852:
It is known that the number of cases in which the differential equations of analytic dynamics can be integrated in closed form, or even reduced to quadratures, is quite limited; and in view of the repeated efforts that the greatest mathematicians have applied to this subject, it is very probable that most of the mechanical problems

whose solution has not yet been achieved in this form are by their nature not susceptible of integration by quadratures and can be effectively handled only through the introduction of other analytic forms. Since Jacobi has recently given a beautiful representation in series form of the rotation of a rigid body on which no accelerative force acts, it seems worthwhile to attempt to give a wider expression to the application of series and use them to handle cases of rotational motion that have not yet been reduced to quadratures. One such case is offered by the rotation of a heavy body, for which reduction to quadratures has been attained only in one special case due to Lagrange. The Academy therefore makes the complete solution of this problem the subject of a competition and poses the problem:

> "Integrate the differential equations for the motion of a body rotating about a fixed point on which the only accelerative force is that of gravity, by means of regular series representing as explicit functions of time all quantities required to determine the motion."

Despite the prize of 100 ducats, however, no entries were received in this competition by the deadline of 1855. The competition was extended for another three years, again without any entries.

4.4. Weierstrass and C. Neumann. Since Weierstrass was in Berlin at this time, he cannot have been unaware of this competition. From a mathematical point of view the problem posed for the competition would have been particularly attractive to him at that time, since he had been one of the pioneers in using the newly-discovered theta functions of two variables to solve the general Jacobi inversion problem. It undoubtedly seemed to him that the time had come to make new advances in the problem of rigid body motion. There were other advances along these lines as well that would have convinced him that this problem could now be solved, the most important of which was the dissertation of Carl Neumann (1832–1925). The problem Neumann posed was *to describe the motion of a point constrained to move on the unit sphere* $x^2+y^2+z^2=1$ *and subject to a force whose potential is* $u(x,y,z)=ax^2+by^2+cz^2$, where a, b, and c are real numbers (not necessarily positive). Neumann found four possible cases of motion, depending on the values of the initial conditions imposed on the motion. Most importantly, he found that by using the Hamilton–Jacobi formulation of the equations of motion, parameterizing the sphere by a pair of coordinates λ_1, λ_2, he obtained the following equations for the trajectory of a solution

$$\varepsilon_1\int_{l_1}^{\lambda_1}\frac{d\lambda}{L}+\varepsilon_2\int_{l_2}^{\lambda_2}\frac{d\lambda}{L}=0,\qquad \varepsilon_1\int_{l_1}^{\lambda_1}\frac{\lambda\,d\lambda}{L}+\varepsilon_2\int_{l_2}^{\lambda_2}\frac{\lambda\,d\lambda}{L}=t$$

where $L=\sqrt{8(\lambda-a)(\lambda-b)(\lambda-c)(\lambda-A)(\lambda-B)}$, A and B are arbitrary constants, $\varepsilon_1=\pm1$ and $\varepsilon_2=\pm1$. (The four possible values of the pair $(\varepsilon_1,\varepsilon_2)$ give the four cases of motion distinguished by Neumann.) These are, of course, *exactly* the equations of the Jacobi inversion problem! Naturally Neumann was able to express the solutions in terms of theta functions of two variables.

Weierstrass was aware of Neumann's work and no doubt intrigued by the application of theta functions of two variables. He must have seen this area as a fruitful one for further investigation. As we shall see below, he suggested this line of work to Kovalevskaya.

4.5. The mathematical mermaid. It is time to return to Jacobi's other contribution to this problem, namely his development of the last-multiplier method of solving a system of ordinary differential equations with zero divergence. The final form of the Euler equations was given by the British mathematician R. B. Hayward in 1858. For a body subject only to gravitational torque having one point fixed at $(0, 0, 0)$ with center of gravity at the point (x_0, y_0, z_0) in a set of mutually perpendicular axes fixed in the body, the system consists of the following six equations for six unknown functions $p(t)$, $q(t)$, $r(t)$, $\gamma_1(t)$, $\gamma_2(t), \gamma_3(t)$ which are respectively the three coordinates of the angular velocity vector of the body and the three coordinates (in body coordinates) of a unit vector pointing downward in space.

$$\begin{aligned} A\frac{dp}{dt} &= (B-C)qr + Mg(y_0\gamma_3 - z_0\gamma_2), & \frac{d\gamma_1}{dt} &= r\gamma_2 - q\gamma_3, \\ B\frac{dq}{dt} &= (C-A)rp + Mg(z_0\gamma_1 - x_0\gamma_3), & \frac{d\gamma_2}{dt} &= p\gamma_3 - r\gamma_1, \\ C\frac{dr}{dt} &= (A-B)pq + Mg(x_0\gamma_2 - y_0\gamma_1), & \frac{d\gamma_3}{dt} &= q\gamma_1 - p\gamma_2. \end{aligned}$$

Here A, B, and C are the three principal moments of inertia (which Euler had denoted a^2, b^2, c^2), M is the mass of the body, and g is the acceleration of gravity. For purposes of analysis, it is important to observe that they really give six different expressions for dt, none of which involves time. That is, they can be written as the set of five equations

$$\frac{dp}{P} = \frac{dq}{Q} = \frac{dr}{R} = \frac{d\gamma_1}{\Gamma_1} = \frac{d\gamma_2}{\Gamma_2} = \frac{d\gamma_3}{\Gamma_3}$$

where none of the equations contains t, and each of the denominators is a (quadratic) polynomial in the six variables. Even more convenient is the fact that the polynomial P is independent of p, Q is independent of q, and so forth, that is, each denominator is independent of the variable that occurs in the differential of its numerator. That means certainly that the divergence of the system is zero. Hence the Jacobi last-multiplier method will yield a complete set of five independent integrals provided we can find four independent integrals.

Now three integrals are immediate:

1. Conservation of energy shows that the function $(Ap^2 + Bq^2 + Cr^2) + Mg(x_0\gamma_1 + y_0\gamma_2 + z_0\gamma_3)/2$ must be constant.
2. Since the force is vertical, the torque is horizontal, and hence the vertical component of the angular momentum must be constant, that is, $Ap\gamma_1 + Bq\gamma_2 + Cr\gamma_3$ must be constant.
3. Since $(\gamma_1, \gamma_2, \gamma_3)$ is a unit vector, $\gamma_1^2 + \gamma_2^2 + \gamma_3^2 = 1$.

Thus the general problem would be solved if only one more independent integral could be found. Where should one look for it? No one knew at the time. That is why the problem was so tempting, and yet so elusive, that the Germans referred to as *die mathematische Nixe* (the mathematical mermaid).

4.6. Kovalevskaya's work. Kovalevskaya worked on this problem while she was studying with Weierstrass, but she was not able to make any progress on it. It was while she was trying to re-establish her mathematical credentials, in the early 1880s that she found herself returning to this problem with renewed hope. In

November 1881 she wrote to Mittag–Leffler,

... The problem involves solving the general case of rotation of a heavy body about a fixed point by means of Abelian functions.... The route I followed consisted in expressing the variables of the problem by theta functions of two variables that for certain values of the constants reduce to the elliptic theta functions that arise in the special case of Lagrange, then trying to choose them so as to be able to integrate the differential equations between the theta functions and time. The calculations which I got into in this way were so difficult and complicated that I cannot yet say if I will reach the desired end by this route. In any case I hope to know in two or three weeks at the most what I should do about it, and M. Weierstrass is consoling me that even in the worst case I could always invert the question and try to find out which forces lead to a rotation whose variables can be expressed by Abelian functions—a poor problem, to be sure, and one far from having the same interest as the one I have set myself, but one which I shall have to settle for if I have bad luck, relying on the example of M. Neumann, who chose an analogous problem for his doctoral dissertation.

Working her ideas into a substantial result took Kovalevskaya another four years. Only in June 1886 was the work sufficiently advanced to present to Hermite and his colleagues in Paris. As is well-known, the Paris Academy, at the instigation of Hermite, arranged a prize competition suitable for this work, which Kovalevskaya entered and won in 1888.

4.7. The result: Part I. The general equations of motion of a heavy body about a fixed point turned out to be too complicated to have meromorphic solutions (as Kovalevskaya indicates Weierstrass had already shown, though he never published this work). Taking as the basic problem the system of five equations, Kovalevskaya asked when such a system could be satisfied by six meromorphic functions of time $p(t)$, $q(t)$, $r(t)$, $\gamma_1(t)$, $\gamma_2(t)$, $\gamma_3(t)$. Thus she assumed the system of equations

$$\begin{aligned}
p(t) &= t^{-n_1}(p_0 + p_1 t + p_2 t^2 + \cdots),\\
q(t) &= t^{-n_2}(q_0 + q_1 t + q_2 t^2 + \cdots),\\
r(t) &= t^{-n_3}(r_0 + r_1 t + r_2 t^2 + \cdots),\\
\gamma_1(t) &= t^{-m_1}(f_0 + f_1 t + f_2 t^2 + \cdots),\\
\gamma_2(t) &= t^{-m_2}(g_0 + g_1 t + g_2 t^2 + \cdots),\\
\gamma_3(t) &= g^{-m_3}(h_0 + h_1 t + h_2 t^2 + \cdots).
\end{aligned}$$

The equations themselves immediately yield the fact that $n_1 = n_2 = n_3 = 1$ and $m_1 = m_2 = m_3 = 2$, that is, the components of the angular velocity have at worst a simple pole, and those of the unit vertical vector at worst a pole of order two.

If the general solution of the system is to be expressed in this way, five of the coefficients of the power series must be arbitrary. The equations provide a recursive system of linear equations for each set of coefficients. In order to get the required five degrees of freedom, the complete set of these systems must have a rank deficiency of five. Kovalevskaya, overlooking the case in which one of the systems might have a rank deficiency greater than one, examined the cases in which five different systems each have rank deficiency one, and found that this is possible only if $A = B = 2C$ and $z_0 = 0$. That is the *Kovalevskaya case* of motion. It means that the body has partial kinetic symmetry and that its center of gravity lies in the plane through the fixed point spanned by the two principal axes about which the moments of inertia are equal. (To exhibit an example of a body having these properties Kovalevskaya appealed to Hermann Amandus Schwarz, another student of Weierstrass, who produced a model consisting of a pair of identical cylinders with parallel axes.) Thus, Kovalevskaya had carried out the modest goal Weierstrass had suggested, having discovered the last case in which the equations of the problem can be expressed as meromorphic functions of time.

4.8. Bypassing the last-multiplier method. After discovering this case, Kovalevskaya also discovered along the way a fourth algebraic integral. Normalizing the units so that $A = B = 2$, $C = 1$, and choosing axes in the plane where all moments of inertia are equal in such a way that the center of gravity lies on one of the axes, one gets $y_0 = 0$ also. For this case the fourth algebraic integral can be written in complex form as

$$\bigl[(p+qi)^2 + Mgx_0(\gamma_1 + i\gamma_2)\bigr]\bigl[(p-qi)^2 + Mgx_0(\gamma_1 - i\gamma_2)\bigr].$$

At this point, *in principle*, Jacobi's last-multiplier method provides a recipe to follow to construct the fifth independent integral, after which the trajectories could be made to appear as (in modern terms) the intersections of the level surfaces of these five functions of the six variables. One can well imagine, however, that the computations involved are going to be very complicated. In fact, Kovalevskaya does not follow this route, nor does she even mention the last-multiplier method. Instead she resorted to an intricate series of changes of variable in order to get back, in the end, to the same Jacobi inversion problem that Neumann had derived. Her system was in fact simply the differentiated form of Neumann's system:

$$0 = \frac{ds_1}{\sqrt{\mathcal{R}(s_1)}} + \frac{ds_2}{\sqrt{\mathcal{R}(s_2)}}, \qquad dt = \frac{s_1\,ds_1}{\sqrt{\mathcal{R}(s_1)}} + \frac{s_2\,ds_2}{\sqrt{\mathcal{R}(s_2)}},$$

in which $\mathcal{R}(s)$ is a polynomial of degree 5.

4.9. The result: Part 2. Complicated as this work was, it was only the beginning of Kovalevskaya's work. The expression of the variables s_1 and s_2 as quotients of theta functions of two variables whose arguments are linear functions of time was by 1886 a standard procedure. It took Kovalevskaya 16 pages to get this far on the problem. The true depth of the problem lies in the recovery of the original variables from the intricate series of changes of variables mentioned above. That part of the work required 50 more pages of very discouraging computation. In the end she was able to express the components of the angular velocity vector as quotients of the following form:

$$p = -i\frac{L_1P_1 + M_1P_2 + N_1P_3}{LP_1 + MP_2 + NP_3}$$

where L, M, N, L_1, M_1, N_1 are constants determined algebraically from the data of the problem, and P_1, P_2, and P_3 are quotients of theta functions of two variables, each of which is a linear function of time.

Even our abbreviated sketch of the paper shows how very complicated the motion of a rigid body is, even for a body possessing the considerable amount of symmetry of the Kovalevskaya case. The analytic description of the general case must necessarily be at least this complicated, and probably far worse. Kovalevskaya completed her work by showing that the components of the rotation matrix that describes the position of the body axes in terms of spatially fixed axes can all be expressed using quotients of theta functions of two variables, each of which is a linear function of time.

4.10. Impact of the work. The report of the jury for the 1888 Bordin Prize showed the aspect of the work that contemporaries considered to be of most importance:
The author has done more than merely adding a result of very high interest to those bequeathed to us by Euler and Lagrange; he has made a profound study of his result in which the resources of the modern theory of theta functions of two independent variables make it possible to give the solution in the most precise and elegant form. The result is a new and memorable example of a mechanical problem involving these transcendental functions, whose applications had previously been confined to pure analysis and geometry.

This point of view was shared by Weierstrass and Hermite. There seemed to be a hope that new advances could be made in mathematical physics by solving more complicated differential equations and that the theory of Abelian functions would play an important role in this advance.

5. Minor Works

In addition to her work on the two topics discussed in detail above, Kovalevskaya published four other mathematical papers, three of which date to her student years in Berlin and one of which was from the early 1880s. Although one of these, on hyperelliptic integrals that degenerate to elliptic integrals, was an elegant piece of work that impressed Poincaré, none of them would be remembered today as important contributions to nineteenth-century analysis. We shall therefore describe each of them in a few sentences.

5.1. Degenerate Abelian integrals. There are certain values of the parameter in a hyperelliptic integral that cause it to become an elliptic integral. Weierstrass had discovered that such would be the case if the theta functions associated with the integral had a certain form. This criterion was a "transcendental" condition, since theta functions are not themselves algebraic functions. Kovalevskaya found an algebraic criterion based on the location of the points where the curve associated with the integral has double tangents. This paper lay unpublished for ten years until Kovalevskaya succumbed to the "publish or perish" ambience of the day and published it in the *Acta Mathematica* in 1884. There it attracted the attention of Poincaré, who found it sufficiently impressive to publish his own extensions of it.

5.2. The shape of Saturn's ring. This paper occupies an anomalous position among Kovalevskaya's works. It is not referred to in any of the letters she received from Weierstrass as a student. The paper was based on Laplace's study of a ring orbiting a planet for which the shape of a cross-section was determined by assuming that a thin layer of liquid on the surface would be in equilibrium. (Not that the ring was actually believed to be liquid. Rather this assumption was a way of saying that no shear stresses were acting on the ring as a result of the gravitational attraction of the planet.) Laplace had assumed an elliptic cross-section of the ring. Kovalevskaya attempted to generalize this assumption by taking the cross section in polar coordinates as the solution of the equation

$$r(\theta) = m_0 + m_1 \cos\theta + m_2 \cos 2\theta + \cdots .$$

Truncating the series by setting $m_2 = m_3 = \cdots = 0$ yields the Laplace case. Kovalevskaya carried the series out to one more term, obtaining an oval shape. This is the only paper in which she made serious efforts at numerical approximation. Like the paper on Abelian integrals, this one formed part of her dissertation and lay unpublished for a decade until her friend the astronomer Hugo Gyldén published it for her in the *Astronomische Nachrichten.*

5.3. The Lamé equations. When Kovalevskaya was trying to re-establish her credentials as a mathematician in 1881, Weierstrass gave her a paper he had written many years earlier containing a surface-integral method of solving certain differential equations. He suggested that she apply this technique to solve the differential equations that Gabriel Lamé (1795–1870) had derived for propagation of a wave in a solid medium. Lamé had hoped to explain light as an elastic disturbance in a medium, but the solutions of his equations could not explain propagation of light from a point source, since the solutions became infinite at the source. Weierstrass hoped that his technique would yield new solutions that did not become infinite at that point. (A philosophical issue hung on the outcome, as Lamé had been driven to the point of introducing a weightless either in order to uphold the application of his beloved elasticity theory in the explanation of light.) Despite enormous personal hardship at the time, Kovalevskaya produced formulas for a solution. Weierstrass, however, wasn't feeling well at the time and delayed his consent to publication of his portion of the paper, which was of extreme importance to Kovalevskaya. In the end, he allowed Kovalevskaya to publish the paper in the *Acta Mathematica* with quotation marks around the portion of it due to him. Although the paper was highly regarded and was proofread by no less a mathematician than Carl Runge (1856–1927), apparently no one attempted to apply it to any particular problem until Vito Volterra (1860–1940) did so in early 1891, just at the time of Kovalevskaya's death. Reading this paper in preparation for giving a course on light propagation, Volterra discovered that in fact Kovalevskaya's formulas do not satisfy the Lamé equations. Thus it was necessary for Mittag-Leffler to print a retraction of this paper simultaneously with his obituary of Kovalevskaya.

5.4. Bruns' theorem. E. H. Bruns (1848–1919) is best known for having proved that the differential equations of the three-body problem have only a ten-dimensional space of algebraic integrals. He also proved that the potential of a body bounded by a surface $W(x, y, z) = 0$, where W is an analytic function, is analytic at each regular point of the surface (each point where the gradient of W does not vanish). The proof of this result involves constructing an analytic solution of a boundary-value problem for Poisson's equation in the region. Kovalevskaya derived this result as a corollary of the Cauchy–Kovalevskaya theorem. There seems to be little doubt that this paper was written while she was working on her dissertation; Weierstrass alludes to such a paper in his letter requesting her degree. However, she did not decide to publish it until 1889, when, no doubt exhausted from the work on the rotation of a heavy body, she needed a paper to give at a conference. Mittag–Leffler published it after her death in the *Acta Mathematica.*

References

1. Niels Henrik Abel, *Mémoire sur une propriété générale d'une classe très étendue de fonctions transcendantes*, Mém. Présentés par Divers Savants **7** (1841).
2. Reinhard Bölling, ed., *Briefwechsel. Karl Weierstraß—Sofja Kowalewskaja*, Akademie Verlag, Berlin, 1993.
3. Augustin Louis Cauchy, *Équations différentielles ordinaires*, Cours inédit, Éditions Études Vivantes, Ltée., Ville Saint-Laurent, Qc, Johnson Reprint Corp., New York, 1981.
4. ______, *Mémoire sur un théorème fondamental dans le calcul intégrale*, Comptes Rendus de l'Académie des Sciences **14** (1842), 1020–1026.
5. ______, *Mémoire sur l'emploi du nouveau calcul des limites dans l'intégration d'un système d'équations différentielles*, Comptes Rendus de l'Académie des Sciences **15** (1842), 44–59.
6. ______, *Mémoire sur l'application du calcul des limites à l'intégration d'un système d'équations aux dérivées partielles*, Comptes Rendus de l'Académie des Sciences **15** (1842), 85–101.
7. ______, *Mémoire sur les systèmes d'équations aux dérivées partielles d'ordre quelconque, et sur leur réduction à des systèmes d'équations linéaires du premier order*, Comptes Rendus de l'Académie des Sciences **15** (1842), 131–138.
8. Roger Cooke, *Abel's theorem*, The History of Modern Mathematics (Poughkeepsie, NY, 1989), vol. 1, Academic Press, Boston, MA, 1989, pp. 389–421.
9. H. R. Dullin, P. H. Richter, and A. P. Veselov, *Action variables of the Kovalevskaya top. J. Moser at* 70, Regul. Chaotic Dyn. **3** (1998), no. 3, 18–31
10. Leonhard Euler, *Du mouvement de rotation des corps solides autour d'un axe variable*, Mém. Acad. Sci. Berlin **14** (1758), 154–193.
11. ______, *Theoria Motus Corporum Solidorum seu Rigidorum*, Rostock, 1765.
12. Jean-Pierre Françoise, *Sur les action-angles de la toupie de Kowalevski*, C. R. Acad. Sci. Paris Sér. I Math. **300** (1985), no. 13, 427–430.
13. Carl Gustav Jacob Jacobi, *De investigando ordine systematis æquationum differentialium vulgarium cujuscunque*, J. Reine Angew. Math. **64** (1865), 297–320.
14. ______, *De æquationum differentialium systemate non normali ad formam normalem revocando*, C. G. J. Jacobi's Gesammelte Werke, vol. 5, 1890, pp. 483–513.
15. ______, *Sur la rotation d'un corps*, C. R. Acad. Sci. **24** (1849), 97–106.
    ______, *Sur la rotation d'un corps*, J. Reine. Angew. Math. **39** (1849), 293–350.
16. ______, *S. V. Kovalevskaya's mathematical legacy: The rotation of a rigid body*, Vita Mathematica (Toronto, 1992; Quebec City, 1992), 1MAA Notes, vol. 40, Math. Assoc. America, Washington, DC, 1996, pp. 77–190.
17. ______, *Sonya Kovalevskaya's place in nineteenth-century mathematics*, Contemp. Mathematics **64** (1987), 17–51.
18. ______, *Considerationes generales de transcendentibus abelianis*, J. Reine Angew. Math. **9** (1832), 394–403.
19. Paul R. Garabedian, *Partial differential equations*, John Wiley & Sons, Inc., New York-London-Sydney, 1964.
20. V. V. Golubev, *The work of S. V. Kovalevskaya on the motion of a rigid body about a fixed point* (Работа С. В. Ковалевской о движении твердого тела вокруг неподвижной точки), Appl. Math. Mech. **14**, (1950), 236–244.
21. ______, *Lectures on the integration of the equations of motion of a rigid body about a fixed point*, National Science Foundation, 1953.
22. R. B. Hayward, *On a direct method of estimating velocities, accelerations, and all similar quantities with respect to axes movable in any manner in space*, Trans. Cambridge Philos. Soc. **10** (1858), 1–22.
23. Lars Hörmander, *The analysis of linear partial differential operators.* I. *Distribution theory and Fourier analysis*, Fundamental Principles of Mathematical Sciences, vol. 256, Springer-Verlag, Berlin, 1983.
24. P. Ya. Kochina, ed., *The correspondence of S. V. Kovalevskaya and G. Mittag-Leffler* (Переписка С. В. Ковалевской и Г. Миттаг-Леффлера), Scientific Heritage, vol. 7, Nauka, Moscow, 1984.
25. S. V. Kowalewsky, *Zur Theorie der partiellen Differentialgleichungen*, J. Reine Angew Math. **80** (1875), pp. 1–32.

26. ———, *Über die Reduction einer bestimmten Klasse abel'scher Integrale dritten Ranges auf elliptische Integrale*, Acta Math. **4** (1884), 51–74.
27. ———, *Über die Brechung des Lichtes in crystallinischen Mitteln*, Acta Math. **6** (1885), 249–304.
28. ———, *Zusätze und Bemerkungen zu Laplace's Untersuchung über die Gestalt der Saturnringe*, Astronom. Nachr. **111** (1885), 37–48.
29. ———, *Mémoire sur un cas particulier de la rotation d'un corps solide autour d'un point fixe*, Acta Math. **12** (1889), 235–244.
30. ———, *Mémoire sur un case particulier du problème de la rotation d'un corps solide autour d'un point fixe, où l'intégration s'éffectue à l'aide de fonctions ultraélliptiques du temps*, Mém. Présentés par Divers Savants **31** (1890), 1–62.
31. ———, *Sur une propriété du système d'équations différentielles qui définit la rotation d'un corps solide autour d'un point fixe*, Acta Math. **14** (1890), 81–93.
32. ———, *Sur un théorème de M. Bruns*, Acta Math. **15** (1891), 45–52.
33. S. V. Kovalevskaya and Nauchnye Raboty, *S. V. Kovalevskaya, scientific works*, USSR Academy of Sciences Press, 1948.)
34. Joseph Louis Lagrange, *Mécanique analytique*, Paris, 1811–1815.
35. Carl Neumann, *De problemate quodam mechanico, quod ad primam integralium ultraellipticorum classem revocatur*, (Dissertation, Königsberg, 1856), J. Reine. Angew. Math. **56** (1860), 46–63.
36. Karl Weierstraß, *Vorlesungen über die Theorie der abelschen Transcendenten*, Math. Werke, vol. 4, 1897.
37. ———, *Definition analytischer Functionen einer Veränderlichen vermittelst algebraischer Differentialgleichungen*, Math. Werke, vol. 1, 1894, pp. 75–84.

Department of Mathematics, University of Vermont, Burlington, VT 05401, USA.
E-mail address: cooke@emba.uvm.edu
URL: http://www.emba.uvm.edu/~cooke

Centre de Recherches Mathématiques
CRM Proceedings and Lecture Notes
Volume **32**, 2002

The KZB Connection: Parametrizations, Flat Sections and q-Deformation

Benjamin Enriquez

Wess–Zumino–Witten (WZW) theories are conformal field theories associated with a simple Lie algebra $\mathfrak{g}$. These theories lie at the heart of the interaction of many different domains of mathematics: representation theory of quantum groups and affine Kac–Moody algebras, hypergeometric theory, 3-dimensional topology, etc. Suitable limits of the WZW theories also give rise to interesting integrable systems, namely the Gaudin system and higher genus analogues (see Kuroki and Takebe's paper in the volume [**44**]).

An important building block of a WZW theory is a certain flat connection. This connection reflects the differential equations satisfied by correlators of a "chiral half" of the WZW theory. These were found by Knizhnik and Zamolodchikov in the genus zero case, and by Bernard in genus ≥ 1. Bernard's approach followed the construction of [**32**]; later, the authors of [**57**] gave a formulation of this construction in the language of stacks. To each family $(\mathfrak{g}, (g, n), (k, \lambda_1, \dots, \lambda_n))$, where $\mathfrak{g}$ is a simple Lie algebra $\mathfrak{g}$, (g, n) is a pair of integers, k is a complex number, and $(\lambda_1, \dots, \lambda_n)$ is a n-uple of weights of $\mathfrak{g}$, we then associate a flat connection (the KZB connection) over the moduli space of complex curves of genus g with n marked points. The underlying bundle is called the *bundle of conformal blocks*.

In this paper, I will first review some aspects of the theory of the Knizhnik–Zamolodchikov–Bernard (KZB) connection, and of its q-deformation. Then I will explain my results in this subject. They all have been obtained in collaboration, either with G. Felder or V. Rubtsov, whom I would like to thank here. In Section 5, I will mention some open problems related to these results.

1. The KZB Connection

This section is a short review of the KZB connection and its developments. I first present the case $g = 0$ explicitly (Subsection 1.1). Then I explain the construction of flat sections for the connection in this case (Subsection 1.2). The representation-theoretic interpretation of the KZB connection and its generalizations in higher genus are presented in Subsection 1.3. In Section 2, we will see how this representation-theoretic interpretation naturally leads to the q-deformed versions of the KZ equation.

2000 *Mathematics Subject Classification.* Primary: 81R10; Secondary: 81R50.
This is the final form of the paper.

A more detailed account of the material reviewed here can be found e.g. in [**19**].

1.1. The Knizhnik–Zamolodchikov connection in genus 0. The KZ connection can be presented as follows. Let n be an integer, $\mathfrak{g}$ be a simple Lie algebra, $V_1, \dots, V_n$ be a collection of highest weight $\mathfrak{g}$-modules, and κ be a nonzero complex number. Let $t \in (S^2(\mathfrak{g}))^{\mathfrak{g}}$ be the split Casimir element of $\mathfrak{g}$. Set $V = \otimes_{i=1}^n V_i$. If $i \neq j$, we denote by $t^{(ij)}$ the endomorphism of V acting as t on the factor $V_i \otimes V_j$ and as the identity on the other factors.

We set $X = (\mathbb{C}P^1)^n$ – diagonals. X should be viewed as the configuration space of n distinct points on $\mathbb{C}P^1$. Then the KZ connection is a connection over the trivial bundle $V \otimes \mathcal{O}_X$ with fiber V over X, given by

$$\nabla_{\mathrm{KZ}} = d - \frac{1}{2\kappa} \sum_{i \neq j} t^{(ij)} d \ln(z_i - z_j).$$

One can check that this connection is flat. It was first introduced in the paper [**38**] as an equation satisfied by correlators in the Wess–Zumino–Witten models.

The connection ∇_{KZ} is obviously invariant under the diagonal action of $\mathfrak{g}$, as well as under the action of the group $\mathrm{PSL}_2(\mathbb{C})$ of homographic transformations of $\mathbb{C}P^1$.

Two limits of this system are of special interest.

Critical level. A corollary of the flatness of ∇_{KZ} is that the operators

$$H_i = \sum_{j \neq i} \frac{t^{(ij)}}{z_i - z_j}$$

(the Gaudin Hamiltonians) pairwise commute. These operators correspond to the critical limit ($\kappa \to 0$) of the KZ connection.

Integrable case. It can be checked that when κ has the form $\kappa = k + h^\vee$, where k is an integer ≥ 0 and $h^\vee$ is the dual Coxeter number of $\mathfrak{g}$, then $V \otimes \mathcal{O}_X$ contains a subbundle (the subbundle of conformal blocks), which is preserved by the connection ∇_{KZ}. It can be described as the subbundle of sections σ which are $\mathfrak{g}$-invariant, and such that $(\sum_i z_i e_\alpha^{(i)})^{k+1}(\sigma) = 0$, where e_α are the positive Chevalley generators of $\mathfrak{g}$. This subbundle is nonzero only when the highest weights satisfy certain integrality conditions (see [**22**]).

Since ∇_{KZ} is integrable, natural questions are the explicit computation of its monodromy and of its solutions. The first problem was treated in [**7, 39**] (for κ formal near ∞) and in [**37**] (when k is an integer > 0): the monodromy representations of the braid group are equivalent to representations arising from quantum groups. The approaches of both papers are category-theoretic. Schechtman and Varchenko proposed another approach to this problem: they expressed explicitly a "large" family of flat sections of ∇_{KZ} by integral formulas. This allowed Varchenko to recover the equivalence result with quantum groups-based representations under genericity assumptions (see [**60**]).

1.2. Flat sections of the KZ connection. In [**50**] (see also the monograph [**60**]), Schechtman and Varchenko gave integral formulas for flat sections of the KZ connection, in the case or arbitrary k and λ_i. Together with Feigin, they proved in [**22**] that when k is a positive integer, their solutions lie in the subbundle of conformal blocks. Let us explain the simplest case $\mathfrak{g} = \mathfrak{sl}_2$ of Schechtman and Varchenko's construction.

Assume that each V_i is a module with the highest weight λ_i. Let v_i be a highest vector of V_i and set $v = \otimes_{i=1}^n v_i$. Then for each integer m, the expression

$$(1.1)\quad \psi_C(z_1,\dots,z_n) = \int_C \prod_{1\le i<j\le n} z_{ij}^{\lambda_i\lambda_j/(2\kappa)} \prod_{\substack{1\le i\le n\\ 1\le k\le m}} (t_k - z_i)^{-\lambda_i/\kappa} \prod_{1\le k<l\le m} t_{kl}^{2/\kappa} \times \prod_{k=1}^m \Big(\sum_{i=1}^n \frac{e^{(i)}}{t_k - z_i} dt_k\Big) v$$

is a flat section of the KZ equation. It takes values in the weight subspace $V\big[\sum_i \lambda_i - 2m\big]$ of V. Here $a_{ij} = a_i - a_j$, and C is a suitable integration contour.

The proof of this result is an integration by parts argument. Let us set $\mathbf{z} = (z_1,\dots,z_n)$ and $\mathbf{t} = (t_1,\dots,t_m)$. We will denote the partial differentials with respect to each set of variable by $d_{\mathbf{z}}$ and $d_{\mathbf{t}}$. Let $\mathrm{KZ}(\mathbf{z})$ be the $U(\mathfrak{g})^{\otimes n}$-valued 1-form such that

$$\nabla_{\mathrm{KZ}} = d_{\mathbf{z}} - \mathrm{KZ}(\mathbf{z}).$$

Let $\omega(\mathbf{z},\mathbf{t})$ be the $U(\mathfrak{g})^{\otimes n}$-valued 1-form acting on v in integral (1.1), so we have

$$\psi_C(\mathbf{z}) = \int_C \omega(\mathbf{z},\mathbf{t})$$

(partial integration over the variable $\mathbf{t}$). Then there exist $U(\mathfrak{g})^{\otimes n}$-valued 1-forms $\alpha(\mathbf{z},\mathbf{t})$ and $\sigma(\mathbf{z},\mathbf{t})$, such that

$$d_{\mathbf{z}}\omega(\mathbf{z},\mathbf{t}) = \mathrm{KZ}(\mathbf{z})\omega(\mathbf{z},\mathbf{t}) + \omega(\mathbf{z},\mathbf{t})\alpha(\mathbf{z},\mathbf{t}) + d_{\mathbf{t}}\sigma(\mathbf{z},\mathbf{t})$$

and $\alpha(\mathbf{z},\mathbf{t})v = 0$.

One can make the dependence in κ and $\boldsymbol{\lambda} = (\lambda_1,\dots,\lambda_n)$ explicit. Then $\omega(\mathbf{z},\mathbf{t})$ has the form

$$\omega(\mathbf{z},\mathbf{t}) = \Phi_{\boldsymbol{\lambda}}(\mathbf{z},\mathbf{t})^{1/\kappa}\widetilde{\omega}(\mathbf{z},\mathbf{t});$$

we have $\mathrm{KZ}(\mathbf{z}) = \kappa^{-1}\widetilde{\mathrm{KZ}}(\mathbf{z})$, and $\alpha(\mathbf{z},\mathbf{t})$ and $\sigma(\mathbf{z},\mathbf{t})$ have the expansions $\alpha(\mathbf{z},\mathbf{t}) = \kappa^{-1}\alpha_{-1}^{\boldsymbol{\lambda}}(\mathbf{z},\mathbf{t}) + \alpha_0^{\boldsymbol{\lambda}}(\mathbf{z},\mathbf{t})$ and $\sigma(\mathbf{z},\mathbf{t}) = \alpha_0^{\boldsymbol{\lambda}}(\mathbf{z},\mathbf{t})$. So the above equations reduce to the system

$$\frac{d_{\mathbf{z}}\Phi_{\boldsymbol{\lambda}}}{\Phi_{\boldsymbol{\lambda}}}(\mathbf{z},\mathbf{t})\widetilde{\omega}(\mathbf{z},\mathbf{t}) = \widetilde{\mathrm{KZ}}(\mathbf{z})\widetilde{\omega}(\mathbf{z},\mathbf{t}) + \widetilde{\omega}(\mathbf{z})\alpha_{-1}^{\boldsymbol{\lambda}}(\mathbf{z},\mathbf{t}) + d_{\mathbf{t}}\sigma_{-1}^{\boldsymbol{\lambda}}(\mathbf{z},\mathbf{t}),$$

$$d_{\mathbf{z}}\widetilde{\omega}(\mathbf{z},\mathbf{t}) = \widetilde{\omega}(\mathbf{z},\mathbf{t})\alpha_0^{\boldsymbol{\lambda}}(\mathbf{z},\mathbf{t}) + d_{\mathbf{t}}\sigma_0^{\boldsymbol{\lambda}}(\mathbf{z},\mathbf{t}).$$

Let us show how this system can be solved. If we set

$$\widetilde{\omega}'(\mathbf{z},\mathbf{t}) = \prod_{k=1}^{l}\Big(\sum_{i=1}^n e^{(i)} d\ln(t_k - z_i)\Big),$$

then $\widetilde{\omega}'$ is closed, so $(d_{\mathbf{z}} + d_{\mathbf{t}})(\widetilde{\omega}') = 0$; if we denote by $\widetilde{\omega}'_i$ the component of $\widetilde{\omega}'$ homogeneous of degree i as a form in $\mathbf{t}$, we get $\widetilde{\omega} = \widetilde{\omega}'_m$, so

$$d_{\mathbf{z}}\widetilde{\omega} + d_{\mathbf{t}}\widetilde{\omega}'_{m-1} = 0,$$

so we set $\alpha_0^{\boldsymbol{\lambda}} = 0$ and $\sigma_0^{\boldsymbol{\lambda}} = -\widetilde{\omega}'_{m-1}$. This gives a solution of the system.

On the other hand, we have $\sigma_{-1}^{\boldsymbol{\lambda}}(\mathbf{z},\mathbf{t}) = -\ln\big(\Phi_{\boldsymbol{\lambda}}(\mathbf{z},\mathbf{t})\big)\widetilde{\omega}(\mathbf{z},\mathbf{t})$, so we can rewrite the first equation as

$$\frac{d\Phi_{\boldsymbol{\lambda}}}{\Phi_{\boldsymbol{\lambda}}}(\mathbf{z},\mathbf{t})\widetilde{\omega}(\mathbf{z},\mathbf{t}) = \widetilde{\mathrm{KZ}}(\mathbf{z})\widetilde{\omega}(\mathbf{z},\mathbf{t}) + \widetilde{\omega}(\mathbf{z})\alpha_{-1}^{\boldsymbol{\lambda}}(\mathbf{z},\mathbf{t}).$$

In particular, if $\mathbf{t}$ is a solution $\mathbf{t}^*$ of

$$d_{\mathbf{t}}\big(\ln \Phi_\lambda(\mathbf{z},\mathbf{t})\big) = 0, \tag{1.2}$$

then $\widetilde{\omega}(\mathbf{z},\mathbf{t}^*)v$ is an eigenvector for the commuting family $H_i(\mathbf{z})$, with eigenvalues $\lambda_i(\mathbf{z},\mathbf{t}^*)$ such that

$$\frac{d_{\mathbf{z}}\Phi_{\boldsymbol{\lambda}}}{\Phi_{\boldsymbol{\lambda}}}(\mathbf{z},\mathbf{t}^*) = \sum_{i=1}^{n} \lambda_i(\mathbf{z},\mathbf{t}^*)\,dz_i.$$

Eq. (1.2) is the system of the Bethe ansatz equations for the Gaudin system [**33**]. There is also a stationary phase argument for deriving Bethe equations from the Schechtman–Varchenko solutions (it involves taking the limit $\kappa \to 0$ in the Schechtman–Varchenko integral). The above argument is its algebraic version.

REMARK 1. Formula (1.1) is a generalization of integral formulas for hypergeometric functions; all hypergeometric functions can be recovered from the case $m = 1$.

1.3. The KZB connection. Let us explain now the general construction of the KZB connection. We will follow the treatment of [**57**].

We first define conformal blocks. Let C be a curve with distinct marked points $P_0,\dots,P_n$. Let k be an integer > 0 and let $\lambda_0,\dots,\lambda_n$ be dominant integral weights of $\mathfrak{g}$. We define $\widehat{\mathfrak{g}}$ as the universal central extension $L\mathfrak{g}\oplus\mathbb{C}K$ of the formal loop algebra $L\mathfrak{g}$ and $\mathbb{V}_{\lambda,k}$ as the integrable representation of $\widehat{\mathfrak{g}}$ with highest weight (λ,k). $\mathbb{V}_{\lambda,k}$ is the unique integrable quotient of the Weyl module $\mathrm{Ind}_{\widehat{L_+\mathfrak{g}}}^{\widehat{\mathfrak{g}}}(V_{\lambda,k})$, where $L_+\mathfrak{g}\subset L\mathfrak{g}$ is the subalgebra of regular loops inside a formal disc, $\widehat{L_+\mathfrak{g}} = L_+\mathfrak{g}\oplus\mathbb{C}K$ is its preimage by $\widehat{\mathfrak{g}}\to L\mathfrak{g}$, and $V_{\lambda,k}$ is the $\mathfrak{g}$-module V_λ, on which $L_+\mathfrak{g}$ acts through the evaluation map $L_+\mathfrak{g}\to\mathfrak{g}$ and K acts by k.

The space $\mathrm{CB}(C,\mathbf{P},\boldsymbol{\lambda})$ may be then be defined as a space of intertwining operators

$$\mathrm{CB}(C,\mathbf{P},\boldsymbol{\lambda}) = \mathrm{Hom}_{\mathfrak{g}^{\mathrm{out}}}\big(\mathbb{V}_{\lambda_0,k},\otimes_{i=1}^n V^*_{\lambda_i}(P_i)\big);$$

here $\mathbf{P} = (P_1,\dots,P_n)$ and $\boldsymbol{\lambda} = (\lambda_1,\dots,\lambda_n)$, we identify $L\mathfrak{g}$ as the algebra of formal loops around P_0 with values in $\mathfrak{g}$, $\mathfrak{g}^{\mathrm{out}}$ is its Lie subalgebra consisting of loops regular outside P_0, $V(Q)$ is the $\mathfrak{g}$-module V, on which $\mathfrak{g}^{\mathrm{out}}$ acts through the evaluation morphism $\mathfrak{g}^{\mathrm{out}}\to\mathfrak{g}$ attached to a point $Q\in C-\{P_0\}$.

The space $\mathrm{CB}(C,\mathbf{P},\boldsymbol{\lambda})$ is symmetric under permutation of the $n+1$ pairs (P_i,λ_i), because it can also be defined as a space of invariant linear forms $(\mathbb{V}^*)^{\mathfrak{g}^{\mathrm{out}}_{\mathbf{P}}}$. Here $\mathfrak{g}^{\mathrm{out}}_{\mathbf{P}}$ is the Lie subalgebra of the central extension of $\oplus_{i=1}^n L\mathfrak{g}$ consisting of Taylor expansions at the points of $\mathbf{P}$ of maps $C\to\mathfrak{g}$, regular outside $\mathbf{P}$, and $\mathbb{V}$ is the integrable representation of $\widehat{\oplus_{i=1}^n L\mathfrak{g}}$, of level k and associated with the family of weights $\boldsymbol{\lambda}$. Its dimension is computed by the Verlinde formula [**61**].

The family of vector spaces $\mathrm{CB}(C,\mathbf{P})$ forms a bundle over the moduli space $\mathcal{M}_{g,n}$. Lift this bundle to the space $\mathcal{M}_{g,1^\infty,n\cdot 1^2}$ of families $m = (C,\mathbf{P},z_\infty,\mathbf{v})$, where z_∞ is a local coordinate at P_0, and $\mathbf{v} = (v_1,\dots,v_n)$ is a collection of nonzero tangent vectors at each point of $\mathbf{P}$. The KZB connection over the pull-back of the bundle CB to the space $\mathcal{M}_{g,1^\infty,n\cdot 1^2}$ is defined by the formula

$$\nabla^{\mathrm{KZ}}_\xi\psi(m) = \partial_\xi\psi(m) - \psi(m)\circ T[\xi]. \tag{1.3}$$

Here $m\mapsto\psi(m)$ is a local section of the bundle of conformal blocks over $\mathcal{M}_{g,1^\infty,n\cdot 1^2}$; a vector field ξ on $\mathcal{M}_{g,1^\infty,n\cdot 1^2}$ may be identified with an element of $\mathbb{C}((z_\infty))\partial/\partial z_\infty$ /

$\Theta_{\rm out}$, where $\Theta_{\rm out}$ is the space of formal vector fields at P_0, regular on C outside P_0 and vanishing at the points of $\mathbf{P}$ (see [**1, 36, 41**]); $\xi \mapsto T[\xi]$ is the linear map from $\mathbb{C}((z_\infty))\partial/\partial z_\infty$ to the local completion of $U(\widehat{\mathfrak{g}})_{\rm loc}$ given by the Sugawara tensor; explicitly, we have

$$T\left[z_\infty^n \frac{\partial}{\partial z_\infty}\right] = \frac{1}{2\kappa} \sum_{k+l=n-1} : I_\alpha[k] I_\alpha[l] : \tag{1.4}$$

where (I_α) is an orthonormal basis of $\mathfrak{g}$ and $I_\alpha[n] = I_\alpha \otimes z_\infty^n \in L\mathfrak{g}$, and $:\ :$ is a normally-ordered product (the product $I_\alpha[k]I_\alpha[l]$ may be reversed according to the sign of k and l). The main property of the Sugawara map is that it implements the action of the Lie algebra of vector fields $\mathbb{C}((z_\infty))\partial/\partial z_\infty$ on $\widehat{\mathfrak{g}}$ by element of (a completion of) $U(\widehat{\mathfrak{g}})$.

The connection defined by (1.3) is projectively flat, i.e., its curvature is scalar. The fact that this connection is not flat can be related with the fact that the normal-ordering convention in (1.4) does not take into account any global information related to the curve C. One can find a "better" normalization of $T[\xi]$ (it differs from (1.4) by an additive constant), for which the corresponding curvature is "almost everywhere" zero; more precisely, this curvature is scalar, equal to a sum of Dirac distributions located at the Weierstrass points of C.

1.4. Trigonometric and elliptic KZ equations. In the case $g = 1$, the KZB system can be described explicitly. This description involves classical r-matrices (possibly dynamical). Such r-matrices can be of rational, trigonometric or elliptic types. KZ equations corresponding to rational r-matrices are discussed in Section 1.1.

There is an equivalent formulation of the KZ equations of trigonometric type. This formulation, and the expression of flat sections by integral formulas, were discussed by Cherednik in [**6**].

In the elliptic case, the KZB equations as well as the precise description of conformal blocks were discussed in [**28**]. In this situation, parameters of the KZB equations are $(\tau, z_1, \dots, z_n)$ (elliptic moduli and location of marked points on the elliptic curve). The evolution equation for variation in the moduli direction is a heat-type equation. Integral formulas for this system were found in [**26**]. In this paper, Bethe ansatz equations for the critical limit of the KZB system (the "standard" elliptic Gaudin system) are also derived from these integral formulas.

In addition to the "standard" elliptic case, there exist a "twisted" elliptic situation. Then $\mathfrak{g} = \mathfrak{sl}_n$, and $\mathfrak{g}^{\rm out}$ is replaced by a twisted form, involving the action of the discrete Heisenberg group by automorphisms of $\mathfrak{g}$. The corresponding KZ-type equation has been formulated by Etingof in [**18**] (see also [**43**]). Separation of variables for the critical limit of this system was obtained in [**53**].

2. q-Deformations of the KZ Connection

In [**30**], I. Frenkel and N. Reshetikhin introduced a q-deformation of the rational and trigonometric KZ equations (these equations appeared earlier in the case of zero central charge in work by Smirnov). In this approach, the affine Kac–Moody Lie algebra $\widehat{\mathfrak{g}}$ is replaced either by its quantization $U_q(\widehat{\mathfrak{g}})$ (in the trigonometric case), or by the double Yangian algebra $DY(\mathfrak{g})$ (in the rational case). Their work was motivated by the computation of correlators in integrable field theory (a natural deformation of conformal field theory).

2.1. Analogues of the Sugawara construction. The most direct approach to quantizing the KZ(B) equation would rely on a quantization of the Sugawara construction. However, such a direct approach does not work: there is no known quantum analogue of the Sugawara construction, outside the critical level (although there are quantum analogues of the Virasoro algebra).

Here is what remains at the quantum level from the Sugawara construction:

1. when the level is critical, the Sugawara construction yields central elements in $U(\widehat{\mathfrak{g}})_{\mathrm{loc}}$. The center of this algebra is "large" and well-understood [**20**]. There is an explicit construction of central elements of a local completion of $U_q(\widehat{\mathfrak{g}})$ (see [**48**]), and the center of this algebra is also well-understood [**31,51**]. So the large center at critical level "survives" quantization;
2. there is a quantization of "one mode" of the Sugawara tensor. This follows from the construction of the quantum Casimir element (see [**7**]), which says that if $(A, m, \Delta, \mathcal{R}, S)$ is a quasitriangular Hopf algebra, then the element $u = \sum_i S^2(b_i)a_i$ is central (we write the R-matrix of A as $\mathcal{R} = \sum_i a_i \otimes b_i$). Applying this construction to $U_q(\widetilde{\mathfrak{g}})$ or to $\widetilde{DY(\mathfrak{g})}$ (the versions "with derivations" of $U_q(\widehat{\mathfrak{g}})$ and $DY(\mathfrak{g})$), one finds a central element of the form
$$u = q^{(K+h^\vee)D}u_0$$
where D is the derivation generator, K is the generator of the center, and u_0 is an invertible element of the initial algebra $U_q(\widehat{\mathfrak{g}})$ (or $DY(\mathfrak{g})$). This means that the adjoint action of u_0 is the same as the action of $q^{-(K+h^\vee)D}$, which is a property of the exponential of one mode of the Sugawara tensor.

2.2. The qKZ connection in the trigonometric case. The approach of [**30**] is a q-deformation of the "intertwiners" approach to the KZ equations (the "coinvariants" approach was treated in [**10**], in the rational case). As before, the general picture for intertwiners is as follows. We have a Hopf algebra inclusion $U_q(\mathfrak{g}^{\mathrm{out}}) \subset U_q(\widehat{\mathfrak{g}})$, quantizing an "adelic" inclusion $\mathfrak{g}^{\mathrm{out}} \subset \mathfrak{g}$. We consider evaluation representations $U_q(\mathfrak{g}^{\mathrm{out}})$ of the "small" algebra $U_q(\mathfrak{g}^{\mathrm{out}})$, attached to points z_i, and a level k representation $\mathbb{V}$ of the "big" algebra $U_q(\widehat{\mathfrak{g}})$. Then the space of intertwiners attached to these data will be
$$\mathrm{Hom}_{U_q(\mathfrak{g}^{\mathrm{out}})}\big(\mathbb{V}, \otimes_{i=1}^n V_i(z_i)^*\big).$$
The underlying geometric picture in the case of the paper [**30**] is the following. On $C = \mathbb{C}P^1$, consider two marked points 0 and ∞, so $\mathbf{P} = (0, \infty)$. The algebra of regular maps from $C - \mathbf{P}$ to $\mathfrak{g}$ is just $L\mathfrak{g} = \mathfrak{g}[t, t^{-1}]$. On the other hand, and up to completions, the adelic embedding $\mathfrak{g}^{\mathrm{out}}_{\mathbf{P}} \subset \widehat{L_0\mathfrak{g} \times L_\infty\mathfrak{g}}$ can be identified with the quotient of the diagonal embedding $\widehat{\mathfrak{g}} \subset \widehat{\mathfrak{g}} \times \widehat{\mathfrak{g}}$ (taking K to $K_1 + K_2$, where $K_1 = (K, 0)$ and $K_2 = (0, K)$) by the quotient maps taking K to 0 on the left side, and $K_1 + K_2$ to 0 on the right side.

The quantization of this embedding is as follows. Define $U_q(L\mathfrak{g})$ as the quotient of $U_q(\widehat{\mathfrak{g}})$ by the ideal generated by K, then the coproduct induces an algebra morphism
$$U_q(L\mathfrak{g}) \hookrightarrow \big(U_q(\widehat{\mathfrak{g}}) \otimes U_q(\widehat{\mathfrak{g}})\big)/(K_1 + K_2)$$
where $K_1 = K \otimes 1$ and $K_2 = 1 \otimes K$ (the right algebra still has a central element $K_1 = -K_2$). We can take for the representation $\mathbb{V}$, the tensor product $\mathbb{V}_{\lambda,k} \otimes \mathbb{V}^*_{\mu,k}$, where λ, μ are weights of $\mathfrak{g}$. Evaluation modules are of the form $V_i(z_i)$, where

$z_i \in \mathbb{C}^\times$. In that case, dualization shows that the space of intertwiners has the form

$$\mathrm{CB}(\mathbf{z}) = \mathrm{Hom}_{U_q(\widehat{\mathfrak{g}})}\big(\mathbb{V}_{\lambda,k}, \widehat{\mathbb{V}}_{\mu,k} \otimes (\otimes_{i=1}^n V_i(z_i)^*)\big),$$

which is the definition considered in [**30**]. Here $\widehat{\mathbb{V}}_{\mu,k}$ means the completion of $\widehat{\mathbb{V}}_{\mu,k}$ with respect to the grading induced by the derivation D, and $\mathbf{z} = (z_1, \dots, z_n)$.

The spaces $\mathrm{CB}(\mathbf{z})$ can be computed when λ, μ are generic and $\mathbb{V}_{\lambda,k}$ and $\mathbb{V}_{\mu,k}$ are Verma modules. In this case, identification of the highest degree terms in these modules sets up an isomorphism

$$\mathrm{CB}(\mathbf{z}) \to (\otimes_{i=1}^n V_i^*)[\lambda - \mu], \tag{2.1}$$

so the family $\mathrm{CB}(\mathbf{z})$ is actually a trivial bundle.

The next main remark is that using the quantum Casimir element, one can define a flat discrete connection on this bundle (the equivalent viewpoint of [**19**] is that any intertwiner modules over the derivation-extended algebra $U_q(\widehat{\mathfrak{g}})$ has to be a flat section of this connection). The translation of this connection via the isomorphism (2.1) is the *quantum KZ connection*

$$\Psi(q^{\kappa\delta_i}\mathbf{z}) = K_i(\mathbf{z})\Psi(\mathbf{z}) \tag{2.2}$$

where δ_i is the ith basis vector of $\mathbb{N}^n$, and

$$K_i(\mathbf{z}) = \prod_{j=i-1}^{i} R^{ij}_{V_i,V_j}(q^\kappa z_i/z_j)(q^{\lambda+\mu+\rho})^{(i)} \prod_{j=n}^{i+} R^{ji}_{V_j,V_i}(z_j/z_i)^{-1}.$$

Here $R_{V,W}(z)$ is the R-matrix for two representations of the quantum affine algebra, and ρ is the half-sum of positive weights of $\mathfrak{g}$.

The connection given by (2.2) is flat, which means that the equation

$$K_j(q^{\kappa\delta_i}\mathbf{z})K_i(\mathbf{z}) = K_i(q^{\kappa\delta_j}\mathbf{z})K_j(\mathbf{z})$$

holds. In particular, when $\kappa = 0$, we get a commuting family of operators, the inhomogeneous XXZ family (the homogeneous case corresponds to the case when all z_i are equal).

When k is an integer > 0 and $\mathbb{V}_{\lambda,k}$ and $\mathbb{V}_{\mu,k}$ are replaced by integrable $U_q(\widehat{\mathfrak{g}})$-modules, the space of intertwiners is a subspace of $(\otimes_{i=1}^n V_i^*)[\lambda - \mu]$, whch can be characterized by conditions quantizing the algebraic conditions of [**22**]. In [**47**], these conditions were made explicit in the rational case, for $\mathfrak{g} = \mathfrak{gl}_n$.

Integral formulas for solutions of the qKZ equations were written in the $\mathfrak{sl}_2$ case in [**30, 46, 59**], and in the $\mathfrak{sl}_n$ case in [**55**].

In [**56**], Tarasov and Varchenko constructed a q-deformation of the Kohno–Drinfeld theorem expressing the monodromy of the KZ equation. The role of the monodromy representation in the case of difference systems is played by "connection matrices", which are then expressed in terms of elliptic quantum groups.

2.3. Rational and elliptic qKZ equations. Replacing in the system (2.2), trigonometric R-matrices by rational R-matrices, and dropping the middle term $(q^{\lambda+\mu+\rho})^{(i)}$, one gets the system of *rational* KZ *equations*. This is a flat discrete connection on $\mathbb{C}^n$. In the critical limit, the operators $K_i^{\mathrm{rat}}(\mathbf{z})$ are nothing but the commuting family of operators of the inhomogeneous XXX system (of the homogenous system when all z_i coincide). This system has a representation-theoretic interpretation, in terms of intertwiners of representations of the double Yangian $DY(\mathfrak{g})$ (the underlying picture is $C = \mathbb{C}P^1$, $\mathbf{P} = (\infty)$).

In the same way, one can replace the R-matrices by elliptic analogues. One can derive in this way elliptic analogues of the KZB connection when $g = 1$, more precisely, of the equations governing the motion of the points z_i, the modulus τ being fixed. In their paper [**27**], Felder and Varchenko also found a q-deformation of the "heat equation part" of the qKZB equation, describing the moduli variation. The system obtained this way at the critical level is the same as a "face-type" model in statistical mechanics.

Integral formulas for solutions of this system were obtained in [**25**].

3. Critical Level

The construction of [**57**] can be carried out with Verma modules replacing integrable modules. When the level is critical, we get a commuting family of differential operators over the moduli space of G-bundles over the curve C, where G is a Lie group with Lie algebra $\mathfrak{g}$. This construction is one of the ingredients of the work of Beilinson and Drinfeld on the geometric Langlands correspondence [**3**]. In [**29**], E. Frenkel proposed that this construction may be q-deformed, and that a "q-deformed Langlands correspondence" should emerge, at least for genus 0, 1, from the qKZ equations at critical level. On the other hand, in [**54**], Stoyanovsky proposed a scheme for "quantizing" the geometric Langlands correspondence; then the deformation parameter is the level k, and the basic object of the correspondence is the "off-critical" KZB connection.

While both programs are still far from being completed, it seems useful to review some works on the critical level.

3.1. KZB equations. When $C = \mathbb{C}P^1$, the critical level limit of the KZ connection is the commuting family of Gaudin operators. A natural question about this system is the computation of its spectrum. There are two main approaches to this problem: Bethe ansatz and the separation of variables. As we have seen, the Bethe ansatz here is a classical limit of the integral formulas of Schechtman and Varchenko (see [**49**] and [**21**]). On the other hand, the separation of variables was treated by Sklyanin ([**52**]). The connection between his work and the Langlands problem was clarified by E. Frenkel in [**29**].

In the elliptic case, Bethe ansatz equations were obtained in [**26**] as a limit of integral formulas for solution of the KZB equations; in the contribution of Kuroki and Takabe to this volume, the same equations are recovered using bosonization. On the other hand, the separation of variables and its relation with geometric Langlands were discussed in [**9**].

3.2. qKZB equations. As we have seen, at the critical level, the qKZ equations can be identified with commuting families of operators arising from models of statistical mechanics (XXX in the rational case, XXZ in the trigonometric case, elliptic face models in the elliptic case, elliptic vertex models in the "twisted" elliptic case associated with $\mathfrak{sl}_n$).

Integration of these systems may be approached by several methods:

1. algebraic Bethe ansatz, which involves applying operators to a distinguished "highest-weight" eigenvector;
2. separation of variables (this method is also called functional Bethe ansatz and is close to Baxter's method of integration);
3. taking limits of integral formulas for the "off-critical" qKZ system.

In [52], the separation of variables was obtained for the XXX system. The elliptic version of this result was obtained in [24], in the situation of a $\mathfrak{sl}_2$ elliptic face model. On the other hand, integral formulas and the critical limit were obtained for the qKZB system in [25].

4. Results in Higher Genus

In this section, I will present the results of [11–13, 17]. In the two first papers, we present explicit parametrizations of the KZB connection. In the next paper, we give explicit formulas for flat sections of this connection, in the $\mathfrak{sl}_2$ case. Finally, in the last paper, we construct q-deformations of the commutative family of operators arising from the KZB connection at the critical level.

4.1. Parametrizations of the KZB connection. In this section, we will describe two parametrizations of the KZB connection. The first one was obtained in [11], and relies on the Feigin and Stoyanovsky's parametrization of conformal blocks [23]. The second parametrization was obtained in [17], and it relies on a parametrization of moduli spaces of vector bundles over curves introduced by Tyurin in [58].

Both papers rely on the fact that the space of conformal blocks of level k associated to the curve C and the Lie algebra $\mathfrak{g}$ may be viewed as the space $H^0(\mathrm{Bun}_G(C), \det^{\otimes k})$ of sections the kth power of the determinant line bundle, over the moduli space of G-bundles over C (see [2]). To each parametrization $\pi\colon P \to \mathrm{Bun}_G(C)$ of $\mathrm{Bun}_G(C)$ is attached a linear map

$$\pi^*\colon H^0(\mathrm{Bun}_G(C), \det^{\otimes k}) \to H^0(P, (\pi^* \det)^{\otimes k}).$$

The question is then whether this map is injective, and if it is, to characterize its image.

Before we proceed with the description of the contents of [11] and [17], we mention the works [34] and [35], where the KZB connection was parametrized explicitly in the case of genus 2 and hyperelliptic curves. To find a connection with the parametrizations of [11] and [17] is an interesting unsolved problem.

4.1.1. *Parametrization by B-bundles.* A complete description of the space of conformal blocks was obtained by Feigin and Stoyanovsky in [23]. They used the parametrization of $\mathrm{Bun}_G(C)$ by twisted forms of the space of N-bundles, where N is the radical of a Borel subgroup of G. When $G = \mathfrak{sl}_n$, such a twisted form is associated to a sequence $(L_1, \dots, L_n)$ of line bundles, whose tensor product is trivial; it can be described as the space of flags $E_1 \subset \cdots \subset E_n$ of vector bundles, together with isomorphisms $E_i/E_{i-1} \to L_i$. For suitable degrees of the L_i, one can show that π^* is injective. Moreover, in the $\mathfrak{sl}_2$ case, P is just the projectivization of an Ext group of line bundles; it has a similar description in the general case (in terms of towers of projective spaces). The space $H^0(P, (\pi^* \det)^{\otimes k})$ has an explicit functional description: this is a space of top degree forms $\omega(z_i^{(\alpha)}, i = 1, \dots, n_\alpha, \alpha = 1, \dots, r)$ on the product $\prod_{\alpha=1}^r C^{(n_\alpha)}$, satisfying regularity and poles conditions ($C^{(n)} = C^n/\mathfrak{S}_n$ is the nth symmetric power of C). Here r is the rank of $\mathfrak{g}$, $(n_1, \dots, n_\alpha)$ are integers related with the degrees of the L_i, and the map π^* may be described as follows.

Let us denote by e_α, $\alpha = 1, \dots, r$ the positive Chevalley generators of $\mathfrak{g}$, and by $e_\alpha(z)\,dz$ the current associated with e_α. $e_\alpha(z)\,dz$ belongs to a completion of

$\widehat{\mathfrak{g}}\otimes\mathbb{C}((z_\infty))\,dz_\infty$ and is defined by $e_\alpha(z)\,dz=\sum_{k\in\mathbb{Z}}e_\alpha[z^k]z^{-k-1}\,dz$. Here z_∞ is a local coordinate at a point $P_0\in C$. If $\mathbb{V}$ is an integrable $\widehat{\mathfrak{g}}$-module (located at P_0), one defines the extremal vectors of $\mathbb{V}$ as the vectors obtained by the action of the affine Weyl group of $\mathfrak{g}$ on the highest weight vector of $\mathbb{V}$. Then if v is a suitable extremal vector, the map π^* has the form

$$\psi\in(\mathbb{V}^*)^{\mathfrak{g}^{\mathrm{out}}}\mapsto\omega_{\mathrm{FS}}(\psi)=\Big\langle\psi,\prod_{\nu=1}^{r}\Big(\prod_{i=1}^{n_\nu}e_\alpha(z_i^{(\nu)})\,dz_i^{(\nu)}\Big)(v)\Big\rangle,$$

so the map π^* is a "correlation functions" map. In [**23**], Feigin and Stoyanosvsky give explicit conditions for π^* to be injective, and they characterize its image.

Actually, this parametrization of conformal blocks does not seem to allow a uniform description of the KZB connection. To obtain such a description, we consider in [**11**] the natural parametrization by B-bundles instead of N-bundles.

In the case $G=\mathrm{SL}_n$, a B-bundle over C is the same as a flag $E_1\subset\cdots\subset E_n$ of bundles, such that each quotient E_i/E_{i-1} is a line bundle. So the twisted space of N-bundle appears as a fiber of the natural map $\mathrm{Bun}_B(C)\to\mathrm{Bun}_T(C)$, where T is the Cartan group $T=B/N$. When the parameter space P is $\mathrm{Bun}_B(C)$, the space $H^0(P,(\pi^*\det)^{\otimes k})$ can also be described in functional terms, provided a choice of A- and B-cycles on C has been fixed. This space identifies with a space of forms

$$\omega(\lambda_a^{(\nu)}\mid z_i^{(\nu)}),\quad \nu=1,\dots,r,\ a=1,\dots,g,\ i=1,\dots,n_\nu,$$

on $(\mathbb{C}^g)^r\times\prod_{\nu=1}^r C^{(n_\nu)}$ with functional properties (g is the genus of C, and we identify $(\mathbb{C}^g)^r$ with $\mathfrak{h}^g$, where $\mathfrak{h}$ is the Cartan subalgebra of $\mathfrak{g}$). The map π^* should be modified as follows. We fix functions $(r_a)_{a=1,\dots,g}$ on the universal cover of C, regular outside the points above P_0 (the marked point corresponding to the representation $\mathbb{V}$), single-valued along A-cycles and such that the action of the deck transformation corresponding to the B-cycle $B_{a'}$ tranforms r_a into $r_a-\delta_{a,a'}$. Then the map π^* associates, to each invariant linear form ψ, the form

$$\omega(\psi)=\Big\langle\psi,e^{\sum_{a=1}^g\sum_{\nu=1}^r\lambda_a^{(\nu)}h_\nu[r_a]}\prod_{\nu=1}^{r}\Big(\prod_{i=1}^{n_\nu}e_\alpha(z_i^{(\nu)})\,dz_i^{(\nu)}\Big)(v)\Big\rangle$$

where h_ν is the Cartan algebra generator associated with the simple root ν. The map π^* does not depend on the choice of the functions r_a. Composing this map with evaluation at $\lambda_a^{(\nu)}=0$, one recovers the Feigin-Stoyanovsky map. The map $\psi\mapsto\omega(\psi)$ is therefore injective when the map $\psi\mapsto\omega_{\mathrm{FS}}(\psi)$ is.

The next question is to give a functional characterization of the image of the map $\psi\mapsto\omega(\psi)$. In [**11**], we make some steps in this direction. We show that the forms $\omega(\psi)$ have a theta-function-like behavior with respect to the $\lambda_a^{(\nu)}$. To fully characterize the image of conformal blocks, one should supplement this condition with the functional translation of invariance under the Weyl group action, which implies vanishing conditions on hyperplanes in $(\mathbb{C}^g)^r$, as in [**28**]. (See Remark 3 in [**11**]).

Let us denote by $\mathcal{M}^{(a)}_{g,n\cdot 1^2}$ the moduli space of curves of genus g with n pairs (P_i,v_i) of a marked point and a nonzero tangent vector, together with a choice of a system of A-cycles. There is a canonical projection map $p\colon\mathcal{M}^{(a)}_{g,n\cdot 1^2}\to\mathcal{M}_{g,n\cdot 1^2}$. Then the map

$$\psi\mapsto\omega(\psi)$$

may be viewed as a morphism from the lift $p^*(\mathrm{CB})$ of the bundle CB of conformal blocks to a functional bundle $\mathcal{F}$ defined on $\mathcal{M}^{(a)}_{g,n\cdot 1^2}$. Then $p^*(\mathrm{CB})$ is naturally equipped with a flat connection $\nabla^{p^*(\mathrm{CB})}$. The main result of [**11**] is then:

THEOREM 4.1. *$\mathcal{F}$ is naturally equipped with a flat connection $\nabla^{\mathcal{F}}$, such that the morphism $p^*(\mathrm{CB}) \to \mathcal{F}$ is compatible with the connections $\nabla^{p^*(\mathrm{CB})}$ and $\nabla^{\mathcal{F}}$. The connection $\nabla^{\mathcal{F}}$ may be described explicitly as follows. If the tangent vector ξ to $\mathcal{M}^{(a)}_{g,n\cdot 1^2}$ at m corresponds to the formal vector field $\hat{\xi} \in \mathbb{C}((z_\infty))\partial/\partial z_\infty$, and $m \mapsto f(m)$ is a local section of $\mathcal{F}$, then*

$$\nabla^{\mathcal{F}}_\xi(f) = (\partial_\xi f)(m) - \mathrm{res}_{z_\infty = P_0}\big(\hat{\xi}(z_\infty)\big(T(z_\infty)f\big)(m)\big)$$

where $T(z_\infty)\colon \mathcal{F} \to \mathcal{F} \otimes \mathbb{C}((z_\infty))(dz_\infty)^2$ is a bundle homomorphism defined using differential-evaluation operators ($\mathbb{C}((z_\infty))(dz_\infty)^2$ is viewed as a trivial bundle over the moduli space). When $\mathfrak{g} = \mathfrak{sl}_2$ and $n_\nu = 0$, the explicit formula for $T(z_\infty)$ is

$$\big(T(z_\infty)f\big)(\boldsymbol{\lambda}) = \frac{1}{2\kappa}\Bigg(\Bigg(\sum_{\nu=1}^{r}\Big(\sum_{a=1}^{g}\omega_a(z_\infty)\partial_{(H_\nu)_a}\Big)^2 + \sum_{\alpha\in\Delta_+}(D^{(\lambda,\alpha)}_{z_\infty}\omega_a)(z_\infty)\partial_{(\alpha^\vee)_a} + k\sum_{\alpha\in\Delta_+}\omega_{(\lambda,\alpha)}(z_\infty)\Bigg)f\Bigg)(\boldsymbol{\lambda})$$

where $\boldsymbol{\lambda} = (\lambda_1, \dots, \lambda_g)$ is a variable in $\mathfrak{h}^g = (\mathbb{C}^r)^g$, $(\omega_a)_{a=1,\dots,r}$ is the canonical basis of regular differentials on C, $(H_\nu)_{\nu=1,\dots,r}$ is an orthonormal basis of the Cartan subalgebra $\mathfrak{h} \subset \mathfrak{g}$, x_a is the image of $x \in \mathfrak{h}$ in the ath copy of $\mathfrak{h}^g$, $D^\zeta_{z_\infty}$ is a connection on the canonical bundle Ω_C, depending on a variable $\zeta \in \mathbb{C}^g$, and $\omega_\zeta(z_\infty)$ is a quadratic differential on C, depending on the same variable.

4.1.2. *Hecke–Tyurin parametrization.* In [**17**], we used another parametrization of $\mathrm{Bun}_G(C)$ to describe conformal blocks. This parametrization is due to Tyurin [**58**], and it relies on the notion of a Hecke modification of a bundle. Tyurin's construction works when $\mathfrak{g} = \mathfrak{gl}_n$ or $\mathfrak{g} = \mathfrak{sl}_n$, but Hecke modifications can be defined for arbitrary G-bundles [**3**]; the connection between the parametrizations of [**23**] and of [**17**] are discussed in the Introduction of [**17**].

Hecke modifications may be defined as follows. Let E be a vector bundle over C. Let P be points of C and let $V \subset E_P$ be a vector subspace of the fiber of E at P. Then the modification $\mathcal{H}(P,V)(E)$ is the vector subbundle of $E(P)$, such that the space of sections over an open $U \subset C$ is $\Gamma(U, E)$ if $P \notin U$, and the subspace of $\Gamma\big(U, E(P)\big)$ of all sections with a simple pole at P, whose polar term at P belongs to V. If $(P_i, V_i), i = 1, \dots, k$ is a collection of pairs of distinct points $P_i \in C$ and $V_i \subset E_{P_i}$, then the Hecke modifications "commute" and we can define a modified bundle $\mathcal{H}\big((P_i, V_i)_{i=1,\dots,k}\big)$. In particular, when E is the trivial bundle $\mathcal{O}^n_C$, $k = (n+1)g$ and the V_i are 1-dimensional, one gets a linear map

$$\mathcal{H}_{(P_i)_i}\colon (\mathbb{C}P^n)^{(n+1)g} \to \mathrm{Bun}_n(C)_{\mathcal{O}(\sum_i P_i)},$$

the family $(P_i)_i$ being fixed. Here $\mathrm{Bun}_n(C)_{\mathcal{O}(\sum_i P_i)}$ is the moduli space of rank n bundles, with an isomorphim of their determinant bundle with the line bundle $\mathcal{O}(\sum_i P_i)$. This map factors through the diagonal action of the group $\mathrm{PGL}_n(\mathbb{C})$ on the left side of the map, and it is close to the Tyurin parametrization.

In [**17**], we show that when $n = 2$, the map $\mathcal{H}_{(P_i)_i}$ is a "good" parametrization of the moduli space. More precisely, we define a function $\mathrm{Den}_{(P_i)_i}\big((\ell_i)_i\big)$ on $(\mathbb{C}P^1)^{3g}$

by the formula

$$\mathrm{Den}_{(P_i)_i}((\ell_i)_i) = \sum_{\substack{\sigma\in\mathfrak{S}_{3g}|\\ \sigma(3a-2)<\sigma(3a-1)<\sigma(3a)\\ \text{for } a=1,\dots,g}} \varepsilon(\sigma)\sigma \cdot \Big(\prod_{a=1}^{g} \omega_a(P_{3a-2})\omega_a(P_{3a-1})\omega_a(P_{3a})\ell_{3a-2;3a-1}\ell_{3a-2;3a}\ell_{3a-1;3a}\Big)$$

where $\ell_{ij} = \ell_i - \ell_j$, and the elements σ of $\mathfrak{S}_{3a}$ act by permuting both the P_i and the ℓ_i; we then show that

1. $\mathcal{H}_{(P_i)_i}$ maps the complement of $\mathrm{Den}^{-1}_{(P_i)_i}(0)$ to the locus of stable bundles of $\mathrm{Bun}_n(C)_{\mathcal{O}(\sum_i P_i)}$,
2. the restriction of $\mathcal{H}_{(P_i)_i}$ to this complement is an etale morphism to its image (Theorem 1.1).

Using this map, we can parametrize the conformal blocks associated to the location at P_0 of the integrable representation with highest weight (λ, k) as follows: we set

$$\mathcal{F}' = \mathrm{Hom}_{\mathfrak{g}}\big(W, \mathbb{C}[\ell_1]_{\le k} \otimes \cdots \otimes \mathbb{C}[\ell_{3g}]_{\le k}\big)$$

where $\mathbb{C}[\ell]_{\le k}$ is the irreducible $(k+1)$-dimensional $\mathfrak{g}$-module, and we construct a linear map $(\mathbb{V}^*)^{\mathfrak{g}^{\mathrm{out}}} \to \mathcal{F}'$.

The main result of [**17**] is then:

THEOREM 4.2. *Consider $\mathcal{F}'$ as a constant bundle over $\mathrm{Bun}_2(C)_{\mathcal{O}(\sum_i P_i)}$. Equip it with the connection $\nabla^{\mathcal{F}'} = d - A$, where if ξ is a vector field on $\mathcal{M}^{(a)}_{g,n\cdots 1^2}$ and $\hat\xi$ is an associated vector field in $\mathbb{C}((z_\infty))\partial/\partial z_\infty$, then $A(\xi)$ is the endomorphism of $\mathcal{F}'$ given by*

$$A(\xi)(f) = \mathrm{res}_{z_\infty=P_0}\big(\xi(z_\infty)(T^{\mathrm{diff}}(z_\infty)f)\big),$$

and $T^{\mathrm{diff}}(z_\infty)$ is the linear map $\mathcal{F}' \to \mathcal{F}' \otimes \mathbb{C}((z_\infty))(dz_\infty)^2$, which can be expressed in the form

$$2\kappa T^{\mathrm{diff}}(z_\infty) = \sum_\alpha \ell^{\mathrm{diff}}_{(P_i)_i, I_\alpha}(z_\infty)^2 + a_{(P_i)_i, I_\alpha}(z_\infty)\ell^{\mathrm{diff}}_{(P_i)_i, I_\alpha}(z_\infty) + s_{(P_i)_i}(z_\infty)$$

where $\sum_\alpha I_\alpha \otimes I_\alpha$ is the split Casimir element of $\mathfrak{sl}_2$, and the maps $x \mapsto \ell^{\mathrm{diff}}_{(P_i)_i,x}(z_\infty)$, $x \mapsto a_{(P_i)_i,x}(z_\infty)$ and $x \mapsto s_{(P_i)_i,x}(z_\infty)$ are linear maps

$$\mathfrak{g} \to \mathbb{C}[\ell_i, \partial_{\ell_i}, i = 1, \dots, 3g][\mathrm{Den}^{-1}_{(P_i)_i}] \otimes \mathbb{C}((z_\infty))\, dz_\infty,$$

and $s_{(P_i)_i}(z_\infty)$ is an element of $\mathbb{C}[\ell_i, \partial_{\ell_i}, i = 1, \dots, 3g][\mathrm{Den}^{-1}_{(P_i)_i}] \otimes \mathbb{C}((z_\infty))(dz_\infty)^2$, defined in terms of Green functions on C (the operators are all of order zero as differential operators in the ℓ_i, except the "L-operator" $\ell^{\mathrm{diff}}_{(P_i)_i, I_\alpha}(z_\infty)$, which has order 1).

When $\kappa = 0$, the family $T^{\mathrm{diff}}_{(P_i)_i}(z_\infty)$ is a regular quadratic differential in the variable z_∞ and yields a $(3g-3)$-dimensional family of commuting differential operators. These operators can be used to construct a differential system on the complement of $\mathrm{Den}^{-1}_{(P_i)_i}(0)$, locally isomorphic to the Beilinson–Drinfeld systems. We also get an explicit expression for the classical limit of this system, which is locally isomorphic to the Hitchin system (the classical limit of the Beilinson–Drinfeld systems).

The Hecke–Tyurin parametrization was recently used by Krichever, who constructed an infinite-dimensional analogue of the Hitchin system. His construction depends on the additional choice of a holomorphic differential ω on C.

4.2. Integral formulas for flat sections of the connection $\nabla^{\mathcal{F}}$. Let us come back to the situation of a family of curves of genus g, with marked points P_i and a Verma module of weight Λ_i sitting at P_i. In [**13**], we give integral formulas for flat sections of the connection $\nabla^{\mathcal{F}}$.

To do this, we proceed in two steps. We first write the solutions in the case $n = 0$ (there are no variables $z_1, \ldots, z_n$). We find that any solution $\varphi(m \mid \lambda_1, \ldots, \lambda_g)$ of the heat equation

$$\partial_\xi \varphi(m \mid \boldsymbol{\lambda}) = \frac{\kappa}{8i\pi} \sum_{a,b} \delta\tau_{ab} \partial^2_{\lambda_a \lambda_b} \varphi(m \mid \boldsymbol{\lambda})$$

(where $\delta\tau_{ab}$ is the variation of periods of C corresponding to the vector ξ, and $\boldsymbol{\lambda} = (\lambda_1, \ldots, \lambda_g)$) yields a flat section $f_\varphi(m, P_i, v_i \mid \lambda_1, \ldots, \lambda_g)$ of the bundle $\mathcal{F}$, when $n = 0$. The formula for this flat section has the form

$$f_\varphi(m, P_i, v_i \mid \lambda_1, \ldots, \lambda_g) = A(\boldsymbol{\lambda}, P_i, v_i) \varphi\Big(m \mid \kappa\boldsymbol{\lambda} + \Delta - (g-1)P_0 - \sum_i \Lambda_i (P_i - P_0)\Big)$$

where the first factor does not depend on φ, Δ is the Riemann vector of constants of C and the points P_i are identified with their images in the Jacobian of C via the Abel–Jacobi map.

To find solutions in the general case, we use an observation of [**6**] and [**26**]: if one adds auxiliary points t_j with Verma modules of highest weight -2 to the picture, the resulting correlation function is a differential in each variable t_j. Integrating it along the proper contour, one again gets a flat section of the KZB connection. What we do then is inserting as many t_j as there are variables z_i, to preserve the weight zero.

We then find:

THEOREM 4.3. *If we set*

$$\begin{aligned} &f_C(m, P_i, v_i)(\boldsymbol{\lambda} \mid z_1, \ldots, z_n) \\ &\qquad = \int_C \big(\Phi(t_1, \ldots, t_n \mid z_1, \ldots, z_n) f_\varphi\big)(m, P_i, v_i, t_1, \ldots, t_n \mid \lambda_1, \ldots, \lambda_g) \end{aligned}$$

where $\Phi(t_1, \ldots, t_n \mid z_1, \ldots, z_n)$ is a suitable differential-evaluation operator, and C is a cycle for the local system defined by the integrand, then $(m, P_i, v_i) \mapsto f_C(m, P_i, v_i)(\boldsymbol{\lambda} \mid z_1, \ldots, z_n)$ is a flat section of the bundle $\mathcal{F}$.

4.3. q-deformed KZB operators at critical level. When $\kappa = 0$, the operator $T(z_\infty)$ introduced in Theorem 4.1 gives rise (by taking Taylor coefficients) to a commutative family of differential-evaluation operator, acting on forms in two groups of variables (the λ's and the z's). When $\mathfrak{g} = \mathfrak{sl}_2$, the general form of the operator $T(z_\infty)$ is (we set $z = z_\infty$)

$$\begin{aligned} T(z) = {}& \frac{1}{2} O(z \mid z_1, \ldots, z_n)^2 + O'(\boldsymbol{\lambda}, z \mid z_1, \ldots, z_n) + \omega(\boldsymbol{\lambda}, z) \\ & - 2\Big(\sum_{i=1}^n G_{2\boldsymbol{\lambda}}(z, z_i) O(z_i \mid z_1, \ldots, z, \ldots, z_n) + \sigma(\boldsymbol{\lambda}, z, z_i)\Big) \circ \mathrm{ev}_z^{(i)} \end{aligned}$$

where $O(w \mid w_1, \dots, w_n)$ is the operator

$$\sum_a \omega_a(w)\partial_{\lambda_a} + 2\sum_i G(w, w_i),$$

$O'(\boldsymbol{\lambda}, w \mid w_1, \dots, w_n)$ is an operator of the form

$$\sum_a \varphi_a(\boldsymbol{\lambda}, w)\partial_{\lambda_a} + 2\sum_i \gamma(\boldsymbol{\lambda}, w, w_i),$$

and $\mathrm{ev}_z^{(i)}$ is the evaluation operator taking $f(\boldsymbol{\lambda} \mid z_1, \dots, z_n)$ to $f(\boldsymbol{\lambda} \mid z_1, \dots, z, \dots, z_n)$ (z in ith position); $G(z, w)$ is the Green function and $G_{\boldsymbol{\lambda}}(z, w)$ is its twisted version.

The purpose of [**12**] is to construct a deformation $\widetilde{T}(z_\infty)$ of this commuting family of operators. The construction of this family is parallel to that of the family $T(z_\infty)$. It relies on quantum group deformations of the algebras $U(\widehat{\mathfrak{g}})$, which we introduced in [**14, 15**]. Such a deformation depends on the choice of a nonzero differential ω on C. We denote it by $U_{\hbar,\omega}\mathfrak{g}$. The construction of [**12**] involves several steps:

1. $U_{\hbar,\omega}\mathfrak{g}$ contains a subalgebra $U_\hbar\mathfrak{g}^{\mathrm{out}}$, such that the inclusion $U_\hbar\mathfrak{g}^{\mathrm{out}} \subset U_{\hbar,\omega}\mathfrak{g}$ is a flat deformation of the inclusion $U(\mathfrak{g}^{\mathrm{out}}) \subset U(\widehat{\mathfrak{g}})$;
2. we construct a deformation of the Sugawara tensor of $U(\widehat{\mathfrak{g}})$, which yields a commuting family of operators in $U_{\hbar,\omega}\mathfrak{g}$;
3. we construct suitable deformations of the critical level representations of $\widehat{\mathfrak{g}}$;
4. then we construct the operators $\widetilde{T}(z)$ in a way parallel to the construction of $T(z)$.

The result of this construction is a family $\widetilde{T}(z)$ of commuting difference-evaluation operators, acting on functions $f(\boldsymbol{\lambda} \mid z_1, \dots, z_n)$; these operators have the following form

$$(4.1)\quad (\widetilde{T}(z)f)(\boldsymbol{\lambda} \mid z_1, \dots, z_n)$$
$$= \sum_{\varepsilon \in \{+,-\}} \left(A^\varepsilon(\boldsymbol{\lambda}, z \mid z_1, \dots, z_n) + \Pi^\varepsilon(z_i) B^\varepsilon(\boldsymbol{\lambda}, z \mid z_1, \dots, z_n) \circ \mathrm{ev}_z^{(i)}\right) \circ e^{\sum_a \omega^\varepsilon(z)\partial/\partial\lambda_a}$$

where $A^\pm(\boldsymbol{\lambda}, z \mid z_1, \dots, z_n)$, $B^\pm(\boldsymbol{\lambda}, z \mid z_1, \dots, z_n)$ and $\omega^\pm(z)$ are functions derived from C and from the vector field inverse to ω. The main result of [**12**] is then

THEOREM 4.4. *The operators $\widetilde{T}(z)$ defined by* (4.1) *form a commutative family, deforming the family $T(z)$.*

5. Open Problems

In this section, we gather some problems that naturally arise from the works reviewed in Section 4. Subsection 5.1 contains problems related to [**11, 13, 17**], and Subsection 5.2 contains problems arising from [**12**]. Other problems are formulated at the end of the book [**19**] (some of them have been solved since the publication of this book).

5.1. KZB equations.

Integral formulas. One would like to generalize the integral formulas obtained in [**13**]. It is also natural to seek "abstract" integral formulas for the KZB connection. The role of variables $\mathbf{t}$ would then be played by a nontrivial bundle over $\mathcal{M}_{g,n}$.

Monodromy problems. The monodromy representation of the KZB connection is a projective representation of the mapping class group. When the (λ_i, k) are integrable (which implies k integer > 0), this representation is conjectured to be isomorphic to a natural representation, defined in terms of modular categories attached to quantum groups at roots of 1 (see [**40, 45**]). This is a natural generalization of the Kohno-Drinfeld statement.

Integral formulas are a natural way to compute monodromy representations. It is therefore natural to ask what kind of monodromy representations arise from the homologies with local coefficients which we found in [**13**], and whether they can be related with the representations conjectured in [**40, 45**].

Separation of variables and geometric Langlands. It is an open problem to generalize the discussion "separation of variables–Bethe ansatz–integral formulas" in higher genus. Parametrizations might be helpful for this.

On the other hand, as a step in the program of [**29**] of relating Bethe ansatz with the geometric Langlands problem, one can try to solve the following problem. According to the general philosophy, the Bethe ansatz equations found in [**26**] (in the elliptic case) should be trivial monodromy conditions for an appropriate $\mathfrak{g}^L$-oper, where $\mathfrak{g}^L$ is the Langlands dual to $\mathfrak{g}$. The problem is to find this $\mathfrak{g}^L$-oper and to prove this statement.

It is also interesting to investigate the "mixed" differential systems (partly vector and partly differential) arising from opers with fake singularities. According to the Langlands program philosophy, infinitely close such opers, related by isomonodromy vector fields, should give rise to isomorphic mixed differential systems. It would be interesting to compute this isomorphism explicitly, and to see whether this picture could be q-deformed.

Field-theoretic analogues. In [**42**], Krichever constructs infinite-dimensional analogues of the Hitchin system. This work leads to natural questions. We have a diagram

	classical system	quantum system
critical level	Hitchin system	BD system
non-critical level	Schlesinger connection	KZB connection

Natural questions raised by Krichever's work are

1. to formulate this system in an "intrinsic" (i.e., parametrization-independent) way, and
2. to construct an "infinite-dimensional" array parallel to the above diagram.

5.2. qKZB equations. One would like to generalize to the systems of [**12**] more constructions of the genus 0, 1 qKZB equations at critical level, in particular, to find the Bethe ansatz equations for these systems. Another question would be to formulate the analogous systems in higher rank (quantum groups in higher genus have been constructed for any simple Lie algebra $\mathfrak{g}$ in [**8**]).

Let us now mention two other problems.

Noncritical level. It would be interesting to find "off-critical-level" qKZB systems, whose critical limits would be the systems of [**12**]. A new feature should emerge with respect to the standard qKZ equations: since the algebra $U_{\hbar,\omega}\mathfrak{g}$ is now quasi-Hopf, the parenthesis order of tensor product of modules at marked points should matter.

Nonformal versions. The algebras obtained in [**8**] make sense only when the deformation parameter is a formal variable. In general, one cannot assign to it numerical values. When $(C,\omega) = (\mathbb{C}P^1, z^N\,dz)$, we managed to define $U_{\hbar,\omega}\mathfrak{g}$ for complex $\hbar$ ([**16**]). What we used there is the fact that there is a Galois cover $\pi\colon C \to \mathbb{C}P^1$, such that $\omega = \pi^*(dz)$. One may hope that this can be generalized to situations where (C,ω) is a Galois cover of either $(\mathbb{C}P^1, dz)$, or $(\mathbb{C}P^1, dz/z)$, or (E, dz), where E is an elliptic curve $\mathbb{C}/L$. If this can be done, the systems of [**12**] could probably make sense for nonformal $\hbar$.

References

1. E. Arbarello, C. de Concini, V. Kac, and C. Procesi, *Moduli space of curves and representation theory*, Comm. Math. Phys. **117** (1988), no. 1, 1–36.
2. A. Beauville and Y. Laszlo, *Conformal blocks and generalized theta-functions*, Comm. Math. Phys. **164** (1994), no. 2, 385–419.
3. A. Beilinson and V. Drinfeld, *Quantization of Hitchin's integrable fibration and Langlands program*, Algebraic and Geometric Methods in Mathematical Physics (Kaciveli, 1993), Math. Phys. Stud., vol. 19, Kluwer Acad. Publ., Dordrecht, 1996, pp. 3–7.
4. D. Bernard, *On the Wess–Zumino–Witten on the torus*, Nuclear Phys. B **303** (1988), 77–93.
5. ———, *On the Wess–Zumino–Witten on Riemann surfaces*, Nuclear Phys. B **309** (1988), 145–74.
6. I. Cherednik, *Integral solutions of trigonometric Knizhnik–Zamolodchikov equations and Kac–Moody algebras*, Publ. Res. Inst. Math. Sci. **27** (1991), 727–744.
7. V. Drinfeld, *Quasi-Hopf algebras*, Algebra i Analiz **1** (1989), no. 6, 114–148 (Russian); English transl. Leningrad Math. J. **1** (1990), no. 6, 1419–1457.
8. B. Enriquez, *Quasi-Hopf algebras associated with semisimple Lie algebras and complex curves*, Selecta Math., N.S. (to appear).
9. B. Enriquez, B. Feigin, and V. Rubtsov, *Separation of variables for the elliptic Gaudin–Calogero systems*, Compositio Math. **110** (1998), no. 1, 1–16.
10. B. Enriquez and G. Felder, *Coinvariants for Yangian doubles and* qKZ *equations*, Internat. Math. Res. Notices **2** (1999), 81–104.
11. ———, *Commuting differential and difference operators associated with complex curves.* I, Compositio Math. **124** (2000), no. 1, 17–56.
12. ———, *Commuting differential and difference operators associated with complex curves.* II, Preprint math/9812152.
13. ———, *Solutions of the* KZB *equations in genus greater than one*, Preprint math/992198.
14. B. Enriquez and V. Rubtsov, *Quantum groups in higher genus and Drinfeld's new realizations method* ($\mathfrak{sl}_2$ *case*), Ann. Sci. École Norm. Sup. (4) **30** (1997), 821–846.
15. ———, *Quasi-Hopf algebras associated with* $\mathfrak{sl}_2$ *and complex curves*, Israel J. Math. **112** (1999), 61–108.
16. ———, *Some examples of quantum groups in higher genus*, math/9608005, Moscow Seminar in Mathematical Physics, Amer. Math. Soc. Transl. Ser. 2, vol. 191, Amer. Math. Soc., Providence, RI, 1999, pp. 33–65.
17. ———, *Hecke–Tyurin parametrization of the Hitchin and* KZB *systems*, Preprint ENS Paris, DMA-99-32, math/9911087.
18. P. Etingof, *Representations of affine Lie algebras, elliptic r-matrix systems, and special functions*, Comm. Math. Phys. **159** (1994), 471–502.
19. P. Etingof, I. Frenkel, and A. Kirillov, Jr., *Lectures on representation theory and Knizhnik–Zamolodchikov equations*, Math. Surveys and Monogr., vol. 58, 1991, Amer. Math. Soc., Providence, RI, 1998.
20. B. Feigin and E. Frenkel, *Affine Kac–Moody algebras at the critical level and Gel'fand–Dikii algebras*, Infinite Analysis, Part A, B (Kyoto, 1991), Adv. Ser. Math. Phys., vol. 16, World Sci. Publishing, River Edge, NJ, 1992, pp. 197–215.
21. B. Feigin, E. Frenkel, and N. Reshetikhin, *Gaudin model, Bethe ansatz and critical level*, Comm. Math. Phys. **116** (1994), no. 1, 27–62.

22. B. Feigin, V. Schechtman, and A. Varchenko, *On algebraic conditions satisfied by hypergeometric correlators of* WZW *models.* I, Comm. Math. Phys. **163** (1994), 173–84; II, Comm. Math. Phys. **170** (1995), 219–247.
23. B. Feigin and A. Stoyanovsky, *A realization of the modular functor in the space of differentials and geometric approximation of the moduli space of G-bundles*, Funct. Anal. Appl. **28** (1994), no. 1, 68–90.
24. G. Felder and A. Schorr, *Separation of variables for quantum integrable systems on elliptic curves*, Preprint math/9905072.
25. G. Felder, V. Tarasov, and A. Varchenko, *Solutions of the elliptic* qKZB *equations and the Bethe ansatz.* I, Amer. Math. Soc. Trans. Ser. 2 **180** (1997), 45–75.
26. G. Felder and A. Varchenko, *Integral representation of solutions of the elliptic* KZB *equations*, Internat. Math. Res. Notices, **5** (1995), 221–233.
27. ———, *The q-deformed* KZB *heat equation*, math/9809139.
28. G. Felder and C. Wieczerkowski, *Conformal blocks on elliptic curves and the* KZB *equations*, Comm. Math. Phys. **176** (1996), no. 1, 133–161.
29. E. Frenkel, *Affine algebras, Langlands duality and Bethe ansatz*, XIth International Congress of Mathematical Physics (Paris, 1994) (D. Iagolnitzer, ed.), Internat. Press, Cambridge, MA, 1995, pp. 606–642.
30. I. Frenkel and N. Reshetikhin, *Quantum affine algebras and holonomic difference equations*, Comm. Math. Phys. **146** (1992), 1–60.
31. E. Frenkel, N. Reshetikhin, and M. Semenov-Tian-Shansky, *Drinfeld–Sokolov reduction for difference operators and deformations of $\mathcal{W}$-algebras.* I. *The case of Virasoro algebra*, Comm. Math. Phys. **192** (1998), vol. 3, 605–629.
32. D. Friedan and S. Shenker, *The analytic geometry of two-dimensional conformal field theory*, Nuclear Phys. B **281** (1987), 509–545.
33. M. Gaudin, *Diagonalisation d'une classe d'hamiltoniens de spin*, J. Physique **37** (1976), 1087–1098.
34. K. Gawędzki and P. Tran-Ngoc-Bich, *Hitchin system at low genera*, Preprint hep-th/9803101; J. Math. Phys. **41** (2000), no. 7, 4965–4712.
35. B. van Geemen and E. Previato, *On the Hitchin system*, Duke Math. J. **95** (1996), no. 3, 659–683.
36. N. Kawamoto, Yu. Namikawa, A. Tsuchiya, and Y. Yamada, *Geometric realzation of conformal field theory and Riemann surfaces*, Comm. Math. Phys. **116** (1988), no. 2, 247–308.
37. D. Kazhdan and G. Lusztig, *Tensor structures arising from affine Lie algebras.* I, J. Amer. Math. Soc. **6** (1993), 905–947; II, **6** (1993), 949–1011; III, **7** (1994), 335–381; IV, **7** (1994), 383–453.
38. V. Knizhnik and A. Zamolodchikov, *Current algebras and Wess–Zumino model in two dimensions*, Nuclear Phys. B **247** (1984), 83–103.
39. T. Kohno, *Monodromy representations of braid groups and Yang–Baxter equations*, Ann. Inst. Fourier (Grenoble) **37** (1987), vol. 4, 139–160.
40. ———, *Topological invariants for 3-manifolds using representations of mapping class groups.* I, Topology **31** (1992), vol. 2, 203–230.
41. M. Kontsevich, *The Virasoro algebra and Teichmüller spaces*, Funktsional. Anal. i Prilozhen. **21** (1987), no. 2, 78–79 (Russian).
42. I. Krichever, *Vector bundles and Lax equations on algebraic curves*, Preprint hep-th/0108110.
43. G. Kuroki and T. Takebe, *Twisted* WZW *models on elliptic curves*, Comm. Math. Phys. **190** (1997), 1–56.
44. ———, *Wess–Zumino–Witten model on elliptic curves at the critical level*, J. Phys. A **34** (2001), 2403–2413.
45. V. Lyubashenko, *Invariants of* 3*-manifolds and projective representations of mapping class groups via quantum groups at roots of* 1, Comm. Math. Phys. **172** (1995), 467–516.
46. A. Matsuo, *Jackson integrals of the Jordan–Pochhammer type and* qKZ *equation*, Comm. Math. Phys. **151** (1993), 263–273.
47. E. Mukhin and A. Varchenko, *Quantization of the space of conformal blocks*, Preprint math/9710039; Lett. Math. Phys. **44** (1998), no. 2, 157–167.
48. N. Reshetikhin and M. Semenov-Tian-Shansky, *Central extensions of quantum currents groups*, Lett. Math. Phys. **19** (1990), 133–142.

49. N. Reshetikhin and A. Varchenko, *Quasiclassical asymptotics of solutions to the* KZ *equations*, Geometry, Topology, and Physics, Conf. Proc. Lecture Notes Geom. Topology, IV, Internat. Press, Cambridge, MA, 1995, pp. 293–322.
50. V. Schechtman and A. Varchenko, *Integral representations of N-point correlation functions in the* WZW *model*, Preprint Max Planck Institute MPI/89-51 (1989), Bonn.
51. M. Semenov-Tian-Shansky and A. Sevostyanov, *Drinfeld–Sokolov reduction for difference operators and deformations of $\mathcal{W}$-algebras.* II. *The semisimple case*, Comm. Math. Phys. **192** (1998) vol. 3, 631–647.
52. E. Sklyanin, *Separation of variables in the Gaudin model*, J. Sov. Math. 47 (1989), 2473–2488.
53. E. Sklyanin and T. Takebe, *Separation of variables in the elliptic Gaudin model*, Comm. Math. Phys. **204** (1999), no. 1, 17–38.
54. A. Stoyanovsky, *On quantization of the geometric Langlands correspondence.* I, Preprint math/9911108.
55. V. Tarasov and A. Varchenko, *Jackson integral representations for solutions of the Knizhnik–Zamolodchikov quantum equation*, (Russian) Algebra i Analiz **6** (1994), no. 2, 90–137 (Russian); English transl. St. Petersburg Math. J. **6** (1995), no. 2, 275–313.
56. ———, *Geometry of q-hypergeometric functions, quantum affine algebras and elliptic quantum groups*, Astérisque no. 246, 1997.
57. A. Tsuchiya, K. Ueno, and Y. Yamada, *Conformal field theory on universal family of stable curves with gauge symmetries*, Adv. Stud. Pure Math. **19** (1989), 459–566.
58. A. Tyurin, *Classification of vector bundles on universal family of stable curves of arbitrary genus*, Amer. Math. Soc. Transl. Ser. 2, **63** (1967), 245–279.
59. A. Varchenko, qKZ *equations, quantum Yang-Baxter equation, and q-hypergeometric functions*, Comm. Math. Phys. **162** (1994), 499–528.
60. ———, *Multidimensional hypergeometric functions and representation of quantum groups*, Adv. Ser. Math. Phys., vol. 21, 1995, World Scientific Publishing Co., Inc., River Edge, NJ, 1995.
61. E. Verlinde, *Fusion rules and modular transformations in* 2D *conformal field theory*, Nuclear Phys. B **300** (1988), 360–376.

Département de mathématiques, Université Louis Pasteur, 7, rue René Descartes, 67084 Strasbourg, France.

E-mail address: enriquez@math.u-strasbg.fr

Centre de Recherches Mathématiques
CRM Proceedings and Lecture Notes
Volume **32**, 2002

Jacobians of Singularized Spectral Curves and Completely Integrable Systems

Lubomir Gavrilov

Abstract. We state two recent results concerning the linearization of integrable systems on generalized Jacobians. Then we apply this to the (complexified) spherical pendulum.

1. Introduction

Let M^J be the affine vector space of complex matrix polynomials $A(x)$ in a variable x, of fixed degree d and dimension r

$$A(x) = Jx^d + A_{d-1}x^{d-1} + \cdots + A_0, \quad A_i \in \mathbf{gl}_r(\mathbb{C})$$

where $J \in \mathbf{gl}_r(\mathbb{C})$ is a fixed matrix. The matricial polynomial Lax equations

$$\frac{d}{dt}A(x) = \left[\frac{A^k(a)}{x-a}, A(x)\right], \quad k \in \mathbb{N},\ a \in \mathbb{C} \tag{1.1}$$

are well known to be Hamiltonian (with respect to several compatible Poisson structures on M^J) and completely integrable. The corresponding Hamiltonian vector fields define a complete set of commuting vector fields on the isospectral manifolds

$$M_P^J = \{A(x) \in M^J \colon \det(A(x) - yI_r) = P(x,y)\}.$$

The corresponding complex flows are not complete, but may be completed by adding a suitable divisor D_∞ called the Painlevé divisor of the system. The new manifold $\overline{M}_P^J = M_P^J \cup D_\infty$ becomes a noncompact complex-analytic commutative group. The most remarkable fact concerning $\overline{M}_P^J$ is that it has a commutative algebraic group structure which is compatible with the structure of an analytic group. This algebraic group is just the generalized Jacobian $J(X')$ of a suitable singularized spectral curve X' (to be defined below). The variety $\overline{M}_P^J$ is noncompact and it admits several algebraic structures. The "right" one is defined by the symmetry group G of the system (1.1). Namely, let $G = \mathbb{P}\mathbf{GL}_r(\mathbb{C}; J)$ be the subgroup of

---

2000 *Mathematics Subject Classification.* Primary: 14H70; Secondary: 37J35, 14H40, 37K10, 70H06.

I am grateful to R. Cushman for sending me the unpublished text [**8**] part of which is reproduced in the last section.

This is the final form of the paper.

the projective group $\mathbb{P}\mathbf{GL}_r(\mathbb{C})$ formed by matrices which commute with J. In the applications G is the symmetry group of the corresponding dynamical system (e.g. a rigid body with an axis of symmetry has a symmetry group $\mathbb{C}^*$ which is the complexified rotation group S^1). The group G acts on M^J by conjugation, the action is Poisson, and the reduced Hamiltonian system is completely integrable too. The action of G on M^J_P can be extended on the completed variety $\overline{M}^J_P$ and it is proper and free. Therefore $\overline{M}^J_P$ can be considered as the total space of a holomorphic principal fibre bundle ξ with base $\overline{M}^J_P/G$, structural group G, and natural projection map

$$\overline{M}^J_P \xrightarrow{\phi} \overline{M}^J_P/G.$$

The fibre bundle ξ is described as follows. When the spectral curve X defined by $\{(x,y) \in \mathbb{C}^2 \colon P(x,y)=0\}$ is smooth, then the partially compactified variety $\overline{M}^J_P$ is smooth and bi-holomorphic to the generalized Jacobian variety $J(X')$. The curve X' is singular and as a topological space it is just X with its "infinite" points $\infty_1, \infty_2, \ldots, \infty_r$ identified to a single point ∞. Thus $J(X')$ is a noncompact commutative algebraic group and it can be described as an extension of the usual Jacobian $J(X)$ by the algebraic group $G = (\mathbb{C}^*)^{s-1} \times \mathbb{C}^{r-s}$, where $s \leq r$ is the number of distinct eigenvalues of the leading term J

$$0 \to G \to J(X') \xrightarrow{\phi} J(X) \to 0. \tag{1.2}$$

As analytic spaces $J(X')$ and $J(X)$ are complex tori

$$J(X') = \mathbb{C}^{p_a}/\Lambda', \quad J(X) = \mathbb{C}^{p_g}/\Lambda$$

where Λ', Λ are lattices of rank $2p_g + s - 1$ and $2p_g$ respectively, p_g is the genus of X, and $p_a = p_g + r - 1$ is the arithmetic genus of X'. The generalized Jacobian $J(X')$ can be also considered as the total space of a holomorphic principal fibre bundle with base $J(X)$, projection ϕ, and structural group G. The group G is then identified with the symmetry group $\mathbb{P}\mathbf{GL}_r(\mathbb{C}; J)$ of (1.1), and the manifold $\overline{M}^J_P/G$ with the usual Jacobian $J(X) = J(X')/G$. The algebraic description of the reduced invariant manifold $\overline{M}^J_P/G$ is a well known result proved by A. Beauville [**5**] and M. R. Adams, J. Harnad, J. Hurtubise [**1**] (see also M. Adler, P. van Moerbeke [**2**], and [**26**, Section 8.2. of the survey] by A. G. Reyman and M. A. Semenov-Tian-Shansky).

The above can be illustrated on the following "simple" example. Let $\tau_1, \tau_2 \in \mathbb{C}$ be generic complex numbers. Consider the $\mathbf{Z}$-module of rank three $\Lambda \subset \mathbb{C}^2$

$$\Lambda = \mathbf{Z}\begin{pmatrix} 2\pi i \\ 0 \end{pmatrix} \oplus \mathbf{Z}\begin{pmatrix} 0 \\ 2\pi i \end{pmatrix} \oplus \mathbf{Z}\begin{pmatrix} \tau_1 \\ \tau_2 \end{pmatrix}. \tag{1.3}$$

The set $\mathbb{C}^2/\Lambda$ is a noncompact *algebraic group* and it can be considered as a (nontrivial) extension of the elliptic curve $\mathbb{C}/\{2\pi i\mathbf{Z} \oplus \tau_1\mathbf{Z}\}$ by $\mathbb{C}^* \sim \mathbb{C}/2\pi i\mathbf{Z}$

$$0 \to \mathbb{C}/2\pi i\mathbf{Z} \to \mathbb{C}^2/\Lambda \xrightarrow{\phi} \mathbb{C}/\{2\pi i\mathbf{Z} \oplus \tau_1\mathbf{Z}\} \to 0, \quad \phi(z_1, z_2) = z_1. \tag{1.4}$$

If $d = 2$, $r = 2$ and $P(x,y)$ is a suitable polynomial, then it may be shown that $\overline{M}^J_P = \mathbb{C}^2/\Lambda$, the symmetry group is $G = \mathbb{C}/2\pi i\mathbf{Z}$ and acts as

$$G \times \mathbb{C}^2/\Lambda \to \mathbb{C}^2/\Lambda : g \times \begin{pmatrix} z_1 \\ z_2 \end{pmatrix} \to \begin{pmatrix} z_1 \\ z_2 + g \end{pmatrix}$$

The quotient group $\{\mathbb{C}^2/\Lambda\}/G$ is the elliptic curve $\mathbb{C}/\{2\pi i\mathbf{Z} \oplus \tau_1\mathbf{Z}\}$ which is the Jacobian of the spectral curve X with an affine equation $\{P(x,y)=0\}$. Finally $\mathbb{C}^2/\Lambda$ is the generalized Jacobian of the singular spectral curve X' which is obtained from the regular curve X by a suitable singularization (a procedure opposite to the usual desingularization of a singular curve). We note that $\mathbb{C}^2/\Lambda$ is also an extension of the elliptic curve $\mathbb{C}/\{2\pi i\mathbf{Z} \oplus \tau_2\mathbf{Z}\}$ and hence it has a second algebraic structure. The "right" algebraic structure is the one related to the symmetry group $G = \mathbb{C}/2\pi i\mathbf{Z}$.

Generalized Jacobians as generic invariant manifolds of integrable systems appeared (implicitly) first in the papers of Jacobi [**19–21**] which is easily seen from the analytic expressions of the solutions (see also Klein and Sommerfeld [**22**]). This motivated the further study of the corresponding generalized Jacobi inversion problem (e.g. Clebsch an Gordan [**7**]). The modern theory of generalized Jacobians, without relation to integrable systems, is due to Rosenlicht [**27, 28**]. In our exposition we shall follow Serre [**29**]. In the modern literature on integrable systems generalized Jacobians appeared again in B. A. Dubrovin [**10**] and E. Previato [**25**]. For more examples see Fedorov [**3, 13, 14**] and Faye [**11, 12**].

In the present note we describe the invariant manifold of the integrable system (1.1) in the case when the leading term J of the Lax matrix $A(x)$ is either regular (Theorem 1), or it is not regular but diagonalizable (Theorem 2). The proofs are given in [**15,30**]. Then we apply Theorem 1 to the complexified spherical pendulum. The description which we obtain of the invariant manifold of the system completes the recent results of M. Audin, F. Beukers and R. Cushman [**4, 6, 8**].

2. Singularized Spectral Curves and Their Jacobians

A polynomial

$$P(x,y) = y^r + s_1(x)y^{r-1} + \cdots + s_r(x)$$

is called spectral, provided that the affine curve $\{(x,y) \in \mathbb{C}^2 : P(x,y)=0\}$ is the spectrum of some polynomial $r \times r$ matrix $A(x)$

$$P(x,y) = \det\big(A(x) - y.I_r\big).$$

In this case $\deg\big(s_i(x)\big) \leq i.d$, where d is the degree of $A(x)$

$$A(x) = A_d x^d + A_{d-1}x^{d-1} + \cdots + A_0, \quad A_i \in \mathbf{gl}_r(\mathbb{C}). \tag{2.1}$$

Consider the weighted projective space $\mathbb{P}^2(d) = \mathbb{C}^3\backslash\{0\}/\mathbb{C}^*$, where the $\mathbb{C}^*$-action on $\mathbb{C}^3$ is defined by

$$t \cdot (x,y,z) \to (tx, t^d y, tz), \quad t \in \mathbb{C}^*.$$

$\mathbb{P}^2(d)$ is a compact complex surface with one singular point $\{[0,1,0]\} = \mathbb{P}^2(d)_{\text{sing}}$. The affine curve $\{(x,y) \in \mathbb{C}^2\colon \det\big(A(x) - y.I_r\big) = 0\}$ is naturally embedded in $\mathbb{P}^2(d)$,

$$\mathbb{C}^2 \to \mathbb{P}^2(d)\colon (x,y) \mapsto [x,y,1],$$

and the condition $\deg\big(s_i(x)\big) \leq i.d$ shows that its closure X is contained in the smooth surface $\mathbb{P}^2(d)_{\text{reg}} = \mathbb{P}^2(d)\backslash\{[0,1,0]\}$. Let x be an affine coordinate on $\mathbb{P}^1$. The surface $\mathbb{P}^2(d)_{\text{reg}}$ is identified with the total space of the holomorphic line bundle $\mathcal{O}_{\mathbb{P}^1}(d)$ with base $\mathbb{P}^1$ and projection

$$\pi\colon \mathbb{P}^2(d)_{\text{reg}} \to \mathbb{P}^1\colon [x,y,z] \to [x,z].$$

The induced projection

$$\pi\colon X \to \mathbb{P}^1 \tag{2.2}$$

is a ramified covering of degree r, and over the affine plane $\mathbb{C}$ it is simply the first projection

$$\pi\colon \{(x,y) \in \mathbb{C}^2\colon P(x,y) = 0\} \to \mathbb{C}\colon (x,y) \to x.$$

DEFINITION 1 (Spectral curve of $A(x)$). We define the spectral curve X of the matrix polynomial $A(x)$ (2.1), to be the closure of the affine curve $\{(x,y) \in \mathbb{C}^2\colon \det(A(x) - y.I_r) = 0\}$ in the total space of the line bundle $\mathcal{O}_{\mathbb{P}^1}(d)$.

From now on we fix the spectral polynomial $P(x,y)$ and suppose that the spectral curve X is *smooth* and irreducible.

We are going now to singularize the curve X. Let $m = \sum_{i=1}^{s} n_i P_i$, $P_i \in X$, $n_i > 0$, be an effective divisor on X. To the pair (X, m) we associate a singular curve $X' = X_{\text{reg}} \cup \infty$, where if $S = \cup_{i=1}^{s} P_i$ is the support of m, then $X_{\text{reg}} = X - S$, and ∞ is a single point. The structure sheaf $\mathcal{O}'$ of $X' \sim (X, m)$ is defined in the following way. Let $\mathcal{O}_{X'}$ be the direct image of the structure sheaf $\mathcal{O} = \mathcal{O}_X$ under the canonical projection $X \to X'$. Then

$$\mathcal{O}'_P = \begin{cases} \mathcal{O}_P, & P \in X_{\text{reg}} \\ \mathbb{C} + i_\infty, & P = \infty \end{cases}$$

where i_∞ is the ideal of $\mathcal{O}_\infty$ formed by the functions f having a zero at P_i of order at least n_i. Thus a regular function f on X' is a regular function f on X, and such that for some $c \in \mathbb{C}$ and any i holds $v_{P_i}(f - c) \geq n_i$, where $v_P(.)$ is the order function. If p_g is the genus of X then the arithmetic genus p_a of the singular curve X' is $p_a = p_g + \deg(m) - 1$.

EXAMPLE. Let $m = P^+ + P^-$ be a divisor on the Riemann surface X. Then in a neighborhood of ∞ the singularized curve X' is analytically isomorphic either to the germ of analytical curve $xy = 0$ ($P^+ \neq P^-$), or to $y^2 = x^3$ ($P^+ = P^-$).

DEFINITION 2 (Singularized spectral curve of $A(x)$). If π is the projection (2.2) and $\infty = [1, 0] \in \mathbb{P}^1$ the "infinite" divisor, then the effective divisor $m = \pi^*(\infty)$ is called modulus of the spectral curve X. We have $\deg(m) = r$ and we denote by X' the singular curve associated to the regular curve X and to the modulus m.

The generalized Jacobian of the singular curve X' is the analytic manifold

$$J(X') = H^0(X, \Omega^1(-m))^* / H_1(X_{\text{reg}}, \mathbf{Z}) = \mathbb{C}^{p_a}/\Lambda'$$

where Λ' is a rank $2p_g + s - 1$ lattice, and $\Omega^1(-m)$ is the sheaf of meromorphic one-forms ω, such that $(\omega) \geq -m$. Similarly, for the usual Jacobian $J(X) \subset J(X')$, we have

$$J(X) = H^0(X, \Omega^1)^* / H_1(X, \mathbf{Z}) = \mathbb{C}^{p_g}/\Lambda$$

where $\Lambda \subset \Lambda'$ is a rank $2p_g$ lattice. It may be shown that both $J(X')$ and $J(X)$ are commutative algebraic groups related as follows. $J(X')$ is a nontrivial *extension* of $J(X)$ by the algebraic group $G = (\mathbb{C}^*)^{s-1} \times \mathbb{C}^{\deg(m)-s}$

$$0 \to G \to J(X') \xrightarrow{\phi} J(X) \to 0 \tag{2.3}$$

where the map ϕ is induced by the natural homomorphisms

$$H_1(X_{\text{reg}}, \mathbf{Z}) \to H_1(X, \mathbf{Z}), H^0(X, \Omega^1) \to H^0(X, \Omega^1(-m)).$$

This means that the sequence (2.3) is exact in the usual sense and moreover the algebraic structure of G (respectively of $J(X)$) is induced (respectively quotient) of the algebraic structure of $J(X')$. Note that $J(X')$ is noncompact. Indeed, while the topological space of $J(X)$ is $(S^1)^{2p_g}$, the one of $J(X')$ is $(S^1)^{2p_g+s-1}\times\mathbb{R}^{2\deg(m)-s-1}$.

Let M_P be the variety of $r \times r$ polynomial matrices of degree d (2.1), which have a fixed spectral polynomial $P(x,y)$

$$M_P = \{A(x)\colon \det(A(x) - yI_r) = P(x,y)\}$$

and let $M_P^J = M_P \cap M^J$ be the isospectral manifold formed by matrices of the form (2.1) with fixed leading term $A_d = J$

$$A(x) = Jx^d + A_{d-1}x^{d-1} + \cdots + A_0, \quad A_i \in \mathbf{gl}_r(\mathbb{C}). \tag{2.4}$$

The stabilizer

$$\mathbb{P}\mathbf{GL}_r(\mathbb{C};J) = \{R \in \mathbb{P}\mathbf{GL}_r(\mathbb{C})\colon RJR^{-1} = J\}$$

of $\mathbb{P}\mathbf{GL}_r(\mathbb{C})$ at $J \in \mathbf{gl}_r(\mathbb{C})$ is a commutative algebraic group isomorphic to $(\mathbb{C}^*)^{s-1}\times \mathbb{C}^{\deg(m)-s}$. It is a well known fact that M_P^J is a smooth manifold, $\mathbb{P}\mathbf{GL}_r(\mathbb{C};J)$ acts freely and properly on M_P^J by conjugation, and the quotient space $M_P^J/\mathbb{P}\mathbf{GL}_r(\mathbb{C};J)$ is a smooth manifold biholomorphic to $J(X) - \Theta$ [**1, 5**].

Consider the holomorphic principal fibre bundle ξ with total space M_P^J, structural group $\mathbb{P}\mathbf{GL}_r(\mathbb{C};J)$, base $M_P^J/\mathbb{P}\mathbf{GL}_r(\mathbb{C};J)$, and natural projection map $\varphi\colon M_P^J \to M_P^J/\mathbb{P}\mathbf{GL}_r(\mathbb{C};J)$. Consider also the associate principal bundle η with base space $J(X) - \Theta$, total space $J(X') - \Theta'$, projection map ϕ, and structural group G (see (2.3)).

We are ready to formulate our first result

THEOREM 1. [**15**, L. Gavrilov] *Suppose that the spectral curve X is smooth. There exists a partial compactification of M_P^J to a noncompact algebraic manifold $\overline{M}_P^J = M_P^J \cup D_\infty$ which is isomorphic (as an algebraic manifold) to the generalized Jacobian $J(X')$ of the singularized spectral curve X'. The Hamiltonian flows* (1.1) *are translation invariant on $J(X')$. The action of the symmetry group $\mathbb{P}\mathbf{GL}_r(\mathbb{C};J)$ on M_P^J extends to an action on $\overline{M}_P^J$ which is free, proper and compatible with the algebraic structure of $\overline{M}_P^J$. The principal algebraic bundles*

$$0 \to \mathbb{P}\mathbf{GL}_r(\mathbb{C};J) \to \overline{M}_P^J \to \overline{M}_P^J\mathbb{P}\mathbf{GL}_r(\mathbb{C};J) \to 0$$

and

$$0 \to G \to J(X') \xrightarrow{\phi} J(X) \to 0$$

are isomorphic. The Painlevé divisor $D_\infty = \overline{M}_P^J \backslash M_P^J$ is the pre-image $\Phi^{-1}(\Theta)$ of the Riemann theta divisor $\Theta \subset J(X)$.

A local analysis (at infinity) shows that the smoothness of the compact spectral curve X implies the regularity of the leading term J of $A(x)$ (a matrix is regular if its minimal and characteristic polynomials coincide). We shall suppose now that J is not regular, and hence X is never smooth (at infinity). We consider only the simplest case when J is diagonal. Equivalently, if $m(\lambda)$ is the minimal polynomial of J, then $m(\lambda) = \Pi_{i=1}^k(\lambda - \lambda i)$, where $\lambda_i \neq \lambda_j$ for $i \neq j$, and $k < r = \dim J$. Geometrically this means that X has only normal crossing singularities at infinity. Without loss of generality we suppose that J is diagonal, so $\{e_i\}_1^r$ are its eigenvectors. If J is not regular the dimension of the stabilizer $\mathbb{P}\mathbf{GL}_r(\mathbb{C};J)$ is strictly bigger than $d-1$. This suggests that, in order to compensate the increase of the dimension of the stabilizer of J, additional restrictions should be imposed on the subleading term A_{d-1}. Let $\oplus_i E_i$ be a spectral decomposition of $\mathbb{C}^r$, $E_i = \mathrm{Ker}(J-\lambda_i I)$. To an endomorphism $u \in \mathrm{End}(\mathbb{C}^r)$ we associate its restriction on E_i denoted $u|_{E_i} \in \mathrm{End}(E_i)$

$$u|_{E_i} : v \to Pr_{E_i} u(v).$$

Let us fix now a basis $\{e_i\}_{i=1}^r$ of eigenvectors. To each matrix A we associate by the above formula its restriction $A|_{E_i}$. If J is diagonal, then $A|_{E_i}$ is a "diagonal" minor of A of dimension $\dim E_i \times \dim E_i$. To each proper subspace $E_i = \mathrm{Ker}(J-\lambda_i I)$ of dimension $m_i > 1$ we associate a diagonalizable regular matrix K_i of dimension m_i. Denote further the set of matrices K_i by K. We replace further the space M^J by the affine space

$$M^{J,K} = \Big\{ A(x) = \sum_{i=1}^{d} A_i x^i \colon A_i \in \mathbf{gl}_r(\mathbb{C}), A_d = J, A_{d-1}|_{E_i} = K_i \quad \forall i \text{ such that } \dim E_i > 1. \Big\}$$

and M_P^J by

$$M_P^{J,K} = \big\{ A(x) \in M^{J,K}, \det\big(A(x) - yI\big) \equiv P(x,y) \big\}.$$

The singularization X' of the *singular* spectral curve X is defined as follows. Let $\widetilde{X}$ be the desingularized X. Then apply Definition 2 to the curve $\widetilde{X}$ and to the divisor m, where m is the pre-image of the points at 'nfinity" on X. It is seen that the topological space of X is $X_{\mathrm{aff}} \cup \infty$, where $X_{\mathrm{aff}} = \big\{(x,y) \in \mathbb{C}^2 \colon P(x,y) = 0\big\}$. Finally we define the stabilizer (or the symmetry group)

$$\mathbb{P}\mathbf{GL}_r(\mathbb{C}; J, K) = \big\{ R \in \mathbb{P}\mathbf{GL}_r(\mathbb{C}, J) \colon (RA_{d-1}R^{-1})\big|_{E_i} = K_i, \quad \forall i \text{ such that } \dim E_i > 1 \big\}.$$

The analogue of Theorem 1 is the following

THEOREM 2. [**30**, O.Vivolo] *Suppose that J is a nonregular diagonal matrix , and let K be a set of regular diagonal matrices associated to each proper space of J of dimension bigger than one. If the affine part X_{aff} of the spectral curve is smooth, then the variety $M_P^{J,K}$ is smooth. There exists a partial compactification of $M_P^{J,K}$ to a noncompact algebraic manifold $\overline{M}_P^{J,K} = M_P^{J,K} \cup D_\infty$ which is isomorphic (as an algebraic manifold) to the generalized Jacobian $J(X')$ of the singularized spectral curve X'. The Hamiltonian flows (1.1) are translation invariant on $J(X')$. The action of the symmetry group $\mathbb{P}\mathbf{GL}_r(\mathbb{C}; J, K)$ on $M_P^{J,K}$ extends to an action on $\overline{M}_P^J$ which is free, proper and compatible with the algebraic structure of $\overline{M}_P^{J,K}$. The principal algebraic bundles*

$$0 \to \mathbb{P}\mathbf{GL}_r(\mathbb{C}; J, K) \to \overline{M}_P^{J,K} \to \overline{M}_P^{J,K} / \mathbb{P}\mathbf{GL}_r(\mathbb{C}; J, K) \to 0$$

and

$$0 \to G \to J(X') \xrightarrow{\phi} J(X) \to 0$$

are isomorphic. The Painlevé divisor $D_\infty = \overline{M}_P^{J,K} \backslash M_P^{J,K}$ is the pre-image $\Phi^{-1}(\Theta)$ of the Riemann theta divisor $\Theta \subset J(X)$.

See [**30**] for applications.

3. The Spherical Pendulum

The complexified spherical pendulum is described by the following system of differential equations

$$\dot{x} = v, \quad \dot{v} = -e_3 + \big(\langle x, e_3\rangle - \langle v, v\rangle\big)x \tag{3.1}$$

constrained on the (invariant) manifold

$$M = \big\{(x, v) \in \mathbb{C}^3 \times \mathbb{C}^3 \mid \langle x, x\rangle = 1, \langle x, v\rangle = 0\big\}$$

where $\langle x, y\rangle = \sum_{i=1}^{3} x_i y_i$ and e_1, e_2, e_3 is the standard basis of $\mathbb{C}^3$. The algebraic manifold M is the tangent bundle TQ of the complexified sphere $Q = \big\{x\colon \langle x, x\rangle = 1\big\}$ and may be also identified to T^*Q via the bilinear form $\langle . \,, .\rangle$ on T_xQ. Therefore TQ is a symplectic manifold and with respect to this symplectic structure (3.1) is a two degrees of freedom completely integrable Hamiltonian system with Hamiltonian function

$$H(x, v) = \frac{1}{2}\langle v, v\rangle + x_3.$$

The second integral of motion

$$K(x, v) = x_1 v_2 - x_2 v_1$$

is associated to the symmetry group $G = \mathrm{SO}(2, \mathbb{C})$ (rotations about the axes e_3). We denote the corresponding Hamiltonian vector fields by X_H, X_K.

Following Cushman [**8**], consider the following automorphism of M

$$\varphi\colon M \to M\colon (x, v) \to (y, u) = (x, x \times v). \tag{3.2}$$

It is straightforward to check that $\varphi^2(x, v) = (x, -v)$, $\varphi^{-1}(y, u) = (y, u \times y)$. The push forward under φ of the Hamiltonian vector field (3.1) is the Hamiltonian vector field $X_{\mathcal{H}}$ corresponding to the Hamiltonian

$$\mathcal{H}(y, u) = (\varphi^{-1})^* H(y, u) = \frac{1}{2}\langle u, u\rangle + \langle y, e_3\rangle.$$

The integral curves of $X_{\mathcal{H}}$ satisfy

$$\dot{y} = \dot{x} = v = u \times y, \quad \dot{u} = \dot{x} \times v + x \times \dot{v} = -x \times e_3 = e_3 \times y. \tag{3.3}$$

Physically the above Hamiltonian system is obtained by reducing the *right* S^1 symmetry given by rotating the Lagrange top about its symmetry axis at its momentum value 0 and then choosing appropriate time and length scales. For more details see [**9**, pp. 191–200].

As in [**17**] we introduce the complex change of variables

$$\begin{aligned} U_1 &= u_3 - iu_2, & U_2 &= y_3 - iy_2,\\ V_1 &= u_1, & V_2 &= y_1,\\ W_1 &= u_3 + iu_2, & W_2 &= y_3 + iy_2. \end{aligned} \tag{3.4}$$

Consider the matrices

$$A(\lambda) = \begin{pmatrix} V_1\lambda + V_2 & \lambda^2 + U_1\lambda + U_2\\ \lambda^2 + W_1\lambda + W_2 & -(V_1\lambda + V_2)\end{pmatrix} \tag{3.5}$$

and

$$B(\lambda) = \frac{A(\lambda) - A(0)}{\lambda} = \begin{pmatrix} V_1 & \lambda + U_1\\ \lambda + W_1 & -V_1\end{pmatrix} \tag{3.6}$$

THEOREM 3. [**8**, R. Cushman] *Equation* (3.1) *can be written in Lax form as*

$$2i\frac{d}{dt}A(\lambda) = [A(\lambda), B(\lambda)] = -\left[A(\lambda), \frac{A(0)}{\lambda}\right]. \tag{3.7}$$

The proof is straightforward. An equivalent Lax pair may be found in the recent preprint of M. Audin [**4**] (she attributes it to A. Reyman). A Lax pair in the Lie algebra ao(3, 1) appeared earlier in [**26**].

A short calculation shows that characteristic polynomial of $A(\lambda)$ is

$$\begin{aligned} \det(\mu I - A(\lambda)) &= \mu^2 + (V_1\lambda + V_2)^2 + (\lambda^2 + U_1\lambda + U_2)(\lambda^2 + W_1\lambda + W_2) \\ &= \mu^2 + \lambda^4 + (U_1 + W_1)\lambda^3 + (U_2 + W_2 + U_1W_1 + V_1^2)\lambda^2 \\ &\quad + (U_1W_2 + U_2W_1 + 2V_1V_2)\lambda + (U_2W_2 + V_2^2). \end{aligned} \tag{3.8}$$

Using (3.4) and (3.2) we find that

$$\begin{aligned} U_1 + W_1 &= 2\langle u, e_3\rangle = 2\langle x \times v, e_3\rangle = 2k \\ U_2 + W_2 + U_1W_1 + V_1^2 &= 2\left(\frac{1}{2}\langle u, u\rangle + \langle y, e_3\rangle\right) \\ &= 2\left(\frac{1}{2}\langle v, v\rangle + \langle x, e_3\rangle\right) = 2h \\ U_1W_2 + U_2W_1 + 2V_1V_2 &= 2\langle u, y\rangle = 2\langle v \times x, x\rangle = 0 \\ U_2W_2 + V_2^2 &= \langle y, y\rangle = \langle x, x\rangle = 1. \end{aligned}$$

Hence we obtain the spectral curve

$$X_{\text{aff}} = \{0 = \mu^2 + F_c(\lambda)\} \tag{3.9}$$

of the Lax equation (3.7) where

$$F_c(\lambda) = \lambda^4 + 2k\lambda^3 + 2h\lambda^2 + 1. \tag{3.10}$$

Let X be the compactified and normalized curve X_{aff}. When X_{aff} is smooth, X is an elliptic curve, $X = X_{\text{aff}} \cup \infty_1 \cup \infty_2$. Let X' be the singularized curve X with respect to the modulus $\infty_1 + \infty_2$ (Definition 2). The topological space of X' is pictured in [**17**, Fig. 2] and [**4**, Fig. 2]. As a corollary of Theorem 1 applied to the Lax pair (3.7), we obtain the following

THEOREM 4. *Suppose that the affine spectral curve X_{aff} of the spherical pendulum is smooth. Then the invariant manifold M_{hk} is also smooth. There exists a partial compactification of M_{hk} to a noncompact algebraic manifold $\overline{M}_{hk} = M_{hk} \cup D_\infty$ which is isomorphic (as an algebraic manifold) to the generalized Jacobian $J(X')$ of the singularized spectral curve X'. The Hamiltonian flows X_H and X_K of the spherical pendulum are translation invariant on $J(X')$. The action of the symmetry group $G = \mathrm{SO}(2,\mathbb{C}) = \mathbb{C}^*$ on M_{hk} extends to an action on $\overline{M}_{hk}$ which is free, proper and compatible with the algebraic structure of $\overline{M}_{hk}$. The principal algebraic bundles*

$$0 \to \mathbb{C}^* \to \overline{M}_P^J \to \overline{M}_{hk}/\mathbb{C}^* \to 0$$

and

$$0 \to \mathbb{C}^* \to J(X') \xrightarrow{\phi} J(X) \to 0$$

are isomorphic. The Painlevé divisor D_∞ is isomorphic to $\mathbb{C}^*$

REMARKS. The complexified spherical pendulum is a limiting case of the complexified Lagrange top and the algebro-geometric structure of the latter was studied in [**17**]. The above theorem is the analogue of this result. A big part of Theorem 4 is proved by Beukers and Cushman [**6,8**]. It is shown in [**6**] that the general invariant manifold

$$M_{hk} = \{(x, v) \in M \colon H(x, v) = h, K(x, v) = k\}$$

is the total space of a principal $\mathbb{C}^*$ bundle with base an elliptic curve. The equations satisfied by the variables U_i, V_i, W_i as well the Lax pair (3.7) are obtained by R. Cushman [**8**].

Our second remark concerns the monodromy of the energy-momentum map

$$M \to \mathbb{R}^2 \colon (x, v) \to \big(H(x, v), K(x, v)\big)$$

restricted to the set of its regular values. This is a "classical" question and we refer the reader to [**9**]. As the regular fibers of the energy-momentum map are real parts of the affine part of the generalized Jacobian $J(X')$, then it suffices to study the monodromy of the flat bundle

$$H_1\big(J(X'), \mathbf{Z}\big) \to (h, k)$$

restricted to the set of real regular (h, k) (and taking into account the real structure of $J(X')$. But $H_1\big(J(X'), \mathbf{Z}\big) = H_1(X_{\mathrm{aff}}, \mathbf{Z})$ and hence the last question is equivalent to the study of the familiar Picard–Lefschetz monodromy of the homology bundle associated to the fibration

$$\{(\lambda, \mu) \in \mathbb{C}^2 \colon \lambda^4 + 2k\lambda^3 + 2h\lambda^2 + 1 + \mu^2 = 0\} \to (h, k).$$

In the case of the Lagrange top this was (implicitly) used by O. Vivolo [**31,32**] and in the case of the spherical pendulum by M. Audin [**4**].

The nondegeneracy of the frequency map is another remarkable feature of the spherical pendulum. This has been first proved by Emil Horozov (as conjecture of J. J. Duistermaat) in relation to applications of KAM theory [**18**]. Another proof, revealing the hidden algebraic geometry of the problem, can be found in [**16**].

References

1. M. R. Adams, J. Harnad, and J. Hurtubise, *Isospectral Hamiltonian flows in finite and infinite dimension*, Comm. Math. Phys. **134** (1990), 555–585.
2. Mark S. Adler and P. van Moerbeke, *Linearization of Hamiltonian systems, Jacobi varieties and representation theory*, Adv. in Math. **38** (1980), 318–379.
3. Mark S. Alber and Yuri N. Fedorov, *Wave solutions of evolution equations and Hamiltonian flows on nonlinear subvarieties of generalized Jacobians*, J. Phys. A **33** (2000), no. 47, 8409–8425.
4. M. Audin, *Monodromy of the spherical pendulum* via *algebraic geometry*, Preprint, Strasbourg, 2001.
5. A. Beauville, *Jacobiennes des courbes spectrales et systèmes hamiltoniens complètement intégrables*, Acta Math. **164** (1990), 211–235.
6. F. Beukers and R. Cushman, *The complex geometry of the spherical pendulum*, Preprint, Utrecht, 2000.
7. A. Clebsch and P. Gordan, *Theorie der Abelschen Functionen*, Teubner, Leipzig, 1866.
8. R. Cushman, Letter to L. Gavrilov dated 18 December 2000.
9. R. Cushman and L. Bates, *Global aspects of classical integrable systems*, Birkhäuser Verlag, Basel, 1997.
10. B. A. Dubrovin, *Geometry of Abelian varieties, Riemann surfaces and nonlinear equations*, Ph.D., Moscow, 1984 (in russian).

11. I. Faye, *Semi-Abelian surfaces and integrable systems*, Proc. Edinburg Math. Soc. **44** (2001), 249–265.
12. ———, *Sur l'intégrabilité algébrique et la non-intégrabilité des systèmes Hamiltoniens*, Ph.D. thesis, University of Toulouse III, 2001.
13. Yu. Fedorov, *Classical integrable systems and billiards related to generalized Jacobians*, Acta Appl. Math. **55** (1999), no. 3, 251–301.
14. ———, *Discrete versions of some algebraic integrable systems related to generalized Jacobians*. SIDE III—*symmetries and integrability of difference equations* (Sabaudia, 1998), CRM Proc. Lecture Notes, vol. 25, Amer. Math. Soc., Providence, RI, 2000, pp. 147–160.
15. L. Gavrilov, *Generalized Jacobians of spectral curves and completely integrable systems*, Math. Z. **230** (1999), 487–508.
16. L. Gavrilov and O. Vivolo, *The real period function of the A_3 singularity and perturbations of the spherical pendulum*, Compositio Math. **123** (2000), 167–184.
17. L. Gavrilov and A. Zhivkov, *The complex geometry of Lagrange top*, Enseign. Math. (2) **44** (1998), 133–170.
18. Emil Horozov, *Perturbations of the spherical pendulum and Abelian integrals*, J. Reine Angew. Math. **408** (1990), 114–135.
19. C. Jacobi, *Über eine neue Methode zur Integration der Hyperelliptischen Differentialgleichungen und über die Rationale Form ihrer Vollständigen Algebraischen Integralgleichungen*, J. Reine Angew. Math. **32** (1846), 220–226.
20. ———, *Sur la rotation d'un corps*, Gesammelte Werke, Bd. 2, Chelsea, 1969, pp. 289-352.
21. ———, *Fragments sur la rotation d'un corps tirés des manuscrits de Jacobi et communiqués par E. Lotner*, Gesammelte Werke, Bd. 2, Chelsea, 1969, pp. 425-514.
22. F. Klein and A. Sommerfeld, *Theorie des Kreisels*, Teubner, Leipzig, pp. 1897-1910.
23. P. van Moerbeke and D. Mumford, *The spectrum of difference operators and algebraic curves*, Acta Math. **143** (1979), 93–154.
24. D. Mumford, *Tata lectures on theta* II, Progress in Mathematics, vol. 43, Birkhäuser Boston, Inc., Boston, MA, 1984
25. E. Previato, *Hyperelliptic quasi–periodic and soliton solutions of the nonlinear Schrödinger equation*, Duke Math. J. **52** (1985), 329–377.
26. A. G. Reyman and M. A. Semenov-Tian-Shansky, *Dynamical systems*. VII. *Integrable systems, nonholonomic dynamical systems*, Enciclopaedia of Mathematical Sciences, Springer-Verlag, Berlin, 1994.
27. M. Rosenlicht, *Generalized Jacobian varieties*, Ann. of Math. (2) **59** (1954), 505–530.
28. ———, *A universal mapping property of generalized Jacobian varieties*, Ann. of Math. (2) **66** (1957), 80–88.
29. J.-P. Serre, *Groupes algébriques et corps de classes*, Hermann, Paris, 1959.
30. O. Vivolo, *Jacobians of singular spectral curves and completely integrable systems*, Proc. Edinburg Math. Soc. (2) **43** (2000), 605–623.
31. ———, *The monodromy of action variables of Lagrange top*, Preprint no. 62, Laboratoire Emile Picard, Iniversité de Toulouse III, 1995.
32. ———, *Systèmes intégrables et courbes algébriques*, Thèse, Université de Toulouse III, 1997.

Laboratoire Emile Picard, UMR 5580, Université Paul Sabatier, 118, route de Narbonne, 31062 Toulouse Cedex, France.

E-mail address: `gavrilov@picard.ups-tlse.fr`

Centre de Recherches Mathématiques
CRM Proceedings and Lecture Notes
Volume **32**, 2002

The q-Hypergeometric Equation, Askey–Wilson Type Solitons and Rational Curves With Singularities

Luc Haine

ABSTRACT. I survey some recent work aiming at describing all families of functions which, besides satisfying a doubly infinite three-term recursion relation, are also eigenfunctions of a q-difference operator of an arbitrary order. Links with the theory of solitons are emphasized and some new results are presented.

1. Introduction

We denote by L a tridiagonal doubly infinite matrix

$$L = a(n)E^{-1} + b(n)I + E, \quad n \in \mathbb{Z} \tag{1.1}$$

where I is the identity operator and E, E^{-1} denote respectively the forward and the backward shift operators acting on an infinite vector $h(n)$, $n \in \mathbb{Z}$, as follows

$$(Eh)(n) = h(n+1), \quad (E^{-1}h)(n) = h(n-1), \quad n \in \mathbb{Z}. \tag{1.2}$$

We look for all the solutions of the following problem:

DISCRETE (TRIDIAGONAL) BISPECTRAL PROBLEM. Find all tridiagonal doubly infinite matrices L, for which some family of eigenfunctions $f(n,z)$ satisfying

$$a(n)f(n-1,z) + b(n)f(n,z) + f(n+1,z) = x(z)f(n,z), \tag{1.3}$$

is also a family of eigenfunctions of a q-difference operator

$$\sum_{\text{finitely many i's} \in \mathbb{Z}} c_i(z)f(n,q^i z) = \theta(n)f(n,z), \tag{1.4}$$

for some appropriate choice of the "spectral variable" $x(z)$.

Thorough the paper, we assume that $|q| \neq 1$. We abbreviate (1.3) and (1.4) as

$$Lf = xf, \quad Bf = \Theta f, \tag{1.5}$$

with Θ the diagonal matrix $\Theta = \operatorname{diag}(\ldots, \theta(-1), \theta(0), \theta(1), \ldots)$.

2000 *Mathematics Subject Classification.* Primary: 33D45; Secondary: 35Q51.

Research Associate of the Belgian National Fund for Scientific Research.

I am grateful to Alberto Grünbaum and Plamen Iliev for their collaboration on various parts of the work reported here. I also would like to thank Vadim Kuznetsov for his unshakable willingness to put this volume together, honoring Sophie Kowalevski.

This is the final form of the paper.

If we impose the additional conditions that $f(-1,z)=0$ and $f(0,z)=1$, then the $f(n,z), n \geq 0$, are polynomials of degree n in the variable $x(z)$. The celebrated Askey–Wilson polynomials [**2**] provide instances of solutions, with B a second-order q-difference operator and $x(z)=(z+z^{-1})/2$. In [**11**], jointly with F. A. Grünbaum, we proved that these polynomials exhaust all the polynomial solutions, with an operator B of order two. Our proof is based on a discrete analogue of an identity which was discovered and used in a pioneering paper of J. J. Duistermaat and F. A. Grünbaum [**9**], where a purely differential version of the problem was considered. This work and its various extensions, are now part of a body of knowledge, referred to as *bispectral problems*. It was the driving force that led us in [**12**] to relax the special conditions leading to the orthogonal polynomial solutions, by allowing for a doubly infinite three-term recursion relation, keeping B of order two. We discovered that the only solutions of this more general problem, are provided by what we call the *associated Askey–Wilson matrices*. The entries of these matrices are obtained by making an *arbitrary* shift in the coefficients of the recursion relations satisfied by the Askey–Wilson polynomials, extending them over *all* integers. Precisely,

$$a(n)=\tilde{a}(n+\varepsilon), \quad b(n)=\tilde{b}(n+\varepsilon), \qquad n\in\mathbb{Z} \tag{1.6}$$

where $\tilde{a}(n)$, $n \geq 1$, and $\tilde{b}(n), n \geq 0$, denote the coefficients of the three-term recursion relation satisfied by the Askey–Wilson polynomials.

The coefficients (1.6) define the the so-called *associated Askey–Wilson polynomials*, when the conditions $f(-1,z)=0$ and $f(0,z)=1$ are imposed. The associated Askey–Wilson polynomials were extensively studied by M. E. H. Ismail and M. Rahman [**19**]. However, as soon as $\varepsilon \neq 0$, the functions $f(n,z)$ that solve (1.3) and (1.4), with a second-order q-difference operator, are *not* given by the associated polynomials, but rather by a *two-dimensional* space of contiguous basic hypergeometric *functions*, that are naturally called the *Askey–Wilson functions*. These functions were in fact already studied in [**19**]. Thus, our result characterizes the Askey–Wilson functions by their bispectral property. This is reviewed in Sections 2 and 3, with several simplifications in the proofs compared with [**12**].

The rest of our paper is concerned with the description of some solutions to (1.3) and (1.4), when both the recursion relation in (1.3) and the q-difference equation in (1.4) are allowed to be of an arbitrary order. It aims mainly at introducing the reader to the beautiful circle of ideas that was opened up by the work of G. Wilson [**26**], connecting the subject with the Burchnall–Chaundy [**4–6**] theory of commutative rings of differential and difference operators. An important invariant of these rings is the *rank*, meaning the greatest common divisor of the orders of all the operators in the ring. Section 4 reviews the theory in the context of rank 1 commutative rings of difference operators. G. Wilson discovered that the spectrum (in the sense of algebraic geometry) of a *bispectral* rank 1 commutative ring of differential operators, is an irreducible rational curve with only *cusp-like singularities*, which completes by adding *one* nonsingular point at infinity. His beautiful idea was that exchanging the role of the "space" and the "spectral" variables in the Baker–Akhiezer function of such a curve, produces a Baker–Akhiezer function of another curve of the same type. Jointly with P. Iliev [**15**], we showed that irreducible rational curves with only *cusp-like singularities*, which complete by adding *two* nonsingular points at infinity, lead to rank 1 bispectral rings of *difference* operators, with a (rank 1) ring of *differential* operators as *dual* ring. Thus, these curves provide solutions to the problem above, when the recursion relation (1.3) is allowed to be of an arbitrary order, and the q-difference equation (1.4) is replaced by a differential equation. The specific curves leading to solutions satisfying a three-term recursion relation were classified in [**16**].

I have chosen to illustrate the ideas on the purely discrete version of the bispectral problem, following [**17**]. An advantage is that in this context, we can (mostly) deal with the simplest possible singular curves, namely *rational curves with ordinary double points.* Section 5 introduces the reader to the geometry of these curves, which are familiar from soliton theory. In Section 6, I describe a class of rational curves with double points, which provide solutions to the purely discrete version of the bispectral problem. I call these solutions "Askey–Wilson type solitons". The generic case of this construction is described in a short paper by F. W. Nijhoff and O. A. Chalykh [**23**], from a different point of view. A related study was done independently in [**14**] and [**18**].

In Section 7, I consider some nongeneric examples, which lead to solutions of the *tridiagonal* discrete bispectral problem posed at the beginning of the introduction. These examples are obtained by iteration of the (matrix) Darboux transformation, starting from the operator $L = E + E^{-1}$, and adding bound states at a sequence of points of the form

$$\pm(q^{k_i/2} + q^{-k_i/2}), \quad k_i \in \mathbb{N} \setminus \{0\}. \tag{1.7}$$

Notice that, when $q \to 1$, these bound states all accumulate at ± 2, i.e. at the boundary of the continuous spectrum of the operator L, which is formed by the interval $[-2, 2]$. This is reminiscent of the work of H. L. Krall in the 1940's, aiming at classifying all families of orthogonal polynomials which are eigenfunctions of a differential operator. In [**21**], he proved that all the orthogonal polynomials, which satisfy a fourth-order differential equation, could be obtained by adding point masses (with an arbitrary weight) at the end of the support of the orthogonality measure of some instances of the Laguerre and the Jacobi polynomials. For a review of Krall's work, including some recent developments, I refer the reader to [**13**]. The paper ends by showing that (1.7) gives the clue about how to build q-analogues of the Krall polynomials, adding bound states to the orthogonality measure of the Askey–Wilson polynomials. A complete account of this result will appear in [**17**].

2. The Basic Operator Identity and Its Integrated Form

In the next two sections, I shall describe all the solutions of the discrete tridiagonal bispectral problem (1.5), when B is a second-order q-difference operator. When B is replaced by a differential operator (i.e. when $q \to 1$), the choice of the spectral parameter $x(z)$ is immaterial. Within the context of q-difference operators, this choice cannot be swept under the rug. We limit ourselves to the case

$$x(z) = \gamma(z + \delta z^{-1}), \tag{2.1}$$

with γ and δ arbitrary parameters. The choices $x(z) = z$ and $x(z) = (z + z^{-1})/2$ are up to scaling the only ones that deserve attention.

There is no loss of generality in assuming that the operator B takes the form

$$Bf \equiv c(z)\big(f(n, qz) - f(n, z)\big) + d(z)\big(f(n, q^{-1}z) - f(n, z)\big). \tag{2.2}$$

We denote by

$$[L, M]_q \equiv LM - qML,$$

the "q-commutator" of two matrices L and M. When $q = 1$, we omit the subscript so that $[\,.\,,\,.\,] = [\,.\,,\,.\,]_1$ is the usual commutator. We have the following identity:

BASIC OPERATOR IDENTITY ([**11, 12**]). Any solution of (1.5) with B as in (2.2), satisfies the matrix identity

$$\left[L, \left[L, [L, \Theta]_{q^{-1}}\right]_q + \gamma^2\delta\frac{(q-1)^2(q+1)^2}{q^2}\Theta\right] = 0. \tag{2.3}$$

PROOF. Put

$$u = -\frac{1+q+q^2}{q}, \qquad v = \gamma^2\delta\frac{(q-1)^2(q+1)^2}{q^2}. \tag{2.4}$$

Expanding the commutators in the left-hand side of (2.3) and applying the resulting operator to some eigenvector f satisfying (1.5), one obtains:

$$\begin{aligned}
&(L^3\Theta - \Theta L^3)f + u(L^2\Theta L - L\Theta L^2)f + v(L\Theta - \Theta L)f \\
&\quad = \big(B(x^3 f) - x^3 B(f)\big) + u\big(xB(x^2 f) - x^2 B(xf)\big) + v\big(B(xf) - xB(f)\big) \\
&\quad = c(z)\big(x(qz) - x(z)\big)\left[x^2(z) + x^2(qz) + (1+u)x(z)x(qz) + v\right]f(qz) \\
&\qquad + d(z)\big(x(q^{-1}z) - x(z)\big)\left[x^2(z) + x^2(q^{-1}z) + (1+u)x(z)x(q^{-1}z) + v\right]f(q^{-1}z).
\end{aligned}$$

One checks that the choice of $x(z)$ in (2.1) cancels the two terms in the square brackets. As z varies, the $f(z)$ are linearly independent vectors, so that the *finite band* operator $(L^3\Theta - \Theta L^3) + u(L^2\Theta L - L\Theta L^2) + v(L\Theta - \Theta L)$ has an infinite dimensional kernel; hence it must vanish identically, proving the identity. □

We use the following notation for the "q-integers":

$$[n]_q = \frac{q^n - 1}{q - 1}.$$

PROPOSITION 2.1 ([**11, 12**]). *The most general solution of* (2.3) *with all* $a(n) \neq 0$ *and all* $\theta(n)$ *distinct, depends on seven arbitrary parameters* λ, μ, ν, ρ, σ, $a(0)$, $b(0)$ *and is explicitly given by the following formulas*:

$$\theta(n) = q^{1-n}\left(\frac{[n]_q[n+1]_q}{q+1}\nu + [n]_{q^2}\rho\right) + q^n\sigma, \tag{2.5}$$

$$\begin{aligned}
&\left[\theta(n) - \theta(n+1)\right]\left[\theta(n-1) - \theta(n)\right]b(n) \\
&\qquad = \left[\theta(0) - \theta(1)\right]\left[\theta(-1) - \theta(0)\right]b(0) + \left[\theta(0) - \theta(n)\right]\mu,
\end{aligned} \tag{2.6}$$

$$\begin{aligned}
&\left[\theta(n+1) - \theta(n-1)\right]\left[\theta(n) - \theta(n-2)\right]a(n) \\
&\quad = \left[\theta(1) - \theta(-1)\right]\left[\theta(0) - \theta(-2)\right]a(0) + \left[\theta(-1) + \theta(0) - \theta(n-1) - \theta(n)\right]\lambda \\
&\qquad + \left[\theta(n-1)\theta(n) - \theta(-1)\theta(0)\right]v + \left[\theta(n) - \theta(n+1)\right]^2 b^2(n) \\
&\qquad - \left[\theta(0) - \theta(1)\right]^2 b^2(0) + \left\{\left[\theta(0) - \theta(1)\right]b(0) - \left[\theta(n) - \theta(n+1)\right]b(n)\right\}\mu.
\end{aligned} \tag{2.7}$$

PROOF. The best way to solve equation (2.3) is to notice that this equation can be "integrated", i.e.

$$(2.3) \Leftrightarrow \left[L, [L, \Theta]_{q^{-1}}\right]_q + v\Theta = \lambda I + \mu L + \nu L^2, \tag{2.8}$$

with v as in (2.4), I the identity operator and λ, μ, ν three arbitrary constants. Equating the diagonals of (2.3) to zero, we obtain at the $(n-1, n+2)$-th entry

$$\theta(n+2) - \frac{[3]_q}{q}\theta(n+1) + \frac{[3]_q}{q}\theta(n) - \theta(n-1) = 0, \tag{2.9}$$

which can be written as

$$\left\{\theta(n+2) - \frac{1+q^2}{q}\theta(n+1) + \theta(n)\right\} - \left\{\theta(n+1) - \frac{1+q^2}{q}\theta(n) + \theta(n-1)\right\} = 0,$$

$$\text{i.e. } \theta(n+1) - \frac{1+q^2}{q}\theta(n) + \theta(n-1) = \nu, \tag{2.10}$$

with ν some constant independent of n. This last equation can be further integrated twice, giving (2.5).

The $(n-1, n+1)$-th entry of (2.3) gives

$$\left(-\frac{q^2+1}{q}\theta(n+1) + \frac{[3]_q}{q}\theta(n) - \theta(n-1)\right)b(n+1) + \big(\theta(n+1) - \theta(n-1)\big)b(n)$$
$$+ \left(\theta(n+1) - \frac{[3]_q}{q}\theta(n) + \frac{q^2+1}{q}\theta(n-1)\right)b(n-1) = 0,$$

which, upon using (2.9), can be written as

$$\big\{\big[\theta(n+1) - \theta(n+2)\big]b(n+1) + \big[\theta(n) - \theta(n-1)\big]b(n)\big\} \tag{2.11}$$
$$- \big\{\big[\theta(n) - \theta(n+1)\big]b(n) + \big[\theta(n-1) - \theta(n-2)\big]b(n-1)\big\} = 0,$$

$$\text{i.e. } \big[\theta(n) - \theta(n+1)\big]b(n) + \big[\theta(n-1) - \theta(n-2)\big]b(n-1) = \mu, \tag{2.12}$$

with μ some constant independent on n. Multiplying this last equation by $\theta(n-1) - \theta(n)$ and summing from $n = 1$ till n, we get a telescoping sum leading to

$$\big[\theta(n-1) - \theta(n)\big]\big[\theta(n) - \theta(n+1)\big]b(n)$$
$$+ \big[\theta(0) - \theta(1)\big]\big[\theta(0) - \theta(-1)\big]b(0) = \big[\theta(0) - \theta(n)\big]\mu,$$

which is precisely equation (2.6).

Finally, upon using (2.5), the $(n-1, n)$-th entry of (2.3) gives

$$\big(\theta(n) - \theta(n+2)\big)a(n+1) + \big(\theta(n+1) + \theta(n) - \theta(n-1) - \theta(n-2)\big)a(n)$$
$$+ \big(\theta(n-3) - \theta(n-1)\big)a(n-1) + v\big(\theta(n) - \theta(n-1)\big)$$
$$- \big(\theta(n) - \theta(n-1)\big)\frac{(qb(n-1) - b(n))(qb(n) - b(n-1))}{q} = 0,$$

with v as in (2.4). Again, using now (2.12) and (2.9) (with n replaced by $n-1$), the latter equation can be integrated once leading to

$$\big[\theta(n-1) - \theta(n+1)\big]a(n) + \big[\theta(n-1) - \theta(n-3)\big]a(n-1) \tag{2.13}$$
$$- \big[\theta(n) - 2\theta(n-1) + \theta(n-2)\big]b^2(n-1) - \mu b(n-1) + v\theta(n-1) = \lambda,$$

with λ some other constant independent of n. Multiplying the equation above by $\theta(n-2)-\theta(n)$ and summing from $n=1$ till n we obtain

$$\begin{aligned}(2.14)\quad &\big[\theta(n-2)-\theta(n)\big]\big[\theta(n-1)-\theta(n+1)\big]a(n)\\ &\qquad\qquad + \big[\theta(-1)-\theta(1)\big]\big[\theta(0)-\theta(-2)\big]a(0)\\ &= \big[\theta(-1)+\theta(0)-\theta(n-1)-\theta(n)\big]\lambda + \big[\theta(n-1)\theta(n)-\theta(-1)\theta(0)\big]v\\ &\quad + \sum_{i=0}^{n-1}\big[\theta(i-1)-\theta(i+1)\big]\big\{\mu+\big[\theta(i-1)-2\theta(i)+\theta(i+1)\big]b(i)\big\}b(i).\end{aligned}$$

In oder to show that this formula precisely agrees with the one announced in (2.7), it remains to prove that the sum which appears in the right-hand side of (2.14) can be evaluated in closed form:

$$\begin{aligned}(2.15)\quad &\sum_{i=0}^{n-1}\big[\theta(i-1)-\theta(i+1)\big]\big\{\mu+\big[\theta(i-1)-2\theta(i)+\theta(i+1)\big]b(i)\big\}b(i)\\ &\qquad = \big[\theta(n)-\theta(n+1)\big]^2 b^2(n) - \big[\theta(0)-\theta(1)\big]^2 b^2(0)\\ &\qquad\quad + \mu\big\{\big[\theta(0)-\theta(1)\big]b(0) - \big[\theta(n)-\theta(n+1)\big]b(n)\big\}.\end{aligned}$$

We establish (2.15) by induction on n. It is obviously true for $n=0$. Assume that it is true for n, we establish it for $n+1$. We have to show that

$$\begin{aligned}(2.16)\quad &\big[\theta(n)-\theta(n+1)\big]^2 b^2(n) - \mu\big[\theta(n)-\theta(n+1)\big]b(n)\\ &\quad + \big[\theta(n-1)-\theta(n+1)\big]\big\{\mu+\big[\theta(n-1)-2\theta(n)+\theta(n+1)\big]b(n)\big\}b(n)\\ &\quad = \big[\theta(n+1)-\theta(n+2)\big]^2 b^2(n+1) - \mu\big[\theta(n+1)-\theta(n+2)\big]b(n+1).\end{aligned}$$

Summing equations (2.12) for n and $n+1$, we get that

$$\begin{aligned}&\big[\theta(n-1)-2\theta(n)+\theta(n+1)\big]b(n)\\ &\qquad = \big[\theta(n-1)-\theta(n-2)\big]b(n-1) + \big[\theta(n+1)-\theta(n+2)\big]b(n+1) - 2\mu,\end{aligned}$$

which, using (2.11), gives that

$$\begin{aligned}&\big[\theta(n-1)-\theta(n+1)\big]\big\{\mu+\big[\theta(n-1)-2\theta(n)+\theta(n+1)\big]b(n)\big\}b(n)\\ &\quad = \big\{\big[\theta(n-1)-\theta(n-2)\big]b(n-1) + \big[\theta(n+1)-\theta(n+2)\big]b(n+1) - \mu\big\}\\ &\qquad \times \big\{\big[\theta(n-2)-\theta(n-1)\big]b(n-1) + \big[\theta(n+1)-\theta(n+2)\big]b(n+1)\big\}\\ &\quad = \big[\theta(n+1)-\theta(n+2)\big]^2 b^2(n+1) - \big[\theta(n-2)-\theta(n-1)\big]^2 b^2(n-1)\\ &\qquad - \mu\big[\theta(n-2)-\theta(n-1)\big]b(n-1) - \mu\big[\theta(n+1)-\theta(n+2)\big]b(n+1).\end{aligned}$$

Using this last equation, establishing (2.16) amounts to show that

$$\begin{aligned}&\big[\theta(n)-\theta(n+1)\big]^2 b^2(n) - \mu\big[\theta(n)-\theta(n+1)\big]b(n)\\ &\qquad - \big[\theta(n-2)-\theta(n-1)\big]^2 b^2(n-1) - \mu\big[\theta(n-2)-\theta(n-1)\big]b(n-1) = 0,\end{aligned}$$

which follows readily from (2.12). One checks easily that (2.10), (2.12) and (2.13) are equivalent to the equation (2.8). Thus this equation can be thought of as an "integrated form" of the operator identity (2.3). This finishes the proof. □

REMARK 2.2. The heart of the matter in the above proof (which is inspired by [**13**] in the case q=1), lies in our ability of evaluating the sum (2.15) in closed form. This seems to indicate that the methods of "A=B" [**24**] could be useful in the study of bispectral problems.

3. The q-Hypergeometric Equation and the Askey–Wilson Functions

In Section 2, we showed that the basic operator identity (2.3) is a necessary condition to be satisfied by all the solutions of (1.5), when B is an operator of order two. We now show that it is sufficient. Lemma 3.1 below will allow us to drastically simplify some of the arguments in [**12**]. For further use, we introduce the operators:

$$D_x^{\pm} f(z) = \frac{f(q^{\pm 1}z) - f(z)}{x(q^{\pm 1}z) - x(z)}, \tag{3.1}$$

with $x(z)$ as in (2.1). We observe that equation (2.8) is invariant under the change $q \curvearrowright q^{-1}$. Thus, if we define

$$M^{\pm} = [L, \Theta]_{q^{\mp 1}} + rI + sL, \tag{3.2}$$

with r, s two extra arbitrary parameters, we immediately deduce from (2.8) that

$$[L, M^{\pm}]_{q^{\pm 1}} + v\Theta = H^{\pm}(L), \tag{3.3}$$

with

$$H^{\pm}(x) = \lambda + \big(\mu + (1 - q^{\pm 1})r\big)x + \big(\nu + (1 - q^{\pm 1})s\big)x^2. \tag{3.4}$$

Let us define

$$K^{\pm}(z) = H^{\pm}(x) + \big(q^{\pm 1}x(z) - x(q^{\pm 1}z)\big)(r + sx). \tag{3.5}$$

With these definitions, we introduce the first-order q-difference operators

$$A^{\pm} f = K^{\pm}(z) D_x^{\pm}. \tag{3.6}$$

LEMMA 3.1. *The pair of equations*

$$Lf = x(z)f, \tag{3.7}$$

$$A^{\pm} f = M^{\pm} f + \big(x(q^{\pm 1}z) - q^{\pm 1}x(z)\big)\Theta f, \tag{3.8}$$

is compatible if and only if equation (3.3) *is identically satisfied.*

PROOF. It is enough to establish the "+ case" of this statement. For convenience, we shall omit the + sign in the notations for the rest of the proof ($A = A^{+}$, etc.). Equations (3.7) and (3.8) will be compatible if and only if the following identity holds:

$$[L, M]_q f = [A, x]_q f - \big(x(qz) - qx(z)\big)[L, \Theta]_q f. \tag{3.9}$$

We show that this identity is identical with (3.3). From the definition of A in (3.6) we compute that

$$[A, x]_q f = H(x)f + \big(x(qz) - qx(z)\big)\big(Af - (r + sx)f\big),$$

and thus, since $H(x)$ is a function of x, using (3.7), (3.9) amounts to

$$[L, M]_q f = H(L)f + \big(x(qz) - qx(z)\big)\big(Af - [L, \Theta]_q f - (rI + sL)f\big).$$

If we now use the definition (3.2) of M, this last equation takes the form

$$\begin{aligned}[L,M]_q f &= H(L)f + \big(x(qz) - qx(z)\big)\big(Af - [L,\Theta]_q f + [L,\Theta]_{q^{-1}} f - Mf\big) \\ &= H(L)f + \big(x(qz) - qx(z)\big)\big(Af - Mf + (q - q^{-1})x(z)\Theta f\big) \text{ (using (3.7))} \\ &= H(L)f + \big(x(qz) - qx(z)\big)\big(x(qz) - q^{-1}x(z)\big)\Theta f \text{ (using (3.8))} \\ &= H(L)f - v\Theta f \text{ (using (2.1) with } v \text{ as in (2.4))},\end{aligned}$$

showing that (3.9) is equivalent to (3.3). This establishes the lemma. □

THEOREM 3.2 ([**12**]). *Let $a(n)$, $b(n)$, $\theta(n)$ be the general solution of (2.3) which is explicitly given by (2.5), (2.6) and (2.7), and depends on the seven free parameters λ, μ, ν, ρ, σ, $a(0)$, $b(0)$. Let $w(z)$ be an ARBITRARY solution of the equation*

$$Bw(z) = \sigma w(z), \tag{3.10}$$

with B the second-order q-difference operator defined by

$$B = \frac{A^+ - A^-}{x(qz) - x(q^{-1}z)}, \tag{3.11}$$

with $A^\pm$ as in (3.6). Let us define $f(n,z)$, $n \in \mathbb{Z}$, to be the unique solution of the doubly infinite three-term recursion relation (1.3), with initial conditions $f(0,z)$ and $f(1,z)$ given by

$$f(0,z) = w(z), \tag{3.12}$$

$$\begin{aligned} f(1,z) = \frac{q}{(q^2-1)\sigma + q(2\rho+\nu)} \times \big\{K^+(z)D_x^+ w(z) + \big[(\rho - s)x(z) - r \\ - \big((1-q^{-1})\sigma + \rho\big)b(0) - \big(x(qz) - qx(z)\big)\sigma\big]w(z)\big\}, \end{aligned} \tag{3.13}$$

with D_x^+ as in (3.1) and $K^+(z)$ as in (3.5). The functions $f(n,z)$ so defined, satisfy

$$Bf(n,z) = \theta(n)f(n,z), \quad \forall n \in \mathbb{Z}. \tag{3.14}$$

PROOF. By Lemma 3.1, we know that (3.7) and (3.8) are compatible. We determine the most general family of functions $f(n,z)$, which satisfy both equations. From (3.7) and (3.8) (with the + sign, for instance), one computes easily that $f(1,z)$ must be defined in term of $f(0,z) \equiv w(z)$ via the formula given in (3.13). Thus it remains to specify $w(z)$, since then all other functions $f(n,z)$ are uniquely determined from the three-term recursion relation.

From the definition (3.2) of $M^\pm$ we have that

$$M^+ - q\Theta L = M^- - q^{-1}\Theta L. \tag{3.15}$$

Using (3.7), we can rewrite (3.8) as

$$A^\pm f = (M^\pm - q^{\pm 1}\Theta L)f + x(zq^{\pm 1})\Theta f.$$

Subtracting the "- equation" from the "+ equation" and using (3.15) we get (3.14), with B as in (3.11). Taking the case $n = 0$ of this equation, we get in particular that the still unknown function $w(z)$ must satisfy (3.10), remembering from (2.5) that $\theta(0) = \sigma$. This completes the proof of the theorem. □

We can always modify trivially the general solution of the operator identity (2.3) by translating or rescaling the $\theta(n)$'s with an arbitrary constant. So in fact we are left with $7 - 2 = 5$ free parameters. Remember also that the parameters r and s in (3.2) can be picked arbitrarily. We now show that, using this freedom, our formulas can be converted into a very nice form. Going back to the definition of $A^{\pm}$ in (3.4), (3.5) and (3.6), a simple computation shows that if we put

$$\lambda = \frac{\gamma^2(q+1)(q-1)^2(s_2-\delta s_4-\delta)}{q^2}, \quad \mu = -\frac{\gamma(q-1)^2(qs_1+s_3)}{q^2},$$
$$\nu = \frac{(q-1)^2(s_4+q)}{q^2}, \quad r = \frac{\gamma(q-1)(s_3-s_1)}{q}, \quad s = \frac{(q-1)(1-s_4)}{q},$$
$$\text{with } s_1 = \delta a+\delta b+c+d, \quad s_2 = \delta^2 ab+\delta ac+\delta ad+\delta bc+\delta bd+cd,$$
$$s_3 = \delta abc+\delta abd+acd+bcd, \quad s_4 = abcd$$

where a, b, c, d denote four arbitrary parameters, the operator B in (3.11) takes the form of the famous second-order q-difference operator of Askey and Wilson [**2**]:

$$Bf(z) = c(z)\big(f(qz)-f(z)\big)+c(\delta z^{-1})\big(f(q^{-1}z)-f(z)\big), \tag{3.16}$$

with

$$c(z) = \frac{(1-az)(1-bz)(\delta-cz)(\delta-dz)}{(\delta-z^2)(\delta-qz^2)}, \; c(\delta z^{-1}) = \frac{(z-\delta a)(z-\delta b)(z-c)(z-d)}{(z^2-\delta)(z^2-\delta q)}.$$

If we define then the four remaining free parameters ρ, σ, $a(0)$ and $b(0)$ in our original construction by the formulas

$$\rho = \frac{(q-1)(q^{\varepsilon+1}+q^{\varepsilon}s_4-q^2-q)}{q^{\varepsilon+2}}, \quad \sigma = \frac{(q^{\varepsilon}-1)(q^{\varepsilon}s_4-q)}{q^{\varepsilon+1}}, \tag{3.17}$$

$$a(0) = \gamma^2(q^{-\varepsilon}-1)(q^{2-\varepsilon}-s_4)$$
$$\times\frac{[(q^{2(1-\varepsilon)}-s_4)^2(\delta(q^{2(1-\varepsilon)}+s_4)-q^{1-\varepsilon}s_2)+q^{2(1-\varepsilon)}(q^{1-\varepsilon}s_1-s_3)(q^{1-\varepsilon}s_3-s_1s_4)]}{(q^{1-2\varepsilon}-s_4)(q^{2(1-\varepsilon)}-s_4)^2(q^{3-2\varepsilon}-s_4)},$$
$$b(0) = \frac{\gamma q^{-\varepsilon}[q^{1-2\varepsilon}(s_3+qs_1)-(q+1)q^{-\varepsilon}(s_1s_4+qs_3)+(s_3+qs_1)s_4]}{(q^{-2\varepsilon}-s_4)(q^{2(1-\varepsilon)}-s_4)},$$

with ε a fifth free parameter besides a, b, c and d introduced above, a straightforward (when performed with some symbolic software!) computation gives that the equations (2.5), (2.6) and (2.7) which determine $\theta(n)$, $a(n)$ and $b(n)$ become

$$\theta(n) = q^{-n-\varepsilon}(1-q^{n+\varepsilon})(1-s_4q^{n+\varepsilon-1}), \tag{3.18}$$

$$a(n) = A(n-1)C(n), \quad b(n) = \gamma(\delta a+a^{-1})-\big(A(n)+C(n)\big), \tag{3.19}$$

with

$$A(n) = \frac{\gamma(q^{-n-\varepsilon}-\delta ab)(q^{-n-\varepsilon}-ac)(q^{-n-\varepsilon}-ad)(q^{1-n-\varepsilon}-abcd)}{a(q^{1-2n-2\varepsilon}-abcd)(q^{-2n-2\varepsilon}-abcd)}, \tag{3.20}$$

$$C(n) = \frac{\gamma a(q^{-n-\varepsilon}-1)(q^{1-n-\varepsilon}-bc)(q^{1-n-\varepsilon}-bd)(\delta q^{1-n-\varepsilon}-cd)}{(q^{2-2n-2\varepsilon}-abcd)(q^{1-2n-2\varepsilon}-abcd)}. \tag{3.21}$$

When $\varepsilon = 0$, equations (3.19), (3.20) and (3.21) reduce to the classical formulas defining the coefficients of the normalized recurrence relations satisfied by the Askey–Wilson polynomials when $\gamma = 1/2$ and $\delta = 1$, or the big q-Jacobi polynomials when $\gamma = 1$ and $\delta = 0$ (see for example [**20**]). Thus, for $\varepsilon \neq 0$, these equations define the *associated Askey–Wilson* or the *associated big q-Jacobi matrices*, as introduced in (1.6). Once we have picked a solution $w(z)$ of (3.10) defining $f(0,z)$ as in (3.12), the family of functions $f(n,z)$ satisfying (3.14), is uniquely determined via the definition of $f(1,z)$ in (3.13), and the recurrence relation defined by (3.19). Thus, it remains to determine the general solution of (3.10), which is a *q-analogue* *of the hypergeometric equation.* This can be done in terms of basic hypergeometric functions using the next theorem, due to N. M. Atakishiyev and S. K. Suslov [**3**]. We need to introduce some notations.

The standard q-shifted factorial is defined by

$$(a;q)_n = 1, \text{ if } n=0 \quad \text{and} \quad (a;q)_n = \prod_{i=0}^{n-1}(1-aq^i), \text{ if } n \geq 1. \tag{3.22}$$

We define $(a;q)_\infty = \prod_{i=0}^{\infty}(1-aq^i)$, for $|q|<1$. When products of q-shifted factorials occur, we use the more compact notations $(a_1,a_2,\ldots,a_m;q)_n = (a_1;q)_n (a_2;q)_n \ldots (a_m;q)_n$ and $(a_1,a_2,\ldots,a_m;q)_\infty = (a_1;q)_\infty(a_2;q)_\infty,\ldots,(a_m;q)_\infty$. The series expansion

$${}_r\phi_s\left[\begin{matrix} a_1,a_2,\ldots,a_r \\ b_1,b_2,\ldots,b_s\end{matrix};q,z\right] = \sum_{k=0}^{\infty}\frac{(a_1,a_2,\ldots,a_r;q)_k}{(q,b_1,\ldots,b_s;q)_k}\left[(-1)^k q^{k(k-1)/2}\right]^{1+s-r} z^k,$$

defines the ${}_r\phi_s$ basic hypergeometric series.

THEOREM 3.3 ([**3,12**]). *Let B as in (3.16) and let $y=1$, $q/\delta ab$, q/ac or q/ad. Then the function*

$$w(z) = \frac{(az,\delta a z^{-1};q)_\infty}{(yaz,\delta yaz^{-1};q)_\infty}\,{}_4\phi_3\left[\begin{matrix} yq^{-\varepsilon}, \frac{\tilde a\tilde b\tilde c\tilde d}{yq^{1-\varepsilon}}, \tilde a z, \frac{\delta\tilde a}{z} \\ \delta\tilde a\tilde b, \tilde a\tilde c, \tilde a\tilde d\end{matrix};q,q\right] \tag{3.23}$$

solves the inhomogeneous equation

$$Bw(z) - \sigma w(z) = \frac{(yq^{-\varepsilon}, yq^{\varepsilon-1}abcd, az, \delta a z^{-1};q)_\infty}{y(\delta yab, yac, yad, yq;q)_\infty}, \quad \sigma \text{ as in } (3.17) \tag{3.24}$$

where $\tilde a = ya$ and, depending on the choice of y above, one must pick

$$\tilde b = b,\ \tilde c = c,\ \tilde d = d, \quad \text{or} \quad \tilde b = \frac{q}{\delta a},\ \tilde c = c,\ \tilde d = d, \quad \text{or}$$
$$\tilde b = b,\ \tilde c = \frac{q}{a},\ \tilde d = d, \quad \text{or} \quad \tilde b = b,\ \tilde c = c,\ \tilde d = \frac{q}{a}.$$

By taking appropriate combinations of the functions in (3.23), satisfying the inhomogeneous equations (3.24), we obtain solutions of the homogeneous equation (3.10) defining $f(0,z)$ in (3.12), namely

$$f(0,z) = {}_4\phi_3\left[\begin{matrix} q^{-\varepsilon}, \frac{abcd}{q^{1-\varepsilon}}, az, \frac{\delta a}{z} \\ \delta ab, ac, ad\end{matrix};q,q\right] \tag{3.25}$$
$$- y\frac{(q^{-\varepsilon}, abcd/q^{1-\varepsilon}, \delta yab, yac, yad, yq, az, \delta a/z;q)_\infty}{(yq^{-\varepsilon}, yabcd/q^{1-\varepsilon}, \delta ab, ac, ad, q, yaz, \delta ya/z;q)_\infty}\,{}_4\phi_3\left[\begin{matrix} yq^{-\varepsilon}, \frac{\tilde a\tilde b\tilde c\tilde d}{yq^{1-\varepsilon}}, \tilde a z, \frac{\delta\tilde a}{z} \\ \delta\tilde a\tilde b, \tilde a\tilde c, \tilde a\tilde d\end{matrix};q,q\right].$$

The functions $f(n,z)$ satisfying the doubly infinite three-term recurrence relation defined by (3.19), with initial conditions $f(0,z)$ and $f(1,z)$ given by (3.25) and (3.13), are called the *Askey–Wilson functions.* Explicit formulas for these functions in terms of very-well-poised ${}_8\phi_7$ series can in fact be extracted from [**19**]. Thus, our result characterizes these functions by their bispectral property. When $\varepsilon = 0$, $f(0,z) = 1$ and $f(1,z) = x(z) - b(0)$, as it is easily seen from (3.25) and (3.13). Then, the functions $f(n,z), n \geq 0$, reduce to the Askey–Wilson polynomials, when $\gamma = 1/2$, $\delta = 1$, or to the big q-Jacobi polynomials when $\gamma = 1$, $\delta = 0$, see [**12**].

4. Rank 1 Commutative Rings of Difference Operators

By definition, the *rank* of a commutative ring of difference operators is the greatest common divisor of the orders of all the operators in the ring. In this section, I shall only consider nonsingular curves. The necessary modifications to deal with the case of singular rational curves will be explained in the next section.

Let C be a nonsingular Riemann surface of genus g, with *two* distinguished points P_∞ and Q_∞ on it. Let $D = \sum_{i=1}^{g} P_i$ be a nonspecial divisor of degree g on C and let z^{-1} denote a local coordinate around the point P_∞. Associated with these data, there is a vector of Baker–Akhiezer functions $\psi(n,t,P)$, $n \in \mathbb{Z}$, depending on parameters $t = (t_1, t_2, t_3, \dots)$, which as functions of P are meromorphic functions on $C \setminus \{P_\infty\}$, and are uniquely characterized by the following two properties:

$$\begin{gathered}\big(\psi(n,t,P)\big) + D - nQ_\infty \geq 0 \text{ on } C \setminus \{P_\infty\},\\ \psi(n,t,P) = z^n \exp(t,z)\big(1 + O(1/z)\big) \text{ near } P_\infty.\end{gathered}$$

In the last equation $\big(\psi(n,t,P)\big)$ denotes the divisor of the function $\psi(n,t,P)$ (on $C \setminus \{P_\infty\}$) and $\exp(t,z)$ is a shorthand notation for

$$\exp(t,z) = \exp(t_1 z + t_2 z^2 + t_3 z^3 + \cdots). \tag{4.1}$$

The vector of Baker functions is the common eigenvector of a rank 1 commutative ring of difference operators. More precisely, for any meromorphic function f on C, which is holomorphic on $C \setminus \{P_\infty, Q_\infty\}$, there is a finite band operator L_f such that $L_f\psi(n,t,P) = f(P)\psi(n,t,P)$. The size of the operator L_f is determined by the order of the poles of f at P_∞ and Q_∞. Namely, if f has a pole of order i at P_∞ and a pole of order j at Q_∞, then the operator L_f, thought of as a finite band matrix, has i diagonals above the main diagonal and j diagonals below it. Conversely, any rank 1 commutative ring of difference operator can be obtained by the above construction, allowing eventually the curve C to become singular, with P_∞ and Q_∞ two nonsingular points, see [**22**] for a full discussion.

The Baker functions $\psi(n,t,P)$ as well as the entries of the matrices L_f can be expressed in terms of a family of tau functions $\tau(n,t)$, $n \in \mathbb{Z}$, depending on the infinitely many time variables $t_1, t_2, t_3, \dots$. For an accessible account of the theory of solitons as developed by the Kyoto school, I refer the reader to [**8**, Chapter 7]. The link with this theory can be done via the introduction of the so-called adjoint Baker functions, first introduced by Cherednik [**7**]. Since D is nonspecial, there exists a unique meromorphic differential form δ on C such that $(\delta) - D + 2P_\infty \geq 0$, normalized so that $\delta = \mathrm{d}z\big(1 + O(1/z)\big)$, near P_∞. Since the degree of the divisor of any differential is $2g-2$, δ determines a divisor $D^* = \sum_{i=1}^{g} P_i^*$ on C such that $(\delta) = -2P_\infty + D + D^*$. The adjoint Baker functions $\psi^*(n,t,P)$ are defined as the unique meromorphic functions on $C \setminus \{P_\infty\}$, satisfying

$$\begin{gathered}\big(\psi^*(n,t,P)\big) + D^* + nQ_\infty \geq 0 \text{ on } C \setminus \{P_\infty\},\\ \psi^*(n,t,P) = z^{-n} \exp^{-1}(t,z)\big(1 + O(1/z)\big) \text{ near } P_\infty.\end{gathered}$$

As long as $n \geq m$, the differentials $\psi(n,t,P)\ \psi^*(m,t',P)\delta$ have no singularities on $C \setminus \{P_\infty\}$, and thus, by the residue theorem, one deduces that

$$\operatorname{res}_{P_\infty} \psi(n,t,P)\psi^*(m,t',P)\delta = 0, \quad \forall n \geq m \quad \text{and} \quad \forall t,t'. \tag{4.2}$$

THEOREM 4.1. *The Baker and the adjoint Baker functions can be expanded as follows in the local coordinate* z^{-1} *around* P_∞

$$\psi(n,t,z) = w(n,t,z) \times (\textit{a factor independent of } n \textit{ and } t),$$
$$\psi^*(n,t,z) = w^*(n,t,z) \times (\textit{a factor independent of } n \textit{ and } t).$$

The functions $w(n,t,z)$ *and* $w^*(n,t,z)$ *are called respectively the wave function and the adjoint wave function. They satisfy the so-called BILINEAR IDENTITIES*

$$\operatorname{res}_{z=\infty} w(n,t,z)w^*(m,t',z)\,\mathrm{d}z = 0, \quad \forall n \geq m \quad \textit{and} \quad \forall t,t'. \tag{4.3}$$

PROOF. I shall only give a sketch of the proof. First, one writes down an explicit formula for the Baker function $\psi(n,t,P)$, in terms of Riemann's theta function $\theta(u_1,\ldots,u_g)$ of the curve C and its prime form $E(Q,P)$:

$$\psi(n,t,P) = \exp\Bigl(\sum_{i=1}^{\infty} t_i \int^P \eta_i\Bigr)\left[\frac{E(Q_\infty,P)}{E(P_\infty,P)}\right]^n \frac{1}{E(Q_\infty,P_\infty)^n} \times \frac{\theta(tA+\xi-\int_{P_\infty}^P\omega - n\int_{P_\infty}^{Q_\infty}\omega)\theta(\xi)}{\theta(tA+\xi-n\int_{P_\infty}^{Q_\infty}\omega)\theta(\xi-\int_{P_\infty}^P\omega)}. \tag{4.4}$$

I refer the reader to Section 3 of [**15**] for a detailed explanation of all the ingredients involved in this formula. The formula for the adjoint Baker function is obtained by replacing t by $-t$, n by $-n$ and ξ by ξ^* in (4.4). When performing the product $\psi(n,t,z)\psi^*(m,t',z)$, the two factors which are independent respectively of n, t and m, t' combine to produce precisely the expansion of the function $\mathrm{d}z/\delta$, that is

$$\frac{\theta(\xi)\theta(\xi^*)}{\theta(\xi-\int_{P_\infty}^P\omega)\theta(\xi^*-\int_{P_\infty}^P\omega)} = \frac{\mathrm{d}z}{\delta}, \quad \text{for } P \text{ near } P_\infty,$$

when both sides are expressed in the local coordinate z^{-1}. Thus (4.2) is equivalent to (4.3). This proves the theorem. □

If we pick $n = m$ in (4.3), we obtain that the functions $z^{-n}w(n,t,z)$ and $z^n w^*(n,t,z)$ satisfy the so-called bilinear identities of the standard KP (Kadomtsev–Petviashvili) hierarchy. Hence, by a fundamental theorem (see for instance [**8**, p. 95]), there exists a tau function $\tau(n,t)$ such that

$$w(n,t,z) = z^n \exp(t,z)\frac{\tau(n,t-[z^{-1}])}{\tau(n,t)}, \tag{4.5}$$

$$w^*(n,t,z) = z^{-n} \exp^{-1}(t,z)\frac{\tau(n,t+[z^{-1}])}{\tau(n,t)}, \tag{4.6}$$

with $\exp(t,z)$ defined as in (4.1) and $[z] = (z, z^2/2, z^3/3, \ldots)$.

In [**1**] (see also [**15**]), we gave a definition of the tau functions above "à la Segal–Wilson", by considering the flag of subspaces

$$\mathcal{V}\colon \quad \cdots \subset V_{n+1} \subset V_n \subset V_{n-1} \subset \cdots, \quad \dim(V_n/V_{n+1}) = 1, \text{ with} \tag{4.7}$$
$$V_n = \bigl\{\text{meromorphic functions } f \text{ on } C\colon (f) + D - nQ_\infty \geq 0 \text{ on } C \setminus \{P_\infty\}\bigr\}.$$

The tau function $\tau(n,t)$ coincides precisely with the tau function $\tau_{V_n}(t)$ associated with the plane V_n, as defined by Segal and Wilson in [**25**]. The dual flag of subspaces

$$(4.8)\quad \mathcal{V}^*:\quad \cdots \supset V^*_{n+1} \supset V^*_n \supset V^*_{n-1} \supset \dots,\quad \dim(V^*_n/V^*_{n-1}) = 1,\ \text{with}$$
$$V^*_n = \big\{\text{meromorphic functions } f \text{ on } C\colon (f) + D^* + nQ_\infty \geq 0 \text{ on } C \setminus \{P_\infty\}\big\},$$

defines the same tau functions, up to a change of sign, $\tau_{V^*_n}(t) = \tau_{V_n}(-t)$.

5. The Case of Rational Curves

In this section, I discuss the necessary modifications needed to deal with the situation of rank 1 commutative rings of difference operators, with spectrum (in the sense of algebraic geometry) an irreducible rational singular curve. We denote by $\mathcal{P} = \mathbb{C}[z]$ the space of polynomials in the variable z, and we denote by $\mathcal{R} = \mathbb{C}(z)$ the space of rational functions of z. A rational function f which admits an expansion around $z = \infty$ of the form

$$f = c_n z^n + c_{n-1}z^{n-1} + \cdots,\quad c_n \neq 0,$$

will be called *an element of order n*. It will be useful to consider the nondegenerate bilinear form on $\mathcal{R}$

$$(5.1)\qquad B(f,g) = \operatorname{res}_{z=\infty} f(z)g(z)\,\mathrm{d}z,\quad f,g \in \mathcal{R}.$$

The next definition introduces the analogue in the context of difference operators, of the so-called rational Grassmannian [**25**] (relevant to the study of rank 1 commutative rings of differential operators).

DEFINITION 5.1. A flag $\mathcal{V}$: $\cdots \subset V_{n+1} \subset V_n \subset V_{n-1} \subset \cdots$ of subspaces $V_n \subset \mathcal{R}$ is called a rational flag if and only if there are polynomials $p(z)$ and $r(z)$ (independent of n) such that

$$z^n p(z)\mathcal{P} \subset V_n \subset z^n r^{-1}(z)\mathcal{P},$$

and the codimension of V_n in $z^n r^{-1}(z)\mathcal{P}$ is equal to the degree of $r(z)$.

There is a one-to-one correspondence between rational flags and affine irreducible rational curves which complete by adding *two* nonsingular points at infinity. The curve is recovered from the flag as the spectrum of the ring

$$(5.2)\quad A_{\mathcal{V}} = \big\{f(z) \in \mathcal{R} \text{ with poles only at } z=0 \text{ and } z=\infty,$$
$$\text{such that } \exists k \in \mathbb{Z},\ \text{for which } f(z)V_n \subset V_{n+k},\ \forall n\big\}.$$

In the following, we shall always make a slight abuse of notation by using $A_{\mathcal{V}}$ to denote the isomorphic ring of difference operators.

The proof of the statements above is beyond the scope of this review article. It can be done as in [**25**, (Section 7)], which deals with the case of affine irreducible rational curves which complete by adding *one* nonsingular point a infinity. Instead, we shall content ourselves by illustrating the notions on the example of the simplest rational singular curves, obtained by identifying $2g$ distinct points of $\mathbb{P}^1(\mathbb{C})$ in pairs $\{\lambda_i,\mu_i\}, i = 1,\dots,g$. We assume that all these points are distinct from $Q_\infty \equiv 0$ and $P_\infty \equiv \infty$. In this case, the role of the nonspecial divisor D of degree g in (4.7) is played by a collection of g nonzero complex numbers $\delta_i \in \mathbb{C}^*$, and the definition (4.7) is adapted as follows

$$(5.3)\quad V_n = \frac{1}{\prod_{i=1}^{g}(z-\lambda_i)}\{\text{meromorphic functions } f \text{ on } \mathbb{P}^1(\mathbb{C}) \text{ such that } (f) - nQ_\infty \geq 0 \text{ on } \mathbb{P}^1(\mathbb{C}) \setminus \{\infty\} \text{ and } f(\lambda_i) = \delta_i f(\mu_i), i = 1, \dots, g\}.$$

The next lemma shows that this definition fits within Definition 5.1 above.

LEMMA 5.2. *The plane V_n defined in* (5.3) *satisfies*

$$z^n \prod_{i=1}^{g}(z-\mu_i)\mathcal{P} \subset V_n \subset z^n \prod_{i=1}^{g}(z-\lambda_i)^{-1}\mathcal{P},$$

and the codimension of V_n in $z^n \prod_{i=1}^{g}(z-\lambda_i)^{-1}\mathcal{P}$ is equal to g.

PROOF. The second inclusion as well as the assertion about the codimension both follow immediately from the definition of V_n in (5.3). In order to establish the first inclusion we introduce the dual flag of subspaces

$$(5.4)\quad V_n^* = \frac{1}{\prod_{i=1}^{g}(z-\mu_i)}\{\text{meromorphic functions } f \text{ on } \mathbb{P}^1(\mathbb{C}) \text{ such that } (f) + nQ_\infty \geq 0 \text{ on } \mathbb{P}^1(\mathbb{C}) \setminus \{\infty\} \text{ and } f(\mu_i) = \delta_i^* f(\lambda_i), i = 1, \dots, g\},$$

with

$$(5.5)\qquad \delta_i^* = -\delta_i \prod_{j\neq i, j=1}^{g} \frac{\mu_i - \mu_j}{\lambda_i - \lambda_j} \prod_{j=1}^{g} \frac{\mu_i - \lambda_j}{\lambda_i - \mu_j},$$

playing now the role of the divisor D^* in (4.8). It follows immediately from (5.3), (5.4) and (5.5) that

$$\{\mathrm{res}_{z=\lambda_i} + \mathrm{res}_{z=\mu_i}\} w(z)w^*(z)\,\mathrm{d}z = 0, \quad \forall w(z) \in V_n \quad \text{and} \quad \forall w^*(z) \in V_m^*.$$

For $w(z) \in V_n$ and $w^*(z) \in V_m^*$ with $n \geq m$, $w(z)w^*(z)$ has no pole at $z = 0$. Thus, by the residue theorem, we deduce that

$$\mathrm{res}_{z=\infty}\, w(z)w^*(z)\,\mathrm{d}z = -\sum_{i=1}^{g}\{\mathrm{res}_{z=\lambda_i} + \mathrm{res}_{z=\mu_i}\} w(z)w^*(z)\,\mathrm{d}z = 0,$$
$$\forall w(z) \in V_n \quad \text{and} \quad \forall w^*(z) \in V_m^* \text{ with } n \geq m,$$

showing that $V_n^* \subseteq V_n^\perp$, with $V_n^\perp$ the orthogonal of V_n with respect to the bilinear form B defined in (5.1). Since V_n has a basis of elements of orders $n, n+1, n+2, \dots$, and V_n^* has a basis of elements of orders $-n, -n+1, -n+2, \dots$, we deduce that $V_n^* = V_n^\perp$. By the definition (5.4) of V_n^*, it implies that

$$V_n^\perp \subset z^{-n}\prod_{i=1}^{g}(z-\mu_i)^{-1}\mathcal{P} \Leftrightarrow V_n \supset z^n \prod_{i=1}^{g}(z-\mu_i)\mathcal{P},$$

as desired. This finishes the proof of the lemma. □

PROPOSITION 5.3. *The tau function $\tau(n,t)$ associated with the rational flag $\mathcal{V}$ defined in* (5.3), *is given by*

$$\tau(n,t) = \prod_{j=1}^{g} \mu_j^{-n} \exp\Big(-\sum_{i=1}^{\infty} t_i \sum_{j=1}^{g} \lambda_j^i\Big) \times \det\big(\lambda_j^{n+i-1}\exp(t,\lambda_j) - \delta_j \mu_j^{n+i-1}\exp(t,\mu_j)\big)_{1\le i,j\le g}. \tag{5.6}$$

PROOF. From the definition (5.3), it follows that the wave function associated with the plane V_n is given by

$$w(n,t,z) = z^n \exp(t,z)\Big\{1 - \sum_{i=1}^{g} \frac{w_i(n,t)}{z-\lambda_i}\Big\}, \tag{5.7}$$

subject to the g conditions

$$\operatorname{res}_{z=\lambda_i} w(n,t,z) + \varepsilon_i w(n,t,\mu_i) = 0, \text{ with } \varepsilon_i = -\delta_i \frac{\prod_{j=1}^{g}(\mu_i - \lambda_j)}{\prod_{j\neq i, j=1}^{g}(\lambda_i - \lambda_j)}. \tag{5.8}$$

It is easy to check that the solution is given by

$$w(n,t,z) = \frac{\exp(t,z)\ \det(\lambda_j^{n+i-1}\exp(t,\lambda_j) - \delta_j \mu_j^{n+i-1}\exp(t,\mu_j); z^{n+i-1})}{\prod_{j=1}^{g}(z-\lambda_j)\det\big(\lambda_j^{n+i-1}\exp(t,\lambda_j) - \delta_j\mu_j^{n+i-1}\exp(t,\mu_j)\big)_{1\le i,j\le g}}. \tag{5.9}$$

The notation in the numerator means the determinant of the $(g+1)\times(g+1)$ matrix with (i,j) entries as indicated there for $1\le i\le g+1, 1\le j\le g$, and entries z^{n+i-1}, $1\le i\le g+1$, in the last column. Elementary row manipulations with this determinant lead to

$$w(n,t,z) = \frac{\exp(t,z)}{\prod_{j=1}^{g}(z-\lambda_j)} z^{n+g} \frac{\prod_{j=1}^{g}(z-\lambda_j)\tau(n,t-[z^{-1}])}{z^g \tau(n,t)},$$

in agreement with (4.5), which defines the tau function. This finishes the proof. □

REMARK 5.4. The motivation for the (irrelevant) factor $\prod_{j=1}^{g}\mu_j^{-n}$ in the definition (5.6) of $\tau(n,t)$ will appear in the next section, see (6.20).

From (5.4) and (5.5), the dual wave function is given by

$$w^*(n,t,z) = z^{-n}\exp^{-1}(t,z)\Big\{1 + \sum_{i=1}^{g}\frac{w_i^*(n,t)}{z-\mu_i}\Big\}, \tag{5.10}$$

with the coefficients $w_i^*(n,t)$ determined by the g conditions

$$\operatorname{res}_{z=\mu_i} w^*(n,t,z) - \varepsilon_i w^*(n,t,\lambda_i) = 0, \quad \varepsilon_i \text{ as in (5.8)}.$$

We now look for rational curves with double points, for which the associated rank 1 commutative rings of difference operators contain a tridiagonal operator.

THEOREM 5.5 ([**17**]). *Let us assume that $\mu_i = \lambda_i^{-1}$, and that the $2g$ points λ_i, μ_i, $1\le i\le g$, are still distinct of each other, i.e. $\lambda_i \neq \pm 1$ and $\lambda_i \neq \lambda_j$, $\lambda_i \neq \lambda_j^{-1}$, $\forall i \neq j$. Then, the rank 1 commutative ring of difference operators $A_{\mathcal{V}}$ contains a tridiagonal operator with one diagonal above and one diagonal below the main diagonal and the spectral curve of the ring has for equation*

$$\operatorname{Spec}(A_{\mathcal{V}})\colon v^2 = (u^2-4)\prod_{i=1}^{g}\big(u - (\lambda_i + \lambda_i^{-1})\big)^2. \tag{5.11}$$

PROOF. Since $\mu_i = \lambda_i^{-1}$, it is clear that the function

$$u = z + z^{-1} \tag{5.12}$$

belongs to the ring $A_{\mathcal{V}}$ as defined in (5.2), with $uV_n \subset V_{n-1}$. This function has a simple pole both at $z = 0$ and $z = \infty$, leading thus to a tridiagonal operator L_u with one diagonal above and one diagonal below the main diagonal, satisfying $L_u w(n,t,z) = uw(n,t,z)$.

Since u has a simple pole at $z = 0$, all other generators of $A_{\mathcal{V}}$ can be taken to be polynomials. The functions

$$f_k = z^{-k}\prod_{i=1}^{g}(z-\lambda_i)(z-\lambda_i^{-1}), \quad k \in \mathbb{Z}, \tag{5.13}$$

which vanish at λ_i and $\mu_i = \lambda_i^{-1}$, obviously belong to $A_{\mathcal{V}}$. For $1 \leq k \leq g-1$, by subtracting an appropriate polynomial of degree k in u, one obtains polynomials (in z) $q_{2g-1}(z), q_{2g-2}(z), \ldots, q_{g+1}(z)$, of degrees $2g-1, 2g-2, \ldots, g+1$, that belong to $A_{\mathcal{V}}$. Combining them with the functions f_k, $k \leq 0$, defined in (5.13), we conclude that $A_{\mathcal{V}}$ contains polynomials $q_k(z)$ of degree k, $\forall k \geq g+1$. On the other hand, any polynomial $q(z) \in A_{\mathcal{V}}$ of degree $k \leq g$, must be identically constant. Indeed, such a polynomial must satisfy $q(\lambda_i) = q(\lambda_i^{-1})$, $\forall 1 \leq i \leq g$, which amounts to a linear homogeneous system $Ac = 0$, for the unknown coefficients $c_1, \ldots, c_g$ of $q(z) = c_0 + c_1 z + \cdots + c_g z^g$. One shows easily that

$$\det A = \prod_{i=1}^{g}\lambda_i^{-g}\prod_{i=1}^{g}(\lambda_i^2-1)\prod_{1\leq i<j\leq g}(\lambda_i-\lambda_j)\prod_{1\leq i<j\leq g}(1-\lambda_i\lambda_j).$$

Since the $2g$ points λ_i, $\mu_i = \lambda_i^{-1}$, $1 \leq i \leq g$, are assumed to be distinct, this determinant is nonzero, implying that $c_1 = c_2 = \cdots = c_g = 0$.

The upshot of the discussion above is that the algebra $A_{\mathcal{V}}$ is generated by u in (5.12) and polynomials of degree k, $k \geq g+1$, or equivalently by u and the functions f_k, $k \in \mathbb{Z}$, introduced in (5.13). We now show that the functions u and

$$v = \frac{z - z^{-1}}{z^g}\prod_{i=1}^{g}(z-\lambda_i)(z-\lambda_i^{-1}),$$

are enough to generate $A_{\mathcal{V}}$, which will establish (5.11). Clearly

$$f_g = \prod_{i=1}^{g}\left(u - (\lambda_i + \lambda_i^{-1})\right) \quad \text{and} \quad v = f_{g-1} - f_{g+1}.$$

Since $uf_k = f_{k-1} + f_{k+1}$, we deduce inductively that $f_{g-1}, f_{g+1}, f_{g-2}, f_{g+2}, \ldots$, belong to the algebra generated by u and v. This finishes the proof. □

6. Askey–Wilson Type Solitons

In this section, I construct some rational curves, with the corresponding wave functions enjoying a bispectral property similar to the one satisfied by the Askey–Wilson functions. It is convenient to introduce the so-called reduced wave and adjoint wave functions, which are obtained by omitting the factors $\exp(t,z)$ and $\exp^{-1}(t,z)$ in (5.7) and (5.10) respectively:

$$\widetilde{w}(n,z) = \exp^{-1}(t,z)w(n,t,z), \quad \widetilde{w}^*(n,z) = \exp(t,z)w^*(n,t,z). \tag{6.1}$$

I also omit to write the explicit dependence on t, which is irrelevant for what follows (we just think of $t_1, t_2, \dots$ as parameters).

THEOREM 6.1 ([**17**]). *Consider a rational curve with double points, obtained by identifying $2g$ distinct points of $\mathbb{P}^1(\mathbb{C})$ in pairs $\{\lambda_i, \mu_i\}$, $i = 1, \dots, g$. Assume that*

$$\lambda_i = q^{k_i}\mu_i, \quad \text{with } k_i \in \mathbb{N} \setminus \{0\}. \tag{6.2}$$

Then the reduced wave function $\widetilde{w}(n, z)$, besides being a common eigenfunction of a rank 1 commutative ring $A_{\mathcal{V}}$ of difference operators (in the variable n)

$$L_f \widetilde{w}(n, z) = f(z)\widetilde{w}(n, z), \quad \forall f \in A_{\mathcal{V}},$$

is also a common eigenfunction of a rank 1 commutative ring of q-difference operators $A_{\mathcal{V}'}$, in the variable z

$$B_\theta \widetilde{w}(n, z) = \theta(n)\widetilde{w}(n, z), \quad \forall \theta \in A_{\mathcal{V}'}. \tag{6.3}$$

We need some preliminaries, before giving the proof of the theorem. First, we introduce the following multiplicative groups of formal operators

$$\mathcal{W} = \left\{1 + \sum_{i,j=1}^{\infty} w_{ij} x^{-j} E^{-i}\right\}, \quad \mathcal{W}^* = \left\{1 + \sum_{i,j=1}^{\infty} w_{ij} x^{-j} E^{i}\right\},$$

with $w_{ij} \in \mathbb{C}$, and E, E^{-1} as in (1.2). We denote by a the adjoint isomorphism

$$a \colon \mathcal{W} \to \mathcal{W}^*, \text{ with } a(W) = (W^{-1})^*, \text{ for } W \in \mathcal{W} \tag{6.4}$$

where * denotes the transposed operator, $(E^i)^* = E^{-i}$.

We define an anti-isomorphisms $b \colon \mathcal{W} \to \mathcal{W}$ by

$$b(x) = E, \; b(E) = x, \text{ i.e. } b\left(1 + \sum_{i,j=1}^{\infty} w_{ij} x^{-j} E^{-i}\right) = 1 + \sum_{i,j=1}^{\infty} w_{ij} x^{-i} E^{-j}, \tag{6.5}$$

as well as an anti-isomorphism $b^* \colon \mathcal{W}^* \to \mathcal{W}^*$ by

$$b^*(x) = E^{-1}, \; b^*(E) = x^{-1}, \quad \text{i.e. } b^*\left(1 + \sum_{i,j=1}^{\infty} w_{ij} x^{-j} E^{i}\right) = 1 + \sum_{i,j=1}^{\infty} w_{ij} x^{-i} E^{j}. \tag{6.6}$$

With these definitions, it is straightforward to check that we have the relation

$$b^* a = ab. \tag{6.7}$$

Assume now that $\lambda_i = q^{k_i}\mu_i$, for some positive integers k_i. Since

$$\begin{aligned} &\lambda_j^{n+i-1} \exp(t, \lambda_j) - \delta_j \mu_j^{n+i-1} \exp(t, \mu_j) \\ &\qquad = \mu_j^n \{x^{k_j} \lambda_j^{i-1} \exp(t, \lambda_j) - \delta_j \mu_j^{i-1} \exp(t, \mu_j)\}, \text{ with } x = q^n, \end{aligned} \tag{6.8}$$

formula (5.9) shows that the (reduced) wave function is a function of $x = q^n$ and z, i.e. $\widetilde{w}(n, z) \equiv \widetilde{w}(x, z)$ with

(6.9)

$$\widetilde{w}(x, z) = \frac{e^{(\log x \log z)/(\log q)}}{\prod_{j=1}^{g}(z - \lambda_j)} \frac{\det(x^{k_j} \lambda_j^{i-1} \exp(t, \lambda_j) - \delta_j \mu_j^{i-1} \exp(t, \mu_j); z^{i-1})}{\det(x^{k_j} \lambda_j^{i-1} \exp(t, \lambda_j) - \delta_j \mu_j^{i-1} \exp(t, \mu_j))_{1 \le i,j \le g}}.$$

Since the k_i are positive integers, it follows immediately from this formula that

$$\lim_{x \to \infty} e^{-(\log x \log z)/(\log q)} \widetilde{w}(x, z) = \frac{\det(\lambda_j^{i-1}; z^{i-1})}{\prod_{j=1}^{g}(z - \lambda_j) \det(\lambda_j^{i-1})_{1 \le i,j \le g}} = 1. \tag{6.10}$$

By definition of the reduced wave function (6.1), we always have that

$$\widetilde{w}(n,z) = z^n\left(1+\frac{w_1(n)}{z}+\frac{w_2(n)}{z^2}+\cdots\right) \quad \text{as } z\to\infty \tag{6.11}$$
$$= \left(1+w_1(n)E^{-1}+w_2(n)E^{-2}+\cdots\right)z^n \equiv W(n)z^n,$$

with

$$W(n) = 1+w_1(n)E^{-1}+w_2(n)E^{-2}+\cdots. \tag{6.12}$$

We shall need the following lemma.

LEMMA 6.2.

$$\widetilde{w}^*(n,z) = (W^{-1})^*(n-1)z^{-n}. \tag{6.13}$$

PROOF. It is enough to check that with the definitions (6.11) and (6.13), the bilinear identities (4.3) are satisfied, see [**15**] for details. □

Equations (6.9) and (6.10) show that, when $\lambda_i = q^{k_i}\mu_i$, $k_i \in \mathbb{N}\setminus\{0\}$, $W(n)$ in (6.12) is a function of $x = q^n$, $W(x) = 1+w_1(x)E^{-1}+w_2(x)E^{-2}+\cdots$, with $\lim_{x\to\infty} w_i(x) = 0$. Thus, it admits the expansion

$$W(x) = 1+\sum_{i,j=1}^{\infty} w_{ij}x^{-j}E^{-i}, \quad \text{as } x\to\infty. \tag{6.14}$$

It follows then from (6.13) that the (reduced) adjoint wave function is also a function of $x = q^n$ and z, with

$$\widetilde{w}^*(x,z) = a(W)(q^{-1}x)z^{-n} \tag{6.15}$$

where $a(W)$ is defined as in (6.4). We are ready for the next crucial lemma.

LEMMA 6.3 ([**17**]). *With the assumptions of Theorem* 6.1, *the new functions* $\widetilde{w}'(x,z)$ *and* $(\widetilde{w}')^*(x,z)$ *defined respectively by*

$$\widetilde{w}'(x,z) = \widetilde{w}(z,x) \quad \textit{and} \quad (\widetilde{w}')^*(x,z) = \widetilde{w}^*(qz,q^{-1}x)q^{-1}xz^{-1}, \tag{6.16}$$

with $\widetilde{w}(x,z)$ *and* $\widetilde{w}^*(x,z)$ *defined as in* (6.9) *and* (6.15), *are the* (*reduced*) *wave and adjoint wave functions of another rank* 1 *commutative ring of difference operators* $A_{\mathcal{V}'}$, *with spectrum a rational curve.*

PROOF. In terms of the anti-isomorphism b introduced in (6.5), $\widetilde{w}'(x,z)$ as defined in (6.16) can be written as

$$\widetilde{w}'(x,z) = b(W)(x)z^n, \tag{6.17}$$

with $W(x)$ as in (6.14). Using (6.15) and the definitions (6.6) and (6.16) of b^* and $(\widetilde{w}')^*(x,z)$ respectively, one checks easily that

$$(\widetilde{w}')^*(x,z) = b^*\big(a(W)\big)(q^{-1}x)z^{-n}.$$

From (6.7), we deduce that

$$(\widetilde{w}')^*(x,z) = a\big(b(W)\big)(q^{-1}x)z^{-n}. \tag{6.18}$$

Let

$$V_n' = \left\{\widetilde{w}'(n,z), \widetilde{w}'(n+1,z), \widetilde{w}'(n+2,z), \dots\right\} \quad \text{and}$$
$$(V_n')^* = \left\{(\widetilde{w}')^*(n,z), (\widetilde{w}')^*(n-1,z), (\widetilde{w}')^*(n-2,z), \dots\right\},$$

with $\widetilde{w}'(n,z)$ and $(\widetilde{w}')^*(n,z)$ the functions obtained when substituting $x = q^n$ in $\widetilde{w}'(x,z)$ and $(\widetilde{w}')^*(x,z)$ respectively. By Lemma 6.2 (with W replaced by $b(W)$), equations (6.17) and (6.18) show that $B\big(\widetilde{w}'(n,z), (\widetilde{w}')^*(m,z)\big) = 0$, $\forall n \geq m$, where B is the residue pairing defined in (5.1). Thus

$$(V_n')^* = (V_n')^{\perp} \tag{6.19}$$

where $(V_n')^{\perp}$ denotes the orthogonal of V_n' with respect to B.

Using (6.8), we can write $\tau(n,t)$ in (5.6) as

$$\tau(x,t) = \exp\Big(-\sum_{i=1}^{\infty} t_i \sum_{j=1}^{g} \lambda_j^i\Big) \times \det\big(x^{k_j}\lambda_j^{i-1}\exp(t,\lambda_j) - \delta_j \mu_j^{i-1}\exp(t,\mu_j)\big)_{1\leq i,j\leq g}, \tag{6.20}$$

with $x = q^n$, and thus $\tau(x,t)$ is a polynomial in x. Equations (4.5) and (4.6) combined with the definition (6.16) of $\widetilde{w}'(x,z)$ and $(\widetilde{w}')^*(x,z)$, show then that

$$V_n' \subset z^n \tau(z,t)^{-1}\mathcal{P} \quad \text{and} \quad (V_n')^* \subset z^{-n}\tau(qz,t)^{-1}\mathcal{P},$$

which, because of (6.19), is equivalent to

$$z^n\tau(qz,t)\mathcal{P} \subset V_n' \subset z^n\tau(z,t)^{-1}\mathcal{P}.$$

Clearly, V_n' contains elements of orders $n, n+1, n+2, \ldots$, and only those, meaning that the codimension of V_n' in $z^n\tau(z,t)^{-1}\mathcal{P}$ is equal to the degree (as a polynomial in z) of $\tau(z,t)$. Thus, the flag $\mathcal{V}'$: $\cdots \subset V_{n+1}' \subset V_n' \subset V_{n-1}' \subset \cdots$, is a rational flag in the sense of Definition 5.1. This is equivalent to the statement that $\widetilde{w}'(n,z)$ is the (reduced) wave function of a rank 1 commutative ring of difference operators, with spectrum a rational curve. The lemma is established. □

The proof of Theorem 6.1 has now been reduced to a tautology.

PROOF OF THEOREM 6.1. From Lemma 6.3, we have that

$$\sum_{\text{finitely many i's } \in \mathbb{Z}} c_i(x)\widetilde{w}'(q^i x, z) = \theta(z)\widetilde{w}'(x,z), \quad \forall\theta(z) \in A_{\mathcal{V}'},$$

which, remembering the definition (6.16) of $\widetilde{w}'(x,z)$, amounts to

$$\sum_{\text{finitely many i's } \in \mathbb{Z}} c_i(z)\widetilde{w}(n, q^i z) = \theta(q^n)\widetilde{w}(n,z),$$

and shows that $\widetilde{w}(n,z)$ is an eigenfunction of a rank 1 commutative ring of q-difference operators, as claimed in (6.3). This establishes the theorem. □

REMARK 6.4. In the special case when $\lambda_i = q\mu_i$, F. W. Nijhoff and O. A. Chalykh [**23**] have obtained the following formula for the reduced wave function

$$\widetilde{w}(x,z) = e^{(\log x \log z)/(\log q)} \det\big\{I - (xI + X)^{-1}[X,Z]_{q^{-1}}(zI+Z)^{-1}\big\},$$

with X and Z some $g \times g$ matrices, so that the q-commutator $[X,Z]_{q^{-1}}$ is a rank 1 matrix. They conjecture that all bispectral rank 1 commutative rings of difference operators (in the sense of Theorem 6.1) can be parametrized in this way. In the limit $q = 1$, corresponding to rank 1 bispectral rings of differential operators, this was proved by G. Wilson [**27**]. P. Iliev [**18**] has obtained similar formulas, in relation with a version of the problem involving only the q-derivative operator.

7. Some Rank 1 and Rank 2 Tridiagonal Solutions

Let L be a tridiagonal operator as in (1.1) and let s be an arbitrary parameter. The Darboux factorization method consists in factorizing $L - sI$ as a product

$$L - sI = QP, \tag{7.1}$$

with Q an upper triangular matrix and P a lower triangular matrix, and to producing a new operator $\hat{L}$ by permuting the factors. Precisely,

$$\hat{L} = sI + PQ. \tag{7.2}$$

The most general solution of the factorization (7.1) is given by

$$(Ph)(n) = h(n) - \frac{\phi(n)}{\phi(n-1)}h(n-1), \quad (Qh)(n) = -a(n)\frac{\phi(n-1)}{\phi(n)}h(n) + h(n+1),$$

with $\phi(n)$ an arbitrary element in the kernel of $L - sI$. Because of the freedom in the choice of $\phi(n)$, *one extra free parameter* appears in the new operator $\hat{L}$. Its entries $\hat{a}(n)$ and $\hat{b}(n)$ are explicitly given by

$$\hat{a}(n) = a(n-1)\frac{\phi(n-2)\phi(n)}{\phi(n-1)^2}, \quad \hat{b}(n) = b(n) + \frac{\phi(n+1)}{\phi(n)} - \frac{\phi(n)}{\phi(n-1)}. \tag{7.3}$$

With suitable assumptions, the operator $\hat{L}$ has the same spectrum as the operator L, except for the addition of a bound state at the point s, see [**10**].

7.1. Some rank 1 tridiagonal solutions. We observe that the tridiagonal operators that we constructed in Theorem 5.5, can be obtained by iteration of the Darboux process, starting from the operator

$$L = E + E^{-1},$$

by adding bound states at the g points $s_i = \lambda_i + \lambda_i^{-1}$. This is the spectral meaning of the curve that we computed in (5.11). Assuming the λ_i's to be real, these g extra bound states sit outside of the purely continuous spectrum of L, which is supported on the interval $[-2, 2]$. The operator L is the simplest nontrivial solution of the operator identity (2.3), with $\gamma = \delta = 1$. The functions $f(n, z) = z^n$ satisfy

$$f(n+1, z) + f(n-1, z) = x(z)f(n, z), \quad \text{with } x(z) = z + z^{-1},$$

as well as

$$f(n, qz) - f(n, q^{-1}z) = (q^n - q^{-n})f(n, z).$$

Coming back to the problem of describing discrete bispectral rank 1 commutative rings of difference operators containing a tridiagonal matrix, we see from (6.2), that the only intersection between Theorem 5.5 and Theorem 6.1 is when

$$\lambda_i = q^{k_i}\mu_i \quad \text{and} \quad \mu_i = \lambda_i^{-1} \Leftrightarrow \lambda_i = \pm q^{k_i/2}, \qquad k_i \in \mathbb{N} \setminus \{0\}.$$

This leads immediately to the next theorem.

THEOREM 7.1 ([**17**]). *Let k_i, $1 \le i \le g$, denote g positive integers. Let us partition these integers into two arbitrary sets J and K of r and s elements respectively, r, $s \ge 0$, $r + s = g$. We build accordingly the r points $q^{k_i/2} + q^{-k_i/2}$, $i \in J$, and the s other points $-(q^{k_i/2} + q^{-k_i/2})$, $i \in K$. The tridiagonal operator $\hat{L}$ which is obtained by iteration of the Darboux process, starting at $L = E + E^{-1}$ and adding bound states at these $g = r + s$ distinct points, is part of a rank 1 commutative ring of difference operators. Its (reduced) wave function $\widetilde{w}(n, z)$ is an eigenfunction of a rank 1 commutative ring of q-difference operators in the spectral variable z.*

REMARK 7.2. When $q \to 1$, the spectral curve (5.11) corresponding to the situation described in Theorem 7.1, becomes

$$v^2 = (u-2)^{2r+1}(u+2)^{2s+1},$$

and it has only cusp-like singularities. At the limit, the dual ring becomes a ring of differential operators. This situation was studied by Haine and Iliev [**16**].

7.2. Some rank 2 tridiagonal solutions, with bound states off the continuous spectrum. We consider the doubly infinite tridiagonal matrix L with $a(n)$ and $b(n)$ defined as in (3.19), (3.20) and (3.21), with the following special choice for the free parameters a, b, c, d, ε, γ and δ:

$$\begin{gathered} a = q^{1/2}, \quad b = q^{\alpha+1/2}, \quad c = -q^{\beta+1/2}, \quad d = -q^{1/2}, \\ \gamma = 1/2, \quad \delta = 1, \quad \varepsilon \text{ arbitrary}. \end{gathered} \tag{7.4}$$

The corresponding semi-infinite matrix, with $\varepsilon = 0$, define the so-called *continuous q-Jacobi polynomials*, which are orthogonal on the interval $[-1,1]$, see [**20**, formula (3.10.14)]. When $q \to 1$, these polynomials tend to the classical Jacobi polynomials with weight function $(1-x)^\alpha(1+x)^\beta\, dx$ on $[-1,1]$.

To give the simplest possible example, we pick $\alpha = 1$ in (7.4). We perform a Darboux transformation of the corresponding L as in (7.1) and (7.2), adding a bound state at

$$s = \frac{q^{1/2}+q^{-1/2}}{2}. \tag{7.5}$$

Using the notation introduced in (3.22) for the q-shifted factorial, one checks that, for the choice of s in (7.5), the kernel of $L - sI$ is given by

$$\begin{aligned} \phi(n) = {} & \frac{(q^{1+\varepsilon};q)_n(-q^{1+\varepsilon};q)_n(q^{\beta+1+\varepsilon};q)_n(-q^{\beta+1+\varepsilon};q)_n}{2^n q^{n/2}(q^{\beta+2+2\varepsilon};q)_{2n}} \\ & \times \left[q^{n+\varepsilon} + t(1-q^{n+1+\varepsilon})(1-q^{n+1+\beta+\varepsilon})\right], \end{aligned}$$

with t an arbitrary free parameter, leading to an operator $\hat{L}$ defined as in (7.3).

Given any basic family of functions $f(n,z)$ as described in Theorem 3.2, with the parameters specified as in (7.4) with $\alpha = 1$, the new family of functions

$$\hat{f}(n,z) \equiv Pf(n,z) = f(n,z) - \frac{\phi(n)}{\phi(n-1)} f(n-1,z), \tag{7.6}$$

satisfies a fourth-order q-difference equation

$$\sum_{i=-2}^{2} c_i(z)\hat{f}(n, q^i z) = \theta(n)\hat{f}(n,z),$$

with

$$\begin{aligned} c_{-2}(z) &= \frac{t(q^{1/2}z-1)(z-q^{5/2})(z+q^{\beta+1/2})(z+q^{\beta+3/2})}{q^\beta(q+1)(z^2-1)(z^2-q^2)[(q^\beta+1)qt-1]}, \\ c_{-1}(z) &= -\frac{(z-q^{1/2})(z+q^{\beta+1/2})}{q^{\beta+1/2}(z^2-1)}, \quad c_1(z) = -\frac{(q^{1/2}z-1)(q^{\beta+1/2}z+1)}{q^{\beta+1/2}(z^2-1)}, \\ c_2(z) &= \frac{t(z-q^{1/2})(q^{5/2}z-1)(q^{\beta+1/2}z+1)(q^{\beta+3/2}z+1)}{q^\beta(q+1)(z^2-1)(q^2z^2-1)[(q^\beta+1)qt-1]}, \\ c_0(z) &= -c_{-2}(z) - c_{-1}(z) - c_1(z) - c_2(z), \end{aligned}$$

and

$$\theta(n) = \frac{(q^{n+\varepsilon}-1)(q^{n+\varepsilon+\beta+1}-1)[q(q^{n+\varepsilon+1}-1)(q^{n+\varepsilon+\beta}-1)t+q^{n+\varepsilon}(q+1)]}{q^{2n+2\varepsilon+\beta+1/2}(q+1)[(q^{\beta}+1)qt-1]}.$$

We notice that $c_i(z) = c_{-i}(1/z)$, $i = 1, 2$, a property similar to the one satisfied by the coefficients of the Askey–Wilson operator, see (3.16). When $\varepsilon = 0$, by taking in (7.6) $f(n,z)$ to be the monic continuous q-Jacobi polynomials (with $\alpha = 1$ in (7.4)), one obtains orthogonal polynomials which are eigenfunctions of a fourth-order q-difference operator. In the limit $q \to 1$, these polynomials tend to the so-called Krall–Jacobi polynomials, with weight function on $[-1,1]$ given by $(1+x)^{\beta} + t\delta(1-x)$, where $\delta(x)$ is the Dirac delta function. This is one of the family of orthogonal polynomials satisfying a fourth-order differential equation that was discovered by H. L. Krall in [**21**], see [**13**] for an overview of Krall's work and some recent developments.

One can generalize the above example as follows. For $\alpha \in \mathbb{N} \setminus \{0\}$ and $\beta \in \mathbb{N} \setminus \{0\}$, we define

$$J = \left\{ \frac{q^{\frac{1}{2}+k}+q^{-\frac{1}{2}-k}}{2}, 0 \le k \le \alpha - 1 \right\}, \quad K = \left\{ -\frac{q^{\frac{1}{2}+k}+q^{-\frac{1}{2}-k}}{2}, 0 \le k \le \beta - 1 \right\}.$$

THEOREM 7.3 ([**17**]). *Let $\alpha \in \mathbb{N} \setminus \{0\}$ or (and) $\beta \in \mathbb{N} \setminus \{0\}$. Then, adding point masses with arbitrary weights to the measure of orthogonality of the continuous q-Jacobi polynomials, at the points of J or (and) K, leads to new orthogonal polynomials which are eigenfunctions of some q-difference operator.*

In [**11**], other q-deformations of the Krall polynomials have been considered, starting from the q-Laguerre and the little q-Jacobi polynomials. These polynomials are expressed in terms of basic ${}_2\phi_1$ functions, see [**20**]. Theorem 7.3 provides the first q-analogues of the Krall polynomials, obtained by iterating the Darboux process, starting with polynomials expressed in terms of basic ${}_4\phi_3$ functions. These examples (as well as their limits when $q \to 1$) are part of *rank two* commutative rings of difference operators. For complete proofs, as well as a discussion of a more general situation than the one described in Theorem 7.3, I refer the reader to [**17**].

References

1. M. Adler, L. Haine, and P. van Moerbeke, *Limit matrices for the Toda flow and periodic flags for loop groups*, Math. Ann. **296** (1993), 1–33.
2. R. Askey and J. Wilson, *Some basic hypergeometric orthogonal polynomials that generalize Jacobi polynomials*, Mem. Amer. Math. Soc. **54**, 1985, no. 319.
3. N. M. Atakishiyev and S. K. Suslov, *Difference hypergeometric functions*, Progress in Approximation Theory (Tampa, 1990) (A. A. Gonchar and E. B. Saff, eds.), Springer Ser. Comput., vol. 19, Springer, New York, 1992, pp. 1–35.
4. J. L. Burchnall and T. W. Chaundy, *Commutative ordinary differential operators*, Proc. London Math. Soc. (3) **21** (1923), 420–440.
5. ———, *Commutative ordinary differential operators*, Proc. Roy. Soc. London Ser. A **118** (1928), 557–583.
6. ———, *Commutative ordinary differential operators. II. The identity $P^n = Q^m$*, Proc. Roy. Soc. London Ser. A **134** (1932), 471–485.
7. I. V. Cherednik, *Differential equations for the Baker–Akhiezer functions of algebraic curves*, Funktsional. Anal. i Prilozhen. **12** (1978), no. 3, 45–54, 96 (Russian); English transl. Functional Anal. Appl. **12** (1978), no. 3, 195–203.
8. L. A. Dickey, *Soliton equations and Hamiltonian systems*, Advanced Series in Mathematical Physics, vol. 12, World Scientific Publishing Co., Inc., River Edge, NJ, 1991.

9. J. J. Duistermaat and F. A. Grünbaum, *Differential equations in the spectral parameter*, Comm. Math. Phys. **103** (1986), 177–240.
10. H. Flaschka and D. McLaughlin, *Some comments on Bäcklund transformations, canonical transformations and the inverse scattering method*, Bäcklund Transformations, the Inverse Scattering Method, Solitons, and Their Applications (Nashville, 1974), Lecture Notes in Math., vol. 515, Springer, Berlin, 1976, pp. 253–295.
11. F. A. Grünbaum and L. Haine, *The q-version of a theorem of Bochner*, J. Comput. Appl. Math. **68** (1996), 103–114.
12. ______, *Some functions that generalize the Askey–Wilson polynomials*, Comm. Math. Phys. **184** (1997), 173–202.
13. L. Haine, *The Bochner–Krall problem: Some new perspectives*, Special Functions 2000: Current Perspective and Future Directions (J. Bustoz, M. E. H. Ismail and S. K. Suslov (eds.), NATO Adv. Sci. Inst. Ser. C Math. Phys. Sci., Kluwer, 2001, pp. 141–178.
14. L. Haine and P. Iliev, *The bispectral property of a q-deformation of the Schur polynomials and the* q-KdV *hierarchy*, J. Phys. A **30** (1997), 7217–7227.
15. ______, *Commutative rings of difference operators and an adelic flag manifold*, Internat. Math. Res. Notices (2000), no. 6, 281–323.
16. ______, *A rational analogue of the Krall polynomials*, J. Phys. A **34** (2001), 2445–2457.
17. ______, *Askey–Wilson type functions, with bound states* (to appear).
18. P. Iliev, q-KP *hierarchy, bispectrality and Calogero–Moser systems*, J. Geom. Phys. **35** (2000), 157–182.
19. M. E. H. Ismail and M. Rahman, *The associated Askey–Wilson polynomials*, Trans. Amer. Math. Soc. **328** (1991), no. 1 201–237.
20. R. Koekoek and R. F. Swarttouw, *The Askey–scheme of hypergeometric orthogonal polynomials and its q-analogue*, Reports no. 98-17, Faculty of Tech. Math and Info., Delft, 1998.
21. H. L. Krall, *On orthogonal polynomials satisfying a certain fourth order differential equation*, Report, no. 6, The Pennsylvania Sate College Studies, 1940.
22. D. Mumford, *An algebro-geometric construction of commuting operators and of solutions to the Toda lattice equation, Korteweg–de Vries equation and related nonlinear equations*, Proceedings of International Symposium on Algebraic Geometry (Kyoto, 1977) (M. Nagata, ed.), Kinokuniya Book Store, Tokyo, 1978, pp. 115–153.
23. F. W. Nijhoff and O. A. Chalykh, *Bispectral rings of difference operators*, Russian Math. Surveys **54** (1999), 644–645.
24. M. Petkovšek, H. S. Wilf, and D. Zeilberger, $A = B$, A. K. Peters, Ltd., Wellesley, MA, 1996.
25. G. Segal and G. Wilson, *Loop groups and equations of* KdV *type*, Inst. Hautes Études Sci. Publ. Math. **61** (1985), 5–65.
26. G. Wilson, *Bispectral commutative ordinary differential operators*, J. Reine Angew. Math. **442** (1993), 177–204.
27. ______, *Collisions of Calogero–Moser particles and an adelic Grassmannian.* With an appendix by I. G. Macdonald, Invent. Math. **133** (1998), 1–41.

Department of Mathematics, Université Catholique de Louvain, Chemin du Cyclotron 2, 1348 Louvain-la-Neuve, Belgium.
E-mail address: haine@math.ucl.ac.be

Centre de Recherches Mathématiques
CRM Proceedings and Lecture Notes
Volume **32**, 2002

Quantum Discrete Soliton Equations

Kazuhiro Hikami

ABSTRACT. Studied is an integrable structure of the discrete soliton equation, especially the discrete affine $\widehat{s\ell_N}$ Toda field theory and the N-reduced KP equation. We construct the integrals of motion explicitly, and give an outline of construction of the Baxter equation. We then clarify a relationship with the Bogoyavlensky lattice (the generalized Lotka–Volterra equation). Furthermore we study the integrable cellular automaton, and we reveal alternating aspects which show the quantum group symmetry of the Bogoyavlensky lattice.

1. Introduction

The conformal field theory (CFT) [**7**] is now the powerful tool, and has been widely used in many areas of theoretical physics. As an extension of the conformal symmetry, there exists the W symmetry [**79**], which is closely connected with the integrable differential equations. It is known [**23, 24**] that (a classical limit of) the W_N algebra coincides with the second Hamiltonian structure of the N-reduced KP hierarchy (see Ref. [**16**] for a review). This Poisson algebra is also called the Adler–Gelfand–Dickey algebra [**32**]. It was also shown [**9**] that the W algebra can be constructed from the Toda field theory (see Refs. [**14, 17**] for the τ-function and the Hamiltonian structure of the KdV type equations).

The correspondence between the CFT and the soliton equation has also revealed in the integrable deformation of the CFT [**80**]. It was shown [**18, 58, 66**] that the affine Toda field theory corresponds to deformation of the CFT with W symmetry, and that the integrals of motion (IM) of the deformed CFT are those of the quantum affine Toda field theory.

We take example from $s\ell_2$ case. As the 2-reduced KP equation, we have the KdV equation

$$\partial_t u = 6uu_x + u_{xxx}, \tag{1.1}$$

which by the Miura transformation, $u = v^2 + \mathrm{i}v_x$, is related with the mKdV equation

$$\partial_t v = 6v^2 v_x + v_{xxx}. \tag{1.2}$$

2000 *Mathematics Subject Classification.* Primary: 81T25; Secondary: 17B37, 68Q80, 81T40.

The author would like to thank R. Inoue-Yamazaki for stimulating discussions and for collaborations. He also thanks T. Tokihiro, K. Sogo, A. Kuniba, T. Takagi, and G. Hatayama.

This is the final form of the paper.

The second Poisson algebra (the Magri bracket) of the KdV equation is given by [**61**]

$$\{u(x), u(y)\} = \left(u(x)\partial + \partial u(x) + \frac{1}{2}\partial^3\right)\delta(x-y), \tag{1.3}$$

and it is nothing but the classical Virasoro algebra. Furthermore the CFT perturbed by the primary field $\Phi_{(1,3)}$is integrable, and is known to coincide with the sine-Gordon theory, which is given by

$$\partial_t \phi_x = \sin(\beta\phi), \quad v = \frac{1}{2}\beta\phi_x. \tag{1.4}$$

Recently the integrable structure of the CFT as the quantum KdV equation was reconsidered [**6**]. Therein the so-called Baxter $\mathcal{Q}$ operator was constructed from the universal $\mathcal{R}$ matrix with the q-oscillator representation, and it was shown that it solves the second order difference equation whose coefficient is the transfer matrix of the CFT. The Baxter $\mathcal{Q}$ operator plays an important role in the integrable models. As was originally applied to the XYZ spin chain [**3**], we can compute the spectrum of the model only from the analytic properties of the transfer matrices. Later Gaudin and Pasquier explicitly constructed the Baxter $\mathcal{Q}$ operator in an integral form for the quantum periodic Toda lattice [**63**], and showed that in the classical limit the logarithm of the $\mathcal{Q}$ operator is a generating function of the Bäcklund transformation, *i.e.*, the Bäcklund transformation is given by the classical limit of the evolution equation, $\mathcal{O} \to \mathcal{Q}\mathcal{O}\mathcal{Q}^{-1}$ (see also Ref. [**54, 60**], and Ref. [**67**] for a review). Since then the explicit form of the Baxter $\mathcal{Q}$ operator has been derived for other integrable models such as the XXX model [**15, 64**], the dimer self-trapping model [**59**],and the Volterra model (discrete KdV equation) [**22, 41, 75**].

In this article we study a discrete analogue of the quantum soliton equations by use of a correspondence with the CFT. Introducing a discretized free field of the CFT, we define the discretized affine Toda field theory and the N-reduced KP hierarchy. We shall construct generating functions of these equations in terms of the quantum dilogarithm function, and discuss on the Baxter equations. It is noted that, when we apply a method of Feigin–Frenkel [**27, 29**], an existence of the IMs of these discrete soliton equations can be proved homologically as was done in Ref. [**19**], but our purpose is to construct the IMs explicitly.

By construction, a differential-difference soliton equation derived here is a "good" discretization of the N-reduced KP hierarchy. Accidentally this hierarchy is shown to be equivalent with a previously known differential-difference equation, called the Bogoyavlensky lattice [**10, 51**]. This result proves the quantum group $U_q(\widehat{s\ell_N})$ symmetry of the Bogoyavlensky lattice, and we shall also study how this correspondence appears in the cellular automaton called the box-ball system [**71,72**] as a $q \to 0$ limit of the quantized soliton equation.

This paper is organized as follows. In Section 2 we study the discrete soliton equations related to $s\ell_2$ *i.e.*, the discrete KdV equation. We review the integrable structure of the Volterra model and the discrete sine-Gordon model, and derive the Baxter equation. We generalize these results to $s\ell_N$ case in Section 3, and discuss on the quantum group symmetry of the discrete affine $\widehat{s\ell_N}$ Toda field theory and the N-reduced KP equation. We shall construct the IMs of the discrete affine Toda field theory in terms of the quantum dilogarithm function, and clarify the duality of our system. Section 4 is for another intriguing relationship between the discrete soliton equation and the quantum group. We demonstrate how to derive the integrable cellular automaton from the discrete soliton equations through the *ultra-discretization* and the *crystallization.* The last section is for the concluding remarks.

2. Discrete Sine–Gordon Equation

2.1. Vertex operator. We define a discrete analogue of the free field ϕ_n (for $n \in \mathbb{Z}$) satisfying

$$[\phi_m, \phi_n] = 2\mathrm{i}\gamma, \quad \text{for } m > n \tag{2.1}$$

where γ is a parameter, and hereafter we set

$$q = \exp(\mathrm{i}\gamma). \tag{2.2}$$

Later when we talk about the classical limit, we mean that the variables commute, and that the commutator is replaced with the Poisson algebra by

$$[\ ,\ ] = -\mathrm{i}\gamma\{\ ,\ \} + O(\gamma^2).$$

The screening charges $Q^{(\pm)}$ is set to be

$$Q^{(\pm)} = \sum_n \exp(\pm\phi_n). \tag{2.3}$$

We see that the screening charges satisfy the Serre relation

$$\begin{aligned}(Q^{(a)})^3 Q^{(b)} - (q^2+1+q^{-2})(Q^{(a)})^2 Q^{(b)} Q^{(a)} \\ + (q^2+1+q^{-2}) Q^{(a)} Q^{(b)} (Q^{(a)})^2 - Q^{(b)} (Q^{(a)})^3 = 0,\end{aligned} \tag{2.4}$$

for $a, b = \pm$. With the grading of algebra $\deg(\mathrm{e}^{\pm\phi_n}) = \boldsymbol{\alpha}_{1,0}$ which are roots of $\widehat{s\ell_2}$, we see that the discrete screening charges are the representation of $U_q(\mathfrak{b}_+)$, which is the positive Borel subalgebra of $U_q(\widehat{s\ell_2})$.

2.2. Volterra model. A discrete analogue of the Virasoro generator is defined as an operator in a kernel of screening operator $Q^{(+)}$ [**28**]. It is explicitly given through the lattice Miura transformation [**26**],

$$S_n^{-1} = q^{-1} + \alpha_n + \alpha_{n+1}^{-1}\alpha_n + q^{-2}\alpha_{n+1}^{-1}, \qquad \alpha_n = \mathrm{e}^{\phi_{n+1}}\mathrm{e}^{-\phi_n}, \tag{2.5}$$

which constitutes the discrete Virasoro algebra (or the Faddeev–Volkov–Takhtajan algebra) [**25**]. In a classical limit $\gamma \to 0$ it is written as

$$\begin{aligned}\{S_n, S_m\} = S_n S_m \big((\delta_{n+2,m} S_{n+1} - \delta_{n,m+2} S_{n-1}) \\ + (\delta_{n+1,m} - \delta_{n,m+1})(S_n + S_m - 1)\big).\end{aligned} \tag{2.6}$$

This algebra corresponds to the second Poisson structure (1.3) of the KdV equation, and gives the second Hamiltonian structure of the Volterra model [**26, 76**]. By setting

$$\mathcal{H}_1 = \frac{1}{2}\sum_n \log S_n, \quad \mathcal{H}_2 = \frac{1}{2}\sum_n S_n, \tag{2.7}$$

we get the equations of motion as follows;

$$\frac{\mathrm{d}S_n}{\mathrm{d}t_1} = \{\mathcal{H}_1, S_n\} = S_n(S_{n+1} - S_{n-1}), \tag{2.8}$$

$$\frac{\mathrm{d}S_n}{\mathrm{d}t_2} = \{2\mathcal{H}_2 - 3\mathcal{H}_1, S_n\} = S_n\big(S_{n+1}(S_n + S_{n+1} + S_{n+2} - 4) - S_{n-1}(S_n + S_{n-1} + S_{n-2} - 4)\big). \tag{2.9}$$

The first is called the Volterra model (the Lotka–Volterra model), which is the fundamental equation in studies of the competition in ecosystem (see Ref. [**33**] for a review). The second is a higher flow of the Volterra model, and in the continuum limit it reduces to the KdV equation.

The Volterra model is known to be integrable both in classical and quantum cases [**22, 25, 76**]. To reveal the integrable structure, we set the Lax matrix as

$$\mathbf{L}_n^{\mathrm{V}}(x) = \frac{1}{\sqrt{S_n}} \begin{pmatrix} x & -S_n \\ 1 & 0 \end{pmatrix}. \tag{2.10}$$

This matrix is gauge-equivalent with

$$\mathbf{L}_n(\lambda) = \begin{pmatrix} \mathrm{e}^{(\phi_{n+1}-\phi_n)/2} & \\ & \mathrm{e}^{-(\phi_{n+1}-\phi_n)/2} \end{pmatrix} \cdot \begin{pmatrix} \mathrm{e}^{\lambda} & 1 \\ 1 & \mathrm{e}^{\lambda} \end{pmatrix} \equiv \mathbf{C}_n \mathbf{X}(\lambda) \tag{2.11}$$

where $x^{-2} + \mathrm{e}^{-2\lambda} = 1$. The transfer matrix and the monodromy matrix are respectively defined by

$$t(\lambda) = \operatorname{Tr} \mathbf{T}(\lambda), \quad \mathbf{T}(\lambda) = \overset{\curvearrowleft}{\prod_n} \mathrm{e}^{\mathrm{i}\gamma/4} \mathbf{L}_n(\lambda) \propto \overset{\curvearrowleft}{\prod_n} \mathbf{L}_n^{\mathrm{V}}(x) \tag{2.12}$$

where an overall factor is included for our later convention. With these definitions we see that the fundamental commutation relation (FCR) is fulfilled;

$$\check{\mathbf{R}}(\lambda - \mu)\mathbf{T}(\lambda) \otimes \mathbf{T}(\mu) = \mathbf{T}(\mu) \otimes \mathbf{T}(\lambda)\check{\mathbf{R}}(\lambda - \mu), \tag{2.13}$$

with

$$\check{\mathbf{R}}(\lambda) = \begin{pmatrix} \operatorname{sh}(\lambda + \mathrm{i}\gamma) & & & \\ & \operatorname{sh}(\mathrm{i}\gamma) & \operatorname{sh}\lambda & \\ & \operatorname{sh}\lambda & \operatorname{sh}(\mathrm{i}\gamma) & \\ & & & \operatorname{sh}(\lambda + \mathrm{i}\gamma) \end{pmatrix}.$$

This proves the Liouville integrability of the Volterra model,

$$\big[t(\lambda), t(\mu)\big] = 0, \tag{2.14}$$

and by expanding the transfer matrix $t(\lambda)$ in a power of a spectral parameter x the IMs of the Volterra model such as $\mathcal{H}_{1,2}$ introduced in Eq. (2.7) are generated.

2.3. Discrete Sine-Gordon model. We introduce the fundamental L operator as an intertwining operator of the vertex operators satisfying

$$(\mathrm{e}^{\pm\phi_n} + \mathrm{e}^{\lambda} \cdot \mathrm{e}^{\pm\phi_{n+1}}) \cdot \mathcal{L}_n(\lambda; \phi) = \mathcal{L}_n(\lambda; \phi) \cdot (\mathrm{e}^{\lambda} \cdot \mathrm{e}^{\pm\phi_n} + \mathrm{e}^{\pm\phi_{n+1}}). \tag{2.15}$$

When we suppose $\mathcal{L}_n(\lambda; \phi) = \mathcal{L}(\lambda; \phi_{n+1} - \phi_n)$, above intertwining relations reduce to the difference equation,

$$\frac{\mathcal{L}(\lambda; \varphi + \mathrm{i}\gamma)}{\mathcal{L}(\lambda; \varphi - \mathrm{i}\gamma)} = \frac{1 + \mathrm{e}^{\lambda+\varphi}}{\mathrm{e}^{\varphi} + \mathrm{e}^{\lambda}}. \tag{2.16}$$

This hyperbolic difference equation is solved by [**20, 26, 65, 77**]

$$\mathcal{L}(\lambda;\varphi)=\frac{\Theta_\gamma(\varphi)}{\Phi_\gamma(\lambda+\varphi)\Phi_\gamma(\lambda-\varphi)} \tag{2.17}$$

where the functions $\Phi_\gamma(\varphi)$ is the quantum dilogarithm function. See Appendix for definitions of $\Phi_\gamma(\varphi)$ and $\Theta_\gamma(\varphi)$. We note that originally a solution of the difference equation (2.16) is given by a q-exponential form [**76**], and that a solution (2.17) is a dualization thereof in a sense of Faddeev [**21**].

When we set the Q-operator by

$$\mathcal{Q}(\lambda)=\prod_n^{\curvearrowleft}\mathcal{L}_n(\lambda;\phi), \tag{2.18}$$

we see that due to Eq. (2.15) the Q-operator commutes with the screening charges;

$$\left[\mathcal{Q}(\lambda),Q^{(\pm)}\right]=0. \tag{2.19}$$

Thus the Q-operator generates the IMs of the Hamiltonians

$$\mathcal{H}_{\mathrm{L}}=Q^{(+)},\quad \mathcal{H}_{\mathrm{SG}}=Q^{(+)}+Q^{(-)}, \tag{2.20}$$

which respectively correspond to the discrete Liouville and sine-Gordon model. With these Hamiltonians, the equations of motion, $\mathrm{d}/(\mathrm{d}t_{\mathrm{L,SG}})\mathcal{O}=\{\mathcal{H}_{\mathrm{L,SG}},\mathcal{O}\}$, are written as

$$\frac{\mathrm{d}}{\mathrm{d}t_{\mathrm{L}}}(\phi_{n+1}-\phi_n)=\mathrm{e}^{\phi_{n+1}}+\mathrm{e}^{\phi_n},$$

$$\frac{\mathrm{d}}{\mathrm{d}t_{\mathrm{SG}}}(\phi_{n+1}-\phi_n)=\mathrm{e}^{\phi_{n+1}}+\mathrm{e}^{\phi_n}-\mathrm{e}^{-\phi_{n+1}}-\mathrm{e}^{-\phi_n},$$

and it is trivial to see that these yield to the well known equations of motion of the Liouville and the sine-Gordon model in the continuum limit. One sees that the screening charge $Q^{(-)}$ corresponds to the integrable deformation of the CFT [**18, 58**].

We clarify the relationship between the SG model and the Volterra model from the integrable structure. The generating function of the IMs for the Volterra and the SG models are respectively constructed from the 2×2 matrix $\mathbf{L}_n(\lambda)$ defined in Eq. (2.11) and the fundamental $\mathcal{L}$ operator (2.17). These two operators satisfy

$$\mathcal{L}_n(\lambda-\mu;\phi)\mathbf{L}_{n+1}(\lambda)\mathbf{L}_n(\mu)=\mathbf{L}_{n+1}(\mu)\mathbf{L}_n(\lambda)\mathcal{L}_n(\lambda-\mu;\phi), \tag{2.21}$$

$$\mathcal{L}_n(\lambda;\phi)\mathcal{L}_{n+1}(\lambda+\mu;\phi)\mathcal{L}_n(\mu;\phi)=\mathcal{L}_{n+1}(\mu;\phi)\mathcal{L}_n(\lambda+\mu;\phi)\mathcal{L}_{n+1}(\lambda;\phi). \tag{2.22}$$

As a result combining (2.13), (2.21), and (2.22), we see that two generating functions commute with each other,

$$\left[t(\lambda),\mathcal{Q}(\mu)\right]=0,\left[\mathcal{Q}(\lambda),\mathcal{Q}(\mu)\right]=0, \tag{2.23}$$

which prove that the IMs of the discrete SG are also IMs of the Volterra model.

We note that these Yang–Baxter type identities can be described in terms of the universal $\mathcal{R}$ matrix for $U_q(\widehat{s\ell_2})$ satisfying

$$\mathcal{R}_{12}\mathcal{R}_{13}\mathcal{R}_{23}=\mathcal{R}_{23}\mathcal{R}_{13}\mathcal{R}_{12}. \tag{2.24}$$

Here the universal $\mathcal{R}$ matrix is written as $\mathcal{R}\in U_q(\mathfrak{b}_+)\otimes U_q(\mathfrak{b}_-)$ where $U_q(\mathfrak{b}_\pm)$ are positive (negative) Borel subalgebras of $U_q(\widehat{s\ell_2})$. Indeed the Lax matrix $\mathbf{L}_n(\lambda)$ can be constructed from $\mathcal{R}$ by representing $U_q(\mathfrak{b}_+)$ in matrices and $U_q(\mathfrak{b}_-)$ in infinite dimensional q-oscillating space [**2**]. Correspondingly the fundamental $\mathcal{L}$ operator is also constructed from the universal $\mathcal{R}$ matrix by the q oscillator [**1**]; precisely speaking the fundamental $\mathcal{L}$ operator (2.17) corresponds to a unification of the universal $\mathcal{R}$ matrix for two dual quantum groups [**21**]. We discuss later on the duality of the $\mathcal{L}$ operator.

2.4. Baxter equation. We have seen that the commutative transfer matrices, $t(\lambda)$ and $\mathcal{Q}(\lambda)$, respectively generate the IMs of the Volterra model and the discrete sine-Gordon model. We show that they are unified in a certain kind of difference equation, which is called the Baxter equation.

We compute a product of these two generating functions defined in Eqs. (2.12) and (2.18). By direct computation we have due to Eq. (2.1)

$$
\begin{aligned}
\mathcal{Q}(\lambda)t(\lambda) &= \operatorname{Tr}\overset{\curvearrowleft}{\prod_n} \mathrm{e}^{\mathrm{i}\gamma/4}\mathcal{L}_n(\lambda;\phi)\mathbf{C}_{n+1}\mathbf{X}(\lambda) \\
&= \operatorname{Tr}\overset{\curvearrowleft}{\prod_n} \mathrm{e}^{\mathrm{i}\gamma/4}\mathbf{C}_{n+1}^{-1}\mathcal{L}_n(\lambda;\phi)\mathbf{C}_{n+1}\mathbf{X}(\lambda)\mathbf{C}_n \\
&= \operatorname{Tr}\overset{\curvearrowleft}{\prod_n} \mathrm{e}^{\mathrm{i}\gamma/4}\mathbf{M}^{-1}\begin{pmatrix} \mathcal{L}(\lambda;\chi_n-\mathrm{i}\gamma)\mathrm{e}^{\lambda+\chi_n/2} & \mathcal{L}(\lambda;\chi_n-\mathrm{i}\gamma)\mathrm{e}^{-\chi_n/2} \\ \mathcal{L}(\lambda;\chi_n+\mathrm{i}\gamma)\mathrm{e}^{\chi_n/2} & \mathcal{L}(\lambda;\chi_n+\mathrm{i}\gamma)\mathrm{e}^{\lambda-\chi_n/2}\end{pmatrix}\mathbf{M}
\end{aligned}
$$

where we have used $\chi_n = \phi_{n+1} - \phi_n$. When we set $\mathbf{M} = \left(\begin{smallmatrix}1 & 0\\ 1 & 1\end{smallmatrix}\right)$ and using Eq. (2.16), we see that $(2,1)$-element vanishes and that above matrices yield to a product of upper triangular matrices. Using identities

$$
\begin{aligned}
(1+\mathrm{e}^{\lambda+\varphi})\cdot\mathcal{L}(\lambda;\varphi-\mathrm{i}\gamma) &= \mathrm{e}^{\varphi/2-\mathrm{i}\gamma/4}\cdot\mathcal{L}(\lambda+\mathrm{i}\gamma;\varphi), \\
\mathrm{e}^{\varphi/2+\mathrm{i}\gamma/4}\cdot\mathcal{L}(\lambda;\varphi-\mathrm{i}\gamma) &= (\mathrm{e}^{\lambda}+\mathrm{e}^{\varphi})\cdot\mathcal{L}(\lambda-\mathrm{i}\gamma;\varphi),
\end{aligned}
$$

diagonal elements can be written in terms of the $\mathcal{Q}$ operator, and we finally obtain

$$
\mathcal{Q}(\lambda)t(\lambda) = \mathcal{Q}(\lambda+\mathrm{i}\gamma) + (\mathrm{e}^{2\lambda}-1\,)^L\cdot\mathcal{Q}(\lambda-\mathrm{i}\gamma) \tag{2.25}
$$

where L is a number of sites. As we know that the operators $\mathcal{Q}(\lambda)$ and $t(\lambda)$ commute with each other (2.23), above operator valued identity can be regarded as a functional relation, which we call the Baxter equation. The Baxter equation for the CFT was studied in Ref. [**6**], and later proposed was a universal procedure to derive the Baxter equation for the integrable models related to $s\ell_2$ from the universal $\mathcal{R}$ matrix by use of the q oscillator representation [**2**]. We note that Sklyanin's method of separation of variables [**68**] is also applicable to derive the Baxter equation. As in the case of the XYZ spin chain Baxter succeeded in computing spectrum by use of this type of the functional relation [**4**], the eigenvalues can essentially be computed from the Baxter equation only from the analytic properties of the transfer matrices.

2.5. Duality. As in the case of the CFT,we have the duality for these discrete systems. The duality algebraically comes from the commutativity of the vertex algebra between $\exp\phi_n$ and $\exp(\pi\phi_m/\gamma)$. The $\mathcal{L}$ operator (2.17) solves a dual equation to Eq. (2.15),

$$
\begin{aligned}
(\mathrm{e}^{\pm\pi\phi_n/\gamma}+\mathrm{e}^{\pi\lambda/\gamma}\cdot\mathrm{e}^{\pm\pi\phi_{n+1}/\gamma})\cdot\mathcal{L}_n(\lambda;\phi) \\
= \mathcal{L}_n(\lambda;\phi)\cdot(\mathrm{e}^{\pi\lambda/\gamma}\cdot\mathrm{e}^{\pm\pi\phi_n/\gamma}+\mathrm{e}^{\pm\pi\phi_{n+1}/\gamma}),
\end{aligned} \tag{2.26}
$$

which shows that the $\mathcal{Q}$ operator commutes with the dual screening operators;

$$[\mathcal{Q}(\lambda), \widetilde{Q}^{(\pm)}] = 0, \quad \widetilde{Q}^{(\pm)} = \sum_n \exp\Big(\pm\frac{\pi}{\gamma}\phi_n\Big). \tag{2.27}$$

Note that the dual screening operators $\widetilde{Q}^{(\pm)}$ satisfy the Serre relation (2.4) replacing a deformation parameter q with $\tilde{q}$,

$$\tilde{q} = \exp\Big(\mathrm{i}\frac{\pi^2}{\gamma}\Big), \tag{2.28}$$

and that they represent the positive Borel subalgebra $U_{\tilde{q}}(\mathfrak{b}_+)$. In this sense, the $\mathcal{Q}$ operator (2.18) unifies two quantum groups which are dual to each other [**21**]. Summarizing our results, we have a duality transformation in the discrete SG model;

$$\gamma \to \frac{\pi^2}{\gamma}, \quad \phi_n \to \frac{\pi}{\gamma}\phi_n, \quad \lambda \to \frac{\pi}{\gamma}\lambda. \tag{2.29}$$

It is now straightforward to see that the duality (2.29) implies the existence of the dual Baxter equation,

$$\mathcal{Q}(\lambda)u(\lambda) = \mathcal{Q}(\lambda + \mathrm{i}\pi) + \left(\mathrm{e}^{2\pi\lambda/\gamma} - 1\right)^L \cdot \mathcal{Q}(\lambda - \mathrm{i}\pi) \tag{2.30}$$

where $u(\lambda)$ is given by the duality transform of the transfer matrix $t(\lambda)$. It is clear that $u(\lambda)$ commutes with both $t(\mu)$ and $\mathcal{Q}(\mu)$. In Ref. [**70**] the dual Baxter equations (2.25) and (2.30) were studied from a view point of the duality between homology and cohomology.

3. Discrete Affine Toda Equation

We shall generalize previous results concerning the Volterra model and the discrete SG model to $\widehat{s\ell_N}$ case. In the soliton theory, this corresponds to the N-reduced KP hierarchy and the affine $\widehat{s\ell}_N$ Toda field theory.

3.1. Quantum group $U_q(\widehat{s\ell_N})$. We define a discrete analogue of the free fields $\phi_n^{(a)}$ (for $a = 1, 2, \ldots, N-1$) satisfying

$$\begin{aligned} &[\phi_m^{(a)}, \phi_n^{(a)}] = 2\mathrm{i}\gamma, \quad \text{for } m > n, \\ &[\phi_n^{(a)}, \phi_m^{(a+1)}] = \begin{cases} \mathrm{i}\gamma, & \text{for } n \le m, \\ -\mathrm{i}\gamma, & \text{for } n > m. \end{cases} \end{aligned} \tag{3.1}$$

We set

$$\phi_n^{(0)} = -\sum_{a=1}^{N-1} \phi_n^{(a)}. \tag{3.2}$$

Discrete analogue of the screening charges (and their dual) are defined by (for $a = 0, 1, \ldots, N-1$)

$$Q^{(a)} = \sum_n \exp\phi_n^{(a)}, \quad \widetilde{Q}^{(a)} = \sum_n \exp\Big(\frac{\pi}{\gamma}\phi_n^{(a)}\Big). \tag{3.3}$$

As in a case of a continuum theory [**12**], the screening charges satisfy the Serre relation

$$(Q^{(a)})^2Q^{(b)} - (q+q^{-1})Q^{(a)}Q^{(b)}Q^{(a)} + Q^{(b)}(Q^{(a)})^2 = 0, \tag{3.4}$$

for $\widehat{\mathbf{C}}_{ab} = -1$ with the extended Cartan matrix of $\widehat{s\ell_N}$. By grading the algebra $\deg(\mathrm{e}^{\phi_n^{(a)}}) = \boldsymbol{\alpha}_a$ with a simple root, we see that the screening charges realize the representation of the positive Borel subalgebra $U_q(\mathfrak{b}_+)$ of the quantum group $U_q(\widehat{s\ell_N})$. The dual screening charge $\widetilde{Q}^{(a)}$ enjoys the same Serre relation replacing q by $\tilde{q}$, and represents $U_{\tilde{q}}(\mathfrak{b}_+)$.

3.2. Discrete W_N algebra and the Bogoyavlensky lattice. We can define the generators of the discreteW_N algebra in terms of the free fields $\phi_n^{(a)}$. We have $N-1$ sets of generators $W_n^{(s)}$ (for $s = 2, 3, \ldots, N$, and $n \in \mathbb{Z}$), and they are in

$$\operatorname{Ker} Q^{(1)} \cap \operatorname{Ker} Q^{(2)} \cap \cdots \cap \operatorname{Ker} Q^{(N-1)}.$$

In a classical limit, they are explicitly defined by [**36, 42**]

$$\begin{aligned}(3.5)\quad W_n^{(s)} = & \frac{1}{\prod_{k=0}^{s-1}\left(1+\sum_{i=1}^{N-1}\exp(\chi_{n+k}^{(i)})\right)} \\ & \times\Bigg(\sum_{1\leq i_1<i_2<\cdots<i_{s-1}\leq N-1} \exp(\chi_{n+s-1}^{(i_1)}+\chi_{n+s-2}^{(i_2)}+\cdots+\chi_{n+1}^{(i_{s-1})}) \\ & \qquad + \sum_{1\leq i_1<i_2<\cdots<i_s\leq N-1} \exp(\chi_{n+s-1}^{(i_1)}+\chi_{n+s-2}^{(i_2)}+\cdots+\chi_{n}^{(i_s)})\Bigg).\end{aligned}$$

Here $\chi_n^{(a)}$ is defined in terms of the free fields $\phi_n^{(a)}$ (for $a = 1, 2, \ldots, N-1$) as

$$\chi_n^{(a)} = \sum_{b=1}^{a} \Delta\phi_n^{(b)}$$

where

$$\Delta\phi_n^{(a)} = \phi_n^{(a)} - \phi_{n+1}^{(a)}.$$

As in the case of the Virasoro algebra, the algebra defined by Eq. (3.1) corresponds to the second Hamiltonian structure of the N-reduced KP hierarchy. By setting the Hamiltonian

$$(3.6)\qquad \mathcal{H}_1 = \frac{1}{N}\sum_n \log W_n^{(N)},$$

we get a set of differential-difference equations [**46**]

$$(3.7)\qquad \frac{\mathrm{d}W_n^{(s)}}{\mathrm{d}t_1} = \{\mathcal{H}_1, W_n^{(s)}\} = W_n^{(s)}(W_{n+s-1}^{(2)} - W_{n-1}^{(2)}) - W_n^{(s+1)} + W_{n-1}^{(s+1)}$$

where $W_n^{(N+1)} \equiv 0$. It is noted that for $N = 2$ this equation coincides with the Volterra model (2.8), and that equation for $N = 3$ case was studied in Refs. [**8, 11, 42**]. By construction we can regard this equation as the first integrable flow of the discretized N-reduced KP hierarchy, and in fact we get the Boussinesq equation from the second flow of this hierarchy in the case of $N = 3$ [**42**]. Interesting is that by setting the W currents as [**46**]

$$\begin{aligned}(3.8)\quad & W_n^{(s)} \\ & = \sum_{0\leq i_1<i_2<\cdots<i_{s-1}\leq N-2} V_{n+i_1/(N-1)} V_{n+1+i_2/(N-1)} \cdots V_{n+s-2+(i_{s-1})/(N-1)},\end{aligned}$$

with the dynamical variables V_n on $n \in \mathbb{Z}/(N-1)$, Eq. (3.7) can be mapped into

$$\frac{\mathrm{d}V_n}{\mathrm{d}t_1} = V_n \sum_{k=0}^{N-1} \left(V_{n+k/(N-1)} - V_{n-k/(N-1)}\right). \tag{3.9}$$

This equation is known as the Bogoyavlensky lattice [**10, 51**]. Later we will study another aspect of the quantum group symmetry of the Bogoyavlensky lattice. We note that in Eq. (3.8) the variables V_n exclude N nearest variables, and it looks like a exclusion statistics [**37**].

The integrable structure of the discrete W_N algebra is formulated as follows. We define the Lax matrix [**36, 42**] as

$$\mathbf{L}_n^{\mathrm{W}}(x) \tag{3.10}$$

$$= \frac{1}{\sqrt[N]{W_n^{(N)}}} \begin{pmatrix} x^{N-1} & -x^{N-2}W_{n+N-2}^{(2)} & x^{N-3}W_{n+N-3}^{(3)} & \cdots & (-1)^{N-2}xW_{n+1}^{(N-1)} & (-1)^{N-1}W_n^{(N)} \\ 1 & 0 & 0 & \cdots & 0 & 0 \\ 0 & 1 & 0 & \cdots & 0 & 0 \\ \vdots & 0 & 1 & \ddots & \vdots & \vdots \\ \vdots & \vdots & \ddots & \ddots & \ddots & \vdots \\ 0 & 0 & \cdots & \cdots & 1 & 0 \end{pmatrix}.$$

This $N \times N$ matrix is transformed into

$$\mathbf{L}_n(\lambda) = \exp\Bigl(\sum_{a=1}^{N-1} \mathbf{h}_a \chi_n^{(a)}\Bigr)\mathbf{X}(\lambda) \tag{3.11}$$

where $x^{-N} + \mathrm{e}^{-N\lambda} = 1$, and

$$\mathbf{h}_a = \mathbf{E}_{a,a} - \frac{1}{N}\mathbb{1}, \quad \mathbf{X}(\lambda) = \sum_{k=0}^{N-1} \mathrm{e}^{-k\lambda}\mathbf{C}^k, \quad \mathbf{C} = \mathbf{E}_{1,N} + \sum_{k=1}^{N-1} \mathbf{E}_{k+1,k}.$$

Here an elementary matrix $\mathbf{E}_{j,k}$ is $(\mathbf{E}_{j,k})_{mn} = \delta_{j,m}\delta_{k,n}$. As usual we define the transfer matrix by

$$t(\lambda) = \operatorname{Tr}\mathbf{T}(\lambda), \quad \mathbf{T}(\lambda) = \overset{\curvearrowleft}{\prod_n} \mathbf{L}_n(\lambda) = \overset{\curvearrowleft}{\prod_n} \mathbf{L}_n^{\mathrm{W}}(x), \tag{3.12}$$

which is proved to satisfy the FCR

$$\check{\mathbf{R}}(\lambda-\mu)\mathbf{T}(\lambda) \otimes \mathbf{T}(\mu) = \mathbf{T}(\mu) \otimes \mathbf{T}(\lambda)\check{\mathbf{R}}(\lambda-\mu), \tag{3.13}$$

with the $\mathbb{Z}_N$ symmetric $\check{\mathbf{R}}$ matrix,

$$\check{\mathbf{R}}(\lambda) = \sum_{k=1}^{N} (\mathrm{e}^{N\lambda} - \mathrm{e}^{\mathrm{i}\gamma})\mathbf{E}_{k,k} \otimes \mathbf{E}_{k,k} + \sum_{k=1}^{N}\sum_{\ell=1}^{N-1} \mathrm{e}^{2(1-\ell/N)\mathrm{i}\gamma}(\mathrm{e}^{N\lambda} - 1)\mathbf{E}_{k,k+\ell} \otimes \mathbf{E}_{k+\ell,k}$$
$$+ \sum_{k=1}^{N}\sum_{\ell=1}^{N-1} (1 - \mathrm{e}^{\mathrm{i}\gamma})\mathrm{e}^{\ell\lambda}\mathbf{E}_{k,k} \otimes \mathbf{E}_{k+\ell,k+\ell}.$$

Here we impose a periodicity of modulo N in indices of matrix $\mathbf{E}_{k,\ell}$. The IMs such as $\mathcal{H}_1$ (3.6) of the discrete N-reduced KP equation (3.7) is generated from the transfer matrix $t(\lambda)$ by expanding in a power of x; some of them are in the classical limit given by

$$
\begin{aligned}
\mathcal{H}_1 &= \frac{1}{N}\sum_n \log W_n^{(N)}, \\
\mathcal{H}_2 &= \frac{1}{2}\sum_n W_n^{(2)}, \\
\mathcal{H}_3 &= \sum_n \Bigl(\frac{1}{2}(W_n^{(2)})^2 + W_n^{(2)}W_{n+1}^{(2)} - W_n^{(3)}\Bigr), \\
\mathcal{H}_4 &= \sum_n \Bigl(\frac{1}{3}(W_n^{(2)})^3 + W_n^{(2)}W_{n+1}^{(2)}(W_n^{(2)} + W_{n+1}^{(2)} + W_{n+2}^{(2)}) \\
&\qquad -W_n^{(3)}(W_{n-1}^{(2)} + W_n^{(2)} + W_{n+1}^{(2)} + W_{n+2}^{(2)}) + W_n^{(4)}\Bigr).
\end{aligned} \tag{3.14}
$$

3.3. Integrals of motion. We define the fundamental L operator for the affine $\widehat{s\ell_N}$ discrete Toda field theory by [**38**]

$$
\begin{aligned}
\mathcal{L}_n(\lambda;\phi) &= \frac{1}{\Phi_\gamma(\lambda+\chi_n^{(N-1)})}\cdot\frac{1}{\Phi_\gamma(2\lambda+\chi_n^{(N-2)})}\cdots\frac{1}{\Phi_\gamma((N-1)\lambda+\chi_n^{(1)})} \\
&\quad\times\frac{\Theta_\gamma(\chi_n^{(1)})}{\Phi_\gamma(\lambda-\chi_n^{(1)})}\cdot\frac{\Theta_\gamma(\chi_n^{(2)})}{\Phi_\gamma(\lambda-\chi_n^{(2)})}\cdots\frac{\Theta_\gamma(\chi_n^{(N-1)})}{\Phi_\gamma(\lambda-\chi_n^{(N-1)})}.
\end{aligned} \tag{3.15}
$$

We can check by use of properties of the quantum dilogarithm function $\Phi_\gamma(\varphi)$ in Appendix that it satisfies the intertwining relations of the (dual) vertex operators (for $a = 0, 1, \ldots, N-1$) [**39**],

$$
(\mathrm{e}^{\phi_n^{(a)}} + \mathrm{e}^{\lambda}\cdot\mathrm{e}^{\phi_{n+1}^{(a)}})\cdot\mathcal{L}_n(\lambda;\phi) = \mathcal{L}_n(\lambda;\phi)\cdot(\mathrm{e}^{\lambda}\cdot\mathrm{e}^{\phi_n^{(a)}} + \mathrm{e}^{\phi_{n+1}^{(a)}}), \tag{3.16}
$$

$$
\begin{aligned}
&(\mathrm{e}^{\pi\phi_n^{(a)}/\gamma} + \mathrm{e}^{\pi\lambda/\gamma}\cdot\mathrm{e}^{\pi\phi_{n+1}^{(a)}/\gamma})\cdot\mathcal{L}_n(\lambda;\phi) \\
&\qquad = \mathcal{L}_n(\lambda;\phi)\cdot(\mathrm{e}^{\pi\lambda/\gamma}\cdot\mathrm{e}^{\pi\phi_n^{(a)}/\gamma} + \mathrm{e}^{\pi\phi_{n+1}^{(a)}/\gamma}).
\end{aligned} \tag{3.17}
$$

Thus when we define the $\mathcal{Q}$ operator by

$$
\mathcal{Q}(\lambda) = \overset{\curvearrowleft}{\prod_n}\mathcal{L}_n(\lambda;\phi), \tag{3.18}
$$

we find that the $\mathcal{Q}$ operator commutes with the (dual)screening operators,

$$
\bigl[\mathcal{Q}(\lambda), Q^{(a)}\bigr] = \bigl[\mathcal{Q}(\lambda), \widetilde{Q}^{(a)}\bigr] = 0, \quad \text{for } a = 0, 1, \ldots, N-1. \tag{3.19}
$$

To conclude the $\mathcal{Q}$ operator is a generating function of the IMs for the discrete (affine) Toda equation whose Hamiltonians are

$$
\mathcal{H}_{\text{Toda}} = \sum_{a=1}^{N-1} Q^{(a)}, \quad \mathcal{H}_{\text{affine}} = \mathcal{H}_{\text{Toda}} + Q^{(0)} = \sum_{a=0}^{N-1} Q^{(a)}. \tag{3.20}
$$

The reason why we can identify the discrete (affine) Toda field theory is that the equations of motion is classically given from above Hamiltonians, $(\mathrm{d}\mathcal{O})/(\mathrm{d}t) = \{\mathcal{H}, \mathcal{O}\}$, as

$$
\frac{\mathrm{d}}{\mathrm{d}t}\Delta\phi_n^{(a)} = \mathrm{e}^{\phi_n^{(a+1)}} - \mathrm{e}^{\phi_n^{(a)}} - \mathrm{e}^{\phi_{n+1}^{(a)}} + \mathrm{e}^{\phi_{n+1}^{(a-1)}}, \tag{3.21}
$$

which reduces to the well-known equations of motion,

$$\partial_t \partial_x \phi^{(a)} = -\sum_b \widehat{\mathbf{C}}_{ab} \exp \phi^{(b)}.$$

The fundamental $\mathcal{L}$ operator (3.15) and the $N \times N$ Lax matrix $\mathbf{L}_n(\lambda)$ (3.11) fulfills same identities with Eqs. (2.21) and (2.22). The YBE (2.22) was proved [**38**]for $N = 3$ by recursive use of the pentagon identity, and later for arbitrary N in Ref. [**39**]. We expect that these identities may be given from the universal $\mathcal{R}$ matrix for $U_q(\widehat{s\ell_N})$ [**55**] with the q-oscillator algebra. To conclude we have a commutativity among the transfer matrix $t(\lambda)$ (3.12) and the $\mathcal{Q}(\lambda)$ operator (3.18).

3.4. Baxter equation. Motivated from results of the Volterra model in previous section, we expect that the $\mathcal{Q}$ operator (3.18) solves the Baxter equation for $U_q(\widehat{s\ell_N})$, which is an Nth order difference equation,

$$\begin{aligned}(3.22)\quad \mathcal{Q}(\lambda + 2\mathrm{i}\gamma) + \sum_{k=1}^{N-1} (-1)^k t_k\Big(\lambda + 2\frac{N-k}{N}\mathrm{i}\gamma\Big)\mathcal{Q}\Big(\lambda + 2\frac{N-k}{N}\mathrm{i}\gamma\Big) \\ + \Delta\Big(\lambda + \frac{2}{N}\mathrm{i}\gamma\Big)\mathcal{Q}(\lambda) = 0.\end{aligned}$$

Here $t_1(\lambda)$ is the transfer matrix (3.12) up to constant, and other matrices $t_{k>1}(\lambda)$ are for the highest weight $(\underbrace{1,\ldots,1}_{k},0,\ldots,0)$ of $s\ell_N$, and should be constructed by the fusion method from $t_1(\lambda)$ [**57**]. Among them, we have

$$t_{N-1}(\lambda) \propto \mathrm{Tr}\Big(\overset{\curvearrowleft}{\prod_n} \exp\Big(-\sum_{a=1}^{N-1} \mathbf{h}_a \chi_n^{(a)}\Big)\overline{\mathbf{X}}(\lambda)\Big), \quad \overline{\mathbf{X}}(\lambda) = 1 - \mathrm{e}^{-\lambda}\mathbf{C}^{-1}.$$

This conjecture was explicitly proved for $N = 3$ [**40**]. A strategy is essentially same with the Volterra model, and the result is read as follows. We set two Lax matrices,

$$\mathbf{L}_n(\lambda) = \begin{pmatrix} \mathrm{e}^{\Delta\phi_n^{(1)}/3 - \Delta\phi_n^{(2)}/3} & & \\ & \mathrm{e}^{\Delta\phi_n^{(1)}/3 + 2\Delta\phi_n^{(2)}/3} & \\ & & \mathrm{e}^{-2\Delta\phi_n^{(1)}/3 - \Delta\phi_n^{(2)}/3} \end{pmatrix} \times \begin{pmatrix} \mathrm{e}^{\lambda} & \mathrm{e}^{-\lambda} & 1 \\ 1 & \mathrm{e}^{\lambda} & \mathrm{e}^{-\lambda} \\ \mathrm{e}^{-\lambda} & 1 & \mathrm{e}^{\lambda} \end{pmatrix},$$

$$\overline{\mathbf{L}}_n(\lambda) = \begin{pmatrix} \mathrm{e}^{-\Delta\phi_n^{(1)}/3 + \Delta\phi_n^{(2)}/3} & & \\ & \mathrm{e}^{-\Delta\phi_n^{(1)}/3 - 2\Delta\phi_n^{(2)}/3} & \\ & & \mathrm{e}^{2\Delta\phi_n^{(1)}/3 + \Delta\phi_n^{(2)}/3} \end{pmatrix} \times \begin{pmatrix} \mathrm{e}^{\lambda} & -1 & 0 \\ 0 & \mathrm{e}^{\lambda} & -1 \\ -1 & 0 & \mathrm{e}^{\lambda} \end{pmatrix},$$

and the two transfer matrices are given by

$$t_1(\lambda) = \mathrm{Tr}\prod_n^{\curvearrowleft} g(\lambda)\mathbf{L}_n(\lambda), t_2(\lambda) = \mathrm{Tr}\prod_n^{\curvearrowleft} \overline{g}(\lambda)\overline{\mathbf{L}}_n(\lambda). \tag{3.23}$$

with $g(\lambda) = \mathrm{e}^{\lambda+\mathrm{i}\gamma/3}$, $\overline{g}(\lambda) = \mathrm{e}^{\mathrm{i}\gamma/3}\big(\mathrm{e}^{3\lambda+2\mathrm{i}\gamma} - 1\big)$. The transfer matrices $t_1(\lambda)$ and $t_2(\lambda)$ respectively correspond to the three-dimensional representation for the highest weight $(1,0,0)$ and $(1,1,0)$. Note,in $s\ell_3$ case we have $\overline{\mathbf{L}}_n(\lambda) \propto \left(\left(\mathbf{L}_n(\lambda)\right)^{-1}\right)^{\mathrm{t}}$. We have a commutativity among the transfer matrices $t_{1,2}(\lambda)$ and the $\mathcal{Q}$ operator (3.18) $(N = 3)$. After lengthy computation we see that the $\mathcal{Q}$ operator solves the third order difference equation [**40**],

$$\begin{aligned}\mathcal{Q}(\lambda + 2\mathrm{i}\gamma) - t_1\left(\lambda + \frac{4}{3}\mathrm{i}\gamma\right)\mathcal{Q}\left(\lambda + \frac{4}{3}\mathrm{i}\gamma\right) + t_2\left(\lambda + \frac{2}{3}\mathrm{i}\gamma\right)\mathcal{Q}\left(\lambda + \frac{2}{3}\mathrm{i}\gamma\right)& \\ - \Delta\left(\lambda + \frac{2}{3}\mathrm{i}\gamma\right)\mathcal{Q}(\lambda) = 0&\end{aligned} \tag{3.24}$$

where

$$\Delta(\lambda) = \left((\mathrm{e}^{3\lambda} - 1)(\mathrm{e}^{3\lambda+2\mathrm{i}\gamma} - 1)\right)^L.$$

This type of the third order difference equation was studied for the quantum Boussinesq equation [**5**], and the separation of variables would be also applied to this type of $s\ell_3$ integrable models [**69**].

From the commutativity of the transfer matrices and the $\mathcal{Q}$ operator, above identities can be regarded as a usual third order difference equation. To get another functional equation, we recall that generally such difference equation has three linearly independent solutions, which we denote $\mathcal{Q}_a(\lambda)$ for $a = -,0,+$. We then define the function $\mathcal{P}_{ab}(\lambda)$ as the quantum Wronskian,

$$\mathcal{P}_{ab} = \begin{vmatrix} \mathcal{Q}_a(\lambda) & \mathcal{Q}_a(\lambda - \frac{2}{3}\mathrm{i}\gamma) \\ \mathcal{Q}_b(\lambda) & \mathcal{Q}_b(\lambda - \frac{2}{3}\mathrm{i}\gamma) \end{vmatrix}, \tag{3.25}$$

for $a, b = 0, \pm$. We thus have three independent functions $\mathcal{P}_{ab}(\lambda)$. We find that this function satisfies the third order difference equation,

$$\begin{aligned}&\mathcal{P}(\lambda + 2\mathrm{i}\gamma) - t_2\left(\lambda + \frac{2}{3}\mathrm{i}\gamma\right)\cdot\mathcal{P}\left(\lambda + \frac{4}{3}\mathrm{i}\gamma\right) \\ &+ \Delta\left(\lambda + \frac{2}{3}\mathrm{i}\gamma\right)t_1\left(\lambda + \frac{2}{3}\mathrm{i}\gamma\right)\cdot\mathcal{P}\left(\lambda + \frac{2}{3}\mathrm{i}\gamma\right) - \Delta(\lambda)\Delta\left(\lambda + \frac{2}{3}\mathrm{i}\gamma\right)\cdot\mathcal{P}(\lambda) = 0.\end{aligned} \tag{3.26}$$

Solving two difference equations (3.24) and (3.26), we see that the transfer matrices $t_a(\lambda)$ is diagonalized as

$$\begin{aligned}t_1(\lambda) = \Delta\left(\lambda - \frac{2}{3}\mathrm{i}\gamma\right)&\frac{\mathcal{P}(\lambda - 2\mathrm{i}\gamma/3)}{\mathcal{P}(\lambda)} + \frac{\mathcal{P}(\lambda + 2\mathrm{i}\gamma/3)}{\mathcal{P}(\lambda)} \\ &\times \frac{\mathcal{Q}(\lambda - 2\mathrm{i}\gamma/3)}{\mathcal{Q}(\lambda)} + \frac{\mathcal{Q}(\lambda + 2\mathrm{i}\gamma/3)}{\mathcal{Q}(\lambda)},\end{aligned} \tag{3.27a}$$

$$\begin{aligned}t_2(\lambda) = \frac{\mathcal{P}(\lambda + 4\mathrm{i}\gamma/3)}{\mathcal{P}(\lambda + 2\mathrm{i}\gamma/3)} &+ \Delta(\lambda)\frac{\mathcal{P}(\lambda)}{\mathcal{P}(\lambda + 2\mathrm{i}\gamma/3)} \\ &\times \frac{\mathcal{Q}(\lambda + 2\mathrm{i}\gamma/3)}{\mathcal{Q}(\lambda)} + \Delta(\lambda)\frac{\mathcal{Q}(\lambda - 2\mathrm{i}\gamma/3)}{\mathcal{Q}(\lambda)}.\end{aligned} \tag{3.27b}$$

As was studied in Ref. [**56**], we can essentially determine the spectrum of the transfer matrices $t_i(\lambda)$ by assuming an analytic property of these functions.

We recall that the $\mathcal{Q}$ operator (3.18)have the duality, and that we have another set of quantum group with a deformation parameter $\tilde{q}$. We then have the dual Baxter equation,

$$\mathcal{Q}(\lambda+2\mathrm{i}\pi)+\sum_{k=1}^{N-1}(-1)^k u_k\Big(\lambda+2\frac{N-k}{N}\mathrm{i}\pi\Big)\mathcal{Q}\Big(\lambda+2\frac{N-k}{N}\mathrm{i}\pi\Big) + \tilde{\Delta}\Big(\lambda+\frac{2}{N}\mathrm{i}\pi\Big)\mathcal{Q}(\lambda)=0 \tag{3.28}$$

where the transfer matrices $u_k(\lambda)$ and a function $\tilde{\Delta}(\lambda)$ are given by the duality transformation (2.29) of $t_k(\lambda)$ and $\Delta(\lambda)$ respectively.

4. Bogoyavlensky Lattice and Cellular Automaton

We have shown that the Bogoyavlensky lattice (3.9) is the first flow of the discrete N-reduced KP hierarchy, and the quantization thereof is formulated by use of the quantum group $U_q(\widehat{s\ell_N})$. In this section we show that a link between the Bogoyavlensky lattice and the quantum group $U_q(\widehat{s\ell_N})$ can be seen in the correspondence between the box-ball system (BBS) and the crystal base. Hereafter we set $M \equiv N-1$, and we rescale the coordinate in Eq. (3.9) and use the equation of motion for V_n (for $n \in \mathbb{Z}$)

$$\frac{\mathrm{d}V_n}{\mathrm{d}t_1}=V_n\sum_{k=0}^{M}(V_{n+k}-V_{n-k}). \tag{4.1}$$

4.1. Soliton solution. We first give the soliton solution of the Bogoyavlensky lattice (4.1). By the mapping (3.8) this gives also the soliton solution of the N-reduced KP equation. When we substitute

$$V_n=\frac{\tau_{n+M+1}\tau_{n-M}}{\tau_{n+1}\tau_n}, \tag{4.2}$$

into Eq. (4.1), we get Hirota's bilinear form (see *e.g.* Ref. [**47**]),

$$D_{t_1}\tau_{n+1}\cdot\tau_n=\tau_{n+M+1}\tau_{n-M}-\tau_{n+1}\tau_n. \tag{4.3}$$

The m-soliton solution is constructed by [**46**]

$$\tau_n=\prod_{i=1}^{m}\big(1+c_iX(p_i)\big)\cdot 1. \tag{4.4}$$

Here c_i is arbitrary parameter, and the vertex operator $X(p)$ is defined by

$$X(p)=\mathrm{e}^{\xi_n(p)}\mathrm{e}^{\hat{\xi}_n(p)} \tag{4.5}$$

where

$$\xi_n(p)=n\log p+\sum_{k=1}^{\infty}\big((x(p))^k-(x(p^{-1}))^k\big)t_k,$$

$$\hat{\xi}_n(p)=-\sum_{k=1}^{\infty}\frac{1}{k}\big((x(p))^{-k}-(x(p^{-1}))^{-k}\big)\frac{\partial}{\partial t_k}.$$

with $x(p) = \sum_{i=1}^{M} p^i$. Note that they satisfy

$$X(p)X(q) = A(p,q) {}^{\circ}_{\circ} X(p)X(q) {}^{\circ}_{\circ},$$
$$A(p,q) = \frac{(x(p)-x(q))(x(p^{-1})-x(q^{-1}))}{(x(p)-x(q^{-1}))(x(p^{-1})-x(q))},$$

and that the higher order bilinear equation for the τ-function (4.4) is written as

$$\left(D_{t_2} - D_{t_1}^{\,2}\right)\tau_{n+M+1}\cdot\tau_n = 0.$$

4.2. Box ball system. We derive the integrable cellular automaton from the Bogoyavlensky lattice (4.1) through the *ultra-discretization* method. The bilinear equation (4.3) of the Bogoyavlensky lattice can be discretized in time-direction, and is given by [**48**]

$$\tau_n^t \tau_{n-1}^{t+1-M} - \delta \tau_{n-1}^{t+1}\tau_n^{t-M} = (1-\delta)\tau_n^{t+1}\tau_{n-1}^{t-M} \tag{4.6}$$

where δ denotes the unit size of time $t \in \mathbb{Z}$, and the variables n and t are transformed as $t \to Mt - n$ and $n \to t$. By substituting

$$\tau_n^{Mt+j} = \exp\left(\frac{Y_{n,j}^t}{\varepsilon}\right), \qquad \delta = -\exp\left(-\frac{1}{\varepsilon}\right), \tag{4.7}$$

and take the ultra-discrete limit $\varepsilon \to +0$ [**74**], we get the ultra-discrete bilinear equation,

$$Y_{n+1,j}^{t+1} + Y_{n,j+1}^t = \max[Y_{n,j}^t + Y_{n+1,j+1}^{t+1}, Y_{n+1,j+1}^t + Y_{n,j}^{t+1} - 1] \tag{4.8}$$

where we have used an identity,

$$\lim_{\varepsilon \to +0} \varepsilon \log(\mathrm{e}^{A/\varepsilon} + \mathrm{e}^{B/\varepsilon}) = \max(A,B).$$

By defining a dynamical variables motivating from Eq. (4.2),

$$u_{n,j}^t = Y_{n,j}^t + Y_{n+1,j+1}^t - Y_{n+1,j}^t - Y_{n,j+1}^t, \tag{4.9}$$

we get the evolution equation,

$$u_{n,j}^{t+1} = \min\left[v_{n,j}^t, 1 - \sum_{i=1}^{j} u_{n,i}^t - \sum_{i=j+1}^{M} u_{n,i}^{t+1}\right], \tag{4.10a}$$

$$u_{n,j}^t + v_{n,j}^t = u_{n,j}^{t+1} + v_{n+1,j}^t. \tag{4.10b}$$

Here we have used a simple formula; $-\min[-A,-B] = \max[A,B]$.

We regard above set of equations as the evolution equation for $u_{n,j}^t$. To relate with the "box-ball system" [**71, 72**] we interpret $u_{n,j}^t$ as the number of ball-j (for $j = 1, \ldots, M$) on the n-th box at time t. Here we suppose that each box can have one ball at most. Then the evolution equation (4.10) can be translated into a rule of the BBS;

(i) Take the leftmost ball-M out of its box, and put it in the first empty box to its right.
(ii) Take the new leftmost ball-M (as long as it has not yet been moved at this time step), and move it to the first empty box to its right.
(iii) Continue the process (ii) until every ball-M has been moved.
(iv) Continue the same processes (i)–(iii) for ball-$(M-1), \ldots,$ ball-2, ball-1. Steps (i)–(iv) represent 1 time step.
(v) Repeat the processes(i)–(iv).

In this interpretation, variable $v_{n,j}^t$ denotes the number of ball-j in a carrier.

We give below some typical examples of evolution of the BBS for $M = 3$ case. Here for brevity we denote an empty box as '0'and a box occupied by ball-j as 'j'.

$t = 0:$	033100320000000000	
$1:$	000033103200000000	
$2:$	000000031033200000	$[331]\times[32]\to[31]\times[332]$
$3:$	000000000310033200	

$t = 0:$	03110220000000000	
$1:$	00003112200000000	
$2:$	00000001132200000	$[311]\times[22]\to[11]\times[322]$
$3:$	00000000011032200	

$t = 0:$	0321000002201000000000000	
$1:$	0000321000020210000000000	
$2:$	0000000321002002100000000	$[321]\times\big([22]\times[1]\big) \to [1]\times[32]\times[221]$
$3:$	0000000000321200021000000	
$4:$	0000000000000132200210000	
$5:$	0000000000000010032002210	

$t = 0:$	0321022000100000000000000000	
$1:$	0000310222010000000000000000	
$2:$	0000003100202210000000000000	
$3:$	0000000031020002210000000000	$\big([321]\times[22]\big)\times[1] \to [1]\times[32]\times[221]$
$4:$	0000000000312000002210000000	
$5:$	0000000000001320000002210000	
$6:$	0000000000000103200000002210	

These examples show that a nonincreasing sequence which is sandwiched by '0' (at $t \to \pm\infty$) has a property of *soliton*; it is stable before and after collisions though the internal degree of freedom can change. Furthermore the out-going state does not depend on the order of the scattering. We denote this nonincreasing sequence as $[c_1 \dots c_p]$ $(c_1 \geq \dots \geq c_p > 0)$, and call it as a soliton of length p. In the right column of above examples we have also written explicitly a scattering rule of solitons.

4.3. Crystallization. We shall show an alternative approach from the Bogoyavlensky lattice (4.1) to the BBS [**45**]. To this end, we recall the bi-Hamiltonian structure of the Bogoyavlensky lattice [**44**]. First of all, we study in detail a case of $M = 1$, *i.e.*, the Volterra model (2.8). Besides the discrete Magri bracket (2.6), we define another Poisson algebra $\{\ ,\ \}_1$ as

$$\{S_n, S_{n+1}\}_1 = 2S_nS_{n+1}, \tag{4.11}$$

which corresponds to the first Hamiltonian structure of the KdV equation [**78**]. With the IMs (2.7) it defines the bi-Hamiltonian structure

$$\frac{\mathrm{d}\mathcal{O}}{\mathrm{d}t_1} = \{\mathcal{H}_1, \mathcal{O}\} = \{\mathcal{H}_2, \mathcal{O}\}_1. \tag{4.12}$$

We investigate again the quantum integrable structure based on this first Hamiltonian structure. By use of the canonically conjugate variables P_n and Q_n satisfying

$$\{P_n, Q_m\}_1 = \delta_{m,n}P_nQ_m, \quad \{P_n, P_m\}_1 = \{Q_n, Q_m\}_1 = 0, \tag{4.13}$$

we can set

$$S_n = (P_n P_{n+1})^{-1} Q_n^{-1} Q_{n+1},$$

and the Lax matrix (2.10) of the Volterra model is transformed to

$$\begin{aligned}\text{(4.14)} \qquad \mathbf{M}_n(x) &= \begin{pmatrix} xP_n & Q_n \\ Q_n^{-1} & 0 \end{pmatrix} \\ &\equiv x|0\rangle\!\rangle\langle\!\langle 0| \otimes P_n + |0\rangle\!\rangle\langle\!\langle 1| \otimes Q_n + |1\rangle\!\rangle\langle\!\langle 0| \otimes Q_n^{-1}.\end{aligned}$$

Quantum integrability of the Volterra model is shown as follows. By quantizing the first Poisson algebra (4.13) as

$$\text{(4.15)} \qquad P_n Q_m = q^{\delta_{n,m}} Q_m P_n, \quad [P_n, P_m] = [Q_n, Q_m] = 0,$$

we get the FCR,

$$\text{(4.16)} \qquad \check{\mathbf{R}}(x/y;q)\mathbf{M}_n(x) \otimes \mathbf{M}_n(y) = \mathbf{M}_n(y) \otimes \mathbf{M}_n(x)\check{\mathbf{R}}(x/y;q),$$

with

$$\check{\mathbf{R}}(x;q) = \begin{pmatrix} x^2 - q^2 & & & \\ & x(1-q^2) & q(x^2-1) & \\ & q(x^2-1) & x(1-q^2) & \\ & & & x^2 - q^2 \end{pmatrix}.$$

It is noted that the quantization method is different from that was done in Section 2; these two quantizations depend on the bi-Hamiltonian structure.

We represent the operators P_n and Q_n as infinite dimensional matrices to give a statistical mechanical interpretation;

$$P_n = \sum_{\ell=-\infty}^{\infty} q^\ell |\ell\rangle_n \cdot {}_n\langle \ell|, \quad Q_n = \sum_{\ell=-\infty}^{\infty} |\ell+1\rangle_n \cdot {}_n\langle \ell|, \quad Q_n^{-1} = \sum_{\ell=-\infty}^{\infty} |\ell\rangle_n \cdot {}_n\langle \ell+1|,$$

Due to the FCR (4.16), the Lax matrix with above representation defines the Boltzmann weight of the integrable vertex model. Furthermore we take a *crystal* limit, *i.e.*, restrict the states of the auxiliary space to nonnegative integer, and take a limit $q \longrightarrow 0$;

$$\text{(4.17)} \quad P_n \to |0\rangle_n \cdot {}_n\langle 0|, \quad Q_n \to \sum_{\ell=0}^{\infty} |\ell+1\rangle_n \cdot {}_n\langle \ell|, \quad Q_n^{-1} \to \sum_{\ell=0}^{\infty} |\ell\rangle_n \cdot {}_n\langle \ell+1|.$$

We see that in this limit Q_n^{-1} is no longer an inverse of Q_n ($Q_n Q_n^{-1} \neq Q_n^{-1} Q_n$), but crucial is that the FCR (4.16)preserves in this crystal limit with a diagonal matrix $\check{\mathbf{R}}(x;q=0)$, and that we have an integrable vertex model. The nonzero Boltzmann weight for the Lax matrix (4.14) with a representation (4.17) is depicted as [**45**]

(4.18)

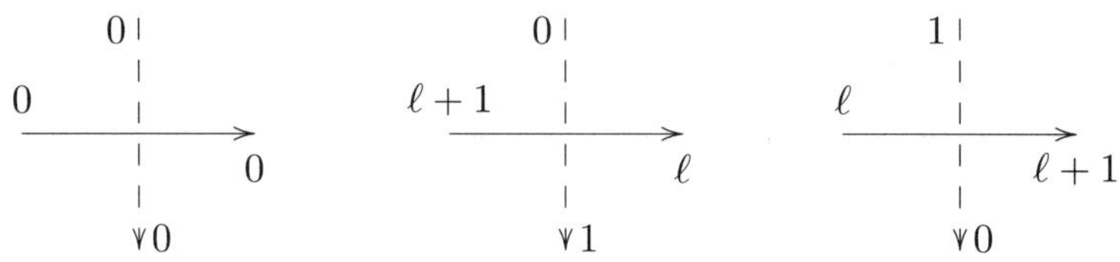

Here the two-dimensional auxiliary space, $|0\rangle\!\rangle$ and $|1\rangle\!\rangle$, is represented by a dashed line, and the Fock space of the operators P and Q is depicted by a solid line. Arrows classify the in-coming and out-going states. Remark that by the crystallization the Fock space is restricted to $\ell \geq 0$.

With the Boltzmann weight (4.18) we can determine a configuration of the integrable vertex model on the two-dimensional square lattice (Fig. 1), where $L \gg 1$ and we fix the boundary on the horizontal lines to be "0". When we fix the boundary state on top end, a unique configuration is allowed. One sees that the configuration on dashed lines coincides with a time evolution of the BBS in Section 4.2 with $M = 1$ (we only have ball-1), once we interpret states $|0\rangle\rangle$ and $|1\rangle\rangle$ as an empty and occupied box respectively. To conclude the time evolution operator of the BBS (with $M = 1$) is given by use of the Lax matrix (4.14) as

$$|\{0,1\}^{\otimes L}\rangle\rangle_{t=n} = \widehat{T}|\{0,1\}^{\otimes L}\rangle\rangle_{t=n-1}$$

where

$$\widehat{T} = \cdots \otimes \mathbf{M}_n(x) \otimes \mathbf{M}_n(x) \otimes \cdots . \tag{4.19}$$

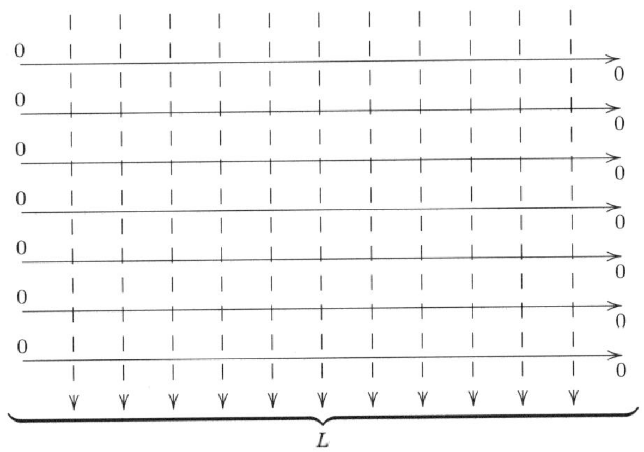

FIGURE 1. Vertex model is defined on the two-dimensional square lattice. The boundary of the horizontal edge on both ends are fixed to be "0". We suppose $L \gg 1$.

We can generalize this correspondence for a case of arbitrary M as follows. We define the first Hamiltonian structure of the Bogoyavlensky lattice by [**49**]

$$\{V_n, V_m\}_1 = 2V_nV_m \sum_{k=1}^{M} (\delta_{m,n+k} - \delta_{m,n-k}), \tag{4.20}$$

which with the Hamiltonians (3.14) gives Eq. (4.12). See that in terms of P_n and Q_n satisfying Eq. (4.13) we can represent the dynamical variable V_n by

$$V_n = (P_nP_{n+1}\cdots P_{n+M})^{-1}Q_n^{-1}Q_{n+M}.$$

As a factorization of the Lax matrix $\mathbf{L}_n^{\mathrm{W}}(x)$ in Eq. (3.10), the $(M+1)\times(M+1)$ Lax matrix for the Bogoyavlensky is then identified with [**49**]

$$\mathbf{M}_n^{\mathrm{B}}(x) = \frac{1}{\sqrt[M+1]{V_n}} \begin{pmatrix} x & 0 & \cdots & 0 & (-1)^M V_n \\ 1 & 0 & \cdots & \cdots & 0 \\ 0 & 1 & \ddots & \ddots & \vdots \\ \vdots & \vdots & \ddots & \ddots & 0 \\ 0 & \cdots & 0 & 1 & 0 \end{pmatrix}, \tag{4.21}$$

which can be gauge-transformed into

$$\mathbf{M}_n(x) = \begin{pmatrix} xP_n & Q_n & 0 & \cdots & 0 \\ 0 & 0 & 1 & \ddots & \vdots \\ \vdots & \ddots & \ddots & \ddots & 0 \\ 0 & \ddots & \ddots & \ddots & 1 \\ Q_n^{-1} & 0 & \cdots & 0 & 0 \end{pmatrix}. \tag{4.22}$$

By use of the q-commuting operators P_n and Q_n (4.15), we see that the FCR (4.16) is satisfied where $\check{\mathbf{R}}(x;q)$ is defined by

$$\begin{aligned} \check{\mathbf{R}}(x;q) = \sum_{j=1}^{M+1}\sum_{k=1}^{M} q(x^{M+1}-1)\mathbf{E}_{j,j+k}\otimes\mathbf{E}_{j+k,j} \\ + \sum_{k=1}^{M+1}(x^{M+1}-q^2)\mathbf{E}_{k,k}\otimes\mathbf{E}_{k,k} + \sum_{j=1}^{M+1}\sum_{k=1}^{M}(1-q^2)x^k\mathbf{E}_{j,j}\otimes\mathbf{E}_{j+k,j+k}. \end{aligned}$$

Here we impose a periodicity in indices of matrix $\mathbf{E}_{j,k}$. We can then directly prove the quantum integrability of the Bogoyavlensky lattice (4.1).

Before taking a crystal limit, we should note that the Lax matrix (4.22) is further gauge-transformed, and that the monodromy matrix $\mathbf{T}(x) = \prod_n^{\curvearrowleft} \mathbf{M}_n(x)$ can also be written in a form by binding M Lax matrices [**45**],

$$\mathbf{T}(x) = \prod_k^{\curvearrowleft}\Big(\prod_{i=1}^{\overset{\curvearrowleft}{M}} \mathbf{K}_{k,i}\Big) \tag{4.23}$$

where

$$\begin{aligned} \mathbf{K}_{k,i}(x) = x|0\rangle\!\rangle\langle\!\langle 0|\otimes P_{kM+i} + |0\rangle\!\rangle\langle\!\langle i|\otimes Q_{kM+i} + |i\rangle\!\rangle\langle\!\langle 0|\otimes Q_{kM+i}^{-1} \\ + \sum_{\substack{j=1\\ j\neq i}}^{M} |j\rangle\!\rangle\langle\!\langle j|\otimes \mathbb{1}. \end{aligned} \tag{4.24}$$

In this case, we can also take the crystal limit (4.17); the FCR is preserved with the matrix $\check{\mathbf{R}}(x;q=0)$. By substituting a representation (4.17) into Eq. (4.24), we obtain an integrable vertex model whose nonzero weight in $\mathbf{K}_{k,i}(x)$ is given by

(4.25)

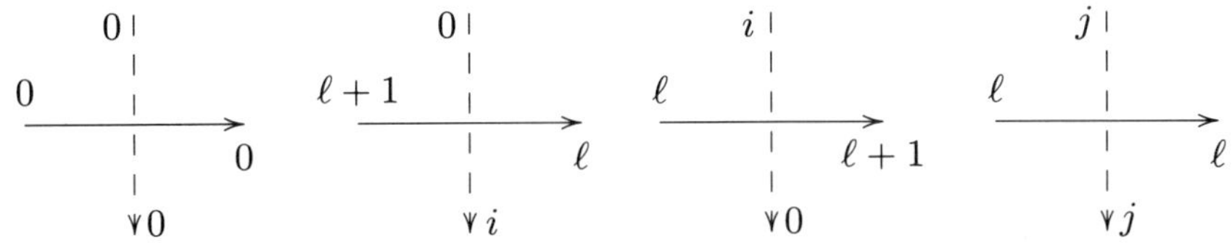

where $1 \le j \le M$, $j \ne i$, and $\ell \ge 0$. Here dashed lines represent an $M+1$ dimensional auxiliary space $|i\rangle\rangle$, while a solid line an infinite dimensional space $|\ell\rangle$ for operators P and Q. As in the case of the Volterra model, a configuration of this vertex model is deterministic once we fix the boundary condition. In addition, when we interpret states $|0\rangle\rangle$ and $|i\rangle\rangle$ (for $1 \le i \le M$) in auxiliary space as an empty box and a box occupied by ball-$(M+1-i)$ respectively, we find that this unique configuration exactly coincides with the evolution rule of the BBS. In this case the evolution of the BBS is written as

$$\left|\{0,1,\dots,M\}^{\otimes L}\right\rangle\!\!\rangle_{t=n} = \widehat{T}\left|\{0,1,\dots,M\}^{\otimes L}\right\rangle\!\!\rangle_{t=n-1}$$

where

$$\widehat{T} = \left(\prod_{i=1}^{\curvearrowleft M} \mathbf{K}_{n,i}(x)\right)^{\otimes L} \tag{4.26}$$

4.4. Crystal base. We have shown that the BBS, which was derived in Section 4.2 as the ultra-discrete limit of the Bogoyavlensky lattice, can also be constructed from a $q \to 0$ limit of the integrable vertex model defined based on the Lax matrix of the Bogoyavlensky lattice. The representation theory of a limit $q \to 0$ of the quantum group has been widely developed as the crystal base [**52**] (see Ref. [**53**] for a review), and in this section we intend to reformulate the BBS using the representation theory of the crystal base.

We set the crystal base $\mathbf{B}$ corresponding to the vector representation of $U_q'(\widehat{s\ell_{M+1}})$ as $\mathbf{B} = \left\{\boxed{i} \mid i \in \{1,2,\dots,M+1\}\right\}$. The action of the Kashiwara operator is summarized in the crystal graph;

(4.27)

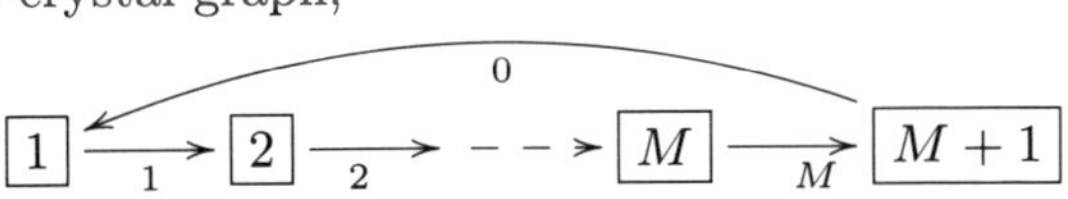

Here as usual, $b \xrightarrow{i} b'$ denotes $\tilde{f}_i b = b'$, and we assign an ordering by $1 < 2 < \cdots < M+1$.

We use the semi-standard tableaux (see, e.g., Ref. [**31**]) to represent the crystal B_ℓ corresponding to the ℓ-fold symmetric tensor representation of the quantum affine algebra $U_q'(\widehat{s\ell_{M+1}})$. Here the "semi-standard" tableaux is defined as a tableaux obtained from a Young diagram by filling the boxes with elements of $\mathbf{B}$ subject to the following two constraints;

(i) the entries in each row are increasing allowing the repetition of elements,
(ii) the entries in each column are increasing, not permitting the repetition of elements.

Then we have

$$B_\ell = \left\{\boxed{m_1 \cdots m_\ell} \mid m_i \in \{1,2,\dots,M+1\}, m_1 \le \cdots \le m_\ell\right\}$$

where we omitted for brevity the $\ell-1$ vertical lines separating the entries in Young diagram. We also denote

$$b=\boxed{m_1\cdots m_\ell}=(x_1,x_2,\dots,x_{M+1})\in B_\ell,\quad x_i=\#\{k\mid m_k=i\}. \tag{4.28}$$

For each element $b=(x_1,\dots,x_{M+1})\in B_\ell$, the Kashiwara operators act as

$$\tilde{e}_0 b=(x_1-1,x_2,\dots,x_{M+1}+1),\qquad \tilde{f}_0 b=(x_1+1,x_2,\dots,x_{M+1}-1),$$
$$\tilde{e}_i b=(x_1,\dots,x_i+1,x_{i+1}-1,\dots,x_{M+1}),\quad \tilde{f}_i b=(x_1,\dots,x_i-1,x_{i+1}+1,\dots,x_{M+1}).$$

We note that the right hand sides should be read as zero if they are not in B_ℓ.

We can define the tensor product of the crystal $B\otimes B'=\{b_1\otimes b_2\mid b_1\in B,b_2\in B'\}$. The Kashiwara operators act on these spaces as

$$\tilde{e}_i(b_1\otimes b_2)=\begin{cases}\tilde{e}_i(b_1)\otimes b_2, & \text{if } \varphi_i(b_1)\geq\varepsilon_i(b_2),\\ b_1\otimes\tilde{e}_i(b_2), & \text{if } \varphi_i(b_1)<\varepsilon_i(b_2),\end{cases}$$

$$\tilde{f}_i(b_1\otimes b_2)=\begin{cases}\tilde{f}_i(b_1)\otimes b_2, & \text{if } \varphi_i(b_1)>\varepsilon_i(b_2),\\ b_1\otimes\tilde{f}_i(b_2), & \text{if } \varphi_i(b_1)\leq\varepsilon_i(b_2).\end{cases}$$

Here we have defined $\varepsilon_i(b)$ and $\varphi_i(b)$ as

$$\varepsilon_i(b)=\max\{n\in\mathbb{Z}_{\geq 0}\mid \tilde{e}_i{}^n b\neq 0\},\quad \varphi_i(b)=\max\{n\in\mathbb{Z}_{\geq 0}\mid \tilde{f}_i{}^n b\neq 0\}.$$

For the tensor product of the crystal we introduce the energy function H_0 for the crystal isomorphism. For a case $b\otimes b'\in B\otimes B'$ being mapped to $\tilde{b}'\otimes\tilde{b}\in B'\otimes B$, the energy function satisfies the following property; for any j such that $\tilde{e}_j(b\otimes b')\neq 0$,

(4.29a)

$$H_0\big(\tilde{e}_j(b\otimes b')\big)=\begin{cases}H_0(b\otimes b')+1, & \text{if } j=0,\ \varphi_0(b)\geq\varepsilon_0(b'),\ \varphi_0(\tilde{b}')\geq\varepsilon_0(\tilde{b}),\\ H_0(b\otimes b')-1, & \text{if } j=0,\ \varphi_0(b)<\varepsilon_0(b'),\ \varphi_0(\tilde{b}')<\varepsilon_0(\tilde{b}),\\ H_0(b\otimes b'), & \text{otherwise.}\end{cases}$$

We further use the generalized energy function H_i (for $i=1,\dots,M$); for any j such that $\tilde{f}_j(b\otimes b')\neq 0$,

$$H_i\big(\tilde{f}_j(b\otimes b')\big)=\begin{cases}H_i(b\otimes b')+1, & \text{if } j=i,\ \varphi_i(b)\leq\varepsilon_i(b'),\ \varphi_i(\tilde{b}')\leq\varepsilon_i(\tilde{b}),\\ H_i(b\otimes b')-1, & \text{if } j=i,\ \varphi_i(b)>\varepsilon_i(b'),\ \varphi_i(\tilde{b}')>\varepsilon_i(\tilde{b}),\\ H_i(b\otimes b'), & \text{otherwise.}\end{cases} \tag{4.29b}$$

Hereafter we use a notation $\boldsymbol{H}(b\otimes b')=\big(H_0(b\otimes b'),\dots,H_M(b\otimes b')\big)$, and for a case of $b\in B_k$ and $b'\in B_\ell$ we normalize H_i as

$$H_i\Big(\underbrace{\boxed{1\cdots 1}}_{k}\otimes\underbrace{\boxed{1\cdots 1}}_{\ell}\Big)=\min(k,\ell).$$

Two crystals, $B\otimes B'$ and $B'\otimes B$, are known to be isomorphic, and there is a unique map $R\colon B\otimes B'\xrightarrow{\sim}B'\otimes B$ which commutes with the Kashiwara operators. This map is called the combinatorial R matrix, and is used to compute a local height probability of the integrable vertex models by the corner transfer matrix method. In a simple case $B_\ell\otimes\mathbf{B}\xrightarrow{\sim}\mathbf{B}\otimes B_\ell$, the action of the combinatorial R matrix is summarized as follows;

- If $\boxed{b\mid a_1}$ is semi-standard, then

$$\boxed{a_1\mid a_2\mid\cdots\mid a_\ell}\otimes\boxed{b}\overset{\sim}{\mapsto}\boxed{a_\ell}\otimes\boxed{b\mid a_1\mid\cdots\mid a_{\ell-1}},$$

- If $\begin{array}{|c|}\hline a_1 \\ \hline b \\ \hline\end{array}$ is semi-standard, then

$$\begin{array}{|c|c|c|c|c|}\hline a_1 & \cdots & a_j & \cdots & a_\ell \\ \hline\end{array} \otimes \boxed{b} \overset{\sim}{\mapsto} \boxed{a_j} \otimes \begin{array}{|c|c|c|c|c|}\hline a_1 & \cdots & b & \cdots & a_\ell \\ \hline\end{array}$$

  where j is the largest integer such that $\begin{array}{|c|}\hline a_j \\ \hline b \\ \hline\end{array}$ is semi-standard.

Using a fact that the admissible reading

$$\begin{array}{ccc} B_\ell & \longrightarrow & \mathbf{B}^{\otimes \ell} \\ \boxed{a_1 \cdots a_\ell} & \mapsto & \boxed{a_\ell} \otimes \cdots \otimes \boxed{a_1} \end{array}$$

induces a crystal structure on B_ℓ, we get a rule to obtain the crystal isomorphism, $B_k \otimes B_\ell \overset{\sim}{\longrightarrow} B_\ell \otimes B_k$, $b_1 \otimes b_2 \overset{\sim}{\mapsto} b'_2 \otimes b'_1$ **[62]**. We suppose$b_1 = (x_1, \ldots, x_{M+1}) \in B_k$ and $b_2 = (y_1, \ldots, y_{M+1}) \in B_\ell$, and represent $b_1 \otimes b_2$ by the two column diagrams; each column has $M+1$ rows, and we put x_i (resp. y_i) dots $\bullet$ in the i-th row of the left (resp. right) column.

$b_1 \otimes b_2 =$ 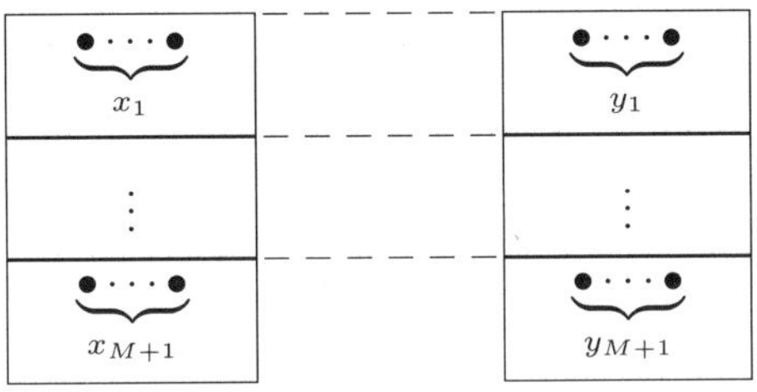

(i) Assume $k \geq \ell$ (resp. $k \leq \ell$). Pick the dot $\bullet_a$ which locates in the highest position in the right (resp. left) column. Connect it with the dot in the left (resp. right)column which has the lowest (resp. highest) position among all dots whose positions are higher (resp. lower) than that of $\bullet_a$. When there is no such dot in the left (resp. right), return to the bottom (resp. top).

(ii) Repeat the previous process for the remaining unconnected dots in the right (resp. left)column.

(iii) The crystal isomorphism, $b_1 \otimes b_2 \overset{\sim}{\mapsto} b'_2 \otimes b'_1$, is obtained by sliding the remaining unpaired dots in the left (resp. right) column to the right (resp. left) one.

(iv) The energy function $H_i(b_1 \otimes b_2)$ is given by a number of lines that cross the $(i+1)$th dotted line.

An example below shows an isomorphism, $\boxed{133} \otimes \boxed{23} \overset{\sim}{\mapsto} \boxed{13} \otimes \boxed{233}$, and $\boldsymbol{H} = (1, 2, 1)$.

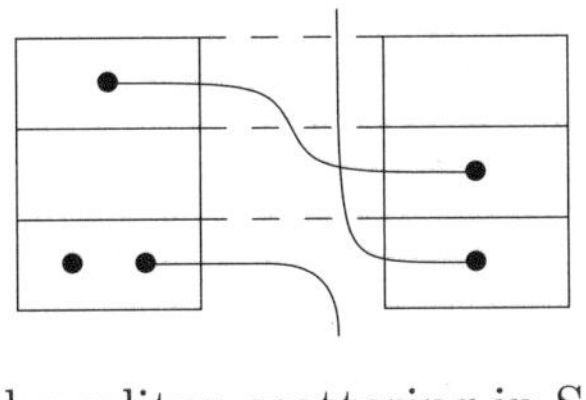

Looking at examples of the soliton scattering in Section 4.2, or especially comparing the first example of soliton scattering with above example of the crystal isomorphism, we notice that the rule of the crystal isomorphism appears **[30, 34, 35, 73]**; by identifying a soliton $[c_1 \ldots c_p]$ with $\boxed{c_p \cdots c_1}$, every soliton scattering is decomposed into a scattering of 2-solitons, and the scattering of 2-solitons with length k and ℓ (we assume $k > \ell$) exactly coincides with the crystal isomorphism $b_1 \otimes b_2 \overset{\sim}{\mapsto} b_2' \otimes b_1'$, where $b_1, b_1' \in B_k$ and $b_2, b_2' \in B_\ell$.

Motivating from this correspondence and result that the time evolution operator $\widehat{T}$ (4.26) of the BBS was identified with a tensor product of the $q \to 0$ limit of the Lax matrix, we redefine the evolution operator T by

$$T = \lim_{\kappa \to \infty} T_\kappa, \tag{4.30}$$

and consider a evolution equation of the crystal $b_n^t \in \mathbf{B}$. Here T_κ (for $\kappa \in \mathbb{Z}$) is defined as

$$T_\kappa \colon b_1^t \otimes b_2^t \otimes \cdots \otimes b_L^t \mapsto b_1^{t+1} \otimes b_2^{t+1} \otimes \cdots \otimes b_L^{t+1} \tag{4.31}$$

where we have the crystal isomorphism,

$$\underbrace{\boxed{1 \cdots 1}}_{\kappa} \otimes b_1^t \otimes \cdots \otimes b_L^t \overset{\sim}{\mapsto} b_1^{t+1} \otimes \cdots \otimes b_L^{t+1} \otimes \underbrace{\boxed{1 \cdots 1}}_{\kappa} \tag{4.32}$$

We assume $L \gg 1$, and $b_n^t = \boxed{1} = b_n^{t+1}$ for sufficiently large n. When we assign $\mathbf{B}$ and B_ℓ for the vertical dotted lines and the horizontal solid lines respectively, the combinatorial R matrix, $R \colon v_n^t \otimes b_n^t \to b_n^{t+1} \otimes v_{n+1}^t$, is depicted as

(4.33)

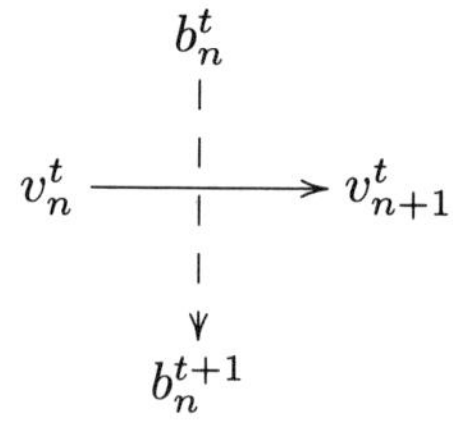

and the evolution operator T_κ (4.31) is schematically drawn as

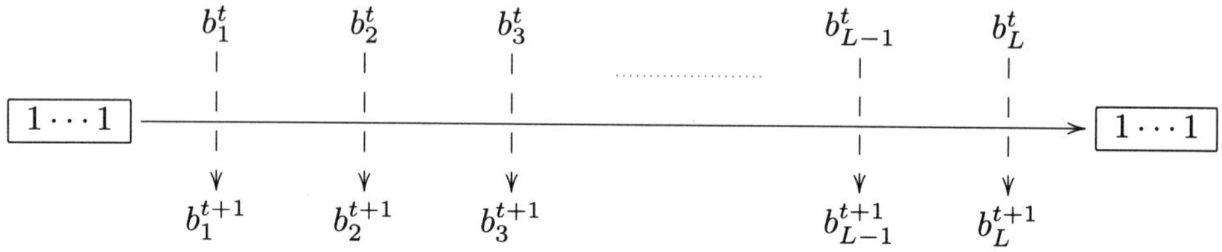

By definition of the combinatorial R matrix, the evolution operators commute with each other for arbitrary positive integers κ;

$$[T_\kappa, T_\eta] = 0. \tag{4.34}$$

When we set (recall the definition (4.28))

$$b_n^t = (u_{n,0}^t, u_{n,1}^t, \dots, u_{n,M}^t), \quad v_n^t = (v_{n,0}^t, v_{n,1}^t, \dots, v_{n,M}^t), \tag{4.35}$$

(note that $u_{n,j}^t$ and $v_{n,j}^t$ denote the number of $j+1$ in semi-standard tableaux b_n^t and v_n^t respectively), we find from a rule of the combinatorial R matrix that the evolution equation by T_κ (4.31) can be written as

$$\begin{aligned} u_{n,j}^{t+1} - v_{n,j}^t = &\max[X_1 - 1, X_2 - 1, \dots, X_{j-1} - 1, X_j - \kappa, \dots, X_M - \kappa, 0] \\ &- \max[X_1 - 1, X_2 - 1, \dots, X_j - 1, X_{j+1} - \kappa, \dots, X_M - \kappa, 0], \end{aligned} \tag{4.36a}$$

$$u_{n,j}^t + v_{n,j}^t = u_{n,j}^{t+1} + v_{n+1,j}^t \tag{4.36b}$$

where $X_\ell = \sum_{i=\ell}^{M} u_{n,i}^t + \sum_{i=1}^{\ell} v_{n,i}^t$. We see that the evolution equation (4.36) exactly coincides with Eq. (4.10) in a limit $\kappa \to \infty$, and that the evolution operator T (4.30) is identified with $\widehat{T}$ (4.26). To conclude the crystal base $\boxed{i} \in \mathbf{B}$ is read in a language of the BBS as

$$\boxed{1}\text{: empty box,}\qquad \boxed{i}\text{: box occupied by ball-}(i-1)(\text{for } i>1).$$

The evolution of the BBS is now identified with the $U_q(\widehat{s\ell_{M+1}})$ crystal isomorphism. The proof that the every soliton scattering is factorized into 2-soliton scattering and that the soliton scattering coincides with the crystal isomorphism of $U_q'(s\ell_M)$ is as follows [**30**,**34**]. The former is based on the commutativity (4.34). As the velocity of soliton $[c_1 \dots c_p]$ propagating rightward is suppressed to be $\min(p,\kappa)$ when we apply the evolution operator T_κ, we can change the order of the soliton scattering due to Eq. (4.34). To prove the latter we recall that the algebra $U_q(s\ell_M)$ is a subalgebra of $U_q'(\widehat{s\ell_{M+1}})$. Thus we need to prove the correspondence between the crystal isomorphism and the 2-soliton scattering only for the highest weight, which can be easily checked.

The m-soliton solution can be essentially constructed from Eq. (4.4) by taking a ultra-discrete limit (4.7). We list some explicit forms below.

1. [1-soliton]. We consider a propagation of the 1-soliton $[c_1 \dots c_p]$, and we set $\boxed{c_p \dots c_1} = (x_1, \dots, x_M) \in B_p$. The τ-function is given by

$$Y_{n,j}^t = \max\bigl(0, C + n - pt - \sum_{i=j}^{M} x_i\bigr) \tag{4.37}$$

where C is arbitrary.

2. [2-soliton]. We consider the soliton scattering $b_1 \otimes b_2 \overset{\sim}{\mapsto} b_2' \otimes b_1'$ with b_r, $b_r' \in B_{p(r)}$, and $p(1) > p(2)$. The τ-function is given by

$$Y_{n,j}^t = \max\bigl(0, \xi_{n,j}^t(1), \xi_{n,j}^t(2), \xi_{n,j}^t(1) + \xi_{n,j}^t(2) - S_j\bigr) \tag{4.38}$$

where

$$\xi_{n,j}^t(r) = C(r) + n - p(r)t - \sum_{i=j}^{M} x_i(r),$$

$$S_j = p(2) + H_{j-1}(b_1 \otimes b_2).$$

We have used $C(r)$ as arbitrary parameters, and have denoted the soliton states as $b_1 = \bigl(x_1(1), \dots, x_M(1)\bigr)$, $b_2' = \bigl(x_1(2), \dots, x_M(2)\bigr)$. We should stress that $x_i(2)$ depends on the out-going state. Interesting is that the generalized energy functions appear as phase shift in the τ-function. See Ref. [**30**] for a discussion on relationship between the energy function H_0 and the phase shift.

5. Concluding Remarks

We have studied the integrable structure of the discrete quantum soliton equations. By use of a discrete analogue of the free fields, we have introduced the discrete affine $\widehat{s\ell_N}$ Toda field theory and the discrete N-reduced KP equation which is equivalent with the Bogoyavlensky lattice. We have further explicitly constructed the $\mathcal{Q}$ operator as a generating function of the integrals of motion in terms of the quantum dilogarithm function. We have conjectured that the $\mathcal{Q}$ operator is related with the universal R matrix of the quantum group $U_q(\widehat{s\ell_N})$ with the q-oscillator representation. It has been shown up to $N = 3$ that the $\mathcal{Q}$ operator solves the Nth order difference equation called the Baxter equation, whose coefficients are given by the transfer matrices of the discrete KP hierarchy.

We have only studied $\widehat{s\ell_N}$ case, but it is well known that the soliton equations such as the Toda field theories and the KdV type equations are associated with the affine Lie algebras. Interesting is to study their integrable discretization, and to construct the Baxter $\mathcal{Q}$ operator explicitly. Some attempts have been made in this direction [**50**].

We have also discussed on another interesting link between the Bogoyavlensky lattice and the quantum group. This is realized in the crystal limit $q \to 0$, and we can formulate the integrable cellular automaton with the Lax matrix of the soliton equations. By using the representation theory of the crystal base, we can generalize the BBS. For instance, we have an inhomogeneous $s\ell_M$-type BBS in which capacities of boxes and carriers are arbitrary [**34**], a BBS associated with other Lie algebras [**35**], a BBS for supersymmetric Lie algebra [**43**]. The inhomogeneous model can be identified with the ultra-discretization of the nonautonomous discrete KP equation [**34**], but to author's knowledge relationship between other BBSs and the integrable solitons equations remains unclear.

Appendix A. Quantum Dilogarithm Function

We briefly review properties of the quantum dilogarithm function defined by

$$\Phi_\gamma(\varphi) = \exp\left(\int_{\mathbb{R}+\mathrm{i}0} \frac{\mathrm{e}^{-\mathrm{i}\varphi x}}{4\,\mathrm{sh}(\gamma x)\,\mathrm{sh}(\pi x)} \frac{\mathrm{d}x}{x} \right). \tag{A.1}$$

This integral was introduced by Faddeev in Ref. [**20**] (see also Refs. [**65, 77**]), and it corresponds to a noncompact analogue of the q-exponential function.

We list below some properties of the Faddeev integral (A.1).

- Duality

$$\Phi_{\pi^2/\gamma}(\varphi) = \Phi_\gamma\left(\frac{\gamma}{\pi}\varphi\right). \tag{A.2}$$

- Zero points,

$$\text{zeros of } \big(\Phi_\gamma(\varphi)\big)^{\pm 1} = \big\{ \mp\mathrm{i}\big((2m+1)\gamma + (2n+1)\pi\big) \mid m, n \in \mathbb{Z}_{\geq 0} \big\}. \tag{A.3}$$

- Inversion relation:

$$\begin{aligned} \Theta_\gamma(\varphi) &= \Phi_\gamma(\varphi) \cdot \Phi_\gamma(-\varphi) \\ &= \exp\left(-\frac{1}{2\mathrm{i}\gamma} \left(\frac{\varphi^2}{2} + \frac{\pi^2 + \gamma^2}{6} \right) \right). \end{aligned} \tag{A.4}$$

- Difference equations:

$$\frac{\Phi_\gamma(\varphi + \mathrm{i}\gamma)}{\Phi_\gamma(\varphi - \mathrm{i}\gamma)} = \frac{1}{1 + \mathrm{e}^{\varphi}}. \tag{A.5}$$

The duality (A.2) gives another type of difference equation,

$$\frac{\Phi_\gamma(\varphi + \mathrm{i}\pi)}{\Phi_\gamma(\varphi - \mathrm{i}\pi)} = \frac{1}{1 + \mathrm{e}^{\frac{\pi}{\gamma}\varphi}}. \tag{A.6}$$

- Pentagon relation [**13, 22**]

(A.7) $$\Phi_\gamma(\hat{p})\Phi_\gamma(\hat{q}) = \Phi_\gamma(\hat{q})\Phi_\gamma(\hat{p}+\hat{q})\Phi_\gamma(\hat{p})$$

where $\hat{p}$ and $\hat{q}$ are the Heisenberg operators satisfying a commutation relation,

(A.8) $$[\hat{p}, \hat{q}] = -2\mathrm{i}\gamma.$$

References

1. A. Antonov, *Universal R-matrix and quantum Volterra model*, Theoret. Math. Phys. **113** (1997), 1520–1529.
2. A. Antonov and B. Feigin, *Quantum group representations and Baxter equation*, Phys. Lett. B **392** (1997), 115–122.
3. R. J. Baxter, *Exactly solved models in statistical mechanics*, Academic Press, London, 1982.
4. ———, *Eight-vertex model in lattice statistics and one-dimensional anisotropic spin chain* I. *Some fundamental eigenvectors*, Ann. Physics **76** (1973), 1–24.
5. V. V. Bazhanov, A. N. Hibberd, and S. M. Khoroshkin, *Integrable structure of $\mathcal{W}_3$ conformal field theory, quantum Boussinesq theory and boundary affine Toda theory*; `hep-th/0105177`(2001).
6. V. V. Bazhanov, S. L. Lukyanov, and A. B. Zamolodchikov, *Integrable structure of conformal field theory, quantum* KdV *theory and thermodynamic Bethe ansatz*, Comm. Math. Phys. **177** (1996), 381–398.
7. A. A. Belavin, A. M. Polyakov, and A. B. Zamolodchikov, *Infinite conformal symmetry in two-dimensional quantum field theory*, Nuclear Phys. B **241** (1984), 333–380.
8. A. A. Belov and K. D. Chaltikian, *Lattice analogue of $\mathcal{W}$-algebras and classical integrable equations*, Phys. Lett. B **309** (1993), 268–274.
9. A. Bilal and J.-L. Gervais, *Systematic approach to conformal systems with extended Virasoro symmetries*, Phys. Lett. B **206** (1988), 412–420.
10. O. I. Bogoyavlensky, *Integrable discretizations of the* KdV *equation*, Phys. Lett. A **134** (1988), 34–38.
11. L. Bonora, L. P. Colatto, and C. P. Constantinidis, *Toda lattice field theories, discrete $\mathcal{W}$ algebras, Toda lattice hierarchies and quantum groups*, Phys. Lett. B **387** (1996), 759–768.
12. P. Bouwknegt, J. McCarthy, and K. Pilch, *Quantum group structure in the Fock space resolutions of $\widehat{sl}(n)$ representations*, Comm. Math. Phys. **131** (1990), 125–155.
13. L. Chekhov and V. V. Fock, *Quantum Teichmüller space*, Theoret. Math. Phys. **120** (1999), 1245–1259.
14. E. Date, M. Kashiwara, M. Jimbo, and T. Miwa, *Transformation groups for soliton equations*, Nonlinear Integrable Systems—Classical Theory and Quantum Theory (Kyoto, 1981), World Sci. Publishing, Singapore, 1983, pp. 39–119.
15. S. Derkachov, *Baxter's Q-operator for the homogeneous* XXX *spin chain*, J. Phys. A **32** (1999), 5299–5316.
16. L. A. Dickey, *Soliton equations and Hamiltonian systems*, Advanced Series in Mathematical Physics, vol. 12, World Scientific Publ. Co., Inc., River Edge, NJ, 1991.
17. V. G. Drinfeld and V. V. Sokolov, *Lie algebras and equations of Korteweg–de Vries type*, Current problems in mathematics, vol. 24, Itogi Nauki i Tekhniki, Akad. Nauk SSSR, Vsesoyuz. Inst. Nauchn. i Tekhn. Inform., Moscow, 1984, pp. 81–180 (Russian).
18. T. Eguchi and S.-K. Yang, *Deformation of conformal field theories and soliton equations*, Phys. Lett. B **224** (1989), 373–378.
19. B. Enriquez and B. L. Feigin, *Integrals of motion of the classical lattice sine-Gordon system*, Theoret. Math. Phys. **103** (1995), 738–756.
20. L. D. Faddeev, *Discrete Heisenberg–Weyl group and modular group*, Lett. Math. Phys. **34** (1995), 249–254.
21. ———, *Modular double of quantum group*, Conférence Moshé Flato 1999 —Vol. I. Quantization, Deformations, and Symmetries (Dijon) (G. Dito and D. Sternheimer, eds.) Math. Phys. Stud., vol. 21, Kluwer Acad. Publ., Dordrecht, 2000, pp. 149–156.
22. L. D. Faddeev, R. M. Kashaev, and A. Y. Volkov, *Strongly coupled quantum discrete Liouville theory* I. *Algebraic approach and duality*, Comm. Math. Phys. **219** (2001), 199–219.

23. V. A. Fateev and S. L. Lukyanov, *The models of two-dimensional conformal quantum field theory with $\mathbb{Z}_n$ symmetry*, Internat. J. Modern Phys. A **3** (1988), 507–520.
24. ———, *Poisson–Lie groups and classical $\mathcal{W}$-algebras*, Internat. J. Modern Phys. A **7** (1992), 853–876.
25. L. D. Faddeev and L. A. Takhtajan, *Liouville model on the lattice*, Field Theory, Quantum Gravity and Strings (H. J. de Vega and N. Sanchez, eds.), Lecture Notes in Phys. **246** (1986), 166–179.
26. L. D. Faddeev and A. Y. Volkov, *Abelian current algebra and the Virasoro algebra on the lattice*, Phys. Lett. B **315** (1993), 311–318.
27. B. Feigin and E. Frenkel, *Free field resolutions in affine Toda field theories*, Phys. Lett. B **276** (1992), 79–86.
28. ———, *Quantization of the Drinfel′d-Sokolov reduction*, Phys. Lett. B **246** (1990), 75–81.
29. ———, *Integrals of motion and quantum groups*, Lecture Notes in Math. **1620** (1995), 349–418.
30. K. Fukuda, M. Okado, and Y. Yamada, *Energy functions in box ball systems*, Internat. J. Modern Phys. A **15** (2000), 1379–1392.
31. W. Fulton, *Young tableaux*, London Mathematical Society Student Texts, no. 35, Cambridge Univ. Press, Cambridge, 1997.
32. I. M. Gelfand and L. A. Dikii, *A family of Hamiltonian structures related to nonlinear integrable differential equations*, (Russian) Akad. Nauk SSSR Inst. Prikl. Mat., Preprint no. 136. 1978.
33. N. S. Goel, S. C. Maitra, and E. W. Montroll, *On the Volterra and other nonlinear models of interacting populations*, Rev. Modern Phys. **43** (1971), 231–276.
34. G. Hatayama, K. Hikami, R. Inoue, A. Kuniba, T. Takagi, and T. Tokihiro, *The $A_M^{(1)}$ automata related to crystals of symmetric tensors*, J. Math. Phys. **42** (2001), 274–308.
35. G. Hatayama, A. Kuniba, and T. Takagi, *Soliton cellular automata associated with crystal bases*, Nuclear Phys. B **577** (2000), 619–645.
36. K. Hikami, *Integrable structure of the classical lattice $\mathcal{W}_N$ algebra*, Chaos, Solitons & Fractals **9** (1998), 1773–1782.
37. ———, *Statistical mechanical interpretation of the inverse scattering method. Level dynamics for the exclusion statistics*, Phys. Rev. Lett. **80** (1998), 4374–4377.
38. ———, *On the fundamental $\mathcal{L}$-operator for the quantum lattice $\mathcal{W}$-algebra*, Chaos, Solitons & Fractals **9** (1998), 853–867.
39. ———, *The $\mathbb{Z}_N$ symmetric quantum lattice field theory—The quantum group symmetry, the Yang–Baxter equation, and the integrals of motion*, J. Phys. Soc. Japan **68** (1999), 55–60.
40. ———, *The Baxter equation for quantum discrete Boussinesq equation*, Nuclear Phys. B **604** (2001), 580–602.
41. K. Hikami and R. Inoue, *The quantum Volterra model and the lattice sine–Gordon system.—Construction of the Baxter Q-operator and the integrals of motion*, J. Phys. Soc. Japan **68** (1999), 376–381.
42. ———, *Classical lattice $\mathcal{W}$ algebras and integrable systems*, J. Phys. A **30** (1997), 6911–6924.
43. ———, *Supersymmetric extension of the integrable box–ball system*, J. Phys. A **33** (2000), 4081–4094.
44. ———, *The Hamiltonian structure of the Bogoyavlensky lattice*, J. Phys. Soc. Japan **68** (1999), 776–783.
45. K. Hikami, R. Inoue, and Y. Komori, *Crystallization of the Bogoyavlensky lattice*, J. Phys. Soc. Japan **68** (1999), 2234–2240.
46. K. Hikami, K. Sogo, and R. Inoue, *Hidden symmetry of the Bogoyavlensky lattice.—The lattice $\mathcal{W}$ algebras and the vertex operators*, J. Phys. Soc. Japan **66** (1997), 3756–3763.
47. R. Hirota, *Direct method in soliton theory*, Solitons (R. K. Bullough and P. J. Caudrey, eds.), Springer, Berlin, 1980, pp. 157–176.
48. R. Hirota and S. Tsujimoto, *Conserved quantities of discrete Lotka–Volterra equation* (Japanese), State of the art and perspectives in studies on nonlinear integrable systems (Japanese) (Kyoto, 1993), Sûrikaisekikenkyûsho Kôkyûroku no. 868 (1994), 31–38.
49. R. Inoue and K. Hikami, *The hungry Volterra model.—The classical and quantum integrable structures*, J. Phys. Soc. Japan **67** (1998), 87–92.
50. ———, *The lattice Toda field theory for simple Lie algebras—Hamiltonian structure and τ-function*, Nuclear Phys. B **581** (2000), 761–775.

51. Y. Itoh, *Integrals of a Lotka–Volterra system of odd number of variables*, Progr. Theoret. Phys. **78** (1987), 507–510.
52. M. Kashiwara, *Crystalizing the q-analogue of universal enveloping algebras*, Comm. Math. Phys. **133** (1990), 249–260.
53. ______, *On crystal bases*, Representations of Groups (B. N. Allison and G. H. Cliff, eds.), CMS Conference Proceedings, vol. 16, Amer. Math. Soc., Providence, 1995, pp. 155–197.
54. S. Kharchev and D. Lebedev, *Integral representations for the eigenfunctions of quantum open and periodic Toda chains from* QISM *formalism*, J. Phys. A **34** (2001), 2247–2258.
55. S. M. Khoroshkin, A. A. Stolin, and V. N. Tolstoy, *Generalized Gauss decomposition of trigonometric $\mathcal{R}$-matrices*, Modern Phys. Lett. A **10** (1995), 1375–1392.
56. P. P. Kulish and N. Y. Reshetikhin, GL_3*-invariant solutions of the Yang–Baxter equation and associated quantum systems*, J. Soviet. Math. **34**, 1948–1971 (1986).
57. P. P. Kulish, N. Y. Reshetikhin, and E. K. Sklyanin, *Yang–Baxter equation and representation theory* I, Lett. Math. Phys. **5** (1981), 393–403.
58. B. A. Kupershmidt and P. Mathieu, *Quantum Korteweg–de Vries like equations and perturbed conformal field theories*, Phys. Lett. B **227** (1989), 245–250.
59. V. B. Kuznetsov, M. Salerno, and E. K. Sklyanin, *Quantum Bäcklund transformation for the integrable* DST *model*, J. Phys. A **33** (2000), 171–189.
60. V. B. Kuznetsov and E. K. Sklyanin, *On Bäcklund transformations for many-body systems*, J. Phys. A **31** (1998), 2241–2251.
61. F. Magri, *A simple model of the integrable Hamiltonian equation*, J. Math. Phys. **19** (1978), 1156–1162.
62. A. Nakayashiki and Y. Yamada, *Kostka polynomials and energy functions in solvable lattice models*, Selecta Math. (N.S.) **3** (1997), 547–599.
63. V. Pasquier and M. Gaudin, *The periodic Toda chain and a matrix generalization of the Bessel function recursion relation*, J. Phys. A **25** (1992), 5243–5252.
64. G. P. Pronko, *On the Baxter's Q-operator for the* XXX *spin chain*, Comm. Math. Phys. **212** (2000), 687–712.
65. S. N. M. Ruijsenaars, *First order analytic difference equations and integrable quantum systems*, J. Math. Phys. **38** (1997), 1069–1146.
66. R. Sasaki and I. Yamanaka, *Virasoro algebra, vertex operators, quantum sine–Gordon and solvable quantum field theories, Conformal Field Theory and Solvable Lattice Models*, Adv. Stud. Pure Math. vol. 16, Academic Press, Boston, MA, 1988, pp. 271–296.
67. E. K. Sklyanin, *Bäcklund transformations and Baxter's Q-operator*, Integrable Systems: From Classical to Quantum (J. Harnad, G. Sabidussi, and P. Winternitz, eds.), CRM Proc. Lecture Notes, vol. 26, AMS, Providence, 2000, pp. 227–250.
68. ______, *Separation of variables. new trends*, Progr. Theoret. Phys. Suppl. **118** (1995), 35–60.
69. ______, *Separation of variables in the quantum integrable models related to the Yangian* $\mathcal{Y}(\mathrm{sl}(3))$, Zap. Nauchn. Sem. S.-Peterburg. Otdel. Math. Inst. Steklov. (POMI) **205** (1993), 166–178.
70. F. A. Smirnov, *Dual Baxter equations and quantization of the affine Jacobian*, J. Phys. A **33** (2000), 3385–3405.
71. D. Takahashi, *One some soliton systems defined by using boxes and balls*, Proc. International Symposium on Nonlinear Theory and Its Applications (NOLTA '93), NOLTA , Tokyo, 1993, pp. 555–558.
72. D. Takahashi and J. Satsuma, *A soliton cellular automaton*, J. Phys. Soc. Japan **59** (1990), 3514–3519.
73. T. Tokihiro, A. Nagai, and J. Satsuma, *Proof of solitonical nature of box and ball systems by means of inverse ultra-discretization*, Inverse Problems **15** (2000), 1639–1662.
74. T. Tokihiro, D. Takahashi, J. Matsukidaira, and J. Satsuma, *From soliton equations to integrable cellular automata through a limiting procedure*, Phys. Rev. Lett. **76** (1996), 3247–3250.
75. A. Y. Volkov, *Quantum lattice* KdV *equation*, Lett. Math. Phys. **39** (1997), 313–329.
76. ______, *Quantum Volterra model*, Phys. Lett. A **167** (1992), 345–355.
77. S. L. Woronowicz, *Quantum exponential function*, Rev. Math. Phys. **12** (2000), 873–920.
78. V. E. Zakharov and L. D. Faddeev, *Korteweg–de Vries equation: A completely integrable Hamiltonian system*, Funct. Anal. Appl. **5** (1971), 280–287.
79. A. B. Zamolodchikov, *Infinite additional symmetries in two-dimensional conformal quantum field theory*, Theoret. Math. Phys. **65** (1985), 1205–1213.

80. ———, *Integrable field theory from conformal field theory*, Integrable Systems in Quantum Field Theory and Statistical Mechanics (M. Jimbo, T. Miwa, and A. Tsuchiya, eds.), Adv. Stud. Pure Math., vol. 19, Academic Press, Boston, MA, 1989, pp. 641–674.

Department of Physics, Graduate School of Science, University of Tokyo, Hongo 7–3–1, Bunkyo, Tokyo 113–0033, Japan.

E-mail address: hikami@phys.s.u-tokyo.ac.jp

Centre de Recherches Mathématiques
CRM Proceedings and Lecture Notes
Volume **32**, 2002

Dual Algebras of Differential Operators

Emil I. Horozov

ABSTRACT. This survey paper is intended to introduce the reader to the subject of bispectral operators. One point of point of view is that the subject is closely connected to certain isomorphisms of noncommutative algebras. This idea helps to drastically simplify both construction and classification of such operators. At the same time we hope that the presentation sheds some light on the nature of the problem, connecting it to by now classical studies of Dixmier on the Weyl algebra.

1. Introduction

The present paper is an attempt to give an account on the present state of the so called **"bispectral problem"**. No doubt it is not realistic to consider all its aspects. For this reason I am forced to choose only the most basic notions and facts, skipping some extremely interesting and (for first acquaintance) unexpected connections and interpretations of the main objects. Before describing the organization of the paper I will give the main definition in its initial form from the pioneering paper [**DG**]. An ordinary differential operator $L(x, \partial_x)$ is called *bispectral* if it has an eigenfunction $\psi(x, z)$, depending on the spectral parameter z, which is at the same time an eigenfunction of another differential operator $\Lambda(z, \partial_z)$, now in the spectral parameter z. In other words one looks for operators L, Λ and a function $\psi(x, z)$ satisfying equations of the form:

$$L\psi = f(z)\psi, \tag{1.1}$$

$$\Lambda\psi = \theta(x)\psi \tag{1.2}$$

with nonconstant functions $\theta(x)$, $f(z)$.

2000 *Mathematics Subject Classification.* Primary: 47E05; Secondary: 47G30, 37K35.

I am grateful to V. Kuznetsov for the invitation to contribute to this volume. Several suggestions of the referee helped me to improve the presentation. Parts of the material were used in my talks in several universities—Nantes, Toulouse, Poitiers, Louvain-la Neuve, Leeds, Loughborough. Many remarks and questions of a lovely and critical audience were incorporated into the text. The main part of the survey is based on joint papers written in collaboration with B. Bakalov, M. Yakimov and T. Milanov. I use the occasion to thank them for this fruitful collaboration and in particular—For their confidence at the time when the subject didn't seem so promising. This research has been supported by a Grant No. MM1003-2000 of the NFSR of Bulgarian Ministry of Education.

This is the final form of the paper.

This problem has naturally arisen in the studies of F. A. Grünbaum on mathematical applications to computer tomography. For more information , see e.g. [**G1, G2**] and the references therein. The goals here are rather to demonstrate the purely mathematical side of the problem.

From the point of view of a pure mathematician the problem may look artificial. But with few steps in the direction of constructing examples, classifying them and in particular—In placing them in the context of other areas of research, the problem becomes very attractive. For example the construction of bispectral operators has revealed connections to integrable systems of KdV type, automorphisms of algebras, etc. [**DG, W1, BHY1, BHY4, KR**]. The classification (up to now) has used the theory of singular points of linear differential operators, algebraic geometry, Dixmier's methods for the study of the structure of the Weyl algebra [**DG, W1, HM, H**]. No doubt the interpretation of various "manifolds" of bispectral operators in the framework of representation theory, "master symmetries" of KdV, noncommutative geometry, etc. [**ZM, BHY2, BHY5, BW1, BW2**] is even more promising.

This review concentrates only on the first two aspects of the problem—Construction and classification of bispectral operators. This has reflected in the division of the paper into two parts.

Next we briefly describe the content of the paper. In Part I we start with a few general properties of bispectral operators. The first fundamental and quite general property is the so-called "ad-*condition*", which says that the operator ad_L acts on θ nilpotently. To the best of my knowledge this is the only way one uses the bispectral property. It seems that the ad-condition (properly formulated) is equivalent to bispectrality of L. This is formulated as a Conjecture 2.3 in Section 2.2. Important consequences of the ad-condition is that after natural normalization the coefficients of the bispectral operator are rational and the function θ is polynomial. All this is the content of Section 2.2. and comes from [**DG**]. More generally one could consider the algebra $\mathcal{B}$ of differential operators, spanned (roughly speaking) by L and θ. Obviously it is (anti)-isomorphic to the algebra, spanned by Λ and f, the concrete isomorphism b playing most important role (which has motivated G. Wilson to give it a special name—"The bispectral involution"). The study of bispectral operators in the framework of two isomorphic algebras $\mathcal{B}$, $\mathcal{B}'$ together with the anti-isomorphism b turns out to be extremely fruitful, in particular for construction of bispectral operators. I call the algebras $\mathcal{B}$, $\mathcal{B}'$ *dual algebras*. The corresponding formalism is introduced in Section 3. Two different applications illustrate the method of dual algebras in Part I. The most widely used procedure for constructing bispectral operators is the Darboux transformation. We demonstrate the power of dual algebras by introducing a version of Darboux transformations which is transparent and very easy to apply (probably in all situations). The point is that now it becomes possible to perform it *simultaneously* to both algebras. This is done in Section 4.1. The next three subsections reproduce the results for rank one algebras [**W1**], the Bessel and Airy cases [**BHY1**]. In Section 5 we describe the other important application for constructing bispectral operators— *Automorphisms of algebras.* Via this method one can get some idea (maybe quite vague) about what should be *a "simple" operators*, used to start the Darboux process. In particular one sees that their class is much broader than the well studied Bessel and Airy cases. Also one can understand easily that they are not so simple. I believe that the two methods are enough to construct all the bispectral operators. Also the automorphism method may give some hint about the classification, which is the subject of Part II. More or less the survey in the first part is quite complete, including essentially all the proofs. This cannot be said for the second part. Here the existing results and methods are not so systematic. I have tried to present few general ideas, which I believe could be useful also in future research. At the same time I have listed more or less the known results but they are either sketched or even without any proof.

Section 6 simply states Wilson's result [**W1**] classifying rank one bispectral operators. In Section 7 I classify the operators with decreasing coefficients at infinity (Section 7.2.) The auxiliary results from Section 7.1 also prepare the classification of operators of prime order. I present a proof of the Conjecture 2.3 (concerning the ad-condition) with the restriction that the coefficients of the operator are bounded at infinity. Section 8 deals with operators of growing coefficients. The main result of Section 9.1 is the classification of operators of prime order. The classical result of [**DG**] is obviously a particular case of this theorem. Some of the results are presented in more general situation. This is done as I believe they (and/or the corresponding techniques) could be useful elsewhere. I would only mention Dixmier's methods in the study of the Weyl algebra or the use of Airy PDO introduced in Section 8.2. (The latter are pseudodifferential operators expanded as Laurent series in Airy operators.) At the end I present some speculations on further possible progress on classification as well as eventual connections with structure of the Weyl algebra.

Part 1. Construction of Bispectral Operators

2. Basic Properties

In this section we have collected some terminology, notations and results relevant for the study of bispectral operators.We introduce the simplest consequences of the ad-condition, rationality of the coefficients, etc.

2.1. Preliminaries. In this subsection we recall some notions from Sato's theory of $\mathcal{K}P$-hierarchy [**S, DJKM, SW**] used in the survey. For a complete presentation of the theory we recommend also [**Di, vM**].

We start with the notion of *the wave operator* $K(x, \partial_x)$. This is a pseudodifferential operator

$$K(x, \partial_x) = 1 + \sum_{j=1}^{\infty} \alpha_j(x)\partial_x^{-j}, \tag{2.1}$$

with coefficients $\alpha_j(x)$ in some reasonable space, e.g. formal (Laurent) series in x^{-1}. The wave operator defines the (stationary) Baker–Akhiezer function $\psi(x, z)$:

$$\psi(x, z) = K(x, \partial_x)e^{xz}. \tag{2.2}$$

From (2.1) and (2.2) it follows that ψ has the following asymptotic expansion:

$$\psi(x, z) = e^{xz}\Big(1 + \sum_{1}^{\infty} \alpha_j(x)z^{-j}\Big), \quad z \to \infty. \tag{2.3}$$

Introduce also the pseudo-differential operator P:

$$P(x, \partial_x) = K\partial_x K^{-1}. \tag{2.4}$$

The following spectral property of P, crucial in the theory of $\mathcal{K}P$-hierarchy, is also very important for the bispectral problem:

$$P\psi(x, z) = z\psi(x, z). \tag{2.5}$$

When it happens that some polynomial of P, say $f(P)$, is a differential operator, we get that $\psi(x, z)$ is an eigenfunction of an ordinary differential operator $L = f(P)$:

$$L\psi = f(z)\psi. \tag{2.6}$$

It is possible to introduce the above objects in many different ways, starting with any of them (and with others, not introduced above). For us it would be important also to start with given *differential operator* L. By change of the variables and conjugation one can bring the operator L to the following form (unique, up to translation of the argument x:

$$L(x, \partial_x) = \partial_x^N + V_{N-2}(x)\partial^{N-2} + \cdots + V_0(x). \tag{2.7}$$

From now on we will work with operators normalized as above. One can define the wave operator K as:

$$LK = Kf(\partial) \tag{2.8}$$

where the first two coefficients of $f(\partial)$ are the same as in L, i.e. $f(\partial) = \partial^N + 0.\partial^{N-1} + \cdots$.

An important notion, connected to an operator L is the algebra $\mathcal{A}_L$ of operators commuting with L (see [**Kr, BC**]). This algebra is commutative one. The wave function $\psi(x, z)$ (defined in (2.2)) is a common wave function for all operators M from $\mathcal{A}_L$:

$$M\psi(x, z) = g_M(z)\psi(x, z). \tag{2.9}$$

We define also the algebra A_L of all functions $g_M(z)$ for which (2.9) holds for some $M \in \mathcal{A}_L$. Obviously the algebras A_L and $\mathcal{A}_L$ are isomorphic. Following [**Kr**] we introduce *the rank of the algebra* $\mathcal{A}_L$ as the greatest common divisor of the orders of the operators in $\mathcal{A}_L$.

2.2. The ad-condition. We start with several simple properties of bispectral operators following [**DG, W1**]. Assuming that L is bispectral means that we have also another operator Λ, a wave function $\psi(x, z)$ and two other nonconstant functions $f(z)$ and $\theta(x)$, such that the equations (1.1) and (1.2) hold. The following lemma, due to [**DG**], has been fundamental for all studies of bispectral operators.

LEMMA 2.1. *Let L be a bispectral operator. Then there exists a number $m \in \mathbb{N}$, such that*

$$(\mathrm{ad}_L)^{m+1}\theta = 0. \tag{2.10}$$

PROOF. Applied to ψ the left-hand side of (2.10) gives

$$(\mathrm{ad}_L)^{m+1}\theta = (-1)^{m+1}(\mathrm{ad}_f)^{m+1}\Lambda.$$

But if m is at least the order of Λ the right-hand side of the last equation is zero (as each application of ad_f reduces the order at any differential operator by one). □

REMARK 2.2. The equation (2.10) is essentially the only property of bispectral operators, used in their studies. Due to its importance it has a special name—The ad-*condition.* It seems that it is equivalent to the bispectrality of L. More precisely the following conjecture is widely believed to be true:

CONJECTURE 2.3. Let L be a differential operator of nonzero order and normalized as in (2.7). If there exists a nonconstant function $\theta(x)$ such that the ad-condition holds for some $m \in \mathbb{N}$ then L is bispectral, i.e. there exist functions $f(z)$, $\psi(x,z)$ and an operator Λ such that (1.1) and (1.2) hold.

This conjecture has been proved in several situations. In Section 6.1 we extract from [**HM, H**] its simple proof for operators with bounded coefficients.

An important corollary of the Lemma 2.1 is the following result.

LEMMA 2.4. *Let the operator L be normalized as in* (2.7). *Then*

(i) *The pseudo-differential operator $\Theta = K^{-1}\theta K$ has polynomial coefficients of order m.*
(ii) *The function $\theta(x)$ is a polynomial.*
(iii) *The coefficients α_j in the expansion* (2.1) *of the wave operator K are rational functions.*

PROOF. We repeat the simple proof following [**W1**]. From the equation (2.10) it follows that

$$(\mathrm{ad}_{\partial_x^N})^{m+1}(K^{-1}\theta K) = 0.$$

On the other hand the kernel of the operator $(\mathrm{ad}_{\partial_x^N})^{m+1}$ consists of all pseudo-differential operators whose coefficients are polynomials in x of degree at most m. This gives that

$$\theta(x)K = K\Theta, \tag{2.11}$$

with a pseudo-differential operator Θ:

$$\Theta = \Theta_0 + \sum_1^\infty \Theta_j \partial_x^{-j} \tag{2.12}$$

whose coefficients Θ_j are polynomials of degree at most m. Hence we get (i). Further $\theta(x) = \Theta_0(x)$. This gives (ii). Comparing the coefficients at ∂_x^{-j} we find that all the coefficients $\alpha_j(x)$ of K are rational functions. □

REMARK 2.5. We notice that at least one of the coefficients of Θ_j has degree *exactly* m, where m from Lemma 2.1 is minimal. This fact will be used later.

The last lemma has as an obvious consequence one of the few general results important in all studies of bispectral operators. Noticing that the coefficients of L are polynomials in the derivatives of $\alpha_j(x)$ we get

LEMMA 2.6. *The coefficients of L are rational functions.*

REMARK 2.7. Obviously the same results hold for the pair Λ, $f(z)$, when imposing the conditions (2.7) on Λ.

3. Dual Algebras

Here, following [**BHY4**] we present a general formulation of the bispectral problem which works not only for differential but also for difference or q-difference modifications of the original problem of [**DG**]. As I restrict myself here only with differential operators I am forced to skip this extremely interesting subject, connected to the classical problem of Bochner–Krall on orthogonal polynomials as eigenfunctions of differential operators, $q-\mathcal{K}dV(\mathcal{K}P)$ hierarchy, q-deformations of Calogero–Moser particle systems, etc. I only can offer an (incomplete) list of references on this recent developments—See, e.g. [**AHvM, GH, GHH1, GHH2, GY, HI, I1, I2**] and the references therein.

Let $\mathcal{R}$ and $\mathcal{R}'$ be associative algebras over a field k and let $\mathcal{M}$ be a left module over both of them. Let $\mathcal{K}$ and $\mathcal{K}'$ be fields such that $k \subset \mathcal{K} \subset \mathcal{R}$ and $k \subset \mathcal{K}' \subset \mathcal{R}'$.

We assume that some $\psi \in \mathcal{M}$ has the property

$$L\psi = 0 \text{ implies } L = 0 \quad \text{for } L \in \mathcal{R} \text{ (or } \mathcal{R}'). \tag{3.1}$$

Let us fix $\psi \in \mathcal{M}$ satisfying (3.1). We are interested in the equation

$$P\psi = Q\psi \tag{3.2}$$

for $P \in \mathcal{R}$, $Q \in \mathcal{R}'$. We put

$$\mathcal{B}_\psi = \{P \in \mathcal{R} \mid \exists Q \in \mathcal{R}' \text{ for which (3.2) is satisfied}\}, \tag{3.3}$$

$$\mathcal{B}'_\psi = \{Q \in \mathcal{R}' \mid \exists P \in \mathcal{R} \text{ for which (3.2) is satisfied}\}. \tag{3.4}$$

We shall assume that the actions of $\mathcal{R}$ and $\mathcal{R}'$ on $\mathcal{M}$ commute and that (3.1) holds. Then $\mathcal{B}_\psi$ and $\mathcal{B}'_\psi$ are associative algebras over k without zero divisors. Obviously (3.2) defines an anti-isomorphism

$$b\colon \mathcal{B}_\psi \to \mathcal{B}'_\psi, \quad b(P) = Q. \tag{3.5}$$

Introduce also the subalgebras

$$\mathcal{K}_\psi = \mathcal{B}_\psi \cap \mathcal{K}, \qquad \mathcal{K}'_\psi = \mathcal{B}'_\psi \cap \mathcal{K}', \tag{3.6}$$

$$\mathcal{A}_\psi = b^{-1}(\mathcal{K}'_\psi), \quad \mathcal{A}'_\psi = b(\mathcal{K}_\psi). \tag{3.7}$$

Then $\mathcal{A}_\psi$ is a commutative algebra isomorphic to $\mathcal{K}'_\psi$ (the *spectral algebra*). With a tilde we shall denote the fields of quotients of these commutative rings, e.g. $\mathcal{K}_\psi \subset \widetilde{\mathcal{K}}_\psi \subset \mathcal{K}$. Then b extends to isomorphisms $\widetilde{\mathcal{K}}_\psi \to \tilde{\mathcal{A}}'_\psi$ and $\tilde{\mathcal{A}}_\psi \to \widetilde{\mathcal{K}}'_\psi$.

DEFINITION 3.1. We call the algebras $(\mathcal{B}_\psi, \mathcal{B}'_\psi)$ *dual algebras* iff both $\mathcal{K}_\psi$ and $\mathcal{K}'_\psi$ contain nonzero elements.

In the above notations the elements of $\mathcal{A}_\psi$ would be *bispectral operators* and the elements of $\mathcal{K}_\psi$ would be *functions*.

REMARK 3.2. We have put the index ψ in all objects in order to stress on the importance of the map $b\colon \mathcal{B}_\psi \to \mathcal{B}'_\psi$. The function ψ plays the role of a kernel of an intertwining operator for two different representations of algebra $\mathcal{B}$. Later, to avoid heavy notation we are going to skip this index. The construction above has appeared in the differential case also in [**KR**].

EXAMPLE 3.3. (i) Let $\mathcal{R} \cong \mathcal{R}'$ be two copies of the rings of differential operators of one variable with rational coefficients and let $\mathcal{M}$ be a module of differentiable functions $\psi(x, z)$ (or formal power series) on which $\mathcal{R}$ acts on the variable x and $\mathcal{R}'$ acts on the variable z. Put $\psi(x,z) = e^{xz}$. Then $\mathcal{B}_\psi$ and $\mathcal{B}'_\psi$ are two copies of the Weyl algebra of the differential operators with polynomial coefficients and b is the anti-involution defined by $b(\partial_x) = z$, $b(x) = \partial_z$. We can put $\mathcal{A}_\psi = \mathbb{C}[\partial_x]$, $\mathcal{K}_\psi = \mathbb{C}[x]$. Obviously $\mathcal{A}'_\psi = \mathbb{C}[\partial_z]$ and $\mathcal{K}'_\psi = \mathbb{C}[z]$.

(ii) [**BHY1, KR**] Let L_α be the (generalized higher) Airy operator

$$L_\alpha = \partial^N + \sum_{i=2}^{N-1} \alpha_i \partial^{N-i} - x, \quad \alpha \in \mathbb{C}^{N-2} \tag{3.8}$$

and let $\Phi(x)$ be a solution of the equation $L_\alpha \Phi = 0$. Put $\psi(x,z) = \Phi(x+z)$. Then $\mathcal{B}_\psi \cong \mathcal{B}'_\psi$ are *again* two copies of the Weyl algebra but now b is defined via

$$b(x) = L_\alpha, \quad b(\partial) = \partial.$$

Then $\mathcal{K}_\psi = \mathbb{C}[x]$, $\mathcal{A}_\psi = \mathbb{C}[L_\alpha]$.

(iii) [**BHY2**] Let L_β be the (generalized) Bessel operator

$$L_\beta = x^{-N}(D-\beta_1)\cdots(D-\beta_N), \quad \beta \in \mathbb{C}^N \tag{3.9}$$

where $D = x\partial$. Let $\Phi(x)$ be a solution of the equation $L_\beta \Phi(x) = \Phi(x)$ and $\psi(x,z) = \Phi(xz)$. Then for "generic" β (see [**BHY1**]) $\mathcal{B}_\psi \cong \mathcal{B}'_\psi$ is generated by x^N, D, L_β; now $b(x^N) = L_\beta$, $b(D) = D$, $b(L_\beta) = x^N$ and $\mathcal{K}_\psi = \mathbb{C}[x^N]$, $\mathcal{A}_\psi = \mathbb{C}[L_\beta]$.

In the next sections we will see how to build new pairs of dual algebras from a given one.

4. Bispectral Darboux Transformations

Up to now the most widely used method for constructing solutions to the bispectral problem is the method of Darboux transformations introduced in the pioneering work of Duistermaat and Grünbaum [**DG**] (see also [**Z, BHY1, KR**]). The main idea of the method is that suitable Darboux transformations applied to "simple" bispectral operators should also produce bispectral operators. In the first subsection we recall one of the standard versions of Darboux transformation. Then we give an abstract version that has the advantage to be performed *on both operators L and Λ*, which automatically assures their bispectrality. In the last three subsections we demonstrate how to obtain the main examples. I hope that they can convince the reader that the procedure is completely algorithmic.

4.1. Darboux transformations for differential operators. In this subsection we recall some notions connected to Darboux transformations for ordinary differential operators in one of its many (*nonequivalent*) forms (see, e.g. [**BHY1, DG, Da**] as well as standard textbooks in quantum mechanics, soliton equations, etc.). Let the ordinary differential operator L of order $N > 0$ be factorized in the form

$$L = Q \circ P \tag{4.1}$$

where Q and P are operators of nonzero orders. Then the operator

$$\bar{L} = P \circ Q \tag{4.2}$$

is called *Darboux transformation of L*. The main property which makes this notion relevant for spectral problems is the fact that if some function ψ is an eigenfunction of L with an eigenvalue λ, then the function $\overline{\psi} = P\psi$ is an eigenfunction for $\overline{\psi} = P\psi$ with the same eigenvalue. In order to describe all Darboux transformations (of the above form) we recall the general scheme for construction of P.

Let L be a differential operator of order N. Then its kernel $\ker L$ is N-dimensional vector space. Choosing any subspace $V \in \ker L$ of dimension $k \leq N$ we can define an operator P (unique, if P is monic) with a kernel—The subspace V. Namely, let $\{f_1, \dots, f_k\}$ be an arbitrary basis of V. Then the monic operator P acts on any function $\psi(x)$ by the formula:

$$P\psi = \frac{Wr(f_1, \dots, f_k, \psi)}{Wr(f_1, \dots, f_k)}. \tag{4.3}$$

Then the operator L factorizes in following way:

$$L = Q \circ P. \tag{4.4}$$

Obviously the operator P does not depend on the basis but *only on the subspace* V. The opposite is also true: Given a factorization (4.4) one can define the subspace $\ker P \subset \ker L$. For details, see e.g. [**I**]. For further use we mention that we will use Darboux transformations also with operators P, which are not monic.

As we see the Darboux transformation (applied to a bispectral operator L preserves the first equation of the group (1.1), (1.2). Nothing guarantees that an equation of the type (1.2) can be found. In general it does not exist. Our next concern is to describe a method to perform Darboux transformations *both on the operators* L *and* Λ.

4.2. Abstract version. Here we give a general definition of bispectral Darboux transformations in the context of bispectral triples introduced in Section 2. We notice that the algebras $(\mathcal{B}_\psi, \mathcal{B}'_\psi)$ being without zero divisors and Noethetian have corresponding skew-fields of fractions $\mathcal{F}$, $\mathcal{F}'$. The antiautomorhpism b naturally extends to $\mathcal{F}$, $\mathcal{F}'$. Below I'm going to use this without mentioning.

DEFINITION 4.1. Let $(\mathcal{B}_\psi, \mathcal{B}'_\psi)$ be dual algebras. Let $L \in \mathcal{A}_\psi$ be a bispectral operator which can be factorized as

$$L = Q\theta^{-1}P \text{ with } P, \quad Q \in \mathcal{B}_\psi, \ \theta \in \mathcal{K}_\psi. \tag{4.5}$$

We call the operator

$$\bar{L} = PQ\theta^{-1} \tag{4.6}$$

a *bispectral Darboux transformation* of L.

THEOREM 4.2. *The operator* $\bar{L}$ *is bispectral. More precisely, if* $f = (L) \in \mathcal{K}'_\psi$ *then*

$$\bar{L}\overline{\psi} = f\overline{\psi}, \tag{4.7}$$

and

$$\bar{\Lambda}\overline{\psi} = \theta\overline{\psi} \tag{4.8}$$

where

$$\bar{\Lambda} = (P)(Q)f^{-1}. \tag{4.9}$$

PROOF. Let $\Lambda = (\theta)$. As $\mathcal{B}_\psi$ has no zero divisors (4.5) implies

$$\theta = PL^{-1}Q, \quad L^{-1} \in \tilde{\mathcal{A}}_\psi.$$

Applying the anti-isomorphism b we obtain

$$\Lambda = (\theta) = (Q)f^{-1}(P).$$

This shows that $\bar{\Lambda}$ is a bispectral Darboux transformation of Λ. From (4.6) and (4.9) it follows th at $\overline{\psi}$ satisfies (4.7) and (4.8). □

Theorem 4.2 is an abstract version of the bispectrality theorems from [**BHY1**]. However, it formalizes only partially the arguments of this theorem. The main difference is that here we do not touch the subtle (in general) question when a factorization of the form (4.5) is possible. In the next subsections we illustrate the power of the theorem on the examples from [**BHY1, W1**].

4.3. Rank one bispectral operators. In this subsection we explain how the method from the previous subsection works on the simplest case—Operators with constant coefficients [**W1**]. This example is important for the rest of the paper as well as for the entire subject of bispectral operators. It serves as a model for bispectral Darboux transformations for the other known examples. At the same time the above method works in this case in a straightforward manner. Finally it should be mentioned that through this particular case the subject was connected to noncommutative geometry and to Calogero–Moser particle systems [**BW1, BW2, W2**].

Consider two copies of the Weyl algebra A_1 in the variables x and z. They will be our algebras $\mathcal{B}$ and $\mathcal{B}'$. We take as an anti-involution b from (3.1), (i), i.e. with $\psi(x, z) = e^{xz}$. The algebra of bispectral operators is $\mathcal{A} = \mathbb{C}[\partial_x]$, i.e. it consists of all operators with constant coefficients. The algebra of functions $\mathcal{K}$ is the algebra of all polynomials $\mathbb{C}[x]$. The corresponding algebras $\mathcal{A}'$ and $\mathcal{K}'$ are the same algebras, but in the variable z. In this case the anti-involution b acts as formal Fourier transform. In order to apply the construction from the previous section we need factorizations of elements of $\mathcal{A}$ in the form required by Theorem 4.2.

Take any operator $h(\partial)$ with constant coefficients. Denote by λ_j, $j = 0, \ldots, k$ the different roots of the polynomial h with corresponding multiplicities d_j, i.e.

$$h(\partial) = \prod_{j=0}^{k} (\partial - \lambda_j)^{d_j}.$$

According to the general theory of linear differential equations with constant coefficients the kernel of the operator $h(\partial)$ is a direct sum of the spaces W_j:

$$\ker L = \bigoplus_{j=0}^{k} W_j$$

where

$$W_j = \operatorname{span}\{e^{\lambda_j x}, xe^{\lambda_1 x}, \ldots, x^{d_j - 1}e^{\lambda_j x}\}.$$

Now we have to pick the subspaces of the kernel of $h(\partial)$ in such a way that the factorization required by Theorem 4.2 exists. It is enough (in this case) to find factorizations of the type (4.1) where the corresponding operators P and Q are with rational coefficients. In view of (4.3) the natural choice is the following one. Let $V_j \subset W_j$ be arbitrary subspaces (they can be also empty). Define

$$V = \bigoplus V_j. \tag{4.10}$$

By the standard procedure explained in Subsection 3.1 we can define the operator P whose kernel is the subspace V. Now it is easy to show that

LEMMA 4.3. *The operator P constructed by (4.3) with $\ker P = V$ has rational coefficients.*

PROOF. The kth column in any of the two determinants in (4.3) contains a polynomial multiplied by one and the same (for each k—Its own) exponent. So in the ratio they cancel. Hence the coefficients of P are rational. □

Now the coefficients of Q can be computed easily from (4.1). As obviously they are differential polynomials of the coefficients of P we get that they are also rational. At the end divide and multiply the operator P from the left by the common divisor of its coefficients to get P in the form required by Theorem 4.2:

$$P = p(x)^{-1} P_1$$

where $p(x)$ is a polynomial and P_1 is an operator with polynomial coefficients. We apply a similar procedure to the operator Q, where first we push all the denominators to the right of the differentiations and present Q in the form:

$$Q = Q_1 q(x)^{-1}$$

where Q_1 is with polynomial coefficients and q is a polynomial. Setting $\theta = p(x)q(x)$ we get the desired form for the factorization of $h(\partial)$:

$$h(\partial) = Q_1 \theta^{-1} P_1$$

with P_1, $Q_1 \in \mathcal{B}$ and $\theta \in \mathcal{A}$.

Now applying Theorem 4.2 we obtain

THEOREM 4.4. *For any space V defined as in (4.10) the operator*

$$\overline{h(\partial)} = P_1 Q_1 \theta^{-1}(x) \tag{4.11}$$

is bispectral.

4.4. Airy algebras. The application of Theorem 4.2 to Airy algebras, defined in (3.1), (ii) essentially repeats the construction from the previous subsection. Here once again the algebras $\mathcal{B}$, $\mathcal{B}'$ are two copies of the Weyl algebra, but the anti-involution is different. We use the notation from (3.1), (ii). Fix an Airy operator A_α of order N. The bispectral operators in this case are polynomials of the Airy operators, i.e. $\mathcal{A} = \mathbb{C}[A]$. As in the previous section we take a polynomial $h(A) \in \mathcal{A}$ and again write its factorization (in $\mathcal{A}$):

$$h(A) = \prod_{j=0}^{k} (A_\alpha - \lambda_j)^{d_j}. \tag{4.12}$$

The description of the kernel of each factor of the above product is quite similar to the case of operators with constant coefficients. We describe it briefly, pointing only the minor differences. Take a basis $\Phi_1(x), \dots, \Phi_N(x)$ of the kernel of the operator A_α. Then it is easy to check that all the functions $\psi_k(x, z) = \Phi_k(x + z)$,

$k = 1, \dots, N$ can play the role of the "intertwiner" ψ from (3.1). More generally we have

$$(A_\alpha - \lambda)^d \psi_j = (z - \lambda)^d \psi_j. \tag{4.13}$$

From this equation one can easily compute the kernel of the factors in (4.12). Take the successive derivatives of (4.13) with respect to z and put $z = \lambda$. This gives that the kernel W_j of any factor $(\mathcal{A}_\alpha - \lambda_j)^{d_j}$ has a basis of the form:

$$\partial_z^l \left(\psi_k(x, z)\right)_{|z=\lambda_j}, \quad l = 0, \dots, l-1; \; k = 1, \dots, N.$$

Again we have to choose our space V to be a direct sum of subspaces $V_j \in W_j$. But now this restriction does not suffice. In order that P has rational conditions it is clear that $\ker P$ should be "symmetric. The following definition takes care of this symmetry.

DEFINITION 4.5. We say that the Darboux transformation connected to a subspace $V \in \ker h(\mathcal{A}_\alpha)$ is *polynomial* iff

(i) the space $V = \bigoplus V_j$, where $V_j \subset W_j$;
(ii) V_j has a basis whose elements are linear combinations of a single function $\psi(x, \lambda_m)$ for some m and its derivatives;
(iii) together with each function $f(x, \lambda_m)$ from (ii) the space V_j contains all the functions $f(x, \lambda_k)$, $k = 1, \dots, N$.

LEMMA 4.6. *If the Darboux transformation is polynomial, then the operator P is with rational coefficients.*

The proof is similar to that of Lemma 4.3. Now one needs to use the fact that the Wronskian of $\psi_1(x, \lambda_k), \dots, \psi_N(x, \lambda_k)$ is a nonzero constant (it could be 1) as well as (ii) and (iii).

The last lemma makes possible to repeat the end of the previous section to obtain the following result [**BHY1**] (see also [**KR**], where it is obtained independently for a subclass of the polynomial Darboux transformations). Namely, again define $P = p(x)^{-1} P_1$, $Q = Q_1 q(x)^{-1}$, with $P_1, Q_1 \in A_1$, $p(x), q(x) \in \mathbb{C}[x]$ and put $\theta = p(x) q(x)$.

THEOREM 4.7. *The operator*

$$\overline{h(A_\alpha)} = P_1 Q_1 \theta^{-1}$$

is bispectral.

4.5. Bessel algebras. Our last application will be to the case of the Bessel algebras, introduced in (3.1), (iii). This reproduces the results of [**BHY1**]. The scheme is essentially the same as in the previous section. The only minor difference here is the description of admissible kernels for the operators P that factorize the initial bispectral operators.

First as in the previous cases we need a description of $\ker h(L_\beta)$. We are going to use the natural condition on β, which implies the restriction on the Bessel operators to be normalized as required in (2.7), i.e.:

$$\sum_{i=1}^{N} \beta_i = \frac{N(N-1)}{2}. \tag{4.14}$$

The following lemma describes the kernel of the operator $h(L_\beta)$ for an arbitrary polynomial h.

LEMMA 4.8. *Let $h(z)$ be a polynomial*

$$h(z) = z^{d_0}(z-\lambda_1^N)^{d_1}\cdots(z-\lambda_r^N)^{d_r}, \quad \lambda_i^N \neq \lambda_j^N,\ \lambda_0 = 0,\ d_i \geq 0. \tag{4.15}$$

Then we have

(i) $\ker h(L_\beta) = \bigoplus_{i=0}^r \ker(L_\beta - \lambda_r^N)^{d_i}$.
(ii) $(L_\beta)^d = L_{\beta^d}$, *where*

$$\beta^d = \big(\beta_1, \beta_1 + N, \ldots, \beta_1 + (d-1)N, \ldots, \beta_N, \ldots, \beta_N + (d-1)N\big). \tag{4.16}$$

(iii) *If* $\{\beta_1, \ldots, \beta_N\} = \{\underbrace{\alpha_1, \ldots, \alpha_1}_{k_1}, \ldots, \underbrace{\alpha_s, \ldots, \alpha_s}_{k_s}\}$ *with distinct* $\alpha_1, \ldots, \alpha_s$*, then*

$$\ker L_\beta = \operatorname{span}\big\{x^{\alpha_i}(\ln x)^k\big\}_{1\leq i\leq s,\ 0\leq k\leq k_i-1}.$$

(iv) *For* $\lambda \neq 0$

$$\ker(L_\beta - \lambda^N)^d = \operatorname{span}\big\{\partial_z^k \Psi_\beta(x,z)\big|_{z=\lambda\varepsilon^j}\big\}_{0\leq k\leq d-1, 0\leq j\leq N-1}$$

where $\varepsilon = e^{2\pi i/N}$ *is an* N*th root of unity.*

The *proof* being obvious is omitted.
Notice that not every subspace of $\ker L$ defines an operator needed to apply Theorem 4.2. Below we give the conditions on the subspaces of the kernel of $h(L_\beta)$ that are needed in order to perform the bispectral Darboux transformations. For further use we are going to introduce some more terminology than needed for this subsection.

DEFINITION 4.9. *Bessel wave function* is called the unique wave function $\Psi_\beta(x,z)$ depending only on xz and satisfying

$$L_\beta(x,\partial_x)\Psi_\beta(x,z) = z^N\Psi_\beta(x,z) \tag{4.17}$$

where the Bessel operator $L_\beta(x,\partial_x)$ is given by (3.9).

Because the Bessel wave function depends only on xz, (4.17) implies

$$D_x\Psi_\beta(x,z) = D_z\Psi_\beta(x,z), \tag{4.18}$$

$$L_\beta(z,\partial_z)\Psi_\beta(x,z) = x^N\Psi_\beta(x,z). \tag{4.19}$$

It is natural to require that $\ker P$ be $\mathbb{Z}_N$-invariant. The $\mathbb{Z}_N$-invariance of $\ker P$ simply means that

$$f(x) \in \ker P \implies f(\varepsilon x) \in \ker P, \quad \varepsilon = e^{2\pi i/N}. \tag{4.20}$$

Now we are ready to give the next definition:

DEFINITION 4.10. We say that the Darboux transformation of $h(L_\beta)$ defined by some operator P is *polynomial* if the kernel of P consists of:

(i) Several groups of elements of $\ker L^{d_0}$ of the form:

$$\partial_y^l\left(\sum_{k=0}^{k_0}\ \sum_{j=0}^{\operatorname{mult}(\beta_i+kN)-1} b_{kj}x^{\beta_i+kN}y^j\right)\Bigg|_{y=\ln x}, \quad 0\leq l\leq j_0 \tag{4.21}$$

where $j_0 = \max\{j | b_{kj} \neq 0 \text{ for some } k\}$;

(ii) Several groups of elements of $\ker(L_\beta) - \lambda_j^N)^{d_j}$ $\lambda_j \neq 0$ of the form:

$$\sum_{k=0}^{k_0} a_k \varepsilon^{ki} \partial_z^k \Psi_\beta(x,z)|_{z=\varepsilon^i \lambda_j}, \quad 0 \le i \le N-1. \tag{4.22}$$

We call the Darboux transformation *monomial* if $h(L_\beta) = L_\beta^m$.

Denote by n_0 the of dimensions of the subspace of $(L_{\beta^{d_0}}$, (i.e. the number of elements of the form (4.21) in the above basis of $\ker P$). For $1 \le j \le r$ denote by n_j the dimension of the number of the elements in the points $\varepsilon^i\lambda_j$, $0 \le i \le N-1$ (i.e. the number of groups of elements of the form (4.22) with $\lambda = \lambda_j$).
It is easy to show that the polynomial Darboux transformations are the ones that we need to apply Theorem 4.2. We again show that factorization (8.2) holds.

LEMMA 4.11. *If the operator P defines a polynomial Darboux transformation it has the form*

$$P(x,\partial_x) = x^{-n} \sum_{k=0}^{n} p_k(x^N)(x\partial_x)^k \tag{4.23}$$

where p_k are rational functions, $p_n \equiv 1$.

The *proof* of this lemma is similar to the corresponding ones in the previous sections. As the reader expects the following result holds:

THEOREM 4.12. *The polynomial Darboux transformations are bispectral.*

5. Automorphisms of Algebras of Differential Operators

In the previous section I described how one can obtain new bispectral operators from "simple" ones by Darboux transformations. The only reason to call them "simple" is that it is trivial to show that they are bispectral.

In this section I will describe following [**BHY4**] entirely different method to construct bispectral operators (again from given ones). This method have changed completely the idea of "*simple*" operators. It is not so easy to show that the examples, illustrating the method, are bispectral. On one hand they are not Darboux transformations of Bessel or Airy operators. On the other hand the machinery of bispectral Darboux transformations can be applied to them very easily. My opinion is that they are candidates for being "simple", a notion which still needs to be clarified (cf. the last section).

In what follows I again use the notations of Section 2. Assume that the algebras $\mathcal{B}$ and $\mathcal{B}'$ consist of differential operators. The main tool for constructing new bispectral operators will be the following simple observation.

PROPOSITION 5.1. *Let $(\mathcal{B}, \mathcal{B}')$ be dual algebras and let $L \in \mathcal{B}$. If ad_L is a locally nilpotent operator $\mathrm{ad}_L\colon \mathcal{B} \to \mathcal{B}$ then $\sigma = e^{\mathrm{ad}_L}$ is an automorphism of $\mathcal{B}$ and $(\mathcal{B}, \mathcal{B}'$ are dual algebras with an automorphism $b \circ \sigma)$.*

(Here *locally nilpotent* means that for each $M \in \mathcal{B}$ there exists an integer n such that $(\mathrm{ad}_L)^n M = 0$.) The *proof* of Proposition 5.1 being obvious is omitted.

An important example of an operator L acting locally nilpotently on $\mathcal{B}$ is any bispectral operator [**DG**]. This can be proved by using the anti-isomorphism b and the fact that the function $f = b(L)$ acts locally nilpotently on $\mathcal{B}'$.

REMARK 5.2. (i) Obviously, exchanging the roles of $\mathcal{B}$ and $\mathcal{B}'$ one can use also the anti-isomorphism $b_2 = e^{\mathrm{ad}_{L'}} \circ b$, where L' acts locally nilpotently on $\mathcal{B}'$.

(ii) In the definition of dual algebras a decisive role is played by the function $\psi(x,z) \in \mathcal{M}$ via which the anti-isomorphism is defined. Proposition 5.1 defines a new anti-isomorphism b_1 in terms of the old function ψ. But to have really dual algebras in the sense of Definition 3.1 we face the difficult task to define a new function $\widetilde{\psi}$ to satisfy $P\widetilde{\psi} = b_1(P)\widetilde{\psi}$. The most natural choice $\widetilde{\psi} = e^{-L}\psi$ needs interpretation (see [**BHY6**]).

Although simple the above Proposition is a source of new bispectral operators.

EXAMPLE 5.3. (i) Let $\mathcal{B}$ and $\mathcal{B}'$ be two copies of the Weyl algebra and let $(x) = \partial_z$, $b(\partial_x) = z$, $\psi = e^{xz}$. For any polynomial $p(\xi)$ the operators $\mathrm{ad}_{p(x)}$ and $\mathrm{ad}_{p(\partial_x)}$ act locally nilpotently on $\mathcal{B}$. Then if we put $\sigma_1 = e^{\mathrm{ad}_{p(x)}}$ and $b_1 = b \circ \sigma_1$ we have

$$b_1(x) = b(x) = \partial_z, \tag{5.1}$$

$$b_1(\partial_x) = \big(-p'(x) + \partial_x\big) = z - p'(\partial_z). \tag{5.2}$$

In this way we get a new dual pair $(\mathcal{B}, \mathcal{B}')$ with a new function $\widetilde{\psi}(x,z) = e^{p(x)}e^{xz}$.

(ii) One can apply the automorphism $\sigma_2 = e^{\mathrm{ad}_{q(\partial_x)}}$ to the above triple $(\mathcal{B}, \mathcal{B}', b_1)$. Then we get $b_2 = b_1 \circ \sigma_2$, which gives

$$b_2(x) = b_1\big(x + q'(\partial_x)\big) = \partial_z + q'\big(z - p'(\partial_z)\big), \tag{5.3}$$

$$b_2(\partial_x) = b_1(\partial_x) = z - p'(\partial_z). \tag{5.4}$$

For the new function $\psi_2(x,z)$ we must have

$$\partial_x\psi_2 = \big(z - p'(\partial_z)\big)\psi_2, \tag{5.5}$$

$$x\psi_2 = \big(\partial_z - q'\big(z - p'(\partial_z)\big)\big)\psi_2. \tag{5.6}$$

This implies

$$\big(\partial_x + p'\big(x - q'(\partial_x)\big)\big)\psi_2 = z\psi_2, \tag{5.7}$$

which together with (5.5) gives a new nontrivial example of bispectral operators when $\deg p \geq 3$, $\deg q \geq 3$, provided ψ_2 exists. One can construct a solution ψ_2 using Fourier transform with respect to both z and x (see [**BHY6**].

Because $\mathcal{K}_{\psi_2} = \mathbb{C}[x]$, $\mathcal{K}'_{\psi_2} = \mathbb{C}[z]$ we see that both spectral curves are isomorphic to $\mathbb{C}$. This shows that the operators defined in (5.6), (5.7) cannot be obtained by the methods from [**BHY1**] (i.e. they are not Darboux transformations from Airy operators).

REMARK 5.4. The automorphisms considered in Example 5.3 generate the entire group of automorphisms of the Weyl algebra [**Dx**].

An anti-isomorphism which can successfully be used for constructing new bispectral operators from old ones is the formal conjugation: $(x) = x$, $(\partial_x) = -\partial_x$. Instead of formulating a general result we shall give an example.

EXAMPLE 5.5. Let $\beta \in \mathbb{C}^N$, $\gamma \in \mathbb{C}^N$, $\beta_i + \gamma_i = N - 1$. Define $\mathcal{B} = \langle x^N, L_\beta, D_x\rangle$, $\mathcal{B}' = \langle z^N, L_\gamma, D_z\rangle$, where the operators L_β, L_γ are Bessel operators. Let b be the formal conjugation: $(x^N) = z^N$, $(D_x) = -D_z - 1$, $(L_\beta) = (-1)^N L_\gamma$. As L_γ acts locally nilpotently on $\mathcal{B}'$ (since it is bispectral) one can define $b_1 = e^{-\operatorname{ad}_{N^{-1}L_\gamma}} \circ b$. Then we have

$$b_1(D_x) = -D_z - 1 + L_\gamma, \quad b_1(L_\beta) = (-1)^N L_\gamma, \quad b_1(x^N) = M_\gamma = e^{-\operatorname{ad}_{N^{-1}L_\gamma}}(z^N).$$

M_γ can be computed easily applying b_1 to both sides of the identity

$$L_\beta x^N = (D_x + N - \beta_1)\cdots(D_x + N - \beta_N).$$

In the simplest nontrivial case $N = 3$ we have

$$(5.8) \quad M_\gamma = -L_\gamma^2 + (3D_z - 9 - \Sigma\gamma_i)L_\gamma - 3D_z^2 + (9 + 2\Sigma\gamma_i)D_z - (9 + 3\Sigma\gamma_i + \Sigma\gamma_i\gamma_j) + z^3.$$

The corresponding function $\psi(x, z)$ satisfies

$$(D_x + D_z + 1)\psi = L_\gamma\psi = (-1)^N L_\beta\psi, \quad x^N\psi = M_\gamma\psi, \quad z^N\psi = M_\beta\psi.$$

As we see the algebra $\mathcal{B}$ is the same for the Bessel operator L_β and for the operator M_β. Therefore for "generic" β (in the sense of [**BHY1**]) the spectral algebra $\mathcal{K}_\Psi = \mathbb{C}[x^N]$, i.e. the spectral curves is again $\mathbb{C}$. Thus we obtain that M_β cannot be constructed by the methods of [**BHY1**].

This construction can be generalized considering $b_1 = e^{\operatorname{ad}_{p(L_\gamma)}} \circ b$ for arbitrary polynomial p.

The results and the examples of this sections exhibit clearly that the classification problem is quite difficult.

Part 2. Classification

The first results on classification of bispectral operators show that different schemes are possible. For example in [**DG**] the authors use classification of operators by order and find the first nontrivial bispectral operators—Those of order two. Another approach could be found in [**W1**]—Classification of *bispectral algebras* of rank one. The examples in **Part I.** show that both classifications by order and by rank may need a more subtle subdivision of the corresponding classes. In [**HM**] a third scheme has been proposed—By the singularity of the operators at infinity. As in the previous schemes the simplest representative has been considered—One can find the complete list of the bispectral operators with regular singular point at infinity. Comparing the constructions of these papers one finds that even restricting the classification to one of the schemes one is forced to use the others. With this point of view in [**H**] I have found the classification of operators of prime order.

In this Part I review the results obtained up to now on the classification of the bispectral operators. Where possible I give some details of the proofs. A brief description of the organization of the review follows.

In Section 6 I list all rank one bispectral operators without proofs. The main classification result of Section 7 concerns operators with decreasing at infinity coefficients (Subsection 7.2). The proofs are sketched but not complete. In contrast to this the auxiliary results are essentially done with details. The reason is that they could be of interest both by themselves and for further research. Section 8 develops some techniques relevant for operators with coefficients growing at infinity. Simple corollary of Sections 7–8 is the classification of operators of prime order. The final result sounds like the classical result of [**DG**] for operators of order two. I hope that apart of the generalization itself, the ideas or their modifications could be of some value to continue the classification of bispectral operators. At the end I present some speculations connecting the classification problem with open questions in the structure of the Weyl algebra.

6. Operators of Rank One

Here I am going only to formulate Wilson's result describing rank one bispectral operators.

THEOREM 6.1. *A rank one operator L is bispectral if and only if it is a Darboux transform of operator with constant coefficients.*

REMARK 6.2. The original formulation in [**W1**] sounds in a different way:

THEOREM 6.3. *A rank one operator L is bispectral if and only if its spectral curve* $\mathrm{Spec}A_L$ *is rational and unicursal. (This means that all the singularities of the curve are cusps.)*

The *idea of the proofs* is roughly speaking the following. The method from the Subsection 3.3. gives operators whose spectral curves sweep all the moduli space of all rational and unicursal curves. One also has to show that a curve of this type defines exactly one maximal algebra of *rank one* bispectral operators. This is achieved by Burchnal–Chaundy–Krichever theory [**BC, Kr**]. The fact is not true always—e.g. for Bessel operators.

7. Operators with Regular Singularity at Infinity

7.1. The ad-condition for operators with bounded coefficients. Probably the most important result of present subsection is the proof of Conjecture 2.3 in the case of operators with bounded coefficients. Also we hope that the construction itself, being simple and natural, could be useful (eventually after some modifications) elsewhere. In what follows we assume that the operator L has bounded at infinity coefficients, i.e. it has the form:

$$L = \partial^p + \sum_{j=0}^{p-2} W_j(x)\partial^j \tag{7.1}$$

where $W_j = c_j + V_j(x)$, c_j are constants and $V_j(x) = \mathcal{O}(x^{-1})$. Next define the polynomial $f(z)$ to be:

$$f(z) = z^p + \sum_{j=O}^{p-2} c_j z^j. \tag{7.2}$$

In this way we can rewrite our operator in the form

$$L = f(\partial) + \sum_{j=0}^{p-2} V_j(x)\partial^j \tag{7.3}$$

where $V_j(x) = \mathcal{O}(x^{-1})$. Let K be the wave operator corresponding to L. In the next lemma, following [**DG**] we find further restrictions on the coefficients of the wave operator K and on L. This will be our main tool to extend the standard involution b in the Weyl algebra to a suitable algebra of PDO.

LEMMA 7.1. (i) *The coefficients* $V_j(x)$, $j = N-2, \ldots, 0$ *of* L *vanish at* ∞ *at least as* x^{-2}.

(ii) *The coefficients* α_j, $j = 1, \ldots$ *of the wave operator* K *vanish at least as* x^{-1}.

PROOF. We are going to prove both statements simultaneously. We use the formula

$$LK = Kf(\partial),$$

Comparing the coefficients at ∂^{N-2} at the both sides of the above identity we get:

$$V_{N-2} + N\alpha_1' = 0.$$

Having in mind that V_{N-2} is equal to the derivative of the rational function α_1 and that it vanishes at ∞ we see that it vanishes at least as x^{-2}. Continuing in the same manner and using induction we get the statement of the lemma. □

Following [**BHY4**] we will introduce an anti-isomorphism b between the algebra $\mathcal{B}$ of pseudo-differential operators $P(x, \partial_x)$ in the variable x and the algebra $\mathcal{B}'$ of pseudo-differential operators $R(z, \partial_z)$ in the variable z. More precisely $\mathcal{B}$ consists of those pseudo-differential operators

$$P = \sum_k^{\infty} p_j(x^{-1})\partial_x^{-j},$$

for which there is a number $n \in \mathbb{Z}$ (depending on P) such that $x^n p_j(x^{-1})$, $j = k$, $k+1, \ldots$ are formal power series in x^{-1}. The involution

$$b\colon \mathcal{B} \to \mathcal{B}'$$

is defined by

$$b(P)e^{xz} = Pe^{xz} = \sum_k^{\infty} z^{-j} p_j(\partial_z^{-1})e^{xz}, \quad \text{for } P \in \mathcal{B}, \tag{7.4}$$

i.e. b is just a continuation of the standard anti-isomorphism between two copies of the Weyl algebra. In what follows we will use also the anti-isomorphism

$$b_1\colon \mathcal{B} \to \mathcal{B}', \quad b_1(P) = b(\mathrm{Ad}_K P). \tag{7.5}$$

Obviously b and b_1 can be considered as involutions of $\mathcal{B}$ and without any ambiguity we can denote the inverse isomorphisms b^{-1}, $b_1^{-1}\colon \mathcal{B}' \to \mathcal{B}$ by the same letters.

We are ready to define the z-counterpart of K and L. Assume also that $\mathrm{ad}_L^{m+1} \theta = 0$. According to Lemma 2.4 there exists a PDO Θ with polynomial coefficients of order at most m. Since the operators K and Θ are from $\mathcal{B}$ we can define S and Λ as follows:

$$S(z, \partial_z) = b\big(K(x, \partial_x)\big), \tag{7.6}$$

$$\Lambda(z, \partial_z) = b(\Theta). \tag{7.7}$$

Explicitly one has for the wave operator S:

$$S = \sum_{j=0}^{\infty} z^{-j}\alpha_j(\partial_z) = \sum_{j=0}^{\infty} a_j(z)\partial_z^{-j}, \quad a_0 = 1, \tag{7.8}$$

and for the differential operator Λ:

$$\Lambda(z, \partial_z) = \sum_{j=0}^{\infty} z^{-j}\Theta_j(\partial_z) = \sum_{i=0}^{m} \Lambda_i(z)\partial_z^i \tag{7.9}$$

where $\Lambda_m = 1$ (see Remark 2.5) and the coefficients Λ_i and a_j should be viewed as formal power series. We are going to see that they are rational functions in z.

LEMMA 7.2. *The coefficients a_j of the operator S are rational functions.*

PROOF. Using that

$$LK = Kf(\partial),$$

we can apply the involution b and find:

$$Sb(L) = f(z)S.$$

Then compute recursively a_j by comparing the coefficients at ∂_z^{-j}. This proves the lemma. □

Now we are ready to show that the operator Λ has coefficients Λ_j, which are rational functions. Denote temporarily by r the degree of the polynomial θ, i.e. if $\theta(x) = z^r + \cdots$, then $\theta_r \neq 0$.

LEMMA 7.3. *With the choice of $f(z)$ as in (7.2) the coefficients Λ_i of the operator Λ are rational functions. The degree of θ $r = m$ and*

$$\Lambda_m = 1, \quad \Lambda_{m-1} = \theta_{m-1} = 0. \tag{7.10}$$

PROOF. Using the definition (7.7) and applying the involution b to the relation $\theta(x)K = K\Theta$ we get:

$$\Lambda S = S\theta(\partial_z).$$

As the coefficients of Λ are expressed as differential polynomials of the coefficients a_j of S we get that Λ_j are also rational functions. Comparing the first two coefficients of the above equality we get also (7.10) . □

Summarizing the above results we obtain the proof of the Conjecture 2.3 for operators with bounded coefficients.

THEOREM 7.4. *Assume the operator L satisfies the* ad*-condition with some function θ. Then L is bispectral.*

We will need two auxiliary results in the classification of bispectral operators. First notice that simple computations give

$$\mathrm{ad}_f^{m-1}(\Lambda) = (-1)^{m-1}m!(f')^{m-1} \cdot \partial_z + cf''(f')^{m-2} \tag{7.11}$$

and

$$\mathrm{ad}_f^{m}(\Lambda) = (-1)^{m}m!(f')^{m}. \tag{7.12}$$

LEMMA 7.5. *If the order of the operator L is equal to its rank, then the coefficients of L vanish at infinity.*

PROOF. Put

$$Q_1 = b_1\big(\mathrm{ad}_f^m(\Lambda)\big).$$

As $\mathrm{ad}_f^{m+1}(\Lambda) = 0$ we get that $[L, Q_1] = 0$, i.e. $Q_1 \in A_L$. Using that the rank of L is equal to its order one obtains

$$Q_1 = q_r L_r + \cdots + q_0, \quad q_j \in \mathbb{C}$$

where $q_r \neq 0$. Again applying b_1 we obtain (using also (7.11))

$$q_r f^r(z) + \sum_{j=0}^{r-1} q_j f^j = m!\big(f'(z)\big)^m. \tag{7.13}$$

Now comparing the powers of the leading terms we get $(N-1)m = Nr$ which implies that $m = Ns$ and $r = s(N-1)$. Next comparing the coefficients at the first two terms of (7.13) we obtain that the second term in $f(z)$ has degree $N-1$, which contradicts to normalization of L. $\square$

We end the section with a version of the so-called "*string equation*"

$$[L, Q] = NL^{n+1}, \tag{7.14}$$

which will play a crucial role. In what follows we would assume that the number m is divisible by N. This is not a restriction as we can always replace Λ by Λ^N.

LEMMA 7.6. *Assume that $f(z) = z^N$. Then there is a natural number n such that:*

$$Q = K^{-1} x \partial_x{}^{nN+1} K, \tag{7.15}$$

is a differential operator. The operator Q is a solution to the string equation (7.14).

PROOF. Using (7.11) with $f = z^N$ we get

$$\mathrm{ad}_f^{m-1}(\Lambda) = c_1 z^{Nl+1}\partial_z + c_2 z^{Nk}.$$

Applying the bispectral involution b_1 we get

$$\mathrm{ad}_L^{m-1}(\theta) = c_1 Q + c_2 L^k, \quad c_1 \neq 0$$

where Q is a differential operator. Having in mind that

$$\mathrm{ad}_f^m(\Lambda) = (-1)^m m! z^{Nl(m-1)},$$

and once again applying b_1 we get (7.14) with suitable $n \geq 0$. $\square$

The string equation is the most important tool in our analysis. I cannot demonstrate all of its power in this review restricting myself only with application to Darboux transformations. But I will need also other consequences, which are formulated below. For their technical proofs see [**HM**].

THEOREM 7.7. *Any rank r bispectral operator L with vanishing at infinity coefficients is $\mathbb{Z}_r$-invariant.*

THEOREM 7.8. *If the bispectral operator L has vanishing at infinity coefficients, then the point $x = \infty$ is a regular singular point.*

At the end denote by $\lambda_1, \ldots, \lambda_N$ the roots of the indicial equation of the point at infinity for L. Then

PROPOSITION 7.9. *Assume that L is a differential operator that solves the string equation. Then we can find roots λ_i and λ_j such that: $\lambda_i - \lambda_j$ is an integer and*

$$n \leq \frac{1}{N}|\lambda_i - \lambda_j|$$

where n is from (7.14).

7.2. Operators with decreasing coefficients. In this subsection we will gradually simplify the operator L by successive applications of Darboux transformations. Our goal is to obtain after a finite number of steps a Bessel operator. According to Theorem 7.8 the point $x = \infty$ is a regular singular point for the operator L. Assume also that L is a rank r differential operator. From Theorem 7.7 we know that in this case L is $\mathbb{Z}_r$-invariant operator. Thus if we represent L as

$$L = \partial^N + V_2\partial^{N-2} + \cdots + V_{N-1}\partial + V_N$$

then the coefficients V_i can be expanded as

$$V_i = \frac{1}{x^i}\sum_{k=0}^{\infty} V_{i,k}x^{-rk}. \tag{7.16}$$

In what follows we need to split the set $M = \{\lambda_1, \lambda_2, \ldots, \lambda_N\}$ of roots of the indicial equation at ∞ for L into subsets of equivalent modulo $\mathbb{Z}$ numbers.

For an arbitrary set M_i denote by λ the number in M_i with minimal real part. The next lemma is a version of a classical result (see e.g. [**I**]) and shows how one can pick an $\mathbb{Z}_r$-invariant function from $\ker L$.

LEMMA 7.10. *If λ is the minimal number of a set M_i, then there is a function ϕ_λ from* $\ker L$ *which can be expanded around ∞ as:*

$$\phi_\lambda(x) = x^\lambda \sum_{k=0}^{\infty} c_k x^{-kr}, \quad c_0 = 1. \tag{7.17}$$

We omit the proof as it repeats the classical one.

Given a function ϕ_λ we construct a first order operator by setting

$$P_\lambda = \partial_x - \frac{\phi'_\lambda}{\phi_\lambda}. \tag{7.18}$$

Then the operator L can be factorized as $L = Q_\lambda P_\lambda$ and after we perform the Darboux transformation

$$L = Q_\lambda P_\lambda \to \tilde{L} = P_\lambda Q_\lambda \tag{7.19}$$

the new operator $\tilde{L}$ will have the following properties:

PROPOSITION 7.11. *Assume that the operator L solves the string equation with a $\mathbb{Z}_r$-invariant wave operator $K = 1 + \alpha_1\partial^{-1} + \cdots, \operatorname{ord}(\alpha_i) \le -i$. Then*

(i) *every operator which is obtained by a Darboux transformation described above also solves the string equation;*

(ii) *if $\{\lambda_1, \lambda_2, \ldots, \lambda_N\}$ are the roots of the indicial equation at ∞ of L and $\lambda = \lambda_{i_0}$ is the number with minimal real part from some M_i then the roots of the indicial equation at ∞ of $\tilde{L}$ are $\tilde{\lambda}_k = \lambda_k - 1$ for $k \ne i_0$ and $\tilde{\lambda}_{i_0} = \lambda_{i_0} + (N-1)$.*

PROOF. Put

$$\widetilde{K} = P_\lambda K \partial^{-1} \quad \text{and} \quad \widetilde{Q} = P_\lambda Q Q_\lambda.$$

Using the equalities $LK = K\partial^N$ and $QK = Kx\partial^{nN+1}$ one easily gets

$$\widetilde{Q}\widetilde{K} = P_\lambda K x \partial^{nN+1}\partial^{N-1} = \widetilde{K}\partial x \partial^{N(n+1)} = \widetilde{K}x\partial^{(n+1)N+1} + \widetilde{K}\partial^{(n+1)N}.$$

Now it is clear that $\widetilde{K}x\partial^{(n+1)N+1}\widetilde{K}^{-1} = \widetilde{Q} - \tilde{L}^{n+1}$ is a differential operator. The second assertion in the lemma is standard. □

After this proposition we are close to our final goal.

PROPOSITION 7.12. *Let L be a bispectral operator with coefficients satisfying (2.7). Then by finitely many $\mathbb{Z}_r$-invariant Darboux transformations we can transform it into a Bessel operator.*

PROOF. We will perform Darboux transformations in the following way: Start with $L_0 = L$. Choose an index i, if there is any, for which the difference between numbers in M_i with maximal real part and with minimal real part exceeds N. Denote by λ the number in M_i with the minimal real part, set $P_1 = P_\lambda$ and factorize L as $L = R_1 P_1$ then the Darboux transformation will be

$$L_0 = L = R_1 P_1 \to L_1 := P_1 R_1.$$

On each step, using Proposition 7.11, we increase the minimal real part of the roots and decrease the real parts of the rest of them.

After finitely many Darboux transformations we obtain an operator L_m for which the difference between the maximal and the minimal parts in each group M_i will be less than N. This is enough to conclude that the number n in the string equation is 0. This will also give that Q_m has the form:

$$Q_m = x\partial_x.$$

Now comparing the coefficients at ∂^j at both sides of the string equation (7.14) with $n = 0$ we obtain the equation

$$-xV_j' + jV_j = NV_j.$$

Integrating it, we obtain that

$$V_j = v_j x^{-N+j}, \quad v_j \in \mathbb{C}.$$

This shows that L_m is a Bessel operator. □

Essentially we have obtained the following theorem.

THEOREM 7.13. *A bispectral operator with decreasing coefficients is monomial Darboux transformation of Bessel one.*

The only thing that we have to prove is that the operators P_λ have rational coefficients. We omit the proof of this technical result referring to [**HM**].

8. Operators with Growing Coefficients

8.1. Normal forms. Let L be a bispectral operator, normalized as in (2.7), i.e.

$$L = \partial_x^N + \sum_{j=0}^{N-2} V_j(x)\partial_x^j. \tag{8.1}$$

We will consider that for some j the corresponding coefficient V_j is growing at infinity, i.e. has Laurent expansion at infinity of the form:

$$V_j = \sum_{m=-\infty}^{r_j} a_{j,m} x^m \tag{8.2}$$

where $r_j > 0$ and $a_{r,j} \neq 0$. We call r_j the order of V_j. In what follows we are going to use several properties shared by the first Weyl algebra A_1 and the larger algebra $\mathcal{R}[\partial]$ of differential operators with rational coefficients. Following [**Dx**] we will prescribe weights to x and ∂ as follows: $wt(x) = \rho$, $wt(\partial) = \sigma$.

DEFINITION 8.1. Assume that $L = V_n\partial^n + V_{n-1}\partial^{n-1} + \cdots + V_0$ is an arbitrary element of $\mathcal{R}[\partial]$. For each term $V(x)\partial_x^i$ define its weight

$$v_{\rho,\sigma}\big(V(x)\partial_x^i\big) = \rho(\operatorname{ord} V) + \sigma i.$$

Then the number

$$v_{\rho,\sigma}(L) := \max_{0\le i\le n} v_{\rho,\sigma}(V_i\partial^i)$$

will be called (ρ,σ)-order of L.

The second definition associates to each differential operator from $\mathcal{R}[\partial]$ a (ρ,σ)-homogeneous polynomial.

DEFINITION 8.2. Assume the notation of the previous definition and denote by $I(L)$ the set $\{i \in \{0,1,\dots,n\} | v_{\rho,\sigma}(V_i\partial^{n-i}) = v_{\rho,\sigma}(L)\}$. The polynomial $p \in \mathbb{C}[X,Y]$ defined as:

$$p(X,Y) = \sum_{i\in I} a_i X^{\operatorname{ord} V_i} Y^i \tag{8.3}$$

where $a_i \in \mathbb{C}$ are uniquely determined from the expansion

$$V_i = a_i x^{\operatorname{ord} V_i} + (\text{lower order terms}),$$

will be called polynomial associated with L. The operator $L_0 =: p(x,\partial_x) :$ will be called *the homogeneous part* of L.

In what follows we assume that ρ and σ are positive integers. Consider again the bispectral operator (8.1). It is always possible to choose them in such a way that the polynomial f associated with L has at least two terms of the kind

$$f = y^N + \alpha x^k y^p + \cdots, \quad \alpha \neq 0 \tag{8.4}$$

where $k > 0$, $p \ge 0$. Denote also by $g(x) = x^l$ the polynomial associated to $\theta(x)$. Our goal is to find severe restrictions on the polynomial f.

We are going to use the following result which is a slight modification of the fundamental from [**Dx**, Proposition 7.4].

LEMMA 8.3. *Suppose the element $L \in \mathcal{R}[\partial]$ acts nilpotently on some $\theta \in \mathcal{R}$, and the associated to L polynomial f contains at least two terms. If $v_{\rho,\sigma}(L) > \rho+\sigma$, then f is either*

$$f = Y^n(Y^r + \mu X)^k \tag{8.5}$$

or

$$f = X^n(X^r + \mu Y)^k \tag{8.6}$$

where $n \ge 0$, $k \ge 0$, $r > 0$.

The proof of this lemma repeats that from [**Dx**] and therefore is omitted. Obviously in our case the polynomial f has the form

$$f = Y^n(Y^r + \mu X)^k, \quad \mu \neq 0. \tag{8.7}$$

Without loss of generality we can assume that $\mu = -1$.

LEMMA 8.4. *The polynomial f has the form*

$$f = (Y^r - X)^k. \tag{8.8}$$

IDEA OF THE PROOF. If we suppose that $n > 0$, then the homogeneous part of L cannot act nilpotently (see for details [**H**]).

REMARK 8.5. Note that the above result gives a normal form for the leading terms of all bispectral operators with growing coefficients of any order.

Now assume that the order N of L is a prime number. This gives that $k = 1$, $N = r$. Using Lemma 8.4 we obtain that:

LEMMA 8.6. *The operator L has the form*

$$L = \partial^N + \sum_{j=1}^{N-2} a_j \partial^j - x + \sum_{j=0}^{N-2} W_j(x)\partial^j \tag{8.9}$$

where $\lim_{x\to\infty} W_j(x) = 0$ *and* $a_j \in \mathbb{C}$.

We will call the operator

$$A = \partial^N + \sum_{j=1}^{N-2} a_j \partial^j - x$$

the *principal part* of L. In this case A is the (generalized) Airy operator.

8.2. Airy PDO's. Define the algebra $\mathcal{B}_1$ of pseudo-differential operators of the type:

$$P(x, \partial_x) = \sum_{j=-m}^{\infty} a_j(x, \partial_x) A^{-j}, \tag{8.10}$$

with operator coefficients of the form:

$$a_j = \sum_{k=0}^{N-1} \alpha_{j,k}(x)\partial^k \tag{8.11}$$

where the functions $\alpha_{j,k}(x)$ are formal Laurent series:

$$\alpha_{j,k}(x) = \sum_{s=r}^{\infty} \beta_{j,k}^{(s)} x^{-s}, \tag{8.12}$$

and the index r depends only on P (*but not on j!*). Then the anti-automorphism b can be continued on $\mathcal{B}_1$ as

$$b(P) = \sum_{j=-m}^{\infty} z^{-j} \sum_{k=0}^{N-1} \sum_{s=r}^{\infty} \beta_{j,k}^{(s)} \partial_z^k A^{-s} = \sum_{s=r}^{\infty} b_s(z, \partial_z) A^{-s}(z). \tag{8.13}$$

Introduce the "wave operator" K as follows:

$$K = 1 + \sum_{j=1}^{\infty} m_j(x, \partial_x) A^{-j} \tag{8.14}$$

where

$$m_j(x, \partial_x) = \sum_{k=1}^{N-1} a_{j,k}(x)\partial_x^k. \tag{8.15}$$

Let L be a differential operator of the form:

$$L = A^l + V_{l-1}A^{l-1} + \cdots \tag{8.16}$$

where

$$V_j(x, \partial) = \sum_{k=0}^{N-1} V_{j,k}(x)\partial^k. \tag{8.17}$$

Then one can find a wave operator (not unique) K of the form (8.14) so that

$$L = KA^lK^{-1}. \tag{8.18}$$

The coefficients $a_{j,k}(x)$ from (8.16) of the expansion of K can be found by induction from the equation:

$$LK = KA^l. \tag{8.19}$$

Multiplying the above equation from the right by powers of A, $A^2, \ldots$ one computes the coefficients $a_{j,k}(x)$ in the expansion (8.14)–(8.16) of K in terms of the functions $V_{j,k}$. We would particularly be interested in the case $l = 1$.

In what follows we assume that the operator L is bispectral. In fact we again assume only that it satisfies the ad-condition. In that case as in Part I and [**DG, W1**] one can prove the following lemma.

LEMMA 8.7. *The coefficients $a_{j,k}$ of the operator K are rational functions.*

8.3. Operators with Airy principal part. Let L be an operator of order N with principal part

$$A = \partial^N + a_{N-2} + \cdots + a_1\partial - x, \tag{8.20}$$

i.e.

$$L = A + \sum_{j=0}^{N-2} V_j(x)\partial^j = A + V(x, \partial_x) \tag{8.21}$$

and

$$\lim_{x\to\infty} V_j(x) = 0. \tag{8.22}$$

Using the techniques of previous subsection we present L in the form

$$L = A + V = KAK^{-1} \tag{8.23}$$

Now we give the statement of the main result.

THEOREM 8.8. *The operator $V \equiv 0$, i.e. $L = A$.*

IDEA OF THE PROOF. Assume that some of the coefficients of V are nonzero. We will come to contradiction. One can obtain recurrent formulas for the *derivatives* of the coefficients of the wave operator K just by comparing the coefficients at the powers of A in (8.23). The expansions at infinity of the first few ones must have x^m with negative degrees m, but the degrees gradually grow to reach -1, which is impossible in view of the rational character of the coefficients of K.

9. Operators of Prime Order

9.1. Classification of operators of prime order. Gathering the patches of the results in the previous sections we have obtained the following theorem.

THEOREM 9.1. *All bispectral operators of prime order are in the following list:*

1. *Generalized Airy operators:*

$$A = \partial^p + \sum_{j=1}^{p-2} a_j \partial^j - x, \quad a_j \in \mathbb{C}; \tag{9.1}$$

2. *Generalized Bessel operators:*

$$B = x^{-p}(x\partial - \beta_1) \cdots (x\partial - \beta_p), \quad \beta_j \in \mathbb{C}; \tag{9.2}$$

3. *Operators with constant coefficients:*

$$C = \partial^p + \sum_{j=1}^{p-2} a_j \partial^j, \quad a_j \in \mathbb{C}; \tag{9.3}$$

4. *Operators, obtained by monomial Darboux transformations from Bessel operators (listed in* 2)
5. *Polynomial Darboux transformations from operators with constant coefficients (listed in* 3).

For $N = 2$ the theorem is just the content of the famous result of J. Duistermaat and F. A. Grünbaum [**DG**].

Sketch of the *proof.* If the coefficients of L are bounded (at infinity) and the order is prime, then using the fact the the rank divides the order we get that either the rank of L is 1 or it is equal to its rank. The latter case is treated in Section 7. The rank one case is the main result of [**W1**] cited in Section 6. Finally the inverse part, i.e. that all the operators listed in Theorem 9.1 are bispectral is the main result of Part I.

10. Comments

It seems that the methods from Sections 4 and 5 would be enough to find all bispectral operators. Assuming that then the classification should be:

1. find all "simple" bispectral operators;
2. show that Darboux transforms reduce any bispectral operator to a "simple" one.

The experience (maybe insufficient) prompts that simple operators L are those that satisfy one of the following "string" equations:

$$[L, Q] = 1, \tag{10.1}$$

$$[L, Q] = L. \tag{10.2}$$

Here I would like to make some speculations about eventual continuation of the classification. It is tempting to consider the above results, namely Section 5, Sections 6–8. as a basis for further progress. The next missing step is the classification of operators with bounded coefficients. One plausible conjecture is the following one.

CONJECTURE 10.1. An operator L with bounded at infinity coefficients is bispectral if and only if it is a polynomial Darboux transformation of a Bessel operator.

This conjecture seems natural in view of the "if" part, obtained in [**BHY1**]. It will be very useful either to give a proof of Wilson's result (Section 5) without using algebraic-geometric arguments or modify his proofs in higher rank situation. Despite of the many interesting results for higher rank solutions of KP-hierarchy I have not found a suitable for our purposes construction. In any case the above conjecture seems to me to be within the reach of the existing tools.

A much more difficult question would be how to treat the operators with unbounded coefficients. A modest restriction presented in Section 6. is the one given for the polynomial, associated to an operator. Nevertheless I would prefer to find first classification of the bispectral operators only in the Weyl algebra. For any operator $L \in A_1$ denote by $C(L)$ its centralizer in A_1.

CONJECTURE 10.2. The centralizer of any bispectral operator $L \in A_1$ is the algebra of polynomials of some operator L_0.

Using the above introduced generator L_0 of the centralizer of L we can conjecture that

CONJECTURE 10.3. An operator $L \in A_1$ is bispectral, if and only if the operator L_0 satisfies the canonical commutation relation $[L_0, Q] = 1$ with some other operator $Q \in A_1$.

A motivation for this and for the next conjecture is the following result from [**BHY4**] (essentially explained in Example 4.3 and Remark 5.4 as well as the entire spirit of the fundamental paper [**Dx**] (see also below).

PROPOSITION 10.4. *If an operator L belongs to the orbits of the operators from $\mathbb{C}[\partial]$ under the action of the automorphism group* $\mathrm{Aut}(A_1)$ *of A_1 then it is bispectral.*

Notice that up to now these are all known bispectral operators in the Weyl algebra. Another conjecture again inspired by the above mentioned result, as well as by Dixmier is the inverse of this result.

CONJECTURE 10.5. If an operator L is bispectral it is in the orbit of $\mathbb{C}[\partial]$.

At the end I have to acknowledge that all the above conjectures are very much connected to the famous conjecture of Dixmier–Kirillov (see [**Dx**]).

CONJECTURE 10.6. Any endomorphism of the Weyl algebra is automorphism.

We finish these speculations by noticing that obviously the Dixmier–Kirillov conjecture follows from the above formulated ones. It seems that the most difficult part is the "if"-part of Conjecture 10.3. A progress in this direction may also be useful for the treatment of Dixmier–Kirillov conjecture. On the other hand Conjecture 10.5 does not seem to be extremely difficult. Also it would be of great importance for the classification problem.

If this program or sufficient parts of it could be achieved one may try to modify some of the analysis above, as well as to try to apply the automorphism group for reduction to simpler cases. Unfortunately this is still insufficient as one needs to perform corresponding studies for broader algebras of differential operators. As shown in [**BHY4**] one can use automorphisms of such algebras as the algebra generated by L_β, D, X^N and obtain again bispectral operators. The study of such automorphisms even in the case $N = 2$, corresponding to certain quotients of the universal enveloping algebra of sl(2) have not been as well studied (to the best of my knowledge) as in the case of the Weyl algebra.

Note added in proof. Now Conjecture 10.5 is proven in [**H1**].

References

[AHvM] M. Adler, E. I. Horozov, and P. van Moerbeke, *The solution to the* q-KdV *hierarchy*, Phys. Lett. A **242** (1998), 139–151.

[AvM] M. Adler and P. van Moerbeke, *Birkhoff strata, Bäcklund transformations, and regularization of isospectral operators*, Adv. Math. **108** (1994), 140–204.

[ASvM] M. Adler, T. Shiota, and P. van Moerbeke, *A Lax representation for the vertex operator and its central extension*, Comm. Math. Phys. **171** (1995), 547–588.

[BHY1] B. Bakalov, E. I. Horozov, and M. Yakimov, *Bispectral algebras of commuting ordinary differential operators*, Comm. Mat. Phys. **190**, (1997) 331–373; q-alg/9602011.

[BHY2] ———, *Tau-functions as highest weight vectors for* $W_{1+\infty}$ *algebra*, J. Phys. A **29** (1996), 5565–5573; hep-th/9510211.

[BHY3] ———, *Bäcklund-Darboux transformations in Sato's Grassmannian*, Serdica Math.J. **22** (1996), 571–588; q-alg/9602010.

[BHY4] ———, *General methods for constructing bispectral operators*, Phys. Lett. A **222** (1996), 59–66.

[BHY5] ———, *Highest weight modules over* $W_{1+\infty}$, *and the bispectral problem*, Duke Math. J. **93** (1998), 41–72.

[BHY6] ———, *Automorphisms of the Weyl algebra and bispectral operators* (Montréal, 1997) (J. Harnad and A. Kasman, eds.), The Bispectral Problem, CRM Proc. Lecture Notes, vol. 14, Amer. Math. Soc., Providence, RI, 1998, pp. 3–10.

[BW1] Yu. Berest and G. Wilson, *Automorphisms and ideals of the Weyl algebra*, Math. Ann. **318** (2000), no. 1, 127–147.

[BW2] ———, *Ideal classes of the Weyl algebra*; math.AG/0104240, (2001).

[BC] J. L. Burchnall and T. W. Chaundy, *Commutative ordinary differential operators*, Proc. London Math. Soc. (3) **21** (1923), 420–440; Proc. Royal Soc. London Ser. A **118** (1928), 557–583; Proc. Royal Soc. London Ser. A **134** (1932), 471–485.

[Da] G. Darboux, *Leçons sur la théorie générale des surfaces*, 2ième partie, Paris, Gauthiers–Villars, 1889 .

[DJKM] E. Date, M. Jimbo , K. Kashiwara, and T. Miwa, *Transformation groups for soliton equations*, Nonlinear Integrable Systems—Classical and Quantum Theory (Kyoto 1981) (M. Jimbo and T. Miwa, eds.), World Sci. Publishing, Singapore, 1983, pp. 39–119.

[Di] L. Dickey, *Soliton and Hamiltonian systems*, Advanced Series in Mathematical Physics, vol. 12, World Scientific Publishing Co., Inc., River Edge, NJ, 1991.

[Dx] J. Dixmicr, *Sur lcs algèbrcs dc Wcyl*, Bull. Soc. Math. France **96** (1968), 209–242.

[DG] J. J. Duistermaat and F. A. Grünbaum, *Differential equations in the spectral parameter*, Comm. Math. Phys. **103** (1986), 177–240.

[F] J. Fastré, *Bäcklund-Darboux transformations and* W*-algebras*, Doctoral Dissertation, Univ. of Louvain, 1993.

[G1] F. A. Grünbaum, *The limited angle reconstruction problem in computer tomography* Computed Tomography (Cincinnati, 1982) (L. Shepp, ed.), Proc. Sympos. Appl. Math., vol. 27, Amer. Math. Soc., Providence, R.I., 1982, pp. 43–61.

[G2] ———, *Time-band limiting and the bispectral problem*, Comm. Pure Appl. Math. **47** (1994), 307–328.

[GH] F. A. Grünbaum and L. Haine, *Bispectral Darboux transformations: An extension of Krall polynomials*, Internat. Math. Res. Notices **8** (1997), 359–392.

[GHH1] F. A. Grünbaum, L. Haine, and E. I.Horozov, *On the Krall–Hermite and Krall–Bessel polynomials*, Internat. Math. Res. Notices **19** (1997), 954–966.

[GHH2] ———, *Some functions that generalize the Krall-Laguerre polynomials*, J. Comput. Appl. Math. **106** (1999), no. 2, 271–297.

[GY] F. A. Grünbaum and M. Yakimov, *Discrete bispectral Darboux transformations from Jacobi operators*, (2000); mat.CA/0012191.

[HI] L. Haine and P. Iliev, *The* q*-bispectral property of* q*-deformations of Schur polynomials and the* q-KdV *hierarchy*, J. Phys. A **30** (1997), 7217–7227.

[BP] J. Harnad, *Bispectral operators, dual isomonodromic deformations and the Riemann–Hilbert dressing method*, The Bispectral Problem (Montréal, 1997) (John Harnad and

Alex Kasman, eds.), CRM Proc. & Lecture Notes, vol. 14. Amer. Math. Soc., Providence, RI, 1998, pp. 67–79.

[H] E. I. Horozov, *Bispectral operators of prime order*, Preprint, Sofia University, 2001.

[H1] ———, *Strictly nilpotent elements and bispectral operators in Weyl algebra*, Preprint, Sofia University, 2002.

[HM] E. I. Horozov and T. Milanov, *Fuchsian bispectral operators*, Prépublication no. 187, Laboratoire de Math. E. Picard, 2000; math.DS/0102093 (2001); Bull Sci. Math. (to appear).

[I1] P. Iliev, *Solutions to Frenkel's deformation of the q-KP hierarchy*, J. Phys. A **31** (1998), L241–L244.

[I2] ———, *The q-KP hierarchy, bispectrality and Calogero–Moser systems*, J. Geom. Phys. **35** (2000), 157–182.

[I] E. L. Ince, *Ordinary differential equations*, New York, Dover, 1944.

[K] A. Kasman, *Bispectral* KP *solutions and linearization of Calogero–Moser particle systems*, Comm. Math. Phys. **172** (1995), 427–448.

[KR] A. Kasman and M. Rothstein, *Bispectral Darboux transformations: The generalized Airy case*, Phys. D **102** (1997), 159–176.

[Kr] I. Krichever, *Commutative rings of linear ordinary differential operators*, Funct. Anal. Appl. **12** (1978), no. 3, 20–31.

[S] M. Sato, *Soliton equations as dynamical systems on infinite-dimensional Grassmann manifolds*, Nonlinear Partial Differential Equations in Applied Science (Tokyo, 1982), North-Holland Math. Stud., vol. 81, North-Holland, Amsterdam, 1983, pp. 259–271.

[SW] G. Segal, and G. Wilson, *Loop groups and equations of* KdV *type*, Inst. Hautes Études Sci. Publ. Math. **61** (1985), 5–65.

[vM] P. van Moerbeke, *Integrable foundations of string theory*, Lectures on Integrable Systems (Sophia-Antipolis, 1991) (O. Babelon et al., eds.), World Sci. Publishing, River Edge, NJ, 1994, pp. 163–267.

[W1] G. Wilson *Bispectral commutative ordinary differential operators*, J. Reine Angew. Math. **442** (1993), 177–204.

[W2] ———, *Collisions of Calogero–Moser particles and an adelic Grassmannian. With an appendix by I. G. Macdonald*, Invent. Math. **133** (1998), 1-41 (1998).

[Z] J. Zubelli, *Differential equations in the spectral parameter for matrix differential operators*, Physica D **43** (1990), 269–287.

[ZM] J. Zubelli and F. Magri, *Differential equations in the spectral parameter, Darboux transformations and a hierarchy of master equations for* KdV, Comm. Math. Phys. **141** (1991), 329–351.

DEPARTMENT OF MATHEMATICS & INFORMATICS, SOFIA UNIVERSITY, 5 J. BOURCHIER BLVD., SOFIA 1126, BULGARIA.

E-mail address: horozov@fmi.uni-sofia.bg

Centre de Recherches Mathématiques
CRM Proceedings and Lecture Notes
Volume **32**, 2002

A Link Between Two Fundamental Contributions of Kowalevski

Ji-Shan Hu and Min Yan

1. Introduction

Among numerous important mathematical contributions of Kowalevski, the following two are fundamental: The first is her doctoral dissertation, "On the theory of partial differential equations", appeared in 1875 [**8**]. The main result is what we now call the Cauchy–Kowalevski theorem. The second is "On the problem of the rotation of a solid body about a fixed point", appeared in 1889 [**9**] and won her the famous Prix Bordin of the French Academy of Sciences. The main result is the discovery of what we now call the Kowalevski top, which completes a program implicit in the works of Euler and Lagrange—to solve in an analytic manner the equations of motion of a rotating rigid body.

It is rather interesting to compare the two contributions from the technical perspective. The first work shows that the formal power series solution to the initial value problem (along a noncharacteristic manifold for PDE) of an analytic differential equation must be convergent (and analytically dependent on the initial data). In particular, Cauchy–Kowalevski theorem justifies the formal power series solutions as being *realistic*. In the second work, Kowalevski determined the integrability of a dynamical system by *formally* testing Laurent series solutions with movable singularity location and enough number (five for a system of six first order equations) of free parameters. The method is what we now call the Painlevé test [**2**] [**11**] and is still the most effective way of detecting integrability. Being the formal process, however, the convergence (i.e., the realism) of the series is not considered in the test.

In this note, we will explain that the following two are equivalent for a differential equation

1. The equation passes the Painlevé test in the most strict sense (see the end of Section 2 for the precise meaning), as the Kowalevski top does;

2000 *Mathematics Subject Classification.* Primary: 35A20, 37K10; Secondary: 34A25, 35A10, 35C10.

Research was supported by Hong Kong Research Grant Council Earmark Grant HKUST6123/00P.

This is the final form of the paper.

2. There is a (usually meromorphic) change of dependent variables that converts the equation to an analytic system of equations, and converts the Laurent series in the Painlevé test to power series.

In particular, if we can apply the Cauchy–Kowalevski theorem to the power series solution, then we conclude the convergence of the Laurent series in the Painlevé test.

An example of such changes of variables already appeared a century ago in Painlevé's proof [**10**] that all solutions of the first Painlevé equation $u'' = 6u^2 + t$ are meromorphic functions on the whole complex plane. However, it did not appear to have been noticed that the existence of such changes of variables is a general property for equations passing the Painlevé test. This note will explain the key ideas for finding the changes of variables. More discussions and examples can be found in our series of papers [**3**] through [**6**].

Both the power series and the Laurent series are local by nature. Therefore it is not reasonable to expect a global link between the two. Even though our change of variables does not *directly* tell us anything about the global questions such as whether singularities accumulate, it can be a potentially useful tool for studying the integrable equations, especially on problems related to singularity analysis. Basically, near regular points of a solution, we may use the original equation to study the behavior of the solution. Near movable pole singularities, we may regularize the situation by applying our changes of variables and then study the system of equations for the new variables. Sometimes the two studies can be combined together and provide us a global understanding of the solution. Painlevé had successfully applied the idea in his proof. We also used the idea in a much better proof of the Painlevé property for certain equations [**7**].

2. Key Idea

We illustrate our key idea through the simple example of Painlevé's first equation:

$$u'' = 6u^2 + t. \tag{2.1}$$

Consider a (regular) initial value problem

$$u(t_0) = r, \quad u'(t_0) = s. \tag{2.2}$$

Formally, through a recursive relation, the initial value problem is equivalent to the following power series solution

$$u = r + s(t - t_0) + \left(3r^2 + \frac{t_0}{2}\right)(t - t_0)^2 + \left(2rs + \frac{1}{6}\right)(t - t_0)^3 + \cdots. \tag{2.3}$$

Cauchy–Kowalevski theorem then concludes that (2.3) is indeed a convergent series, and u is analytically dependent on t_0, r, s.

Now consider pole-like solutions of (2.1), with movable pole location. By dominant balance, the order of the pole must be 2. Therefore we postulate $u = u_0(t - t_0)^{-2} + u_1(t - t_0)^{-1} + \cdots$ and get a recursive relation for the coefficients. The result is a formal Laurent series solution

$$u = (t - t_0)^{-2} - \frac{t_0}{10}(t - t_0)^2 - \frac{1}{6}(t - t_0)^3 + r(t - t_0)^4 + \frac{t_0^2}{18}(t - t_0)^6 + \cdots \tag{2.4}$$

with r arbitrary. Note that the existence of Laurent series solution together with one free parameter r (in general, the number is the degree of the equation minus one) in the series imposes a highly nontrivial condition on the shape of the equation. For example, the equation $u'' = 6u^2 + t^2$ has no Laurent series solutions. Moreover, the Chazy's equation $u''' = 2uu'' - 3u'^2$ has two Laurent series solutions and they can never contain free parameters. Through a similar but much more complicated process, Kowalevski discovered her top by studying the conditions derived from the existence of five free parameters in Laurent series solutions in the system of six equations. More discussion can be found on [**1**] and [**2**].

Comparing the two solutions (2.3) and (2.4), we have the following observations:

1. In addition to t_0, there are two free parameters in (2.3). Such parameters are exactly the initial data in (regular) initial value problems (2.2). The number of such parameters is *always* the order of the equation. Moreover, they *always* appear explicitly as the initial coefficients in (2.3);
2. In addition to t_0, there is one free parameter in (2.4). Such parameters can be considered as the "initial data" in some imaginary singular initial value problems. The number of such parameters is *always* at most the order of the equation minus one, and may be fewer. Moreover, they often do not appear as the initial coefficients in (2.3), but rather buried deep in the series.

In order to convert Laurent series solutions such as (2.4) to power series solutions such as (2.3), therefore, we need to bring the free parameters "buried deep" in the Laurent series "to the front". The most naive way to bring r to the front is to truncate at the place where r first appears

$$u = (t - t_0)^{-2} - \frac{t_0}{10}(t - t_0)^2 - \frac{1}{6}(t - t_0)^3 + v(t - t_0)^4, \tag{2.5}$$

by introducing a new variable

$$v = r + \frac{t_0^2}{18}(t - t_0)^2 + \cdots . \tag{2.6}$$

However, it is not clear what the transformation between u and v is. It is also not clear what equation and the corresponding initial value problem for v is.

Thus we need to do the truncation in a more clever way, and we also need to get a transformation between the old and the new variables at the end. We consider the equation (2.1) as a system for the variables u and u'. Then we introduce a new variable θ through *indicial normalization*

$$u = \theta^{-2}, \tag{2.7}$$

so that θ behaves just like $t - t_0$:

$$\begin{aligned} \theta &= (t - t_0) + \frac{t_0}{20}(t - t_0)^5 + \frac{1}{12}(t - t_0)^6 - \frac{r}{2}(t - t_0)^7 - \frac{t_0^2}{36}(t - t_0)^9 + \cdots \\ &= (t - t_0) + \frac{t}{20}(t - t_0)^5 + \frac{1}{30}(t - t_0)^6 - \frac{r}{2}(t - t_0)^7 - \frac{t^2}{36}(t - t_0)^9 + \cdots . \end{aligned} \tag{2.8}$$

On the other hand, we have the Laurent series for (another variable) u':

$$
\begin{aligned}
(2.9)\qquad u' &= -3(t-t_0)^{-2} - \frac{t_0}{5}(t-t_0) - \frac{1}{2}(t-t_0)^2 + 4r(t-t_0)^3 \\
&\qquad + \frac{t_0^2}{3}(t-t_0)^5 + \cdots \\
&= -3(t-t_0)^{-2} - \frac{t}{5}(t-t_0) - \frac{3}{10}(t-t_0)^2 + 4r(t-t_0)^3 \\
&\qquad + \frac{t^2}{3}(t-t_0)^5 + \cdots .
\end{aligned}
$$

From (2.8), we may rewrite $t - t_0$ as a series in powers of θ. Then we substitute this series into (2.9) and obtain

$$
(2.10)\qquad u' = -2\theta^{-3} - \frac{t}{2}\theta - \frac{1}{2}\theta^2 + 7r\theta^3 + \frac{t^2}{16}\theta^5 + \cdots .
$$

Now we may truncate at the first place where the free parameter r appears

$$
(2.11)\qquad u' = -2\theta^{-3} - \frac{t}{2}\theta - \frac{1}{2}\theta^2 + \xi\theta^3,
$$

by introducing a new variable ξ, which must satisfy

$$
(2.12)\qquad \xi = 7r + \frac{t^2}{16}\theta^2 + \cdots = 7r + \frac{t_0^2}{16}(t-t_0)^2 + \cdots .
$$

The equations (2.7) and (2.11) form a change of variables between (u, u') and (θ, ξ). This converts the original equation (2.1) into a *regular* system

$$
\theta' = 1 + \frac{t}{4}\theta^4 + \frac{1}{4}\theta^5 - \frac{1}{2}\xi\theta^6, \quad \xi' = \frac{t^2}{8}\theta + \frac{3t}{8}\theta^2 + \left(\frac{1}{4} - t\xi\right)\theta^3 - \frac{5}{4}\xi\theta^4 - \frac{3}{2}\xi^2\theta^5.
$$

Moreover, the formal Laurent series for (u, u') are converted to the formal power series (2.8) and (2.12) for (θ, ξ). Note that we may apply Cauchy–Kowalevski theorem to the initial value problem

$$
\theta(t_0) = 0, \quad \xi(t_0) = 7r
$$

to conclude the convergence of the series (2.8) and (2.12). Then we may apply the transformation to see that the Laurent series (2.4) is also convergent.

We summarize the key steps in the computation above:

1. First introduce a new variable θ through the indicial normalization of some variable, so that θ behaves like (a nonzero multiple of) $t - t_0$.
2. Use θ to substitute $t - t_0$, so that t_0 will not appear in the Laurent series for the other variables. Because the change of variables we try to find should be independent of specific solutions, we should eliminate any trace of t_0, which depends on specific solutions.
3. Introduce more new variables by successively truncating at the free parameters in the θ-Laurent series obtained in the second step. This brings the hidden free parameters to the front—the initial term in the power series for the new variables.

Finally, note that in order to get enough number of new variables (so that we have a change of equal number of variables at the end), the number of free parameters appearing in the Laurent series must be the order of the equation minus one. The condition is exactly what we mean by passing the Painlevé test in the most strict sense.

3. The Algorithm

The change of variables (2.7) and (2.11) already appeared in Painlevé's proof [**10**] that all solutions of (2.1) are meromorphic functions on the whole complex plane. In fact, the method we used to derive them in the last section closely resembles Painlevé's original. Next we present a more direct and systematic way of deriving such changes of variables, for equations of the form

$$u^{(n)} = f(t, u, u', \dots, u^{(n-1)}).$$

We will still illustrate our algorithm with the example of the first Painlevé equation (2.1).

STEP 1. Indicial normalization.

This is the same as the Painlevé test. We use the dominant balance to find the order k of the pole. Then we introduce the indicial normalization $u = \theta^{-k}$. For the first Painlevé equation, we have $k = 2$ and $u = \theta^{-2}$.

STEP 2. Laurent series.

We compute the Laurent series *in powers of* θ for the derivatives of u. First postulate a θ-power series

$$\theta' = A = a_0 + a_1\theta + a_2\theta^2 + \cdots \tag{3.1}$$

where a_* are functions of t. Thinking of A as functions of t and θ, we denote

$$\begin{aligned}
A_t &= a_0' + a_1'\theta + a_2'\theta^2 + \cdots, \\
A_\theta &= a_1 + 2a_2\theta + 3a_3\theta^2 + \cdots, \\
A_{t^2} &= a_0'' + a_1''\theta + a_2''\theta^2 + \cdots, \\
A_{t\theta} &= a_1' + 2a_2'\theta + 3a_2'\theta^2 + \cdots, \\
A_{\theta^2} &= 3a_2 + 6a_3\theta + 12a_4\theta^2 + \cdots, \\
&\vdots
\end{aligned}$$

Then we have

$$\begin{aligned}
u' &= -k\theta^{-k-1}A \\
&= -ka_0\theta^{-k-1} - ka_1\theta^{-k} - ka_2\theta^{-k+1} + \cdots, \\
u'' &= (-k)(-k-1)\theta^{-k-2}A^2 - k\theta^{-k-1}A_t - k\theta^{-k-1}AA_\theta \\
&= k(k+1)a_0^2\theta^{-k-2} + \big[k(2k+1)a_0a_1 - ka_0'\big]\theta^{-k-1} \\
&\qquad + [2k^2a_0a_2 - ka_1' + k^2a_1^2]\theta^{-k} + \cdots, \\
&\vdots \\
u^{(n)} &= (-k)\cdots(-k-n+1)\theta^{-k-n}A^n + \cdots \\
&= (-1)^n k\cdots(k+n-1)a_0^n\theta^{-k-n} + \cdots.
\end{aligned}$$

These are θ-Laurent series, with functions of t, a_*, and derivatives of a_* as coefficients. Moreover, these series actually do not depend on the equation. They are formal series depending only on k.

Now we substitute $u = \theta^{-k}$ and the θ-Laurent series of $u', u'', \dots, u^{(n-1)}$ to the right side of the equation, so that the right side becomes a θ-Laurent series. Compared with the θ-Laurent series of $u^{(n)}$ on the left side, we get a recursive relation. Solving the recursive relation will determine the coefficients a_* and the θ-Laurent series of the derivatives of u.

For the first Painlevé equation, the θ-Laurent series on the right side is simply $6\theta^{-4} + t$. Therefore the recursive relation becomes

$$2(2+1)a_0^2 = 6, \quad 2(4+1)a_0a_1 - 2a_0' = 0, \quad 8a_0a_2 - 2a_1' + 4a_1^2 = 0, \dots .$$

We fix a solution $a_0 = 1$ of the first equation and then recursively determine the later coefficients:

$$a_0 = 1, \quad a_1 = 0, \quad a_2 = 0, \dots .$$

As in the Painlevé test, we found that a_6 can be arbitrary, and the θ-power series for θ' is

$$\theta' = 1 + \frac{t}{4}\theta^4 + \frac{1}{4}\theta^5 + a_6\theta^6 - \frac{t^2}{32}\theta^8 + \cdots .$$

This easily gives rise to the θ-Laurent series (2.10) for u'.

STEP 3. New Variables.

We introduce the new variables (in addition to θ introduced in the first step) by truncating the derivatives of u at the places where free coefficients of the θ-power series of θ' appear. Let $a_i, a_j, \dots$, with $i < j < \dots$, be free coefficients in (3.1). Then we introduce a new variable ξ by truncating the θ-Laurent series of u' at θ^{-k+i-1}:

$$u' = u_{1,0}\theta^{-k-1} + u_{1,1}\theta^{-k} + \cdots + u_{1,i-1}\theta^{-k+i-2} + \xi\theta^{-k+i-1}.$$

Note that since i is the first index where a free coefficient appears (i.e., i is the smallest *resonance*), the coefficients $u_{1,0}, u_{1,2}, \dots, u_{1,i-1}$ are functions of t only.

For the first Painlevé equation, we simply truncate (2.10) at θ^3 to obtain (2.11), and the whole algorithm is complete.

If $n > 2$, then we need to further introduce a new variable η by truncating the θ-series of u'' at θ^{-k+j-2}. However, we cannot do this immediately because the θ-series of u'' we have at the moment has a_i appearing in the coefficients. We need to update the θ-series of u'' by replacing a_i with the "equivalent" variable ξ. Here we say a_i and ξ are equivalent because we have brought the hidden arbitrary choice a_i to the front, as the initial coefficient of ξ in a power series expansion. Specifically, from

$$\xi = u_{1,i} + u_{1,i+1}\theta + u_{1,i+2}\theta^2 + \cdots, \qquad u_{1,i} = -ka_i + \cdots,$$

we may rewrite a_i as a θ-series, with $-k^{-1}\xi$ as the initial term and functions of t, ξ, a_j, ..., as coefficients of the subsequent terms. Then we substitute the θ-power series for a_i into the θ-Laurent series for u'' and introduce a new variable η by truncating at θ^{-k+j-2}:

$$u'' = u_{2,0}\theta^{-k-2} + u_{2,1}\theta^{-k-1} + \cdots + u_{2,j-1}\theta^{-k+j-3} + \eta\theta^{-k+j-2}.$$

The coefficients $u_{2,0}$, $u_{2,2}, \dots, u_{2,i-1}$ are functions of t only, and the coefficients $u_{2,i}, u_{2,i+1}, \dots, u_{2,j-1}$ are functions of t and ξ only.

The process continues until we introduce the final new variable by truncating the (updated) θ-series of $u^{(n-1)}$.

4. Comments

The first two steps in our algorithm closely resembles the Painlevé test. Basically the only new ingredient is the use of θ instead of $t-t_0$ as the basis of expansion. The extra computation this causes is not much: whenever taking derivatives, we need to substitute θ' by (3.1). It is quite easy to set up an equivalence between the coefficients of the θ-power series of θ' in our algorithm and the coefficients of the ordinary Laurent series of u' in the Painlevé test. We emphasis that the algorithm does not require any computational result from the Painlevé test. The first two steps involve about the same amount of computation as the Painlevé test and is equivalent to the Painlevé test.

Because of the equivalence, the free coefficients in our algorithm correspond to the free coefficients in the Painlevé test, and their numbers are the same. Note that in order for the number of new variables (θ and the ones introduced in the third step) to be n, the number of free coefficients must be $n-1$. In other words, the third step of our algorithm produces a change of variables if and only if the equation passes the Painlevé test (in the sense that there are enough number of nonnegative integral resonances).

The new system for $\theta, \xi, \eta, \ldots,$ must be regular, due to the fact that we have brought all the free parameters to the front. See [**6**] for a complete and rigorous proof. Here we only mention that the essence of the proof is the following general fact: Suppose a first order system of differential equations has functions meromorphic in θ and analytic in ξ, η, $\cdots$ on the right side. Suppose the system has formal power series solutions of the form

$$\begin{aligned}
\theta &= a(t-t_0) + b(t-t_0)^2 + \cdots \\
\xi &= \xi_0 + \xi_1(t-t_0) + \xi_2(t-t_0)^2 + \cdots \\
\eta &= \eta_0 + \eta_1(t-t_0) + \eta_2(t-t_0)^2 + \cdots \\
&\vdots
\end{aligned}$$

where a is some fixed nonzero number and $t_0, \xi_0, \eta_0, \ldots,$ can be arbitrary. Then the functions on the right side are also analytic in θ (near $\theta = 0$).

Our algorithm is also applicable to systems of equations as well as partial differential equations [**3, 4**]. We have also demonstrated that, for Hamiltonian systems, it is possible construct the change of variables so that it is canonical [**5**]. All these have been theoretically established under suitable conditions [**6**].

There are still many open questions. One particular question is to find suitable changes of variables in less than ideal situation, especially in case some resonances are negative. These include the so called lower balances in the Painlevé test.

References

1. M. J. Ablowitz and P. A. Clarkson, *Solitons, nonlinear evolutions and inverse scattering*, Cambridge University Press, Cambridge, 1991.
2. M. Ablowitz, A. Ramani, and H. Segur, *A connection between nonlinear evolution equations and ordinary differential equations of P-type* I., J. Math. Phys. **21** (1980) 715–721; II, J. Math. Phys. **21** (1980), 1006–1015.
3. J. S. Hu and M. Yan, *Singularity analysis for integrable systems by their mirrors*, Nonlinearity **12** (1999), 1531–1543.
4. ______, *The mirror systems of integrable equations*, Stud. Appl. Math. **104** (2000), 67–90.
5. ______, *Symplectic structure of the Painlevé test*, J. Nonlinear Math. Phys. **8** (2001), 145–148.

6. ———, *Analytical aspects of the Painlevé test*, Preprint, 2000.
7. ———, *An elementary and direct proof of the Painlevé property for the Painlevé equations* I, II *and* IV, Preprint 2001.
8. S. Kowalevski, *Zur Theorie der partiellen Differentialgleichungen*, J. Reine Angew. Math. **80** (1875), 187–207.
9. ———, *Sur le problème de la rotation d'un corps solide autour d'un point fixé*, Acta Math. **12** (1889), 117–232.
10. P. Painlevé, *Mémoires sur les équations différentielles dont l'intégrale générale est uniforme*, Bull. Soc. Math. France, **28** (1900), 201–261.
11. J. Weiss, M. Tabor and G. Carnevale, *The Painlevé property for partial differential equations*, J. Math. Phys. **24** (1983), 522–526.

Department of Mathematics, The Hong Kong University of Science and Technology, Clear Water Bay, Kowloon, Hong Kong, China.

E-mail address, J. Hu: majhu@ust.hk

E-mail address, M. Yan: mamyan@ust.hk

Centre de Recherches Mathématiques
CRM Proceedings and Lecture Notes
Volume **32**, 2002

Monodromy Deformation Approach to the Scaling Limit of the Painlevé First Equation

Andrei A. Kapaev

ABSTRACT. The isomonodromy deformation equation for a 2×2 matrix linear ODE with a large parameter can be locally reduced to a (hyper)elliptic equation. To globalize this result, we apply the isomonodromy deformation method and obtain the modulation equations for the asymptotic algebraic curve. The method is applied to the degenerate solution of the Painlevé first equation.

1. Introduction

The classical Painlevé equations are found by P. Painlevé and B. Gambier [**27, 73**] as the only up to Möbius transformations irreducible second-order ODEs $y_{xx} = R(x, y, y_x)$ with the rational in y_x, algebraic in y and analytic in x right-hand side whose general solutions are free from movable critical points. The last condition is usually referred to as the Painlevé property while the equations which enjoy this property are called the Painlevé-type equations. The term *irreducible* means that the generic solutions of the Painlevé equations can not be expressed in terms of elementary or classical special functions and hence deserve the name of *Painlevé transcendents*.

The main tool in the classical works [**27, 73**] is the α-test of Painlevé based on the assertion that the limiting form of the scaled Painlevé-type equation, or the *reduced form*, also possesses the Painlevé property [**33**]. E.g., the Painlevé second equation PII,

$$y_{xx} = 2y^3 + xy - \alpha, \tag{1.1}$$

scaled in accord with the relations

$$x = \delta^{-2/3}x_0 + \delta^{1/3}t, \quad y = \delta^{-1/3}v, \quad y_x = \delta^{-2/3}v_t, \quad \alpha = \delta^{-1}\beta_0 + \beta_1 \tag{1.2}$$

where t is the new independent variable, v is the new dependable, and x_0, β_0, β_1 are some constant parameters, becomes

$$v_{tt} = 2v^3 + x_0v - \beta_0 + \delta(tv - \beta_1). \tag{1.3}$$

2000 *Mathematics Subject Classification.* Primary: 34M55; Secondary: 34M60, 34M40, 34M30.

This work was partially supported by RFBR under grant number 99–01–00687. The author is grateful to a referee for his valuable comments and to A. Kitaev for remarks.

This is the final form of the paper.

At $\delta = 0$, the latter is known to have the Painlevé property since it can be integrated by the use of elliptic functions. If $x_0 = -6$ and $\beta_0 = -4$, the additional scaling changes $t = \delta^{-1/5}\tau$, $v = 1 + \delta^{2/5}z$, $\beta = -\delta^{-1/5}\gamma$, turn equation (1.3) into $z_{\tau\tau} = 6z^2 + \tau + \gamma + \delta^{2/5}(2z^3 + \tau z)$, which, at $\delta = 0$, coincides with the Painlevé first equation PI.

The scaling reduction of the Painlevé equation to the elliptic equation means that the Painlevé function is described *locally* by the elliptic function. This fact was justified for equations PI and PII by P. Boutroux [**5**] and later by P. Doran-Wu and N. Joshi [**10, 40, 41**]. In our terms, P. Doran-Wu and N. Joshi studied the local behavior of the function $v(t)$ ($x_0 = 0$) with the initial condition $(t, v, v_t) = (t_0, \eta, \eta')$ where t_0, η, η' are such constants that $D := \eta'^2 - \eta^4 + 2\beta_0\eta \neq 3(\beta_0/2)^{4/3}e^{i(2\pi)/(3)k}$, $k \in \mathbb{Z}$. They proved that, in the region $|t - t_0| < |\ln\delta|$, in the leading order with respect to $\delta \to 0$, the function $v(t)$ is uniformly approximated by the elliptic function $V(t)$, $V'^2 = V^4 - 2\beta_0 V + D$.

Any extension of the above result beyond the indicated boundary involves *modulation* of the elliptic curve, i.e. certain change of D when x_0 in (1.2) varies. For the first time, the modulation equations for the elliptic asymptotics as $|x| \to \infty$ of the Painlevé first and second functions were found by P. Boutroux [**5**]. Namely, letting in (1.1) $y = \sqrt{x}v$, $t = 2/3x^{3/2}$ and integrating, he has found the equation $(v')^2 = v^4 + v^2 + D(t)$, where $D(t)$ is a transcendent function. Using integral estimates, P. Boutroux obtained the variation of $D(t)$ on a period $\omega_k = \oint_{\mathcal{L}}(V^4 + V^2 + D_k)^{-1/2}\,dV$ of the local elliptic ansatz, $(V')^2 = V^4 + V^2 + D_k$, $D_k = \text{const.}$, and has shown that, along the unbounded "line of periods", there exists $\lim_{k\to\infty} D_k = D_\infty$. The latter satisfies the transcendent *Boutroux equation*

$$\oint_{\mathcal{L}} \sqrt{V^4 + V^2 + D_\infty}\,dV = 0. \tag{1.4}$$

Nowadays it is clear that (1.4) is strongly related to the Hamiltonian structure associated to the Painlevé equation and appears as the asymptotic form of the *relative Poincaré invariant* [**2**] for PII.

Another approach to the modulation equations is provided by the method of G. Kuzmak [**62**] who proposed to expand v in an ascending series in δ, $v = V + \delta v_1 + \cdots$ with a periodic leading term V, and to introduce, aside from the "quick" variable t, another, formally independent, "slow" variable $T = \delta t$ for description of the variation of the period and phase shift in $V(t, T)$. Eliminating "secular" terms from the linear differential equation for v_1, one finds an integral condition for $V(t, T)$ which is usually interpreted as the constancy of the Lagrange's action. This condition is the desired modulation equation for $D(T)$. Using this way, N. Joshi and M. Kruskal [**42**] described the modulation of the elliptic asymptotic solutions of PI and PII with respect to $\varphi = \arg x$ as $|x| \to \infty$.

Applying the Kuzmak's method and using the connection formulae of M. Ablowitz and H. Segur [**1, 78**] for the class of the physically interesting decreasing as $x \to +\infty$ solutions of PII ($\alpha = 0$), $y(x, a) \sim aAi(x) \sim a/(2\sqrt{\pi})x^{-1/4}e^{-2/3x^{3/2}}$, J. Miles [**67**] described the asymptotic dependence of the solution on the amplitude a as $a \to i\infty$ and $a \to 1 - 0$.

Using the same ideas, O. Kiselev [**57**] constructed the imaginary on the real axis solution of PII ($\alpha \to i\infty$) which is monotonic as $x \to +\infty$ and oscillates as $x \to -\infty$. The oscillating tail is described by the modulated elliptic function, $V'^2 = V^4 + x_0 V^2 + 2iV + D$, where D depends on x_0 in such a way that $\oint \sqrt{V^4 + x_0 V^2 + 2iV + D}\, dV \equiv 2\pi$.

An equivalent way leading to the modulation equations in a differential form is described by J. Whitham [**88**]. This approach utilizes averaging with respect to the "fast" variable rather than considers a single period of the elliptic function. This idea was applied by V. Novokshenov [**89, 90**] and V. Vereshchagin [**84–86**] to the asymptotic as $|x| \to \infty$ description of equations PII, PIII and PIV, and to description of the scaling limits in the classical Painlevé equations. The differential modulation equations for the scaling limits are found in the explicit form $\partial_T D = f(D, K, E)$, where T is the "slow" variable, D is the module, and K, E are the complete elliptic integrals.

In the most systematic way, the averaging method for the integrable "soliton" equations was developed by H. Flaschka, M. Forest and D. McLaughlin [**18**] and I. Krichever [**60**]. Applications of their ideas to equation PI $y_{xx} = 6y^2 + x$ are presented in the papers of I. Krichever [**61**], S. Novikov [**70**] and F. Fucito, A. Gamba, M. Martellini and O. Ragnisco [**26**]. In particular, the differential Whitham equations for PI imply the integral conditions for the asymptotic elliptic curve [**61**],

$$\mathrm{Im} \oint_{a,b} w(\lambda)\, d\lambda = h_{a,b} = \text{const.}, \qquad w^2 = 4\lambda^3 + 2x\lambda - g_3, \tag{1.5}$$

which yield the invariant g_3 in the ansatz $y(x) = \wp\big(x - x_0; -2x, g_3(x)\big)$ as a transcendent function of x.

Possibly, the most effective approach to the Painlevé equations is based on their isomonodromy interpretation. R. Fuchs [**25**] constructed a linear scalar second-order ODE with four Fuchsian singularities and one apparent singular point whose isomonodromy deformation is governed by the Painlevé sixth equation PVI. R. Garnier [**28**] has found the linear equations associated to the lower Painlevé equations imposing on the equation of R. Fuchs the scaling changes of variables of P. Painlevé [**74**] for the cascade of the successive reductions of the Painlevé equations to each other. These linear equations also form a cascade of reductions arising in some neighborhoods of the merging singularities and turning points. Comments and some technical details missed in the original paper of R. Garnier can be found in the article of P. Boutroux [**5**]. L. Schlesinger [**77**] developed the isomonodromy deformation theory for generic linear matrix equations with Fuchsian singular points, while R. Garnier [**30**] transformed his scalar equations to the matrix form and discovered the link between the isomonodromy and "isospectral" deformations [**29**]. In the papers of M. Jimbo, T. Miwa and K. Ueno [**39**] and H. Flaschka and A. Newell [**19**] the isomonodromy deformation theory was developed for the equations with non-Fuchsian singularities.

The asymptotics of the Painlevé functions as x tends to one of its fixed singularities were studied by M. Jimbo, A. Its, V. Novokshenov, B. McCoy, Sh. Tang, B. Suleymanov, A. Kitaev, A. A. Kapaev and others, see [**36,37,43,58,64,65,79**]. The modulation equations for the elliptic asymptotic solutions as $|x| \to \infty$ were discussed from the isomonodromy deformation viewpoint in [**46–48, 54, 89, 90**]. The majority of the mentioned authors used the WKB method as the basic ingredient of the asymptotic investigation of the linear equation.

H. Flaschka and A. Newell [**20**] observed that the WKB approximation to a solution of the λ-equation in the Lax pair, $\Psi_\lambda = \delta^{-1} A \Psi$, $d\Psi = U\Psi$, satisfies as $\delta \to 0$ the "isospectral" problem, $A\psi = \mu\psi$, $d\psi = U\psi$. Making use of this fact, K. Takasaki [**80**] avoided the averaging procedure and proposed the differential modulation equations simply comparing the WKB approximation for the first equation in the Lax pair and the Baker–Akhiezer function for the second equation.

The scaling limits in Painlevé equations find important applications in topological field theories and quantum gravity [**6, 11, 31, 68, 75**], in the theory of semi-classical orthogonal polynomials and random matrices [**24, 63, 66, 69, 76**]. In the series of papers [**21–23**], A. Fokas, A. Its and A. Kitaev described the continuous limit in the matrix model of quantum gravity as the isomonodromy scaling limit induced by the sequence of the Bäcklund transformations in PIV. Similarly, A. A. Kapaev and A. Kitaev [**53**] described the isomonodromy scaling limit from PII to PI, P. Bleher and A. Its [**4**] computed the asymptotics of the semi-classical orthogonal polynomials with the quartic weight related to the isomonodromy scaling limit in PIV, while J. Baik, P. Deift and K. Johansson [**3**] found the distribution law of the increasing subsequence of random permutations related to the isomonodromy scaling limit from PIII to PII. The elliptic asymptotic solutions for the isomonodromy scaling limits in PII and the corresponding modulation equations are found in [**49**]. All the modulation equations which previously appeared in the isomonodromy deformation context, see [**46–49, 54, 89, 90**], have the integral form (1.5) with the proper integrands $w(\lambda)$ and the trivial right-hand side, $h_a = h_b = 0$.

In Section 2, we recall the basic notions of the isomonodromy deformation theory of the linear systems with rational coefficients, introduce the scaling changes of variables which lead to the "isospectral" problem, and define the WKB approximation. In Section 3, we introduce the spectral curve and find the modulation equations for the (hyper)elliptic asymptotic solutions of the isomonodromy deformation equation, describe the domain of their validity and discuss their solvability. In Section 4, we apply the method to equation PI.

2. Isomonodromy Deformations and Scaling Limits

Here, we recall the basic facts of the theory of the matrix equations with rational coefficients following the text books of W. Wasow [**87**], F. Olver [**72**] and M. Fedorjuk [**16**], and of the isomonodromy deformation theory following M. Jimbo, T. Miwa and K. Ueno [**38, 39**] and H. Flaschka and A. Newell [**19**].

Consider a first order 2×2 matrix equation with rational coefficients,

$$\frac{d\Psi}{d\lambda} = A(\lambda)\Psi, \quad A(\lambda) = \sum_{\nu=1}^{n}\sum_{k=0}^{r_\nu} \frac{A_{\nu,-k}}{(\lambda - a^{(\nu)})^{k+1}} - \sum_{k=1}^{r_\infty} A_{\infty,-k}\lambda^{k-1}, \qquad \operatorname{Tr} A(\lambda) = 0. \tag{2.1}$$

Without loss of generality, $\lambda = \infty$ is the singular point of the highest Poincaré rank, i.e. $r_\infty \geq r_\nu$, $\nu = 1, \ldots, n$. We call equation (2.1) *generic* if the eigenvalues of $A_{\nu,-r_\nu}$ are distinct for $r_\nu \neq 0$ and if they are distinct modulo integers for $r_\nu = 0$. Let the generic equation (2.1) be diagonal at infinity, i.e. $A_{\infty,-r_\infty} = x^{(\infty)}_{-r_\infty}\sigma_3$ for $r_\infty \geq 1$, or $A_{\infty,0} = -\sum_{\nu=1}^{n} A_{\nu,0} = x^{(\infty)}_0\sigma_3$ for $r_\infty = 0$, where $\sigma_3 = \operatorname{diag}(1, -1)$.

The generic equation (2.1) has the formal solution near the singularity $a^{(\nu)}$,

$$\begin{aligned}&\Psi_\nu(\lambda) = W^{(\nu)}\widehat{\Psi}^{(\nu)}(\lambda)e^{\theta^{(\nu)}(\lambda)\sigma_3}, \quad \nu = 1, \dots, n, \infty,\\ &\widehat{\Psi}^{(\nu)}(\lambda) = I + \sum_{j=1}^{\infty} \psi_j^{(\nu)}\xi^j, \quad \theta^{(\nu)}(\lambda) = \sum_{j=1}^{r_\nu} x_{-j}^{(\nu)}\frac{\xi^{-j}}{(-j)} + x_0^{(\nu)}\ln\xi\end{aligned} \tag{2.2}$$

where $\xi = \lambda - a^{(\nu)}$ for a finite singular point $a^{(\nu)}$ and $\xi = 1/\lambda$ otherwise. The coefficients $\psi_j^{(\nu)}$, $x_{-j}^{(\nu)}$ of the formal expansion (2.2) are determined uniquely by the eigenvector matrix $W^{(\nu)}$ of $A_{\nu,-r_\nu}$. The formal solution $\Psi_\infty(\lambda)$ is fixed by the normalization condition $W^{(\infty)} = I$. The eigenvector matrix $W^{(\nu)}$ for the finite singularity $a^{(\nu)}$ is unique modulo permutation and scaling of the eigenvectors. Thus, $\Psi_\nu(\lambda)$ (2.2) is unique modulo the right constant diagonal and permutation matrix multipliers. In practice, this ambiguity does not make any difficulty, and we assume below that the eigenvector matrices $W^{(\nu)}$ are chosen in some convenient way.

If $r_\nu = 0$, the singularity is called *Fuchsian.* For Fuchsian singular point, the series $\widehat{\Psi}^{(\nu)}(\lambda)$ converges in a disk centered at $a^{(\nu)}$ [**87**], and the rational terms in the sum for the diagonal matrix $\theta^{(\nu)}(\lambda)$ are absent. If $r_\nu \neq 0$, expansion (2.2) gives an asymptotic representation for $\Psi(\lambda)$ in one of the *Stokes sectors* near $\xi = 0$, i.e. for $\arg\xi \in \big(\pi/(r_\nu)(k-1)+\varphi+\varepsilon, \pi/(r_\nu)(k+1)+\varphi-\varepsilon\big)$, where $\varphi = 1/(r_\nu)(\arg x_{-r_\nu}^{(\nu)} - \pi/2)$, $k = 1, \dots, 2r_\nu$, $\varepsilon > 0$.

A comprehensive description of the fundamental solutions of the nongeneric equation can be found in [**87**].

The ratio $\widetilde{\Psi}^{-1}(\lambda)\Psi(\lambda)$ of any two solutions Ψ and $\widetilde{\Psi}$ of (2.1) does not depend on λ. Ratios of the fundamental solutions normalized by (2.2) are usually called the *monodromy data.* The latter include the monodromy matrices M_ν which describe the branching of $\Psi(\lambda)$ near the singularities, the Stokes matrices $S_\nu^{(k)}$, $k = 1, \dots, 2r_\nu$, which describe the Stokes phenomenon near the non-Fuchsian singularities, and the connection matrices E_ρ^ν which are the ratios of the fundamental solutions normalized at distinct singular points. At this stage, we do not need more detailed description of the monodromy data, see [**19, 39**].

The set of *deformation parameters* is specified for generic equation (2.1) in [**39**]. These include the positions $a^{(\nu)}$ of the singular points and the coefficients $x_{-j}^{(\nu)}$, $j \neq 0$, for $\theta^{(\nu)}(\lambda)$ in (2.2), and form together the vector x (the extension of this definition to some nongeneric equations is given in [**17**]). Remaining parameters $x_0^{(\nu)}$ are usually called the *formal monodromy exponents.*

Using the linear transformation of the complex λ-plane, it is always possible to fix two of the deformation parameters, e.g. to move two of the finite singular points $a^{(\nu)}$ to 0 and 1. Alternatively, if $r_\infty \geq 2$, one may put $x_{-r_\infty}^{(\infty)} = 1$, $x_{1-r_\infty}^{(\infty)} = 0$. Thus the number of the actual deformation parameters is equal to

$$m_c = n + r - 2, \quad r = \sum_{\nu=1,\dots,n,\infty} r_\nu. \tag{2.3}$$

Let d denote the exterior differentiation with respect to entries of the vector of deformation parameters x. The monodromy data of (2.1) do not depend on x if and only if there exist 1-forms Ω and $\Theta^{(\nu)}$ such that the fundamental solutions above additionally satisfy equations [**39**]

$$d\Psi = \Omega\Psi, \quad dW^{(\nu)} = \Theta^{(\nu)}W^{(\nu)}, \qquad \nu = 1, \dots, n, \infty. \tag{2.4}$$

The compatibility condition of (2.1), (2.4),

$$dA = \frac{\partial\Omega}{\partial\lambda} + [\Omega, A], \quad d\Omega = \Omega\wedge\Omega, \tag{2.5}$$

is the completely integrable differential system [**39**] whose fixed singularities are the planes $a^{(\nu)} = a^{(\rho)}$, $\nu \neq \rho$, and $x^{(\nu)}_{-r_\nu} = 0$, $r_\nu \neq 0$ [**39**]. The generic system (2.5) corresponding to $m_c = 1$ is equivalent to one of the classical Painlevé equations [**19, 38**].

An extension of the isomonodromy deformation theory to the nongeneric equations is given in [**17**]. We also mention an exhaustive study [**15**] of a quite important particular nongeneric equation (2.1) with the matrix $A(\lambda)$ given by

$$A(\lambda) = \sum_{\nu=1}^{3} \frac{A_\nu}{\lambda - a^{(\nu)}} \tag{2.6}$$

where the residue matrices A_ν, $\nu = 1, 2, 3$, are all nilpotent and satisfy the normalization condition $A_1 + A_2 + A_3 = -\alpha\sigma_3$. If $a^{(1)} = 0$, $a^{(2)} = 1$, $a^{(3)} = x$, then $\Omega = -A_3/(\lambda - x)\,dx$, and the Schlesinger system which governs the isomonodromy deformations of (2.1), (2.6) is reduced to the particular case of the PVI equation,

$$\begin{aligned} y_{xx} = \frac{1}{2}\Big(\frac{1}{y} + \frac{1}{y-1} + \frac{1}{y-x}\Big)y_x^2 - \Big(\frac{1}{x} + \frac{1}{x-1} + \frac{1}{y-x}\Big)y_x \\ + \frac{y(y-1)(y-x)}{2x^2(x-1)^2}\Big((2\alpha-1)^2 + \frac{x(x-1)}{(y-x)^2}\Big) \end{aligned} \tag{2.7}$$

where y is the only zero of the entry $\big(A(\lambda)\big)_{12}$. Physical applications of (2.7) include the self-dual Bianchi IX model [**7, 71, 83**] and 2D-topological field theories [**12**].

Let us make in (2.1) the changes of variables,

$$\begin{aligned} &\lambda = \delta^{\varkappa}\zeta, \quad a^{(\nu)} = \delta^{\varkappa}(b^{(\nu)} + \delta c^{(\nu)}), \quad \varkappa = \text{const.}, && \nu = 1, \dots, n, \\ &A_{\nu,-k} = \delta^{k\varkappa-1}B_{\nu,-k}, \quad A_{\infty,-k} = \delta^{-k\varkappa-1}B_{\infty,-k}, && k = 0, \dots, r_\nu, \end{aligned} \tag{2.8}$$

so that, for generic equation (2.1),

$$x^{(\nu)}_{-k} = \delta^{k\varkappa-1}(t^{(\nu)}_{-k} + \delta\tau^{(\nu)}_{-k}), \quad x^{(\infty)}_{-k} = \delta^{-k\varkappa-1}(t^{(\infty)}_{-k} + \delta\tau^{(\infty)}_{-k}). \tag{2.9}$$

The constants $b^{(\nu)}$, $t^{(\nu)}_{-k}$, $k \neq 0$, form the vector t which mark the center of the asymptotic domain as $\delta \to +0$. Inequalities $|c^{(\nu)}|$, $|\tau^{(\nu)}_{-k}| <$ const., or, in the vector form, $\|\tau\| <$ const., give the range of the "local" deformation. The entries of t are called the "slow" variables, while the entries of τ are usually called the "quick" or "fast" variables. The parameters $t^{(\nu)}_0$ and $\tau^{(\nu)}_0$ form the vectors α and β.

REMARK 1. If $r_\infty \neq 0$ and $x^{(\infty)}_{-r_\infty}$ is fixed by the use of the proper change of the independent variable λ then equation (2.9) yields $\varkappa = -1/r_\infty$. If all the singularities are Fuchsian, i.e. $r_\nu = 0$, $\nu = 1, \dots, n, \infty$, one may put for simplicity $\varkappa = 0$ which yields the famous autonomous Garnier system [**29**]. E.g., using in (2.7) the changes $x = t + \delta\tau$, $\alpha = \delta^{-1}\alpha_0 + \alpha_1$, $\delta \to 0$, we obtain the reduced autonomous equation $y_{\tau\tau} = \frac{1}{2}(1/y + 1/(y-1) + 1/(y-t))y_\tau^2 + 2\alpha_0^2(y(y-1)(y-t))/(t^2(t-1)^2)$, which has the first integral

$$D_0 = \frac{t^2(t-1)^2}{4y(y-1)(y-t)}y_\tau^2 - \alpha_0^2 y, \tag{2.10}$$

and therefore is solvable in terms of elliptic functions.

REMARK 2. Interpreting the change of the formal monodromy exponents $x_0^{(\nu)}$ as the result of the Schlesinger transformation action which preserve all the monodromy data [**21–23, 38**], we arrive at the quantization condition, $x_0^{(\nu)} = \delta^{-1}t_0^{(\nu)} + \tau_0^{(\nu)} = \rho^{(\nu)} + l^{(\nu)}$ with the complex constants $\rho^{(\nu)}$ and integers $l^{(\nu)}$, $\nu = 1, \ldots, n, \infty$. Hence $l^{(\nu)}$ and δ^{-1} are large and comparable, $\lim_{\delta\to+0} l^{(\nu)}\delta = t_0^{(\nu)} = \text{const.}$, while $\tau_0^{(\nu)}$ remain bounded as $\delta \to +0$. Furthermore, since $\delta > 0$ and $l^{(\nu)} \in \mathbb{Z}$,

$$t_0^{(\nu)} \in \mathbb{R}. \tag{2.11}$$

Thus the quantized real part of the vector α of the formal monodromy exponents becomes the vector of discrete deformation parameters with the step $\mathcal{O}(\delta)$. Redefining δ, we normalize the vector $\operatorname{Re}\alpha$ which thus contains n free parameters.

Using (2.8) in (2.1), (2.4), we obtain

$$\begin{gathered}\frac{d\Psi}{d\zeta} = \delta^{-1}B(\zeta)\Psi, \quad d\Psi = \omega\Psi, \\ B(\zeta) = \sum_{\nu=1}^{n}\sum_{k=0}^{r_\nu} \frac{B_{\nu,-k}}{(\zeta - b^{(\nu)} - \delta c^{(\nu)})^{k+1}} - \sum_{k=1}^{r_\infty} B_{\infty,-k}\zeta^{k-1}.\end{gathered} \tag{2.12}$$

Compatibility of (2.12) reads

$$dB = [\omega, B] + \delta\frac{\partial\omega}{\partial\zeta}, \quad d\omega = \omega\wedge\omega. \tag{2.13}$$

The method of constructing the approximate solution of the first of equations (2.12) is comprehensively described in [**16, 87**]. Let $\mathfrak{R} \subset \mathbb{C}$ be a closed simply connected domain satisfying the following conditions: $B(\zeta)$ is holomorphic in $\mathfrak{R}$, and eigenvalues $\mu_j(\zeta)$ of $B(\zeta)$, $j = 1, 2$, have no zero in $\mathfrak{R}$. This domain can be obtained by removal of independent of δ neighborhoods of all the singularities and turning points, i.e. zeros of $\mu_1(\zeta) - \mu_2(\zeta) = 2\mu(\zeta)$, supplemented by certain cuts of the complex ζ-plane. Given $N \in \mathbb{N}$, equation (2.12) can be diagonalized in $\mathfrak{R}$ up to the order $\mathcal{O}(\delta^N)$. Integrating the leading diagonal part, one finds the WKB approximation, which is holomorphic and invertible in $\mathfrak{R}$ [**16**],

$$\Psi_{\text{WKB}}(\zeta) = W(\zeta)\Big[I + \sum_{k=1}^{N-1}\delta^k W^{(k)}(\zeta)\Big]\exp\Big\{\delta^{-1}\int^{\zeta}\Lambda(\zeta)\,d\zeta\Big\}, \tag{2.14}$$

$$\begin{gathered}\Lambda(\zeta) = \mu\sigma_3 - \delta\operatorname{diag}(W^{-1}W_\zeta) + \sum_{k=2}^{N-1}\delta^k\Lambda^{(k)}(\zeta), \\ W = (W_1, W_2), \quad BW_j = \mu_j W_j, \qquad j = 1, 2, \ \mu = \mu_1 = -\mu_2.\end{gathered} \tag{2.15}$$

Here the off-diagonal $W^{(k)}(\zeta)$ and diagonal $\Lambda^{(k)}(\zeta)$ matrices are recursively determined by the eigenvector matrix $W(\zeta)$.

The contour $\gamma_j(a, b) \subset \mathfrak{R}$, $j = 1, 2$, connecting a and b is called the *j-canonical path* if $\operatorname{Re}\int_a^\zeta \mu_j(\zeta)\,d\zeta$, $\zeta \in \gamma_j(a, b)$, does not decrease as ζ runs from a to b along $\gamma_j(a, b)$. A closed simply connected domain $\mathcal{R} \subset \mathfrak{R}$ is called the *j-canonical domain* if there exists such a point a_j that $\forall b \in \mathcal{R}$ the contour $\gamma(a_j, b) \subset \mathcal{R}$ connecting a_j and b is homotopy equivalent to a j-canonical path. If $\mathcal{R}$ is j-canonical for both $j = 1$ and $j = 2$, it is called the *canonical domain.* In the canonical domain $\mathcal{R}$, the

WKB approximation (2.14) approaches the true solution of (2.12) with the uniform in $\zeta \in \mathcal{R}$ accuracy $\mathcal{O}(\delta^N)$ [**16**].

Any small enough neighborhood of a given point $\zeta_0 \in \mathfrak{R}$ is a canonical domain. This domain can be extended as follows. Let $\gamma_0(a_1, a_2) = \{\zeta \in \mathfrak{R}\colon \operatorname{Im}\int_{\zeta_0}^{\zeta} \mu(\zeta)\, d\zeta = 0\}$ be a segment of the *anti-Stokes level line* which passes through ζ_0 and has endpoints a_1 and a_2. The numbering of the endpoints is chosen in such a way that the oriented contour $\gamma_0(a_1, a_2)$ is the 1-canonical path (i.e. for $\mu_1 = \mu$) while $\gamma_0(a_2, a_1)$ is the 2-canonical path (i.e. for $\mu_2 = -\mu$). Consider an arbitrary point of the segment, $\zeta_* \in \gamma_0(a_1, a_2)$, and introduce the contour $\gamma_* = \{\zeta \in \mathfrak{R}\colon \operatorname{Re}\int_{\zeta_*}^{\zeta} \mu(\zeta)\, d\zeta = 0\}$, which is the segment of the *Stokes level line*. By construction, the union $\mathcal{R} = \cup_{\zeta_* \in \gamma_0(a_1, a_2)} \gamma_*$ of all the contours γ_* is the canonical domain.

REMARK 3. Near the non-Fuchsian singularity, the canonical domain $\mathcal{R}$ can be extended beyond the boundary of $\mathfrak{R}$ to fill out one of the Stokes sectors [**16**].

The approximate solution of the system (2.12) in the canonical domain can be constructed multiplying the WKB approximation (2.14) in a right diagonal matrix $Q(\tau)\big(I + \sum_{n=1}^{N-1} Q_n(\tau)\delta^n + \mathcal{O}(\delta^N)\big)$ independent of ζ. Below, we assume that the diagonal matrix is absorbed into the eigenvector matrix $W(\zeta)$, so that the approximate solution of the system (2.12) is given by (2.14).

3. Modulation of the Spectral Curve

In the canonical domain, integration of the system (2.12) is reduced in the leading order with respect to δ to the eigenvalue problem. Equations (2.13) at $\delta = 0$ with the eigenvalue problem (2.15) are basic ingredients of the algebro-geometric integration of the "soliton" equations, see [**13, 14, 35, 59**]. The "spectral curve" Γ is determined by the characteristic equation

$$\det\big(B(\lambda) - \mu I\big) = 0. \tag{3.1}$$

In our 2×2 traceless case, this is the hyper-elliptic curve $\mu^2 = -\det B(\lambda)$ of genus $g = n + r - 2$. The spectral curve of the original system (2.1) is clearly not stationary since $d(\det A) = -d(A^2) = -A(\partial\Omega)/(\partial\lambda) - (\partial\Omega)/(\partial\lambda)A$, but if scaled via (2.8), it varies "slowly",

$$d(\det B) = -\delta\Big(B\frac{\partial\omega}{\partial\zeta} + \frac{\partial\omega}{\partial\zeta}B\Big). \tag{3.2}$$

We adopt the term *singular* for the curve Γ whose topological properties as $\delta \neq 0$ differ from those at $\delta = 0$, e.g. when the singularities of (2.12) coalesce, or turning points merge with each other or with the singular points. Given a parameterization of the curve Γ, we define the *discriminant set* in the total space of the parameters as the set determining the singular curve.

Let Γ_{as} be the limiting $\delta = 0$ form of the spectral curve Γ (3.1) for generic equation (2.12),

$$\mu_{as}^2 = -\lim_{\delta \to 0} \det B = \frac{P_{2g+2}(\zeta)}{\prod_{\nu=1}^{n} (\zeta - b^{(\nu)})^{2(r_\nu + 1)}} \tag{3.3}$$

where $P_{2g+2}(\zeta)$ is a polynomial of degree $2g + 2$. Totally, the curve (3.3) depends on $2g + n + 1$ complex parameters which form the space $\mathcal{T} \otimes \mathcal{D}$. Here, $\mathcal{T}$ is the space of g complex deformation parameters t and of n real normalized deformation parameters $\operatorname{Re}\alpha$, see Remark 2. Below, we use the notation t for the combined vector $(t, \operatorname{Re}\alpha)$. $\mathcal{D}$ is the space of g complex invariants and of $n+1$ imaginary parts $\operatorname{Im}\alpha$. We call the asymptotic spectral curve Γ_{as} (3.3) *generic* if

1. $b^{(\nu)} \neq b^{(\rho)}$ for $\nu \neq \rho$,
2. $\det B_{\nu,-r_\nu} \neq 0$, and,
3. all the asymptotic turning points are simple.

Two first conditions hold true for the generic equation (2.12) with the parameters lying apart from the fixed singularities of (2.5) while the third condition can be violated. However, because the nongeneric curves Γ_{as} are localized on a union of algebraic surfaces of co-dimension one, a small variation of the parameters turns the nongeneric spectral curve into generic one. In contrast, the generic curve Γ_{as} is stable with respect to the small variation of the parameters, and the corresponding curve Γ is not singular at $\delta = 0$.

To illustrate the introduced notions, let us consider the spectral curve ($g = 1$) for the nongeneric equation (2.1), (2.6),

$$\mu^2 = \frac{\lambda\alpha^2 + H}{\lambda(\lambda-1)(\lambda-x)}, \qquad H = \frac{(x(x-1)dy/dx - y(y-1))^2}{4y(y-1)(y-x)} - \alpha^2 y. \tag{3.4}$$

Letting $x = t + \delta\tau$, $\alpha = \delta^{-1}\alpha_0 + \alpha_1$, $\alpha_0 \neq 0$, we find that $D = \delta^2 H \to D_0$ (2.10) as $\delta \to 0$. The asymptotic spectral curve for (3.4), $\mu_{as}^2(\lambda) = (\alpha_0^2(\lambda - \lambda_0))/(\lambda(\lambda - 1)(\lambda - t))$, $\lambda_0 = -D_0/\alpha_0^2$, depends on the only deformation parameter t. The set of the formal monodromy exponents includes the parameter α_0, where $\operatorname{Re}\alpha_0 = 1$ or $\operatorname{Re}\alpha_0 = 0$. The complex module coincides with λ_0. In the space of pairs (t, λ_0), the discriminant set for (3.4) is given by the union of planes $(t = 0) \vee (t = 1) \vee (\lambda_0 = 0) \vee (\lambda_0 = 1) \vee (\lambda_0 = t)$.

The generic curve (3.3) satisfies the following condition:

THEOREM 3.1. *Let the asymptotic spectral curve* (3.3) *for generic equation* (2.12) *be generic at the initial point* $(t_0, D_0) \in \mathcal{T} \otimes \mathcal{D}$. *Then there exists an open neighborhood* $\mathcal{U} \subset \mathcal{T}$ *of the point* t_0 *such that, for any closed path* $\ell \in \Gamma_{as}$,

$$J_\ell(t, D) := \operatorname{Re} \oint_\ell \mu_{as}(\zeta)\, d\zeta = h_\ell, \quad \forall t \in \mathcal{U}, \qquad h_\ell = J_\ell(t_0, D_0). \tag{3.5}$$

PROOF. Let $\ell \in \Gamma$ be an arbitrary closed contour with the base point ζ_0. Consider the solution $\Psi(\zeta)$ of (2.12) as a multivalued mapping of $\mathbb{C}\backslash\{b^{(1)}+\delta c^{(1)}, \dots, b^{(n)} + \delta c^{(n)}\}$ into $\mathrm{SL}(2, \mathbb{C})$ and the function $\Psi_{\mathrm{WKB}}(\zeta)$ (2.14) as a multivalued mapping of the punctured Riemann surface of the curve Γ into $\mathrm{SL}(2, \mathbb{C})$.

The projection $\pi(\ell)$ of the contour ℓ on the complex ζ-plane passes through a finite number of the overlapping canonical regions $\mathcal{R}_k$, $k = 1, \dots, s$, $\mathcal{R}_{s+1} = \mathcal{R}_1$. In each region $\mathcal{R}_k$, the function (2.14) gives a uniform in ζ approximation for the exact solutions $\Psi_k(\zeta)$ of the system (2.12). The exact solutions for the adjacent canonical regions $\mathcal{R}_k$, $\mathcal{R}_{k+1}$ differ from each other in a right multiplier G_k which is independent of ζ and $t \in \mathcal{T}$, i.e. $\Psi_{k+1}(\zeta) = \Psi_k(\zeta)G_k$. Thus

$$m_\ell\big(\Psi_{s+1}(\zeta)\big) = \Psi_1(\zeta) M_\ell G_1 \cdots G_s$$

where m_ℓ is the operator of the analytic continuation along ℓ, and M_ℓ is the monodromy matrix for $\Psi_1(\zeta)$ along ℓ. Using for $\Psi_k(\zeta)$ the WKB approximations, we find

$$\exp\Big\{\delta^{-1} \oint_\ell \Lambda(\zeta)\, d\zeta\Big\} = \big(I + \mathcal{O}(\delta)\big)G(\delta). \tag{3.6}$$

Due to continuous dependence of $\Psi_k(\zeta)$ and of the generic curve (3.3) on all the parameters, the topology of the curve and of the contour $\ell \in \Gamma$ preserve as soon as the point (t, D) is close enough to the initial point (t_0, D_0). Therefore the right-hand side of (3.6) preserves while the deformation parameters t vary in a neighborhood of t_0. Equating the leading orders of the left-hand side for (3.6) at the initial point t_0 and nearby points t, we arrive at (3.5). $\square$

Theorem 3.1 immediately provides us with the following assertions:

COROLLARY 3.2. *Let $\mathcal{U} \subset \mathcal{T}$ be an open domain and let (3.5) holds true. Let the curve Γ_{as} be generic at the point $t_1 \in \partial\mathcal{U}$. Then there exists an open domain $\mathcal{W} \subset \mathcal{T}$ such that $\mathcal{U} \subset \mathcal{W}$, $t_1 \in \mathcal{W}$, and (3.5) is valid $\forall t \in \mathcal{W}$.*

COROLLARY 3.3. *Let $\mathcal{U}, \mathcal{U}' \subset \mathcal{T}$ be contiguous open domains and $\Sigma \subset \{\partial\mathcal{U} \cap \partial\mathcal{U}'\} \neq \emptyset$. Let $J_\ell(t, D) = h_\ell$ $\forall t \in \mathcal{U}$, $J_\ell(t, D) = h'_\ell$ $\forall t \in \mathcal{U}'$, and the integral $J_\ell(t, D)$ be continuous across Σ. Then $h_\ell = h'_\ell$.*

In accord with Corollaries 3.2 and 3.3, equation (3.5) is valid in a subdomain $\mathcal{U} \subset \mathcal{T}$ bounded by the points where the curve Γ_{as} degenerates. In other words, if $\mathcal{I}$ is the integral manifold for (3.5) parameterized by $t \in \mathcal{T}$, then the boundary $\partial\mathcal{U}$ is the projection on $\mathcal{T}$ of the intersection of $\mathcal{I}$ with the discriminant set of Γ_{as}. Beyond this boundary, the equation (3.5) may be not valid.

Varying the contour ℓ in (3.5), we obtain the system of equations $J_{\ell_j} = h_j$, $j = 1, \dots, 2g+n+1$, where the contours ℓ_j form a homology basis of the punctured Riemann surface of Γ_{as}. For instance, taking a small circle c_ν around $\zeta = b^{(\nu)}$ as the integration path ℓ, we obtain the equation for entries of $\operatorname{Im}\alpha$,

$$\operatorname{Im} t_0^{(\nu)} = -\frac{h_{c_\nu}}{2\pi} = \text{const.}, \qquad \nu = 1, \dots, n, \infty, \tag{3.7}$$

which contains the reality condition (2.11) as a particular case.

To discuss (3.5) further, it is convenient to impose the conditions (3.7) from the very beginning and to remove $n+1$ small circles from the "sufficient" set of contours. The remaining contours ℓ_j, $j = 1, \dots, 2g$, form a homology basis of the Riemann surface of the algebraic curve $w^2 = P_{2g+2}(\zeta)$.

REMARK 4. We refer to the systems of $2g$ real transcendent equations (3.5) over the homology basis of the curve $w^2 = P_{2g+2}(\zeta)$ as the *Krichever's system.* Its particular case with the trivial right-hand side $h_j = 0$, $j = 1, \dots, 2g$, is called the *Boutroux' system.*

The real dimension $2g$ of $\mathcal{D}$ coincides with the number of the real equations in the Krichever's system (3.5). The functions $J_{\ell_j}(t, D)$, $j = 1, \dots, 2g$, are the independent first integrals of the completely integrable Pfaffian system $dJ = 0$,

$$\omega \begin{pmatrix} dD \\ d\bar{D} \end{pmatrix} = -\Omega \begin{pmatrix} dt \\ d\bar{t} \end{pmatrix} \tag{3.8}$$

where $\omega = \{\omega_{ij}, \overline{\omega}_{ij}\}$ and $\Omega = \{\Omega_{ik}, \bar{\Omega}_{ik}\}$ are the $2g \times 2g$ and $2g \times (2g+n)$ matrices of the partial derivatives, $\omega_{ij} = \partial J_{\ell_i}/\partial D_j$, $\Omega_{ik} = \partial J_{\ell_i}/\partial t_k$, $i = 1, \dots, 2g$, $j = 1, \dots, g$, $k = 1, \dots, g+n$. Here, the bar means the complex conjugation. The Pfaffian system (3.8) provides us with the modulation equation in the differential form. If contracted to a real plane $\operatorname{Im} t = 0$, (3.8) yields the system which is equivalent to the modulation equations of Whitham type.

The scaling limits in the classical Painlevé equations are described generically by elliptic equations, $g = 1$. If $g = 1$ and $\theta := \partial/\partial D\mu_{as}(\zeta, t, D)\, d\zeta$ is a holomorphic 1-form on the Riemann surface of Γ_{as}, then the classical inequality $\operatorname{Im}(\overline{\oint_a \theta} \cdot \oint_b \theta) > 0$ provides us with the condition $\det \omega \neq 0$. Since the curve Γ_{as} is generic, the integral manifold for the Pfaffian system (3.8) passing through the initial point (t_0, D_0) is well parameterized by $2g + n$ real deformation parameters t, $\bar{t}$, and the Krichever's system (3.5) determines a differentiable in the real sense complex function $D(t, \bar{t})$. Below, we omit the dependence of D on $\bar{t}$.

In some scaling limit problems, the Painlevé function is fixed by the choice of the monodromy data rather then by initial conditions. The dependence of the unknowns on the additional parameter δ means that we consider a fiber space $\mathcal{E} = (\mathcal{P}, \pi, B)$ where π is the projection of the total space $\mathcal{P}$ on the base space B of the scaling parameter δ. Each fiber $\mathcal{P}(\delta)$ is the functional space of the Painlevé functions which is isomorphic to the monodromy surface $\mathcal{M}$ for the associated linear system. The function $y(., \delta)$ appears as a section of the fiber space $\mathcal{E}$. For instance, the sections of the fiber space $\mathcal{E}$ for the isomonodromy scaling limits yield a trivial, "horizontal" foliation. In this respect, it is quite important to relate the quantities h_ℓ in (3.5) with the monodromy data of the associated linear system.

Let us introduce the following notions [**16**]. The *Stokes line* is a Stokes level line emanating from a turning point. The *Stokes graph* is the union of all the Stokes lines. The *Stokes complex* is a connected component of the Stokes graph. The *Stokes chain* is a broken line whose links are the Stokes lines with the common turning points. We associate the singular point to a Stokes complex if there is a Stokes chain which approaches or encircles this point. The singular point can be associated to more than one Stokes complex. If the Stokes complex approaches or encircles the only singular point then this singularity is non-Fuchsian and there exist at least two Stokes lines of the Stokes complex approaching this point.

In our next assertion, we assume the generosity of the monodromy data. The assumption is specified in [**46–49, 54, 89**] for particular asymptotic problems in equations PII and PIV as the nontriviality of some of the Stokes multipliers and their combinations. Below, the assumption means that the ratios of the fundamental solutions normalized in accord with (2.2) at the singular points associated to the same Stokes complex are neither diagonal nor off-diagonal. It is clear that a small admissible variation of the monodromy data, i.e. the variation preserving the monodromy surface $\mathcal{M}$, turn the nongeneric monodromy data into generic ones while generic monodromy data are stable with respect to the small admissible variation.

THEOREM 3.4. *Let the parameters in generic equation* (2.12) *lie apart from the fixed singularities of* (2.5). *Let the corresponding monodromy data be generic and do not depend on δ as $\delta \to +0$. Then, for any closed path $\ell \in \Gamma_{as}$, the asymptotic curve satisfies the condition*

$$J_\ell(t, D) := \operatorname{Re} \oint_\ell \mu_{as}(\zeta)\, d\zeta = 0. \tag{3.9}$$

PROOF. We give here a sketch proof. Assume for a moment that for a small circle c_ν around $b^{(\nu)}$ taken as the integration path ℓ in (3.5) the equation (2.11) is violated. Then the formula $M_\nu S_\nu^{(1)} \dots S_\nu^{(2r_\nu)} = e^{2\pi i x_0^{(\nu)} \sigma_3}$, which relates the formal monodromy exponent $x_0^{(\nu)} = \delta^{-1} t_0^{(\nu)} + \tau_0^{(\nu)}$ with the monodromy and the

Stokes matrices of the fundamental solution normalized at $b^{(\nu)}$ by (2.2), shows the exponential dependence of the product on δ^{-1} which is a contradiction.

Let $J_\ell = h_\ell \neq 0$ for a contour ℓ which encircles two turning points. Thus the Stokes graph has at least two connected components $\mathcal{S}_j$ $j = 1, 2$. Without loss of generality, the domain $\mathcal{R}$ separating $\mathcal{S}_j$, $j = 1, 2$, does not contain neither Stokes lines nor singular points, and its boundary consists of two Stokes chains $\partial\mathcal{R}_j \subset \mathcal{S}_j$, $j = 1, 2$. Choose the turning points $a_1 \in \partial\mathcal{R}_1$ and $a_2 \in \partial\mathcal{R}_2$. By construction, $2\,\mathrm{Re}\int_{a_1}^{a_2} \mu_{as}(\zeta)\,d\zeta = h_\ell \neq 0$.

First compute the ratio of two fundamental solutions normalized by (2.2) at the singular points $b^{(j)}$ associated to the Stokes complexes $\mathcal{S}_1$ and $\mathcal{S}_2$. With this aim, choose a Stokes chain from a neighborhood of $b^{(j)}$ to a_j, $j = 1, 2$. Since the subdomain of $\mathcal{R}$ bounded by the Stokes level lines is a canonical domain for the WKB approximation (2.14) [**16**], we find the ratio as the product

$$G_{12} = e^{F_1\sigma_3} E_1 \exp\left\{\delta^{-1}\int_{a_2}^{a_1} \mu_{as}(\zeta)\,d\zeta\sigma_3\right\} E_2^{-1} e^{-F_2\sigma_3}$$

where $F_j = \left\{\delta^{-1}\int_{\zeta_j}^{\zeta} \mu\,d\zeta - \theta^{(j)}(\lambda)\right\}_{\lambda\to b^{(j)}}$. Modulo the terms which may behave like a power of δ^{-1}, the matrices E_j, $j = 1, 2$, can be represented by the products of the monodromy data of the reduced linear equations approximating (2.12) near the turning points and of the exponents of the phase integrals $\exp\{\delta^{-1}\int_{\zeta_k}^{\zeta_l} \mu\,d\zeta\sigma_3\}$ between the successive turning points of the Stokes chain, see [**36,50**]. Therefore, if neither of the matrices E_j is diagonal nor anti-diagonal, then, in the leading order, G_{12} exponentially depends on δ^{-1} which is a contradiction. If both or one of E_j is diagonal or anti-diagonal then $\mathrm{Re}\,F_j \neq 0$ for one of $j = 1, 2$. In the latter case, compute the ratio of two fundamental solutions normalized at the singular points of the same Stokes complex $\mathcal{S}_j$,

$$G_j = e^{F_j\sigma_3} N e^{-F_k\sigma_3}$$

where N may behave like a power of δ^{-1}. Due to assumption on G_j, the matrix N is neither diagonal nor anti-diagonal, thus, in the leading order, G_j exponentially depends on δ^{-1} that is a contradiction. □

REMARK 5. Equation (3.9) implies that the asymptotic as $\delta \to 0$ Stokes graph for (2.12) is connected.

REMARK 6. The assumption of Theorem 3.4 can be relaxed to the condition of boundedness or slow dependence of the monodromy data on δ^{-1}.

Absence of the initial condition for the Pfaffian system in the assumptions of Theorem 3.4 does not prevent the solvability of (3.9) for D. Indeed, the degeneration of the spectral curve due to merging of the turning points encircled by the contour ℓ allows us to find a point (t_0, D_0) on the discriminant set which could serve as the initial point for (3.8). For equation PII, this idea was used in [**49**] to prove the existence of the differentiable in the real sense function $D(t)$ satisfying (3.9). There is no principal obstacle to extend the proof for other classical Painlevé equations. The preliminary considerations for PIII and PIV which include the description of the corresponding discriminant sets can be found in [**51, 52**].

To illustrate the said above, consider again equation (2.12), (2.6) and its spectral curve (3.4). Assuming (3.9), we describe its Stokes graph as the union of three finite Stokes lines emanating from λ_0 and terminating at the simple poles $0, 1, t$.

The discriminant set, which is the union of the planes $t = 0$, $t = 1$, $t = \lambda_0$, $\lambda_0 = 0$, $\lambda_0 = 1$, intersects with the integral manifold for (3.9) along the real axis $\operatorname{Im} t = 0$, moreover the values $t < 0$, $0 < t < 1$ and $t > 1$ correspond to the planes $\lambda_0 = 0$, $t = \lambda_0$ and $\lambda_0 = 1$, respectively. Thus if the initial data for (2.7) at $t = t_0$, $\operatorname{Im} t_0 < 0$ satisfy the conditions $\lambda_0 \neq 0, 1, t_0$, then the asymptotic as $\delta \to 0$ Painlevé function is elliptic in the lower half of the complex t-plane and degenerates on the real line $\operatorname{Im} t = 0$. Whether the asymptotics is elliptic or remains degenerate above the real line depends on the chosen initial data, or, equivalently, on the corresponding monodromy data.

4. Scaling Limits in PI

In this section, we present the monodromy deformation approach to the scaling limits in the Painlevé first equation PI,

$$y_{xx} = 6y^2 + x. \tag{4.1}$$

Equation PI governs the isomonodromy deformations of the linear system (2.1) with

$$A(\lambda) = (4\lambda^4 + x + 2y^2)\sigma_3 - i(4y\lambda^2 + x + 2y^2)\sigma_2 - \left(2y_x\lambda + \frac{1}{2\lambda}\right)\sigma_1. \tag{4.2}$$

Here $\sigma_3 = \begin{pmatrix} 1 & 0 \\ 0 & -1 \end{pmatrix}$, $\sigma_2 = \begin{pmatrix} 0 & -i \\ i & 0 \end{pmatrix}$, $\sigma_1 = \begin{pmatrix} 0 & 1 \\ 1 & 0 \end{pmatrix}$. The set of the monodromy data for equation (4.2) consists of the Stokes matrices $S_k = \Psi_k^{-1}(\lambda)\Psi_{k+1}(\lambda)$ where $\Psi_k(\lambda) \sim \exp\big[(4/5\lambda^5 + x\lambda)\sigma_3\big]$ as $|\lambda| \to \infty$, $\arg\lambda \in \big(\pi/5(k-3/2), \pi/5(k+1/2)\big)$. The Stokes matrices satisfy the cyclic relation $S_1S_2S_3S_4S_5 = i\sigma_1$ [**44, 45**], thus the Painlevé function set is parameterized by the points of the 2-dimensional complex monodromy surface:

$$\begin{aligned} &\text{if } 1 + s_2s_3 \neq 0 \text{ then } s_1 = \frac{i - s_3}{1 + s_2s_3}, \quad s_4 = \frac{i - s_2}{1 + s_2s_3}, \quad s_5 = i(1 + s_2s_3), \\ &\text{if } 1 + s_2s_3 = 0 \text{ then } s_2 = s_3 = i, \quad s_5 = 0, \quad s_1 + s_4 = i. \end{aligned} \tag{4.3}$$

Physically interesting solutions of equation (4.1) have the nonoscillating asymptotic behavior $y \sim \pm\sqrt{-x/6} + \mathcal{O}(x^{-2})$ as $x \to -\infty$. Their existence has been proved by Ph. Holmes and D. Spence [**32**]. The solution with the asymptotics $y \sim -\sqrt{-x/6} + \mathcal{O}(x^{-2})$ is unique and corresponds to the values $s_2 = s_3 = 0$, $s_1 = s_4 = s_5 = i$, while the solutions with the asymptotics $y \sim \sqrt{-x/6} + \mathcal{O}(x^{-2})$ form a 1-parametric family distinguished by the condition $1 + s_2s_3 = 0$ [**44, 45**]. In more details,

$$y(x) \sim y_-(x) + a(-x)^{-1/8}e^{-8/5(3/2)^{1/4}(-x)^{5/4}}, \quad x \to -\infty \tag{4.4}$$

where $y_-(x) = \sqrt{-x/6} + \mathcal{O}(x^{-2})$ does not contain any free parameter, and

$$a = -\frac{(2/3)^{1/8}}{2\sqrt{2\pi}}\frac{s_1 - s_4}{2}, \quad s_1 + s_4 = i. \tag{4.5}$$

On the complex x-plane, the Painlevé function for $s_2 = s_3 = i$, $s_5 = 0$, $s_1 + s_4 = i$ has the asymptotics (4.4) as $|x| \to \infty$, $\arg x \in (3\pi/5, 7\pi/5)$ with a piece-wise constant amplitude a which shows a jump across the negative real axis [**44, 45, 81**]; on the rays $\arg x = \pm\pi/5, 3\pi/5, 7\pi/5$, the asymptotics is trigonometric; in the interior of the sectors between the indicated rays, the asymptotics is elliptic, see [**44, 45, 54**].

The scaling change of variables

$$x=\delta^{-4/5}(t_0+\delta t),\quad y=\delta^{-2/5}v,\quad y_x=\delta^{-3/5}v_t, \tag{4.6}$$

transforms (4.1) into

$$v_{tt}=6v^2+t_0+\delta t. \tag{4.7}$$

The reduced form of (4.1), i.e. equation (4.7) at $\delta=0$, is easy to integrate,

$$v_t^2=4v^3-g_2v-g_3,\quad g_2=-2t_0,\quad g_3=\text{const.}. \tag{4.8}$$

Thus, at $\delta=0$, $v(t)=\wp(t-\varphi;g_2,g_3)$ is the elliptic function of Weierstraß. Let $v_0=\sqrt{-t_0/6}$. For the particular value $g_3=-8v_0^3$, the elliptic function degenerates,

$$v(t)=v_0+\frac{3v_0}{\sinh^2(\sqrt{3v_0}(t-\varphi))}. \tag{4.9}$$

Below, we find the asymptotics of the degenerate Painlevé function, $s_2=s_3=i$, $s_5=0$, $s_1+s_4=i$, when $|s_4|$ is large. Also, we find the asymptotics of the "almost degenerate" Painlevé functions when $s_5\neq 0$ is small. In order to solve these problems, let us introduce the elliptic curve

$$w^2=(z-z_1)(z-z_3)(z-z_5)=z^3+\frac{t_0}{2}z+\frac{D_0}{4}. \tag{4.10}$$

The Riemann surface Γ_{as} for (4.10) consists of two complex z-planes pasted along the cuts $[-\infty,z_5]$ and $[z_1,z_3]$. Let $\arg w(z)\to 0$ as $z\to+\infty$ on the upper sheet. Let the a-cycle encircles counter-clockwise the cut $[z_1,z_3]$, and the upper half of the b-cycle connects z_5 and z_3. Introduce the integrals over a- and b-cycles

$$\omega_a=\frac{1}{2}\oint_a\frac{dz}{w(z)},\quad \omega_b=\frac{1}{2}\oint_b\frac{dz}{w(z)},\quad \tau=\frac{\omega_b}{\omega_a}, \tag{4.11}$$
$$I_\ell=\oint_\ell w(z)\,dz,\quad \Omega_\ell=\oint_\ell\frac{z\,dz}{w(z)},\quad \Omega_\ell^{(j)}=\Omega_\ell-2z_j\omega_\ell,\qquad \ell\in\{a,b\}.$$

Introduce the system of the Krichever's equations for the function $D_0(t_0)$,

$$2\operatorname{Re}I_a=0,\quad 2\operatorname{Re}I_b=h_b,\quad h_b=-\frac{8}{5}(3/2)^{1/4}. \tag{4.12}$$

Due to triviality of h_a, the system (4.12) admits the degeneration of the curve (4.10), $z_1=z_3=(-t_0/6)^{1/2}$, $D_0=8(-t_0/6)^{3/2}$. Thus the intersection of the discriminant set with the integral manifold for (4.12) satisfies equation $\operatorname{Re}(-t_0)^{5/4}=1$. Among five lines satisfying the latter equation, we denote by symbol γ the line passing through the point $t_0=-1$ and asymptotic to the rays $\arg t_0=\pm 3\pi/5$.

Finally, let us introduce the perforated region $\mathcal{R}(c_0,\varepsilon_1)$,

$$\mathcal{R}(c_0,\varepsilon_1)=\{t\in\mathbb{C}\colon |t|<c_0, |t-n\omega_a-m\omega_b|>\varepsilon_1>0, n,m\in\mathbb{Z}\}. \tag{4.13}$$

THEOREM 4.1. *Let $|s_4|\gg 1$ and $0<c'\le|s_1/s_4|\le c''$ for some constants c', c''. Then there exist the positive constant c_0 and the positive small constants ε, ε_1 such that, to the right of γ, the Painlevé function is given by*

$$y(x)=\delta^{-2/5}\wp(t-\varphi;-2t_0,-D_0),\quad t=\delta^{-1/5}x-\delta^{-1}t_0 \tag{4.14}$$

where $D_0(t_0)$ is determined by (4.12), and

$$\delta^{-1} = -\frac{\ln|s_4|}{h_b}, \tag{4.15}$$

$$t - \varphi = \delta^{-1}\frac{4}{5}t_0 + t + \frac{\omega_a}{2\pi i}\ln(is_4) - \frac{\omega_b}{2\pi i}\ln\frac{s_1}{s_4} + \mathcal{O}(\delta^{1/2-\varepsilon}) \in \mathcal{R}(c_0, \varepsilon_1). \tag{4.16}$$

PROOF. It follows from the integral estimates similar to presented in [**5, 10, 40, 41**] that there exists the solution $y(x) = \delta^{-2/5}\wp(t-\varphi; -2t_0, -D_0)$, $|t| <$ const., of the Cauchy problem for equation PI with the initial data $x_0 = \delta^{-4/5}t_0$, $y_0 = \delta^{-2/5}v_0$, $y_0' = \delta^{-3/5}v_0'$, where $|\delta| \ll 1$ and $D_0 = v_0'^2 - 4v_0^3 - 2t_0v_0$. Using (4.6) and (2.8) with $\varkappa = -1/5$ (see Remark 1) in (2.1), (4.2), we obtain equation (2.12), $\Psi_\zeta\Psi^{-1} = \delta^{-1}B$, where

$$B(\zeta) = (4\zeta^4 + t_0 + 2v^2 + \delta t)\sigma_3 - i(4v\zeta^2 + t_0 + 2v^2 + \delta t)\sigma_2 - \big(2v_t\zeta + \frac{\delta}{2\zeta}\big)\sigma_1. \tag{4.17}$$

The corresponding spectral curve is given by

$$\mu^2 = \mu_{as}^2 + \delta r(\zeta), \quad \mu_{as}^2 = 4\zeta^2(4\zeta^6 + 2t_0\zeta^2 + D_0) = 16\zeta^2 w^2(\zeta^2), \tag{4.18}$$

$$D := v_t^2 - 4v^3 - 2t_0 v = D_0 + \delta D_1, \tag{4.19}$$

$$r(\zeta) = 4\zeta^2 D_1 + 8t\zeta^2(\zeta^2 - v) + 2v_t + \frac{\delta}{4\zeta^2}.$$

Theorem 3.1 implies that the system (3.5), $2\operatorname{Re} I_a = h_a$, $2\operatorname{Re} I_b = h_b$, holds true, and the leading order $D_0(t_0)$ of the parameter D does not depend on t. Thus the definition of D (4.19) is consistent with the elliptic asymptotic ansatz (4.8),

$$v_t^2 = 4v^3 + 2t_0 v + D_0 = 4w^2(v), \quad v(t) = \wp(t - \varphi; -2t_0, -D_0). \tag{4.20}$$

To find the phase shift φ, we apply the isomonodromy deformation method [**36**]. Let us introduce notations $a_3 = (B)_{11}$, $a_+ = (B)_{12}$, $a_- = (B)_{21}$ for the entries of the matrix B (4.17) and assume the following:

1. $|v|$, $|v_t| \le c$ for some constant $c > 0$, and
2. zeros of $a_+(\zeta)$, or, alternatively, of $a_-(\zeta)$, lie apart from the paths of integration below.

Let us assume that the parameters t_0, D_0 determining the elliptic curve (4.10) and (4.20) are such that the asymptotic at $\delta = 0$ Stokes lines for the system (2.12), (4.17), emanating from $\zeta_1 = z_1^{1/2}$ and $\zeta_3 = z_3^{1/2}$ form the continuous chains $e^{-i3\pi/10}\infty \to \zeta_1 \to e^{-i\pi/10}\infty$, $e^{-i\pi/10}\infty \to \zeta_1 \to \zeta_3 \to e^{i\pi/10}\infty$ and $e^{i\pi/10}\infty \to \zeta_3 \to e^{i3\pi/10}\infty$. Then the result of [**50**] implies the expressions for the Stokes multipliers as $\operatorname{Re}(\delta^{-1}) \to +\infty$, $|\operatorname{Im}(\delta^{-1})| <$ const.,

$$\begin{aligned} \ln s_4 &= 2\delta^{-1}\left[\int_{\zeta_1^{(\sigma)}}^{\zeta} \nu_\sigma\, d\zeta - \theta + \sigma\delta/2 \ln\frac{a_\sigma}{\mu}\right]_{\zeta \to e^{-i\pi/10}\infty} - \sigma\ln(-2i) + \mathcal{O}(\delta), \\ \ln s_1 &= 2\delta^{-1}\left[\int_{\zeta_3^{(\sigma)}}^{\zeta} \nu_\sigma\, d\zeta - \theta + \sigma\frac{\delta}{2}\ln\frac{a_\sigma}{\mu}\right]_{\zeta \to e^{i3\pi/10}\infty} - \sigma\ln(-2i) + \mathcal{O}(\delta) \end{aligned} \tag{4.21}$$

where $\zeta_k^{(\sigma)}$ are zeros of $\nu_\sigma(\zeta)$, $\zeta_k^{(\sigma)}|_{\delta=0} = z_k^{1/2}$, $k = 1, 3$,

$$\nu_\sigma^2 = \mu_{as}^2 + \delta r + \sigma\delta\left(a_3' - a_3\frac{a_\sigma'}{a_\sigma}\right), \quad \theta = \frac{4}{5}\zeta^5 + (t_0 + \delta t)\zeta, \quad \sigma \in \{+, -\}. \tag{4.22}$$

Since the Stokes multipliers do not depend on t, and, at the critical point $t = t_5$, $v_t = 0$, $v = z_5$, the assumptions above are not violated for the non-degenerate curve (4.10), we may compute the right-hand side of (4.21) at this point. Using the change $\zeta^2 = z$, we transform the elliptic integrals in (4.21) as follows (look for similar details in [**46, 49**]):

$$\begin{aligned} \ln(-is_4) &= -2\delta^{-1}I_b - \frac{t_5}{2}\Omega_b^{(5)} - \frac{D_1(t_5)}{2}\omega_b + \mathcal{O}(\delta^{1/2-\varepsilon}), \\ \ln\frac{s_1}{s_4} &= -2\delta^{-1}I_a - \frac{t_5}{2}\Omega_a^{(5)} - \frac{D_1(t_5)}{2}\omega_a + \mathcal{O}(\delta^{1/2-\varepsilon}) \end{aligned} \tag{4.23}$$

where $\varepsilon > 0$ is small. The conditions $\ln|s_4| \gg 1$ and $c' \le |s_1/s_4| \le c''$ are consistent with (4.23) if $h_a = 0$ and $h_b < 0$. Substitutions $\delta \mapsto \alpha^{5/4}\delta$, $t_0 \mapsto \alpha t_0$, $t \mapsto \alpha^{-1/4}t$, $v \mapsto \alpha^{1/2}v$, $v_t \mapsto \alpha^{3/4}v_t$, which leave x, y and y_x invariant, allow us to demand for h_b the value in (4.12). The conditions on the Stokes graph imposed above are consistent with the reality condition, $D_0(t_0) \in \mathbb{R}$ for $t_0 > -1$. The latter implies that the chosen branch of $D_0(t_0)$ yields degeneration of the curve (4.10) on γ, and the validity domain of the elliptic asymptotic solution is located to the right of γ.

Using in (4.23) the identities

$$\omega_a I_b - \omega_b I_a = \frac{4\pi i}{5}t_0, \quad \omega_a\Omega_b - \omega_b\Omega_a = 4\pi i, \tag{4.24}$$

which follow from the Legendre identity, we find the critical point t_5,

$$t_5 = -\delta^{-1}\frac{4}{5}t_0 - \frac{\omega_a}{2\pi i}\ln(-is_4) + \frac{\omega_b}{2\pi i}\ln\frac{s_1}{s_4} + \mathcal{O}(\delta^{1/2-\varepsilon}),$$

and the expression for the argument of the $\wp$-function of Weierstrass (4.16). □

REMARK 7. Because ω_a and ω_b are periods of the elliptic function, the asymptotics (4.14) is invariant with respect to the choice of the argument of the Stokes multipliers.

THEOREM 4.2. *Let $0 < |s_5| \ll 1$, $0 < c' \le |s_2| \le c''$ for some constants c', c''. Then there exist positive constant c_0 and positive small constants ε, ε_1 such that, to the left of γ, the Painlevé function is given by* (4.14) *where $D_0(t_0)$ is determined by* (4.12) *and*

$$\delta^{-1} = \frac{\ln|s_5|}{h_b}, \tag{4.25}$$

$$t - \varphi = \delta^{-1}\frac{4}{5}t_0 + t - \frac{\omega_a}{2\pi i}\ln(is_5) + \frac{\omega_b}{2\pi i}\ln(is_2) + \mathcal{O}(\delta^{1/2-\varepsilon}) \in \mathcal{R}(c_0, \varepsilon_1). \tag{4.26}$$

PROOF. In essential, we repeat the proof of Theorem 4.1. Existence of the elliptic asymptotic solution of the Cauchy problem follows from the integral estimates similar to presented in [**5, 10, 40, 41**]. Using (4.6) in (2.1), (4.2), we obtain equation (2.12), $\Psi_\zeta\Psi^{-1} = \delta^{-1}B$, with the matrix B (4.17) whose spectral curve is given by (4.18). Applying Theorem 3.1, we arrive at the system (3.5) for the function $D_0(t_0)$, $2\,\mathrm{Re}\,I_a = h_a$, $2\,\mathrm{Re}\,I_b = h_b$. The definition of D_0 (4.19) is consistent with the elliptic asymptotic ansatz (4.8), (4.20). Assume that $|v|$, $|v_t| \le c$ and zeros of $a_+(\zeta)$ or $a_-(\zeta)$ lie apart from the paths of integration below. Let us assume also that the parameters t_0, D_0 determining the elliptic curve (4.10) and (4.20) are such that the asymptotic at $\delta = 0$ Stokes lines for the system (2.12), (4.17), emanating from $\zeta_k = z_k^{1/2}$, $k = 1, 3, 5$, form the continuous chains $e^{-i\pi/10}\infty \to \zeta_1 \to e^{i\pi/10}\infty$,

$e^{i\pi/10}\infty \to \zeta_1 \to \zeta_3 \to e^{i3\pi/10}\infty$ and $e^{i3\pi/10}\infty \to \zeta_5 \to e^{i\pi/2}\infty$. Then, using the result of [**50**], we find the Stokes multipliers as $\mathrm{Re}(\delta^{-1}) \to +\infty$, $|\mathrm{Im}(\delta^{-1})| <$ const.,

$$(4.27)\quad \begin{aligned} \ln s_2 &= -2\delta^{-1}\left[\int_{\zeta_5^{(\sigma)}}^{\zeta} \nu_\sigma\, d\zeta - \theta + \sigma\frac{\delta}{2}\ln\frac{a_\sigma}{\mu}\right]_{\zeta\to e^{i\pi/2}\infty} \\ &\qquad + \sigma\ln(2i) + \mathcal{O}(\delta), \\ \ln s_5 &= -2\delta^{-1}\left[\int_{\zeta_1^{(\sigma)}}^{\zeta} \nu_\sigma\, d\zeta - \theta + \sigma\frac{\delta}{2}\ln\frac{a_\sigma}{\mu}\right]_{\zeta\to+\infty} \\ &\qquad + \sigma\ln(2i) + \mathcal{O}(\delta), \quad \sigma\in\{+,-\}. \end{aligned}$$

We compute the right-hand side of (4.27) at the critical point $t = t_3$, $v_t = 0$, $v = z_3$ since the Stokes multipliers do not depend on t and the assumptions above are not violated for the nondegenerate curve (4.10). Using the method of [**46, 49**], we transform the elliptic integrals above as follows,

$$(4.28)\quad \begin{aligned} \ln(-is_2) &= 2\delta^{-1}I_a + \frac{t_3}{2}\Omega_a^{(3)} + \frac{D_1(t_3)}{2}\omega_a + \mathcal{O}(\delta^{1/2-\varepsilon}), \\ \ln(-is_5) &= 2\delta^{-1}I_b + \frac{t_3}{2}\Omega_b^{(3)} + \frac{D_1(t_3)}{2}\omega_b + \mathcal{O}(\delta^{1/2-\varepsilon}) \end{aligned}$$

where $\varepsilon > 0$ is small. The condition $|s_5| \ll 1$ and the boundedness of s_2 are consistent with (4.28) if $h_a = 0$ and $h_b < 0$. Substitutions $\delta \mapsto \alpha^{5/4}\delta$, $t_0 \mapsto \alpha t_0$, $t \mapsto \alpha^{-1/4}t$, $v \mapsto \alpha^{1/2}v$, $v_t \mapsto \alpha^{3/4}v_t$, which leave x, y and y_x invariant, allow us to scale h_b to the value in (4.12). The conditions on the Stokes graph imposed above are consistent with $D_0(t_0) \in \mathbb{R}$ for $t_0 < -1$. Thus, it is chosen the branch of $D_0(t_0)$ which yields the degeneration of the curve (4.10) on γ, and the validity domain of the elliptic asymptotic solution is located to the left of γ.

Using in (4.28) the identities (4.24), we find the critical point t_3,

$$t_3 = -\delta^{-1}\frac{4}{5}t_0 + \frac{\omega_a}{2\pi i}\ln(-is_5) - \frac{\omega_b}{2\pi i}\ln(-is_2) + \mathcal{O}(\delta^{1/2-\varepsilon}),$$

and the argument of the $\wp$-function of Weierstrass (4.26). □

The description of the Painlevé function on the line γ is given by the following assertion.

THEOREM 4.3. *Let $t_0 \in \gamma$ where the asymptotic spectral curve (4.18) degenerates via $z_1 = z_3$. Let the functions v, v_t introduced in (4.6) satisfy the estimates $v - z_1, v_t = \mathcal{O}(\delta^\varepsilon)$, $\varepsilon \in [0, 1/2]$. Then the Stokes multipliers of the associated linear system (2.1), (4.2) are as follows:*

(i) *if $0 \le \varepsilon < 1/3$ and $v_t - \hat\sigma 2(v - z_1)\sqrt{v + 2z_1} = \mathcal{O}(\delta^{1-\varepsilon})$, $\hat\sigma^2 = 1$, then*

$$(4.29)\quad \begin{aligned} s_4 &= -i\frac{\sqrt{v+2z_1} - \hat\sigma\sqrt{3z_1}}{\sqrt{v+2z_1} + \hat\sigma\sqrt{3z_1}}\sqrt{2\pi}\frac{e^{i\pi\rho_b/2}}{\Gamma(1/2+\rho_b)}e^{F_b}\big(1 + \mathcal{O}(\delta^{1/4-3/4\varepsilon})\big), \\ s_5 &= -i\frac{\sqrt{v+2z_1} + \hat\sigma\sqrt{3z_1}}{\sqrt{v+2z_1} - \hat\sigma\sqrt{3z_1}}\sqrt{2\pi}\frac{e^{i\pi\rho_b/2}}{\Gamma(1/2-\rho_b)}e^{-F_b}\big(1 + \mathcal{O}(\delta^{1/4-3/4\varepsilon})\big), \\ s_1 &= e^{-2\pi i\rho_b}s_4\big(1 + \mathcal{O}(\delta^{1-\varepsilon/2})\big), \\ 1 + s_4s_5 &= -e^{2\pi i\rho_b}\big(1 + \mathcal{O}(\delta^{1-\varepsilon/2})\big), \quad \arg(-1 - s_4s_5) \in \left(-\frac{2\pi}{3}, \frac{2\pi}{3}\right) \end{aligned}$$

where

$$(4.30)\quad \begin{aligned} F_b &= \delta^{-1}\frac{8}{5}(3/2)^{1/4}(-t_0)^{5/4} + \rho_b \ln\bigl[i\delta^{-1}2^{19/4}3^{5/4}(-t_0)^{5/4}\bigr] \\ &\qquad - 2(3/2)^{1/4}(-t_0)^{1/4}t, \\ \rho_b &= \frac{1}{4\sqrt{3z_1}}\Bigl(\delta^{-1}\bigl(v_t^2 - 4(v-z_1)^2(v+2z_1)\bigr) - 2t(v-z_1) + \frac{v_t}{v-z_1}\Bigr), \\ &\qquad \operatorname{Re}\rho_b \in \Bigl(-\frac{1}{3},\frac{1}{3}\Bigr); \end{aligned}$$

(ii) *if* $1/3 < \varepsilon \le 1/2$ *then*

$$(4.31)\quad \begin{aligned} s_4 &= \frac{3^{1/8}2^{7/8}(-t_0)^{1/8}\delta^{1/2}}{v_t - 2\sqrt{3z_1}(v-z_1)}\sqrt{2\pi}\frac{e^{i\pi\rho_c/2}e^{-i\pi/4}}{\Gamma(\rho_c)} \\ &\qquad \times e^{F_c}\bigl(1+\mathcal{O}(\delta^{1/4-3/8\varepsilon}) + \mathcal{O}(\delta^{3\varepsilon-1})\bigr), \\ s_5 &= \frac{3^{1/8}2^{7/8}(-t_0)^{1/8}\delta^{1/2}}{v_t + 2\sqrt{3z_1}(v-z_1)}\sqrt{2\pi}\frac{e^{i\pi\rho_c/2}e^{-i\pi/4}}{\Gamma(-\rho_c)} \\ &\qquad \times e^{-F_c}\bigl(1+\mathcal{O}(\delta^{1/4-3/8\varepsilon}) + \mathcal{O}(\delta^{3\varepsilon-1})\bigr), \\ s_1 &= -e^{-2\pi i\rho_c}s_4\bigl(1+\mathcal{O}(\delta^{1-\varepsilon/2})\bigr), \\ 1+s_4s_5 &= e^{2\pi i\rho_c}\bigl(1+\mathcal{O}(\delta^{1-\varepsilon/2})\bigr), \quad \arg(1+s_4s_5) \in \Bigl(-\frac{\pi}{3},\frac{\pi}{3}\Bigr) \end{aligned}$$

where

$$(4.32)\quad \begin{aligned} F_c &= \delta^{-1}\frac{8}{5}(3/2)^{1/4}(-t_0)^{5/4} + \rho_c\ln\bigl(i\delta^{-1}2^{19/4}3^{5/4}(-t_0)^{5/4}\bigr) \\ &\qquad - 2\sqrt{3z_1}\,t + i\frac{\pi}{4}, \\ \rho_c &= \frac{\delta^{-1}}{4\sqrt{3z_1}}\bigl(v_t^2 - 12z_1(v-z_1)^2\bigr) + \mathcal{O}(\delta^{3\varepsilon}), \quad \operatorname{Re}\rho_c \in \Bigl(-\frac{1}{6},\frac{1}{6}\Bigr). \end{aligned}$$

In other notations, this assertion was obtained in [**44, 45**] following the conventional arguments of the isomonodromy deformation method [**36**]. Applying the method of [**50**], it is possible to justify the expressions above. It is also possible to improve the estimates for the error terms. To avoid the repetition and to save space, we skip this known derivation.

The inversion of (4.29) yields the asymptotics of the Painlevé function (4.9) with $v_0 = z_1 = \sqrt{-t_0/6}$ and the phase given in terms of the Stokes multipliers by

$$(4.33)\quad \sqrt{3v_0}\,(t-\varphi) = -\frac{F_b}{2} - \frac{1}{2}\ln A, \quad A = -i\sqrt{2\pi}\frac{e^{i\pi\rho_b/2}}{s_4\Gamma(1/2+\rho_b)}$$

where F_b is defined in (4.30), and

$$\rho_b = \frac{1}{2\pi i}\ln(-1-s_4s_5), \quad \arg(-1-s_4s_5) \in \Bigl(-\frac{2\pi}{3},\frac{2\pi}{3}\Bigr).$$

For our purposes, the inversion of (4.31) is more interesting. In the leading order, this yields

$$
\begin{aligned}
v &= \sqrt{-t_0/6} + \delta^{1/2}(Ae^{-F_c} - Be^{F_c}),\\
AB &= 3^{-1/4}2^{-7/4}(-t_0)^{-1/4}\rho_c\big(1+\mathcal{O}(\delta^{1/4-3/8\varepsilon})+\mathcal{O}(\delta^{3\varepsilon-1})\big),\\
\rho_c &= \frac{1}{2\pi i}\ln(1+s_4s_5)+\mathcal{O}(\delta^{1-\varepsilon/2}), \quad \arg(1+s_4s_5)\in\Big(-\frac{\pi}{3},\frac{\pi}{3}\Big),\\
A &= 3^{-1/8}2^{-7/8}(-t_0)^{-1/8}\sqrt{2\pi}\frac{e^{i\pi\rho_c/2}e^{-i\pi/4}}{s_5\Gamma(-\rho_c)}\\
&\qquad\times\big(1+\mathcal{O}(\delta^{1/4-3/8\varepsilon})+\mathcal{O}(\delta^{3\varepsilon-1})\big),\\
B &= 3^{-1/8}2^{-7/8}(-t_0)^{-1/8}\sqrt{2\pi}\frac{e^{i\pi\rho_c/2}e^{-i\pi/4}}{s_4\Gamma(\rho_c)}\\
&\qquad\times\big(1+\mathcal{O}(\delta^{1/4-3/8\varepsilon})+\mathcal{O}(\delta^{3\varepsilon-1})\big),
\end{aligned}
\tag{4.34}
$$

and F_c is defined in (4.32).

Now, we are prepared to discuss the problems announced earlier.

(i) The large amplitude separatrix solution.

Let us consider the Painlevé function $y(x)$ corresponding to the Stokes multipliers (4.3) satisfying $1+s_2s_3=0$, i.e. $s_2=s_3=i$, $s_5=0$, $s_1+s_4=i$, when $|s_4|\gg 1$. The asymptotics (4.4), (4.5) explain the used term the *large amplitude separatrix solution.* Introduce t_0, t and v as in (4.6) and consider the boundary γ for the domain of validity of (4.12). Then applying Theorem 4.1, we see that to the right of γ, the asymptotics of the Painlevé function is elliptic and is described by the equations (4.14), (4.15), (4.16). The corresponding elliptic curve is determined by the system of the Krichever's equations (4.12).

On γ, the elliptic curve degenerates, and, since $1+s_4s_5=1$, the asymptotics of the Painlevé function is described by the limiting as $\rho_c\to 0$ form of (4.34),

$$
\begin{aligned}
v &= \sqrt{-t_0/6} + \delta^{1/2}Ae^{F}, \quad \delta^{-1} = -\frac{\ln|s_4|}{h_b}, \quad h_b = -\frac{8}{5}(3/2)^{1/4},\\
A &= 3^{-1/8}2^{-7/8}(-t_0)^{-1/8}\frac{s_4}{\sqrt{2\pi}}\big(1+\mathcal{O}(\delta^{1/16})\big),\\
F &= -\delta^{-1}\frac{8}{5}(3/2)^{1/4}(-t_0)^{5/4} + 2^{3/4}3^{1/4}(-t_0)^{1/4}t.
\end{aligned}
\tag{4.35}
$$

To the left of γ, the curve (4.10) remains degenerate, and the Krichever's system is not applicable. Observing however, that the complete series for the solution with the leading terms (4.35) contains the positive degrees of e^F only, we may conjecture that the asymptotics of the Painlevé function to the left of γ, i.e. for $\mathrm{Re}(-t_0)^{5/4}>1$, is given by the analytic continuation of (4.35). This conjecture can be confirmed by the use of the Riemann–Hilbert problem approach, see [**9, 34**]. An alternative investigation of such series based on the Borel summation method is given in [**8, 81**].

(ii) The small perturbation of the degenerate solution.

Let $0<|s_5|\ll 1$, $c'\le |s_2|, |s_1/s_4|\le c''$ for some positive finite constants c', c'', e.g., when $s_2=i(1+\varepsilon)$, $s_3=i(1+\varepsilon\varkappa)$, for $0<|\varepsilon|\ll 1$, $c'\le|\varkappa|\le c''$. Using Theorem 4.2, we see that, to the left of the line $\gamma_l=\gamma$ defined above, the asymptotics of $y(x)$ is described by (4.14), (4.25), (4.26). The elliptic curve (4.10) is determined by the Krichever's system (4.12) with $h_a = 0$, $h_b = -8/5(3/2)^{1/4}$.

Due to Theorem 4.1, the right boundary γ_r of the discriminant set consists of the points satisfying the equation $\operatorname{Re}(-t_0)^{5/4} = -\ln|s_4|/\ln|s_5| < 1$ and is asymptotic to the rays $\arg t_0 = \pm 3\pi/5$. If s_4 is bounded as $\varepsilon \to 0$, the right boundary γ_r coincides with these rays. To the right of γ_r, the asymptotic solution $y(x)$ is given by (4.14), (4.16) where the elliptic curve (4.10) is determined by the Krichever's system with the parameters $h'_a = 0$ and $h'_b = -h_b \ln|s_4|/\ln|s_5|$. If s_4 is bounded, then $h'_a = h'_b = 0$, and the description of the Painlevé function can be obtained by the use of changes (4.6) in the formulae of [**44, 45, 54**] for $y(x)$ as $|x| \to \infty$. On the boundaries γ_l and γ_r, we may use (4.34) with $\delta^{-1} = \ln|s_5|/h_b$, $h_b = -8/5(3/2)^{1/4}$, $\rho_c \to 0$. As easy to see, both the nonconstant terms in (4.34) decrease as $\varepsilon \to 0$ for $-\ln|s_4|/\ln|s_5| < \operatorname{Re}(-t_0)^{5/4} < 1$. Because the complete expansion of the Painlevé function involves the positive degrees of these terms only, we may conjecture the validity of (4.34) between γ_l and γ_r as well.

In particular, the discussion above provides us with the possibility to give the analytic interpretation for the formal 2-parametric "instanton type" series with the initial terms given by (4.34) and extensively studied in [**55, 56, 81, 82**].

References

1. M. J. Ablowitz and J. Segur, *Exact linearization of a Painlevé transcendent*, Phys. Rev. Lett. **38** (1977), 1103–1106.
2. V. I. Arnol'd, *Mathematical methods of classical mechanics*, 2nd ed., Grad. Texts in Math., vol. 60, Springer-Verlag, New York, 1989.
3. J. Baik, P. Deift, and K. Johansson, *On the distribution of the length of the longest increasing subsequence of random permutations*, math. CO/9810105, 1998.
4. P. Bleher and A. Its, *Semiclassical asymptotics of orthogonal polynomials, Riemann–Hilbert problem, and universality of the matrix model*, Preprint no. 97-2, Dept. of Mathematical Sciences, IUPUI, 1997.
5. P. Boutroux, *Recherches sur les transcendantes de M. Painlevé et l'étude asymptotique des équations différentielles du second ordre*, Ann. Sci. École Norm. Sup. (4) **30** (1914), 255–376; **31** (1913), 99–159.
6. E. Brézin and V. A. Kazakov, *Exactly solvable field theories of closed strings*, Phys. Lett. B **236** (1990), 144–150.
7. S. Chakravarty, *A class of integrable conformally self-dual metrics*, Classical Quantum Gravity **11** (1994), no 1, L1–L6.
8. O. Costin, *Correlation between pole location and asymptotic behavior for Painlevé* I *solutions*, MSRI preprint no. 997-094, 1997.
9. P. A. Deift and X. Zhou, *A steepest descent method for oscillatory Riemann–Hilbert problems. Asymptotics for the* MKdV *equation*, Ann. of Math. (2) **137** (1995), 295–368.
10. P. Doran-Wu and N. Joshi, *Direct asymptotic analysis of the second Painlevé equation: Three limits*, J. Phys. A **30** (1997), 4701–4708.
11. M. Douglas and S. Shenker, *Strings in less than one dimension*, Rutgers preprint RU-89-34, 1989.
12. B. Dubrovin, *Geometry of* 2D *topological field theories*, Springer Lecture Notes Math. **1620** (1995), 120–348.
13. B. A. Dubrovin, I. M. Krichever, and S. P. Novikov, *Topological and algebraic geometry methods in contemporary mathematical physics.* II, Soviet Sci. Rev. Sect. C Math. Phys. Rev. **3** (1982), 1–150.
14. B. A. Dubrovin, V. B. Matveev, and S. P. Novikov, *Nonlinear equations of Korteweg-de Vries type, finite band operators and Abelian varieties*, Russian Math. Surveys **31** (1976), 59–146.
15. B. A. Dubrovin and M. Mazzocco, *Monodromy of certain Painlevé* VI *transcendents and reflection groups*, Invent. Math. **141** (2000), 55–147.

16. M. V. Fedorjuk, *Asymptotic methods for linear ordinary differential equations*, Moscow, Nauka, 1983.
17. ———, *Isomonodromy deformations of equations with an irregular singularity*, Diff. Uravnen. **22** (1986), 961–967.
18. H. Flaschka, M. G. Forest, and D. W. McLaughlin, *Multiphase averaging and the inverse spectral solution of the Korteweg-de Vries equation*, Comm. Pure Appl. Math. **33** (1980), 739–784.
19. H. Flaschka and A. C. Newell, *Monodromy- and spectrum-preserving deformations.* I, Comm. Math. Phys. **76** (1980), 65–116.
20. ———, *Multiphase similarity solutions of integrable evolution equations*, Phys. D **3** (1981), 203–221.
21. A. S. Fokas, A. R. Its, and A. V. Kitaev, *Isomonodromic approach in the theory of two-dimensional quantum gravity*, Uspekhi Mat. Nauk **45** (1990), 135–136.
22. ———, *Discrete Painlevé equations and their appearance in quantum gravity*, Comm. Math. Phys. **142** (1991), 313–344.
23. ———, *The isomonodromy approach to matrix models in 2D quantum gravity*, Comm. Math. Phys. **147** (1992), 395–430.
24. G. Freud, *On the coefficients in the recursion formulae of orthogonal polynomials*, Proc. Roy. Irish Acad. Sect. A **76** (1976), 1–6.
25. R. Fuchs, *Über lineare homogene Differentialgleichungen zweiter Ordnung mit drei im Endlichen gelegenen wesentlich singulären Stellen*, Math. Ann. **63** (1906), 301–321.
26. F. Fucito, A. Gamba, M. Martellini, and O. Ragnisco, *Nonlinear* WKB *analysis of the string equations*, Internat. J. Modern Phys. B **6** (1992), 2123–2148.
27. B. Gambier, *Sur les équations différentielles du second ordre et du premier degré dont l'intégrale générale est a points critiques fixes*, Acta Math. **33** (1910), 1–55.
28. R. Garnier, *Sur les équations différentielles du troisième ordre dont l'intégrale générale est uniforme et sur une classe d'èquations nouvelles d'ordre supérieur dont l'intégrale générale a ses points critique fixes*, Ann. Sci. École Norm. Sup. (4) **29** (1912), 1–126.
29. ———, *Étude de l'intégrale générale de l'équation* VI *de M. Painlevé*, Ann. Sci. École Norm. Sup. (4) **34** (1917), 239–253.
30. ———, *Solution du problème de Riemann pour les systèmes différentielles linéaires du second ordre*, Ann. Sci. École Norm. Sup. (4) **43** (1926), 177–307.
31. D. J. Gross and A. A. Migdal, *Nonperturbative two-dimensional quantum gravity*, Phys. Rev. Lett. **64** (1990), 127–130.
32. Ph. Holmes and D. Spence, *On a Painlevé-type boundary-value problem*, Quart. J. Mech. Appl. Math. **37** (1984), 525–538.
33. E. L. Ince, *Ordinary differential equations*, Dover, New York, 1956.
34. A. R. Its, A. S. Fokas, and A. A. Kapaev, *On the asymptotic analysis of the Painlevé equations via the isomonodromy method*, Nonlinearity **7** (1994), 1291–1325.
35. A. R. Its and V. B. Matveev, *A class of solutions of the Korteweg-de Vries equation*, Problems in Mathematical Physics, vol. 8, Izdat. Leningrad. Univ., Leningrad, 1976, pp. 70–92 (Russian).
36. A. R. Its and V. Yu Novokshenov, *The isomonodromic deformation method in the theory of Painlevé equations*, Lecture Notes in Math., vol. 1191, Springer-Verlag, Berlin-Heidelberg-New York-Tokyo, 1986.
37. M. Jimbo, *Monodromy problem and the boundary condition for some Painlevé equations*, Publ. Res. Inst. Math. Sci. **18** (1982), no. 3, 1137–1161.
38. M. Jimbo and T. Miwa, *Monodromy preserving deformation of linear ordinary differential equations with rational coefficients.* II, Phys. D **2** (1981), 407–448.
39. M. Jimbo, T. Miwa, and K. Ueno, *Monodromy preserving deformation of linear ordinary differential equations with rational coefficients*, Phys. D **2** (1981), 306–352.
40. N. Joshi, *The second Painlevé equation in the large-parameter limit.* I. *Local asymptotic analysis*, Stud. Appl. Math. **102** (1999), 345–373.
41. ———, *The second Painlevé equation in the large-parameter limit* I. *Local asymptotic analysis*, Stud. Appl. Math. **102** (1999), 345–373.
42. N. Joshi and M. D. Kruskal, *An asymptotic approach to the connection problem for the first and the second Painlevé equations*, Phys. Lett. A **130** (1988), 129–137.

43. A. A. Kapaev, *Singular solutions of the Painlevé* II *equations*, Lecture Notes in Math., vol. 1191, Springer, Berlin, 1986, pp. 261–283.
44. ———, *Asymptotics of solutions of the Painlevé equation of the first kind*, Differ. Uravn. **24**, 1684–1695.
45. ———, *Global asymptotics of the first Painlevé transcendent*, Inst. Nonlinear Studies, Preprint INS no. 225, 1993.
46. ———, *Essential singularity of the Painlevé function of the second kind and the nonlinear Stokes phenomenon*, Zap. Nauchn. Sem. Leningrad. Otdel. Mat. Inst. Steklov. (LOMI) **187** (1991), 139–170 (Russian); English transl., J. Math. Sci. **73** (1995), no. 4, 500–517.
47. ———, *Global asymptotics of the second Painlevé transcendent*, Phys. Lett. A **167** (1992), 356–362.
48. ———, *Global asymptotics of the fourth Painlevé transcendent*, POMI Preprint no. 6/1996, 1996; Steklov Math. Inst. and IUPUI Preprint no. 96-5, 1996.
49. ———, *Scaling limits in the second Painlevé transcendent*, Zap. Nauchn. Sem. Leningrad. Otdel. Mat. Inst. Steklov. (LOMI) **209** (1994), 60–101.
50. ———, WKB *method for* 2×2 *systems of linear ordinary differential equations with rational coefficients*, Steklov Math. Inst. and IUPUI Preprint no. 96-6, 1996.
51. ———, *Scaling limits in the fourth Painlevé transcendent*, Steklov Math. Inst. Preprint POMI no. 15/1996, 1996.
52. ———, *Discriminant set for the scaling limits in the third Painlevé transcendent*, Steklov Math. Inst. Preprint PDMI no. 21/1997, 1997.
53. A. A. Kapaev and A. V. Kitaev, *The limit transition* $\mathrm{P}_2 \to \mathrm{P}_1$, Zap. Nauchn. Sem. Leningrad. Otdel. Mat. Inst. Steklov. (LOMI) **187** (1991), 75–87 (Russian); English transl., J. Math. Sci. **73** (1995), no. 4, 460–467.
54. ———, *Connection formulae for the first Painlevé transcendent in the complex domain*, Lett. Math. Phys. **27** (1993), 243–252.
55. T. Kawai and Y. Takei, WKB *analysis of Painlevé transcendents with a large parameter*. I, Adv. in Math. **118** (1996), 1–33.
56. ———, WKB *analysis of Painlevé transcendents with a large parameter*. III. *Local equivalence of* 2*-parameter Painlevé transcendents*, Adv. in Math. **134** (1998), 178–218.
57. O. M. Kiselev, *Hard loss of stability in Painlevé-*2 *equation*, J. Nonlinear Math. Phys. **8** (2001), no. 1, 65–95.
58. A. V. Kitaev, *Asymptotic description of solutions of the fourth Painlevé equation on analogues of Stokes's rays*, Zap. Nauchn. Sem. Leningrad Otdel. Mat. Inst. Steklov. (LOMI) **169** (1988), Voprosy Kvant. Teor. Polya i Statist. Fiz. **8**, 84–90 (Russian); English transl., J. Soviet Math. **54** (1991), no. 3, 916–920.
59. I. M. Krichever, *Integration of nonlinear equations by the method of algebraic geometry*, Funct. Anal. Appl. **11** (1977), 12–26.
60. ———, *Method of averaging for two-dimensional "integrable" equations*, Funct. Anal. Appl. **22** (1988), 200–213.
61. ———, *On Heisenberg relations for the ordinary linear differential operators*, ETH preprint, Zürich, IHES preprint, Bur-Sur-Yvette, 1990.
62. G. E. Kuzmak, *Asymptotic solutions of nonlinear second-order differential equations with variable coefficients*, J. Appl. Math. Mech. **23** (1959), 730–744.
63. A. P. Magnus, *Painlevé-type differential equations for the recurrence coefficients of semi-classical orthogonal polynomials*, J. Comput. Appl. Math. **57** (1995), 215–237.
64. B. M. McCoy and Sh. Tang, *Connection formulae for Painlevé* V *functions*, Phys. D **19** (1986), 42–72.
65. ——— *Connection formulae for Painlevé functions*. II. *The* δ *function Bose gas problem*, Phys. D **20** (1986), 187–216.
66. M. L. Mehta, *Painlevé transcendents in the theory of random matrices*, An Introduction to Methods of Complex Analysis and Geometry for Classical Mechanics and Nonlinear Waves (Chamonix, 1993), Frontières, Gif-sur-Yvette, 1994, pp. 197–208.
67. J. W. Miles, *On the second Painlevé transcendent*, Proc. Roy. Soc. London Ser. A **361** (1978), 277–291.
68. G. Moor, *Matrix models of* 2D *gravity and isomonodromic deformations*, Progr. Theoret. Phys. Suppl. **102** (1990), 255–285.

69. P. Nevai, *Géza Freud, orthogonal polynomials and Christoffel functions. A case study*, J. Approx. Theory **48** (1986), 3–167.
70. S. P. Novikov, *Quantization of finite-gap potentials and nonlinear quasiclassical approximation in nonperturbative string theory*, Funct. Anal. Appl. **24** (1990), 296–306.
71. S. Okumura, *The self-dual Einstein-Weyl metric and classical solution of Painlevé* VI, Lett. Math. Phys. **46** (1998), no 3, 219–232.
72. F. W. J. Olver, *Asymptotics and special functions*, Computer Science and Applied Mathematics, Academic Press, New York-London, 1974.
73. P. Painlevé, *Mémoire sur les équations différentielles dont l'intégrale générale est uniforme*, Bull. Soc. Math. France **28** (1900), 201–261.
74. ______, *Sur les équations differentielles du second ordre à points critiques fixes*, Comptes Rendus **143** (1906), 1111–1117.
75. L. D. Paniak and R. J. Szabo, *Fermionic quantum gravity*, Nuclear Phys. B **593** (2001), 671–725; `hep-th/0005128`.
76. L. Pastur, *On the universality of the level spacing distribution for some ensembles of random matrices*, Lett. Math. Phys. **25** (1992), 259–265.
77. L. Schlesinger, *Über eine Klasse von Differntiallensystemen belibiger Ordnung mit festen kritischen Punkten*, J. für Math. **141** (1912), 96–145.
78. H. Segur and M. J. Ablowitz, *Asymptotic solutions of nonlinear evolution equations and a Painlevé transcendent*, Phys. D **3** (1981), 165–184.
79. B. I. Suleimanov, *On asymptotics of regular solutions for a special kind of Painlevé* V *equation*, Lecture Notes in Math. **1191** (1986), 230–261.
80. K. Takasaki, *Spectral curves and Whitham equations in isomonodromic problems of Schlesinger type*, Asian J. Math. **4** (1998), 1049–1478.
81. Y. Takei *On the connection formula for the first Painlevé equation—from the viewpoint of the exact* WKB *analysis*, Painlevé Functions and Asymptotic Analysis (Japanese) (Kyoto, 1995), Sūrikaisekikenkyūsho Kōkyūroku, no. 931, 1995, pp. 70–99.
82. ______, *An explicit description of the connection formula for the first Painlevé equation*, Toward the Exact WKB Analysis of Differential Equations, Linear or Nonlinear (Kyoto, 1998), Kyoto Univ. Press, Kyoto, 2000, pp. 271–296.
83. K. P. Tod, *Self-dual Einstein metrics from the Painlevé* VI *equation*, Phys. Lett. A **190** (1994), 221–224.
84. V. L. Vereschagin, *Global asymptotics for the fourth Painlevé transcendent*, Mat. Sbornik **188** (1997), 11–32.
85. ______, *Asymptotic behavior of solutions, singular at zero. of the sine-Gordon equation*, Mat. Zametki **67** (2000), 329–342.
86. ______, *Nonlinear quasiclassics and the Painlevé equations*, 1997; `solv-int/9707004`.
87. W. Wasow, *Asymptotic expansions for ordinary differential equations*, Pure Appl. Math., vol. 14, Wiley & Sons, Inc., New York-London-Sydney, 1965.
88. J. B. Whitham, *Linear and nonlinear waves*, Wiley & Sons, Inc., New York-London-Sydney 1974.
89. V. Yu. Novokshenov, *The Boutroux ansatz for the second Painlev equation in the complex domain*, Izv. Akad. Nauk SSSR Ser. Math. **54** (1990), 1229-1251 (Russian); English transl., Math. USSR-IZV **37** (1991), no. 3, 587–609.
90. ______, *Radial-symmetric solution of the* cosh-*Laplace equation and the distribution of its singularities*, Russian J. Math. Phys. **5** (1998), 211–226.

V. A. Steklov Institute of Mathematics at St.Petersburg, 27, Fontanka St. Petersburg, 191011, Russia.

E-mail address: `kapaev@pdmi.ras.ru`

Centre de Recherches Mathématiques
CRM Proceedings and Lecture Notes
Volume **32**, 2002

Kowalevski top revisited

Vadim B. Kuznetsov

ABSTRACT. We review the separation of variables for the Kowalevski top and for its generalization to the algebra o(4). We notice that the corresponding separation equations allow an interpretation of the Kowalevski top as a $B_2^{(1)}$ integrable lattice. Consequently, we apply the quadratic r-matrix formalism to construct a new 2×2 Lax matrix for the top, which is responsible for its separation of variables.

Introduction

In 1889 Sophie Kowalevski [**21**] found and integrated new integrable case of rotation of a heavy rigid body around a fixed point which since then carried her name, *the Kowalevski top* (KT). In modern terms, this is an integrable system on the e(3) algebra with a quadratic and a quartic (in angular momenta) integrals of motion.

Her original transformation to new variables led to the solution in terms of quadratures and, eventually, to the separation of variables for this complicated top [**18, 22, 41**]. No separation which is alternative to her original separation of variables is known for this system at the moment, even though there is a large body of literature dedicated to the problem (see, for instance, [**5–7, 9, 10, 28**] to name just a few).

The aim of the present work is to design and construct a new Lax matrix for the Kowalevski top which implies Kowalevski's separation of variables and which has a proper algebraic (r-matrix) structure. The principal existence of such (2×2) Lax matrix was predicted in author's MSc dissertation in 1985 [**22**] (see also [**18**]), after he rewrote the formula defining the separation variables for the top given in [**41**], and this new formula took a form of the spectral curve of the future Lax matrix. In what follows I shall explain how this Lax matrix can be reconstructed in somewhat regular way, without much of a guesswork.

In 1981 the Kowalevski top was generalized to the case of o(4) algebra [**14**]. In 1990 for this o(4) KT the separation of variables was found in [**19, 23**] which generalized the one for the e(3) KT. In the present paper we will always consider

---

2000 *Mathematics Subject Classification.* Primary: 70E40; Secondary: 17A45, 17B63.

I want to thank Igor Komarov, Evgueni Sklyanin and Pol Vanhaecke for interesting discussions of the Kowalevski top. I acknowledge the support of the EPSRC.

This is the final form of the paper.

this one-parameter, o(4) extension of the Kowalevski top. We refer the reader to the work [**17**] for the survey of the Kowalevski top, its generalizations and other related tops.

The structure of the paper is following. In Section 1 we recall the integration in quadratures of the o(4) Kowalevski top. In Sections 2 and 3 we describe its separation of variables and, in particular, write down the separation equations in the form which is best for deducing the ansatz for the future Lax matrix. As a result, we conclude that, as it follows from the separation equations, the Kowalevski top (together with its o(4) version) is an integrable system with boundary conditions of $B_2^{(1)}$-type which means that the corresponding Lax matrix should factorize and factors obey the reflection equation (quadratic r-matrix algebra). In Section 4 we introduce the corresponding quadratic algebra $\mathcal{B}$. Section 5 contains separation representation for this algebra, while Section 6 defines and solves the $B_2^{(1)}$-type integrable system. In Section 7 the 2×2 Lax matrix $T(u)$ for the o(4) Kowalevski top is given which follows by comparing the separation data for two integrable systems from the previous Sections (KT and the integrable system on the quadratic algebra $\mathcal{B}$). The final result is given in terms of initial variables of the top.

1. Integration in Quadratures of the o(4) Kowalevski Top

In this section we collect the results from [**19, 23**] about integration in quadratures of the Kowalevski top on the o(4) algebra.

The Poisson brackets for the o(4) generators J_k, x_k, $k = 1, 2, 3$, are defined in the standard way:

$$\{J_i, J_k\} = \varepsilon_{ikl} J_l, \quad \{J_i, x_k\} = \varepsilon_{ikl} x_l, \quad \{x_i, x_k\} = -\mathcal{P} \varepsilon_{ikl} J_l \tag{1.1}$$

where ε_{ikl} is the completely anti-symmetric tensor, $\varepsilon_{123} = 1$, and $\mathcal{P}$ is a complex (or real) parameter. In fact, one can think of the algebra (1.1) as a complex $o(4, \mathbb{C})$ algebra with the triple of real algebras appearing when the parameter $\mathcal{P}$ is specialized as follows:

$$\mathcal{P} = \begin{cases} -1, & o(4) \\ 0, & e(3) \\ 1, & o(3,1) \end{cases} . \tag{1.2}$$

The Casimirs of the bracket (1.1) have the form

$$\ell = x_1 J_1 + x_2 J_2 + x_3 J_3, \quad a^2 = x_1^2 + x_2^2 + x_3^2 - \mathcal{P}(J_1^2 + J_2^2 + J_3^2). \tag{1.3}$$

The o(4) Kowalevski top has the Hamiltonian H,

$$H = J_1^2 + J_2^2 + 2J_3^2 - 2bx_1, \tag{1.4}$$

and the second integral K,

$$\begin{gathered} K = (J_+^2 + 2bx_+ - \mathcal{P}b^2)(J_-^2 + 2bx_- - \mathcal{P}b^2), \\ J_\pm = J_1 \pm iJ_2, \quad x_\pm = x_1 \pm ix_2, \qquad (i^2 = -1), \end{gathered} \tag{1.5}$$

which are Poisson commuting:

$$\{H, K\} = 0. \tag{1.6}$$

When $\mathcal{P} = 0$, this Liouville integrable system becomes the usual e(3) Kowalevski top. The o(4) version was proved to be integrable in [**14**] and it was integrated in [**19**], where also the separation of variables was performed for this system, with separation equations being put in the form which is best suited for the purpose of this paper.

The integration of the o(4) KT in [**19**] was inspired by the paper [**9**] about integration of the e(3) case where it was done by mapping the KT into the Neumann system, thereby giving another way of looking at the original Kowalevski's transformation. For the o(4) case it goes as follows. Introduce new variables:

$$p_1=\frac{1+J_+J_-}{2(J_+-J_-)},\quad p_2=\frac{J_++J_-}{2i(J_+-J_-)},\quad p_3=\frac{1-J_+J_-}{2i(J_+-J_-)},\tag{1.7}$$

$$\begin{aligned} l_1&=\frac{J_3(1-J_+J_-)-bx_3(J_++J_-)}{2(J_+-J_-)},\quad l_2=\frac{ibx_3}{J_+-J_-},\\ l_3&=\frac{J_3(1+J_+J_-)+bx_3(J_++J_-)}{2(J_+-J_-)}. \end{aligned}\tag{1.8}$$

Under the flow governed by the Kowalevski Hamiltonian,

$$\dot{(\cdot)}\equiv\frac{d(\cdot)}{dt}=\frac{1}{2}\{H,(\cdot)\},\tag{1.9}$$

these new variables, p_k, l_k, $k=1,2,3$, combined into two vectors $\vec{p}$ and $\vec{l}$, evolve as the "e(3) Neumann system":

$$\dot{\vec{p}}=2i\vec{l}\times\vec{p},\quad \dot{\vec{l}}=2iQ\vec{p}\times\vec{p},\tag{1.10}$$

with the symmetric matrix Q depending on Casimirs and both integrals:

$$Q=\begin{pmatrix}-\frac{1}{4}+b^2\tilde{a} & ib\ell & i(\frac{1}{4}+b^2\tilde{a})\\ ib\ell & -\frac{H}{2} & -b\ell\\ i(\frac{1}{4}+b^2\tilde{a}) & -b\ell & \frac{1}{4}-b^2\tilde{a}\end{pmatrix},\tag{1.11}$$

$$\tilde{a}=a^2-K/(4b^2)+\mathcal{P}H/2+b^2\mathcal{P}^2/4.\tag{1.12}$$

The Neumann system (1.10)–(1.12) has four integrals of motion:

$$\begin{aligned} I_1&=\vec{l}^{\,2}+(Q\vec{p},\vec{p}\,),\quad I_2=(Q\vec{l},\vec{l}\,)-(Q^\wedge\vec{p},\vec{p}\,),\\ C_1&=(\vec{l},\vec{p}\,)=0,\qquad C_2=\vec{p}^{\,2}=-\frac{1}{4} \end{aligned}\tag{1.13}$$

where $Q^\wedge$ denotes the adjoint matrix. It is usually formulated together with the canonical e(3) Poisson bracket for the generators p_k and l_k. Unfortunately, the transformation (1.7)–(1.8) is noncanonical, so it does not bring the o(4) bracket of the initial variables into this e(3) bracket of the Neumann system. This means we can use the equations (1.10) only keeping in mind that we have a new Poisson structure on the variables p_k and l_k imposed by the Kowalevski o(4) Poisson structure.

As was already mentioned, the mapping of the e(3) Kowalevski dynamics ($\mathcal{P}=0$) into Neumann system's dynamics was first found in [**9**]. Algebraic geometric arguments proving the existence of such (noncanonical) maps in other 'genus 2' situations was also studied in detail around that time (cf. [**5**] and references therein) producing many other maps between various systems, in particular, tops.

Now, let us proceed further to the integration in quadratures. Since the problem has been reduced to the dynamics of the Neumann system, its integration is well-known (see, for instance, [**33**]). One introduces two new variables λ_1 and λ_2 as zeros of the polynomial:

$$\lambda^2 + \lambda\left(\frac{H}{2} - 4(Q\vec{p},\vec{p}\,)\right) - 4(Q^\wedge\vec{p},\vec{p}\,) = (\lambda-\lambda_1)(\lambda-\lambda_2). \tag{1.14}$$

Then one checks that they satisfy the following equations:

$$\dot{\lambda_1}(\lambda_1-\lambda_2) = 2\sqrt{-R_5(\lambda_1)}, \quad \dot{\lambda_2}(\lambda_2-\lambda_1) = 2\sqrt{-R_5(\lambda_2)}, \tag{1.15}$$

$$R_5(\lambda) = ((\lambda - \mathcal{P}b^2/2)^2 - K/4)\left[\left(\lambda + \frac{H}{2}\right)(\lambda^2 + b^2\tilde{a}) - b^2\ell^2\right]. \tag{1.16}$$

After rewriting these equations in the form

$$\frac{d\lambda_1}{\sqrt{-R_5(\lambda_1)}} + \frac{d\lambda_2}{\sqrt{-R_5(\lambda_2)}} = 0, \quad \frac{\lambda_1 d\lambda_1}{\sqrt{-R_5(\lambda_1)}} + \frac{\lambda_2 d\lambda_2}{\sqrt{-R_5(\lambda_2)}} = 2\,dt, \tag{1.17}$$

one uses Abel–Jacobi map to integrate the dynamics in terms of theta-functions. This is exactly what Kowalevski did in 1889 (for $\mathcal{P} = 0$): she found the variables λ_1 and λ_2 (1.14), she derived the equations (1.15)–(1.16), now known as the Kowalevski equations, and she found theta-function formulas for the initial variables of the top, J_k and x_k, as functions of time t. Of course, she was not aware of the 'simplifying map' to the Neumann system discovered 100 years later, this is why her original calculations looked so complicated, mysterious and attractive for many generations of mathematicians, and they still do!

Formulas (1.14)–(1.17) for arbitrary $\mathcal{P}$ were derived in [**19**].

2. Separation of Variables for the o(4) Kowalevski Top

In this Section we collect the results from [**19**] about separation of variables (SoV) for the Kowalevski top on the o(4) algebra.

In Hamiltonian dynamics, separation variables have been generally used to construct analytic expressions for the action variables (in terms of Abelian integrals) or in order to get a separated representation for the action function. Therefore, the method of separation of variables for a long time served an important, but technical role in solving Liouville integrable systems of classical mechanics. A new, and much more exciting, application of the method came with the development of quantum integrable systems. Because of the fact that quantization of the action variables seems to be a rather formidable task, quantum separation of variables became an inevitable refuge. In fact, it was successfully performed for many families of integrable systems (see, for instance, survey [**39**]).

We should start from a (working) definition of the SoV. Historically, several different definitions have been given, each depending on the context. We will always mean the context of finite-dimensional integrable Hamiltonian dynamics or, in other words, our definition will only be valid for the Liouville integrable systems.

By *separation of variables* we mean a canonical transformation to new variables u_j, v_j, $j = 1, \ldots, n$, which satisfy the following *separation equations*

$$\sum_{j=1}^{n} a_{ij} H_j = b_i, \quad i = 1, \ldots, n, \tag{2.1}$$

$$a_{ij} = a_{ij}(u_i, v_i), \quad b_i = b_i(u_i, v_i), \tag{2.2}$$

or, in other words, in terms of which the integrals of motion H_j acquire the following 'separated form':

$$\vec{H} = A^{-1}\vec{B}, \quad (A)_{ij} = a_{ij}, \quad (\vec{B})_i = b_i, \quad (\vec{H})_i = H_i. \tag{2.3}$$

The conditions (2.2) that the functions a_{ij} and b_i in (2.1) depend on the new (separation) variables with the index i *only*, is crucial. It indeed means that the n equations in (2.1) are really separated from one another.

The above definition includes, as a special canonical transform, the (classical) coordinate separation of variables, when the new coordinates, say u_j, are the functions of the old coordinates (q_j) only, and they do not depend on the momenta (p_j). See the book [**13**] about the history of this sub-class of transformations with many examples of such situation.

General (separating) canonical transforms, however, result in new (separation) variables being nontrivial functions of all $2n$ initial canonical variables,

$$u_j = u_j(q_1, \ldots, q_n, p_1, \ldots, p_n), \quad v_j = v_j(q_1, \ldots, q_n, p_1, \ldots, p_n), \quad j = 1, \ldots, n, \tag{2.4}$$

$$\{u_j, u_k\} = \{v_j, v_k\} = \{u_j, v_k\} = 0, \; j \neq k, \quad \{v_j, u_j\} = 1, \; j = 1, \ldots, n. \tag{2.5}$$

The examples of such separating canonical transforms are usually much more sophisticated than those of the coordinate ones. As far as I know, the first explicit example was given by van Moerbeke in [**32**] concerning separation of variables for the Toda lattice (see also [**8**]). The method was further developed by Komarov in a series of works on tops, including quantum separation of variables, see [**15, 16**]. Many further examples have been produced since 1982, with the theory benefiting mostly from the developments of the algebraic geometric and r-matrix understanding of the method of separation of variables. This led to a rather satisfying picture of the present state-of-art of noncoordinate separation of variables. See [**39**] and [**11, 12**] for more details.

It is interesting to remark that many explicit noncoordinate separations were already produced by the classics, such as those for the Goryachev–Chaplygin and Kowalevski top, *but* they never calculated the Poisson brackets, so that many modern developments involved checking the canonicity of the transformations, proposed by mathematicians of the 19th century! Hence, classics *knew* about noncoordinate separation of variables (sometimes more than we do now). It is a pity that we lack a good review of this subject which would cover, say, the separation of variables for the Clebsch, Euler–Manakov, Steklov and many other tops, including modern results and those obtained by classics.

Two pairs of separation variables for the o(4) Kowalevski top are as follows:

$$s_i = 2\lambda_i + H, \quad p_i = \frac{1}{2\sqrt{-2s_i}} \ln \frac{\xi_i + \sqrt{\xi_i^2 + d_i^2}}{d_i}, \qquad i = 1, 2, \tag{2.6}$$

$$\xi_i = 2\sqrt{y_i^2 + d_i y_i}, \quad y_i = (s_i - H - \mathcal{P}b^2)^2 - K, \qquad d_i = 4b^2\left(\frac{\mathcal{P}s_i}{2} + a^2 - \frac{2\ell^2}{s_i}\right). \tag{2.7}$$

They are canonical variables,

$$\{s_1, s_2\} = \{p_1, p_2\} = \{s_1, p_2\} = \{s_2, p_1\} = 0, \tag{2.8}$$

$$\{p_1, s_1\} = 1, \quad \{p_2, s_2\} = 1, \tag{2.9}$$

and they satisfy the following separation equations:

$$s_i^3 - (2H + \mathcal{P}b^2)s_i^2 + \kappa s_i - 4b^2\ell^2 = 2b^2\left(\frac{\mathcal{P}s_i^2}{2} + a^2 s_i - 2\ell^2\right)\cos(2\sqrt{2s_i}p_i), \tag{2.10}$$

$$\kappa = (H + \mathcal{P}b^2)^2 - K + 2a^2b^2. \tag{2.11}$$

Notice that the role of the H_1 and H_2 from the definition (2.1) is played here by the integrals H and κ. The functions a_{ij} and b_i can be directly read from the formulas (2.10). Also notice that the formulas (2.10) are equivalent to two formulas for p_i's from (2.6).

This result, i.e. the separation of variables, was given for $\mathcal{P} = 0$ in [**41**][1] and for the general case in [**19**]. We will give here a shorter proof than the one in [**19**], but before doing that let us reformulate the statement in terms of other variables, the s_i and y_i.

The variables s_i and y_i, $i = 1, 2$, are 'almost canonical',

$$\{s_1, s_2\} = \{y_1, y_2\} = \{s_1, y_2\} = \{s_2, y_1\} = 0, \tag{2.12}$$

$$\{y_1, s_1\} = -4\sqrt{-2s_1y_1(y_1 + d_1)}, \quad \{y_2, s_2\} = -4\sqrt{-2s_2y_2(y_2 + d_2)}, \tag{2.13}$$

and they satisfy the following simple separation equations:

$$y_i = (s_i - H - \mathcal{P}b^2)^2 - K, \quad i = 1, 2. \tag{2.14}$$

In order to bring the equations (2.12)–(2.14) into the equations (2.8)–(2.11) one needs to make a transformation from y_i's to the canonically conjugated variables p_i's which is easily found by taking the integral,

$$p_i = -\frac{1}{4}\int^{y_i} \frac{dy}{\sqrt{-2s_i y(y + d_i)}} = \frac{1}{2\sqrt{-2s_i}} \ln\left(1 + \frac{2y_i}{d_i} + 2\sqrt{\frac{y_i}{d_i}\left(\frac{y_i}{d_i} + 1\right)}\right). \tag{2.15}$$

The separation equations (2.14) are then transformed into the equations

$$s_i\left(y_i + \frac{d_i}{2}\right) = \frac{s_i d_i}{2}\cos(2\sqrt{2s_i}p_i), \tag{2.16}$$

which are equivalent to the separation equations (2.10). Notice here that $R_5(\lambda_i)$ from (1.15)–(1.16) have the following expressions in terms of the variables s_i, y_i and d_i:

$$R_5(\lambda_i) = \frac{s_i y_i}{32}(y_i + d_i), \quad i = 1, 2. \tag{2.17}$$

Finally, the brackets (2.12)–(2.13) are checked by a direct computation[2].

[1] without a proof

[2] on a computer: checking (2.12) takes about 2 hours

3. Hyperelliptic Prymian and Ansatz for the Lax Matrix

In this Section we change the separation variables s_i and p_i to new separation variables u_i and $m_i^{\pm}$, which respect the symmetry of the problem, and while doing that we will derive the proper ansatz for the future 2×2 Lax matrix of the o(4) Kowalevski top.

First of all, notice that in the definition of p_i's in (2.6) there appears $\sqrt{s_i}$, which is very strange taking into account that $\lambda = -H/2$ (which is equivalent to $s = 0$ according to the definition of s-variables (2.6)) is *not* a branching point of the Kowalevski curve $\mu^2 = R_5(\lambda)$ (cf. (1.16)).

This 'paradox' was explained in [**19**] by stating that the s-variables are *not* the proper variables and one should introduce new, u-variables, which respect the additional symmetry present in the problem[3]. This problem of s-variables passed unnoticed in [**41**] in the e(3) case.

The proper new variables u_i and $\widehat{y}_i$, $i = 1, 2$, are introduced as follows:

$$(3.1) \qquad u_i := \sqrt{\frac{s_i}{2}}, \quad \widehat{y}_i := (u_i^2 - H/2 - \mathcal{P}b^2/2)^2 - K/4, \qquad i = 1, 2.$$

Another pair of variables, which are functions of u-variables and Casimirs, will be used:

$$(3.2) \qquad \widehat{d}_i := b^2\left(\mathcal{P}u_i^2 + a^2 - \frac{\ell^2}{u_i^2}\right), \quad i = 1, 2.$$

The variables u_i and $\widehat{y}_i$ are 'almost canonical':

$$(3.3) \qquad \{u_1, u_2\} = \{\widehat{y}_1, \widehat{y}_2\} = \{u_1, \widehat{y}_2\} = \{u_2, \widehat{y}_1\} = 0,$$

$$(3.4) \qquad \{\widehat{y}_1, u_1\} = -2\sqrt{-\widehat{y}_1(\widehat{y}_1 + \widehat{d}_1)}, \qquad \{\widehat{y}_2, u_2\} = -2\sqrt{-\widehat{y}_2(\widehat{y}_2 + \widehat{d}_2)}.$$

Defining new variables $m_i^{\pm}$,

$$(3.5) \qquad m_i^{\pm} := 1 + 2\frac{\widehat{y}_i}{\widehat{d}_i} \pm 2\sqrt{\frac{\widehat{y}_i}{\widehat{d}_i}\left(\frac{\widehat{y}_i}{\widehat{d}_i} + 1\right)}, \quad i = 1, 2,$$

we finally obtain the separation equations in the form

$$(3.6) \quad u_i^6 - (H + \mathcal{P}b^2/2)u_i^4 + \frac{1}{4}((H + \mathcal{P}b^2)^2 - K + 2a^2b^2)u_i^2 - \frac{b^2\ell^2}{2} = \frac{b^2}{4}(\mathcal{P}u_i^4 + a^2u_i^2 - \ell^2)(m_i^+ + m_i^-).$$

The final set of separation variables u_i and $m_i^{\pm}$ satisfy two separation equations above and the algebra below:

$$(3.7) \qquad \{u_1, u_2\} = 0, \quad \{u_j, m_k^{\pm}\} = 0, \qquad j \neq k,$$

$$(3.8) \qquad \{m_j^{\pm}, u_j\} = \mp 2im_j^{\pm}, \quad m_j^+ m_j^- = 1.$$

As was already mentioned, this was a result of [**18**] and [**19**] in the e(3) and o(4) case, respectively. In [**18, 22**] these separation variables and corresponding action variables were used for constructing the quasiclassical spectrum of the integrals of motion of the Kowalevski top.

[3]for the usual e(3) Kowalevski top, i.e. in the $\mathcal{P} = 0$ case, the original explanation was presented much earlier in author's MSc dissertation in 1985, see [**18, 22**]

If we exclude the variables m_i^+ using the condition that $m_i^+m_i^- = 1$ and if we rescale the variables m_i^-,

$$m_i^+ = \frac{1}{m_i^-}, \quad \widetilde{m}_i^- = \frac{b^2}{4}(\mathcal{P}u_i^4 + a^2u_i^2 - \ell^2)m_i^-, \tag{3.9}$$

the separation equations (3.6) will acquire the following form:

$$(\widetilde{m}_i^-)^2 - P_3(u_i^2)\widetilde{m}_i^- + \big(P_2(u_i^2)\big)^2 = 0 \tag{3.10}$$

where two polynomials $P_3(u)$ and $P_2(u)$ are

$$P_3(u) = u^3 - (H + \mathcal{P}b^2/2)u^2 + \frac{1}{4}((H + \mathcal{P}b^2)^2 - K + 2a^2b^2)u - \frac{b^2\ell^2}{2}, \tag{3.11}$$

$$P_2(u) = \frac{b^2}{4}(\mathcal{P}u^2 + a^2u - \ell^2). \tag{3.12}$$

Therefore, separation equations (3.6) amount to having two points, $(u_1, \widetilde{m}_1^-)$ and $(u_2, \widetilde{m}_2^-)$, on the algebraic curve Γ:

$$\Gamma\colon m^2 - P_3(u^2)m + \big(P_2(u^2)\big)^2 = 0, \tag{3.13}$$

$$(u_i, \widetilde{m}_i^-) \in \Gamma, \quad i = 1, 2. \tag{3.14}$$

The curve Γ (3.13) is a hyperelliptic curve with the involution $u \mapsto -u$. This is the curve that replaces Kowalevski's genus 2 hyperelliptic curve $\mu^2 = R_5(\lambda)$ if one takes into account the Poisson structure and the symmetry of the problem. The Kowalevski dynamics is therefore linearized on the corresponding hyperelliptic Prymian of the curve Γ[4]. We can bring the curve Γ (3.13) into the standard hyperelliptic form by shifting the variable m:

$$m \mapsto \widehat{m} = m - \frac{1}{2}P_3(u^2), \tag{3.15}$$

thus getting the curve,

$$\widehat{m}^2 = \frac{1}{4}\big(P_3(u^2) - 2P_2(u^2)\big)\big(P_3(u^2) + 2P_2(u^2)\big) \equiv \frac{1}{4}R_6(u^2), \tag{3.16}$$

$$R_6(u) = uR_5(u - H/2), \tag{3.17}$$

with Kowalevski's polynomial $R_5(\lambda)$ given in (1.16).

Now we can formulate the problem: *to find a 2×2 Lax matrix for which the curve Γ (3.13) is the spectral curve or, in other words, to find the proper 2×2 Lax matrix for the Kowalevski top which is related to Kowalevski's separation of variables.*

We must mention here the known Lax matrices for the Kowalevski top. Let us consider only the e(3) case, as nothing is known about Lax matrices for the o(4) case. Previous attempts to construct Lax matrices for the KT included:

(i) 2×2 matrix in [**7**],
(ii) 3×3 matrix in [**9**],
(iii) 4×4 and 6×6 matrices in [**5**], and
(iv) 5×5 (or 4×4) matrix in [**35**].

---

[4]see the survey by Markushevich in this volume (and also his work [**31**]) about interrelations between different curves for the Kowalevski top in the case of e(3) algebra and zero Casimir $\ell = 0$

The first three do not respect the Poisson structure of the problem in contrast to the last one, which satisfies a linear r-matrix algebra. Unfortunately, no separation of variables is known which can be related to the Lax matrix of [**35**]. In fact, as I said before, no separation is known which is alternative to the original Kowalevski's separation of the problem!

What can be guessed about the ansatz for the required 2×2 Lax matrix $T(u)$? From the form of the spectral curve Γ (3.13) one can conclude that the entries of $T(u)$ will be polynomials in u of order nor higher then 6, the $\operatorname{tr} T(u)$ will be equal to $P_3(u^2)$ and the $\det T(u)$ will be equal to $\left(P_2(u^2)\right)^2$ and, hence, will depend only on Casimirs. In the next sections we will construct (and solve) an integrable system whose Lax matrix has exactly these properties. This system will be formulated within the framework of the quadratic r-matrix algebra for integrable systems with boundary conditions. The o(4) Kowalevski top, as we will see, will correspond to a system with $B_2^{(1)}$-type boundary conditions.

4. Quadratic Algebra $\mathcal{B}$

Starting from about 1982 the method of separation of variables gets connected with the r-matrix formalism of the quantum inverse scattering method, developed during that time by the Leningrad School. It was noticed by Komarov (see the footnote in [**36**] and a full credit in [**39**]) that for the 2×2 L-operators (Lax matrices) the separation variables ought to be the zeros of the off-diagonal element of the L-operator. This observation was fully exploited by Sklyanin in [**36, 37**] who developed a beautiful (pure algebraic) setting for the method within the framework of the r-matrix technique. Since then this approach took off and led to separations for many families of integrable systems. Sklyanin also generalized the approach to include higher rank L-operators and nonstandard normalizations. See the review [**39**] where many of the examples were exposed. An alternative, algebraic geometric approach, which dates back to Adler and van Moerbeke [**3, 4**] and Mumford [**33**] and includes many researchers, have been developed starting from about the same time (see, for instance, [**1, 2, 10–12, 34, 40**]). It also led to many important new separations for complicated integrable systems and tops. Unfortunately, we can not review this another approach, as it would require much larger scope than the pure algebraic one we adopted here.

In [**36**] the Goryachev–Chaplygin top and in [**28**] the (symmetric) Neumann system and so-called Kowalevski–Chaplygin–Goryachev top were related to special representations of the quadratic r-matrix algebras. In the present paper we show how the o(4) Kowalevski top is related to a special representation of the quadratic r-matrix algebra for integrable systems with boundary conditions. We start by introducing the corresponding quadratic algebra $\mathcal{B}$.

Let $\delta \in \mathbb{C}$. Consider the following L-operator:

$$L(u) = \begin{pmatrix} A(u) & B(u) \\ C(u) & D(u) \end{pmatrix}, \tag{4.1}$$

$$A(u) = A_4u^4 + A_3u^3 + A_2u^2 + A_1u + \delta, \tag{4.2}$$

$$D(u) = A(-u), \tag{4.3}$$

$$B(u) = B_3u^3 + B_1u, \tag{4.4}$$

$$C(u) = u^5 + C_3u^3 + C_1u, \tag{4.5}$$

$$\deg_u L(u) = \begin{pmatrix} 4 & 3 \\ 5 & 4 \end{pmatrix}, \quad L(-u) = L^{\wedge}(u). \tag{4.6}$$

Introduce B_2-type quadratic Poisson algebra $\mathcal{B}$ with eight generators, $A_1, \ldots, A_4$, B_1, B_3, C_1, C_3, and the following Poisson brackets:

$$\{A_4, A_3\} = 2iB_3, \quad \{A_4, A_2\} = 0, \quad \{A_4, A_1\} = 2iB_1, \tag{4.7}$$

$$\{A_3, A_2\} = -2iB_1, \quad \{A_3, A_1\} = 0, \quad \{A_2, A_1\} = 2i(B_1C_3 - B_3C_1), \tag{4.8}$$

$$\{B_3, B_1\} = 0, \qquad \{C_3, C_1\} = 0, \tag{4.9}$$

$$\{B_3, C_3\} = -8iA_3A_4, \qquad \{B_3, C_1\} = -8iA_1A_4, \tag{4.10}$$

$$\{B_1, C_3\} = -8iA_1A_4, \qquad \{B_1, C_1\} = -8i(-\delta A_3 + A_1A_2), \tag{4.11}$$

$$\{B_3, A_4\} = 0, \qquad \{B_3, A_3\} = 4iB_3A_4, \tag{4.12}$$

$$\{B_3, A_2\} = 0, \qquad \{B_3, A_1\} = 4iB_1A_4, \tag{4.13}$$

$$\{B_1, A_4\} = 0, \qquad \{B_1, A_3\} = 4iB_1A_4, \tag{4.14}$$

$$\{B_1, A_2\} = 4i(B_1A_3 - B_3A_1), \qquad \{B_1, A_1\} = 4i(B_1A_2 - \delta B_3), \tag{4.15}$$

$$\{C_3, A_4\} = 4iA_3, \qquad \{C_3, A_3\} = 4i(A_2 - C_3A_4), \tag{4.16}$$

$$\{C_3, A_2\} = 4iA_1, \qquad \{C_3, A_1\} = 4i(\delta - C_1A_4), \tag{4.17}$$

$$\{C_1, A_4\} = 4iA_1, \qquad \{C_1, A_3\} = 4i(\delta - C_1A_4), \tag{4.18}$$

$$\{C_1, A_2\} = 4i(A_1C_3 - A_3C_1), \qquad \{C_1, A_1\} = 4i(\delta C_3 - C_1A_2). \tag{4.19}$$

This is a rank 2 algebra because there are four Casimirs $Q_1, \ldots, Q_4$, which are the coefficients of the $\det L(u)$:

$$\det L(u) = Q_4u^8 + Q_3u^6 + Q_2u^4 + Q_1u^2 + \delta^2, \tag{4.20}$$

$$Q_4 = A_4^2 - B_3, \tag{4.21}$$

$$Q_3 = 2A_2A_4 - A_3^2 - B_3C_3 - B_1, \tag{4.22}$$

$$Q_2 = A_2^2 + 2\delta A_4 - 2A_1A_3 - B_3C_1 - B_1C_3, \tag{4.23}$$

$$Q_1 = 2\delta A_2 - A_1^2 - B_1C_1. \tag{4.24}$$

In the matrix notations, this algebra looks very compact,

$$\{L_1(u), L_2(v)\} = \big[r(u-v), L_1(u)L_2(v)\big] + L_1(u)r(u+v)L_2(v) - L_2(v)r(u+v)L_1(u), \tag{4.25}$$

with the r-matrix $r(u)$ being as follows:

$$r(u) = \frac{-2i}{u}\begin{pmatrix} 1 & 0 & 0 & 0 \\ 0 & 0 & 1 & 0 \\ 0 & 1 & 0 & 0 \\ 0 & 0 & 0 & 1 \end{pmatrix}. \tag{4.26}$$

In (4.25) we use the standard notations for the tensor products,

$$L_1(u) = L(u) \otimes \begin{pmatrix} 1 & 0 \\ 0 & 1 \end{pmatrix}, \quad L_2(v) = \begin{pmatrix} 1 & 0 \\ 0 & 1 \end{pmatrix} \otimes L(v). \tag{4.27}$$

Also, in the left-hand-side of (4.25) we have a 4×4 matrix with the entries $\{L_1(u), L_2(v)\}_{ij,kl} \equiv \{(L(u))_{ij}, (L(v))_{kl}\}$ and in the right-hand-side of (4.25) there are (4×4) matrix commutator and products. For more details on these notations and on quadratic r-matrix algebras see [**20, 24–30, 36–39**].

The Poisson brackets for the polynomials $A(u)$, $B(u)$ and $C(u)$ (the polynomial $D(u) \equiv A(-u)$) read

$$\{A(u), A(v)\} = \frac{-2i}{u+v}\big(B(u)C(v) - B(v)C(u)\big), \tag{4.28}$$

$$\{B(u), B(v)\} = 0, \quad \{C(u), C(v)\} = 0, \tag{4.29}$$

$$\{B(u), A(v)\} = \frac{-2i}{u-v}\big(B(u)A(v) - B(v)A(u)\big) + \frac{2i}{u+v}\big(A(v)B(u) + B(v)A(-u)\big), \tag{4.30}$$

$$\{C(u), A(v)\} = \frac{-2i}{u-v}\big(A(u)C(v) - A(v)C(u)\big) - \frac{2i}{u+v}\big(C(u)A(v) + A(-u)C(v)\big), \tag{4.31}$$

$$\{B(u), C(v)\} = \frac{-2i}{u-v}\big(A(-u)A(v) - A(-v)A(u)\big) - \frac{2i}{u+v}\big(A(u)A(v) - A(-v)A(-u)\big). \tag{4.32}$$

5. Separation Representation of the Algebra $\mathcal{B}$

Let us realize the algebra $\mathcal{B}$ (4.7)–(4.19) in terms of 'separation variables' $\widehat{u}_j$, $\widehat{m}_j^{\pm}$, $j = 1, 2$.

First, introduce two new Poisson commuting variables $\widehat{u}_1$ and $\widehat{u}_2$ as zeros of the polynomial $C(u)$:

$$C(u) = u^5 + C_3u^3 + C_1u = u(u^2 - \widehat{u}_1^2)(u^2 - \widehat{u}_2^2), \tag{5.1}$$

$$\widehat{u}_1^2 + \widehat{u}_2^2 = -C_3, \quad \widehat{u}_1^2\widehat{u}_2^2 = C_1, \tag{5.2}$$

$$\{C_3, C_1\} = 0 \iff \{\widehat{u}_1, \widehat{u}_2\} = 0. \tag{5.3}$$

The 'conjugated' variables $\widehat{m}_j^{\pm}$ are introduced as corresponding values of the polynomial $A(\mp u)$ at these $\widehat{u}_j$:

$$\widehat{m}_j^{\pm} = A(\mp\widehat{u}_j), \quad j = 1, 2. \tag{5.4}$$

It is easy to prove that these new variables obey the following (separated) algebra of Poisson brackets:

$$(A) \quad \widehat{m}_j^{+}\widehat{m}_j^{-} = \det L(\widehat{u}_j), \tag{5.5}$$

$$(B) \quad \{\widehat{m}_j^{\pm}, \widehat{u}_j\} = \mp 2i\,\widehat{m}_j^{\pm}, \qquad j = 1, 2, \tag{5.6}$$

$$(C) \quad \{\widehat{m}_k^{\pm}, \widehat{u}_j\} = \{\widehat{m}_k^{\pm}, \widehat{m}_j^{\pm}\} = \{\widehat{u}_k, \widehat{u}_j\} = 0, \qquad k \neq j. \tag{5.7}$$

Let us prove, for instance, that $\{\widehat{u}_j, \widehat{m}_k^{-}\} = -2i\delta_{jk}\widehat{m}_k^{-}$ (cf. [**24, 36**]). Notice first that by the differentiation property of the Poisson bracket,

$$0 \equiv \{C(\widehat{u}_j), A(v)\} = C'(\widehat{u}_j)\{\widehat{u}_j, A(v)\} + \{C(u), A(v)\}_{|u=\widehat{u}_j}. \tag{5.8}$$

Hence,

$$\{\widehat{u}_j, A(v)\} = -\frac{1}{C'(\widehat{u}_j)}\{C(u), A(v)\}_{|u=\widehat{u}_j}. \tag{5.9}$$

Now, substituting $u=\widehat{u}_j$ into the algebraic relation (4.31), we obtain

$$\{C(u),A(v)\}_{|u=\widehat{u}_j} = \frac{-2i}{\widehat{u}_j - v}\widehat{m}_j^- C(v) - \frac{2i}{\widehat{u}_j + v}\widehat{m}_j^+ C(v). \tag{5.10}$$

Therefore,

$$\{\widehat{u}_j,\widehat{m}_k^-\} = 0, \quad j\neq k, \tag{5.11}$$

and

$$\{\widehat{u}_j,\widehat{m}_j^-\} = -2i\widehat{m}_j^-, \quad j=1,2. \tag{5.12}$$

The rest of the relations (A), (B), (C) can be proved in the similar way.

Now we can realize the four polynomials $A(u)$, $B(u)$, $C(u)$ and $D(u)$, whose coefficients are the generators of the algebra $\mathcal{B}$, in terms of $\widehat{u}_j$, $\widehat{m}_j^{\pm}$, $j=1,2$, from the following data:

- $C_3=-\widehat{u}_1^2-\widehat{u}_2^2$, $C_1=\widehat{u}_1^2\widehat{u}_2^2 \Rightarrow C(u)$.
- $A(\pm\widehat{u}_j)=\widehat{m}_j^{\mp}$, $A(0)=\delta \Rightarrow A(u)$.
- $D(u)=A(-u)$.
- $B(u)=\frac{A(u)D(u)-\det L(u)}{C(u)}$, $C(u)$ and the numerator have common zeros at $u=0$ and at $u=\pm\widehat{u}_{1,2}$, so that $C(u)$ divides the numerator.

The result is given by the formulas

$$A(u) = \frac{(u^2-\widehat{u}_1^2)(u^2-\widehat{u}_2^2)}{\widehat{u}_1^2\widehat{u}_2^2}\delta + \frac{u(u^2-\widehat{u}_1^2)((u-\widehat{u}_2)\widehat{m}_2^+ + (u+\widehat{u}_2)\widehat{m}_2^-)}{2\widehat{u}_2^2(\widehat{u}_2^2-\widehat{u}_1^2)} + \frac{u(u^2-\widehat{u}_2^2)((u-\widehat{u}_1)\widehat{m}_1^+ + (u+\widehat{u}_1)\widehat{m}_1^-)}{2\widehat{u}_1^2(\widehat{u}_1^2-\widehat{u}_2^2)}, \quad D(u)=A(-u), \tag{5.13}$$

$$C(u)=u(u^2-\widehat{u}_1^2)(u^2-\widehat{u}_2^2), \quad B(u)=B_3u^3+B_1u, \tag{5.14}$$

$$B_3=A_4^2-q_4, \quad B_1=2A_2A_4-A_3^2+(\widehat{u}_1^2+\widehat{u}_2^2)(A_4^2-q_4)-q_3 \tag{5.15}$$

where q_4 and q_3 are fixed values of the Casimirs from the determinant (cf. (4.20)),

$$\det L(u) = q_4u^8+q_3u^6+q_2u^4+q_1u^2+\delta^2. \tag{5.16}$$

6. $B_2^{(1)}$-Type Integrable System

Let us now introduce an integrable system (of $B_2^{(1)}$-type) on the algebra $\mathcal{B}$. Denote by $K(u)$ the following constant representation of the algebra (4.25) with the r-matrix (4.26):

$$K(u)=\begin{pmatrix}1 & u\\ 0 & 1\end{pmatrix}. \tag{6.1}$$

Define the integrable system by its Lax matrix $T(u)$,

$$T(u) := K(u)L(u). \tag{6.2}$$

Its integrals of motion are the coefficients of the trace of the Lax matrix $T(u)$ ($\det T(u)$ is a Casimir),

$$\operatorname{tr} T(u) = u^6+\mathcal{H}_1u^4+\mathcal{H}_2u^2+2\delta, \tag{6.3}$$

$$\mathcal{H}_1=2A_4+C_3, \quad \mathcal{H}_2=2A_2+C_1, \quad \{\mathcal{H}_1,\mathcal{H}_2\}=0. \tag{6.4}$$

The variables $\widehat{u}_j$, $\widehat{m}_j^{\pm}$, $j = 1, 2$, from the previous section *are* the separation variables for the integrable system (6.4). Indeed, because $\operatorname{tr} T(u) = A(u) + D(u) + uC(u)$, one obtains the following separation equations:

$$\operatorname{tr} T(\widehat{u}_j) = \widehat{u}_j^6 + \mathcal{H}_1 \widehat{u}_j^4 + \mathcal{H}_2 \widehat{u}_j^2 + 2\delta = \widehat{m}_j^+ + \widehat{m}_j^-, \quad j = 1, 2. \tag{6.5}$$

This is the separation of variables for the $B_2^{(1)}$-type integrable system on the algebra $\mathcal{B}$ with the integrals of motion $\mathcal{H}_1$ and $\mathcal{H}_2$ (6.4).

Notice here that a general constant representation of the algebra (4.25) with the r-matrix (4.26) is given by the full matrix,

$$\widehat{K}(u) = \begin{pmatrix} \alpha u + \delta & \beta u \\ \gamma u & -\alpha u + \delta \end{pmatrix}, \tag{6.6}$$

which is used to generate integrable systems with more general boundary conditions, of BC-type (see, for instance, the examples of Toda lattice in [**38**] and Kowalevski–Chaplygin–Goryachev top in [**28**]). The D-type boundary conditions for the Toda lattice are described by a nonconstant matrix $\widehat{K}(u)$ depending on the dynamical variables (cf. [**20, 24–27, 29, 30**]).

7. Lax Matrix for the o(4) Kowalevski Top

Now we have everything to write down the 2×2 Lax matrix $T(u)$ for the o(4) KT. It will have the factorization (6.2),

$$T(u) = \begin{pmatrix} 1 & u \\ 0 & 1 \end{pmatrix} \begin{pmatrix} A(u) & B(u) \\ C(u) & D(u) \end{pmatrix}, \tag{7.1}$$

and we only have to find the expressions for the polynomials $A(u)$, $B(u)$, $C(u)$ and $D(u)$ in terms of the o(4) variables J_k and x_k, $k = 1, 2, 3$.

Notice that two integrable systems considered above, the o(4) KT and the $B_2^{(1)}$-type integrable system on the quadratic algebra $\mathcal{B}$, look the same in terms of the separation variables, so that one can identify their separation variables and separation equations as follows:

$$\widehat{u}_j = u_j, \quad \widehat{m}_j^{\pm} = P_2(u_j^2) m_j^{\pm}, \qquad j = 1, 2, \tag{7.2}$$

$$P_3(u^2) = \operatorname{tr} T(u) = u^6 + \mathcal{H}_1 u^4 + \mathcal{H}_2 u^2 + 2\delta, \tag{7.3}$$

$$\left(P_2(u^2)\right)^2 = \det T(u) = q_4 u^8 + q_3 u^6 + q_2 u^4 + q_1 u^2 + \delta^2 \tag{7.4}$$

where polynomials $P_3(u)$ and $P_2(u)$ are given in (3.11)–(3.12). To see that, compare the separation equations (6.5) and the ones for the o(4) KT (3.6).

As a direct consequence, we can restore the whole algebra, i.e. get a new representation of the algebra $\mathcal{B}$ in terms of the o(4) variables. The result is as follows:

$$A_4 = \frac{1}{2}(u_1^2 + u_2^2 - H - \mathcal{P}b^2/2) = \frac{\widehat{y}_1 - \widehat{y}_2}{2(u_1^2 - u_2^2)} + \frac{\mathcal{P}b^2}{4}, \tag{7.5}$$

$$A_2 = \frac{1}{8}\left((H + \mathcal{P}b^2)^2 - K + 2a^2b^2 - 4u_1^2u_2^2\right) = \frac{\widehat{y}_2 u_1^2 - \widehat{y}_1 u_2^2}{2(u_1^2 - u_2^2)} + \frac{a^2b^2}{4}, \tag{7.6}$$

$$A_0 = -\frac{b^2\ell^2}{4}, \quad A_3 = \frac{i}{8}\{H, u_1^2 + u_2^2\}, \quad A_1 = -\frac{i}{8}\{H, u_1^2u_2^2\}, \tag{7.7}$$

$$C_3 = -u_1^2 - u_2^2, \qquad C_1 = u_1^2 u_2^2, \tag{7.8}$$

$$B_3 = \frac{1}{4}(u_1^2+u_2^2-H)(u_1^2+u_2^2-H-\mathcal{P}b^2) = \frac{\widehat{y}_1-\widehat{y}_2}{4(u_1^2-u_2^2)}\left(\frac{\widehat{y}_1-\widehat{y}_2}{u_1^2-u_2^2}+\mathcal{P}b^2\right), \tag{7.9}$$

$$B_1 = 2A_2A_4 - A_3^2 + (u_1^2+u_2^2)B_3 - \mathcal{P}a^2b^4/8. \tag{7.10}$$

Or, explicitly,

$$A_4 = -\frac{1}{2}(X^2+J_3^2-\mathcal{P}b^2/2), \tag{7.11}$$

$$A_2 = \frac{J^2}{2}(X^2+J_3^2) + bx_2J_3X - bx_1J_3^2 + \frac{b^2}{2}\left(\mathcal{P}J_3^2 - x_3^2 + \frac{a^2}{2}\right), \tag{7.12}$$

$$A_0 = -\frac{b^2\ell^2}{4}, \quad A_3 = -\frac{i}{2}(X^3 + (J_3^2+bx_1-\mathcal{P}b^2)X + bx_2J_3), \tag{7.13}$$

$$\begin{aligned} A_1 = \frac{i}{2}\big(J^2X^3 &+ 2bx_2J_3X^2 + \big(J_3^2(J^2-bx_1) + b^2(x_2^2-x_3^2-\mathcal{P}J_1^2) \\ &+ bJ_1(x_1J_1+x_2J_2)\big)X \\ &+ bx_2J_3(J_2^2+J_3^2-bx_1) + J_1J_2J_3(bx_1-\mathcal{P}b^2)\big), \end{aligned} \tag{7.14}$$

$$C_3 = X^2 - J^2 + 2bx_1 - \mathcal{P}b^2, \tag{7.15}$$

$$C_1 = -J^2X^2 - 2b(x_2J_3-x_3J_2)X - 2b\ell J_1 - b^2(x_2^2+x_3^2-\mathcal{P}J_1^2), \tag{7.16}$$

$$B_3 = \frac{1}{4}(X^2+J_3^2)(X^2+J_3^2-\mathcal{P}b^2), \tag{7.17}$$

$$\begin{aligned} B_1 = -\frac{1}{4}\big(J^2X^4 &+ 2bx_2J_3X^3 + \big(2J_3^2(J^2-bx_1+\mathcal{P}b^2/2) \\ &+ a^2b^2 - b^2(2x_3^2+x_1^2)\big)X^2 + 2bx_2J_3(J_3^2-bx_1)X \\ &+ J_3^4(J^2-2bx_1) - b^2J_3^2(x_3^2-x_1^2+\mathcal{P}(J_1^2+J_2^2)) + \mathcal{P}b^4x_3^2\big) \end{aligned} \tag{7.18}$$

where

$$X := \frac{J_1J_3+bx_3}{J_2}, \quad J^2 := J_1^2+J_2^2+J_3^2. \tag{7.19}$$

Recall that

$$\ell = x_1J_1+x_2J_2+x_3J_3, \quad a^2 = x_1^2+x_2^2+x_3^2-\mathcal{P}J^2. \tag{7.20}$$

The formulas (7.11)–(7.20) give the polynomials $A(u)$, $B(u)$, $C(u)$ and $D(u)$ (cf. (4.2)–(4.5)), which together with the formula (7.1) define the 2×2 Lax matrix $T(u)$ for the o(4) Kowalevski top. This Lax matrix satisfies the quadratic (r-matrix) algebra, corresponding to the B_2-type boundary conditions, and it is the one which is responsible for the separation of variables for the Kowalevski top.

Recall that the spectral curve Γ of the Lax matrix $T(u)$ (7.1) has the following form:

$$\Gamma\colon \det\big(T(u)-m\big) \equiv m^2 - \operatorname{tr}T(u)m + \det T(u) = 0, \tag{7.21}$$

$$\begin{aligned} \operatorname{tr}T(u) = P_3(u^2) = u^6 &- (H+\mathcal{P}b^2/2)u^4 \\ &+\frac{1}{4}((H+\mathcal{P}b^2)^2 - K + 2a^2b^2)u^2 - \frac{b^2\ell^2}{2}, \end{aligned} \tag{7.22}$$

$$\det T(u) = \big(P_2(u^2)\big)^2 = \frac{b^4}{16}(\mathcal{P}u^4+a^2u^2-\ell^2)^2. \tag{7.23}$$

Knowing the r-matrix algebraic structure (4.25)–(4.26) of the found Lax matrix, it is not difficult to derive the full Lax pair, namely: the Hamiltonian flow given by the Kowalevski Hamiltonian H (cf. (1.9)),

$$\dot{J}_1 = J_2 J_3, \quad \dot{J}_2 = -J_3 J_1 - b x_3, \quad \dot{J}_3 = b x_2, \tag{7.24}$$

$$\begin{gathered} \dot{x}_1 = 2x_2 J_3 - x_3 J_2, \quad \dot{x}_2 = x_3 J_1 - 2x_1 J_3 + b\mathcal{P} J_3, \\ \dot{x}_3 = x_1 J_2 - x_2 J_1 - b\mathcal{P} J_2, \end{gathered} \tag{7.25}$$

has the following Lax pair, $T(u)$ and $M(u)$,

$$\dot{T}(u) = -i\big[T(u), M(u)\big], \quad M(u) = \begin{pmatrix} u & 2A_4 \\ 2 & -u \end{pmatrix}. \tag{7.26}$$

References

1. M. R. Adams, J. Harnad, and J. Hurtubise, *Coadjoint orbits, spectral curves and Darboux coordinates*, The Geometry of Hamiltonian Systems (Berkeley, 1989), Math. Sci. Res. Inst. Publ., vol. 22, Springer, New York, 1991, pp. 9–21.
2. ———, *Darboux coordinates and Liouville–Arnol'd integration in loop algebras*, Comm. Math. Phys. **155** (1993), no. 2, 385–413.
3. M. Adler and P. van Moerbeke, *Completely integrable systems, Euclidean Lie algebras, and curves*, Adv. in Math. **38** (1980), no. 3, 267–317.
4. ———, *Linearization of Hamiltonian systems, Jacobi varieties and representation theory*, Adv. Math. **38** (1980), no. 3, 318–379.
5. ———, *The Kowalewski and Hénon–Heiles motions as Manakov geodesic flows on* SO(4)*— A two-dimensional family of Lax pairs*, Comm. Math. Phys. **113** (1988), no. 4, 659–700.
6. A. I. Bobenko, A. G. Reyman, and M. A. Semenov-Tian-Shansky, *The Kowalewski top* 99 *years later: A Lax pair, generalizations and explicit solutions*, Comm. Math. Phys. **122** (1989), no. 2, 321–354.
7. L. D. Fairbanks (1988), *Lax equation representation of certain completely integrable systems*, Compositio Math. **68** (1988), no. 1, 31–40.
8. H. Flaschka and D. W. McLaughlin, *Canonically conjugate variables for the Korteweg-de Vries equation and the Toda lattice with periodic boundary conditions*, Progr. Theoret. Phys. **55** (1976), no. 2, 438–456.
9. L. Haine and E. Horozov, *A Lax pair for Kowalevski's top*, Phys. D **29** (1987), no. 1-2, 173–180.
10. E. Horozov and P. van Moerbeke, *The full geometry of Kowalewski's top and* $(1,2)$*-Abelian surfaces*, Comm. Pure Appl. Math. **42** (1989), no. 4, 357–407.
11. J. C. Hurtubise, *The algebraic geometry of the Kostant–Kirillov form*, J. London Math. Soc. (2) **56** (1997), no. 3, 504–518.
12. J. C. Hurtubise and M. Kjiri, *Separating coordinates for the generalized Hitchin systems and the classical* r*-matrices*, Comm. Math. Phys. **210** (2000), no. 2, 521–540.
13. E. G. Kalnins, *Separation of variables for Riemannian spaces of constant curvature*, Longman Scientific & Technical, Harlow, 1986.
14. I. V. Komarov, *The Kovalevskaja basis for the hydrogen atom*, Teoret. Mat. Fiz. **47** (1981), no. 1, 67–72.
15. ———, *The Goryachev–Chaplygin top in quantum mechanics*, Teoret. Mat. Fiz. **50** (1982), no. 3, 402–409.
16. ———, *Exact solution of the Goryachev–Chaplygin problem in quantum mechanics*, J. Phys. A **15** (1982), no. 6, 1765–1773.
17. ———, *Remarks on Kowalevski's top*, Kowalevski Workshop on Mathematical Methods of Regular Dynamics (Leeds, 2000), J. Phys. A **34** (2001), no. 11, 2111–2120.
18. I. V. Komarov and V. B. Kuznetsov, *Quasiclassical quantization of the Kovalevskaya top*, Teoret. Mat. Fiz. **73** (1987), no. 3, 335–347 (Russian); English transl., Theoret. and Math. Phys. **73** (1987), 1255–1263.
19. ———, *Kowalewski's top on the Lie algebras* o(4), e(3) *and* o(3, 1), J. Phys. A **23** (1990), no. 6, 841–846.

20. T. H. Koornwinder and V. B. Kuznetsov, *Gauss hypergeometric function and quadratic R-matrix algebras*, Algebra i Analiz **6** (1994), no. 3, 161–184.
21. Sophie Kowalevski, *Sur le problème de la rotation d'un corps solide autour d'un point fixe*, Acta Math. **12** (1889), 177–232.
22. V. B. Kuznetsov, *Quasiclassical quantization of the Kowalevski top*, M.Sc. Thesis, Leningrad University, 1985.
23. ———, *Applications of the inverse scattering method to 2-dimensional classical and quantum integrable systems*, Ph.D thesis, Leningrad University, 1989.
24. ———, *Generalized polyspheroidal periodic functions and the quantum inverse scattering method*, J. Math. Phys. **31** (1990), no. 5, 1167–1174.
25. ———, ${}_3F_2(1)$ *hypergeometric function and quadratic R-matrix algebra*, Symmetries and Integrability of Difference Equations (Estérel, 1994), CRM Proc. Lecture Notes, vol. 9, Amer. Math. Soc., Providence, RI, 1996, pp. 185–197.
26. ———, *Separation of variables for the D_n-type periodic Toda lattice*, J. Phys. A **30** (1997), no. 6, 2127–2138.
27. V. B. Kuznetsov, M. F. Jørgensen, and P. L. Christiansen, *New boundary conditions for integrable lattices*, J. Phys. A **28** (1995), no. 16, 4639–4654.
28. V. B. Kuznetsov and A. V. Tsiganov, *A special case of Neumann's system and the Kowalewski–Chaplygin–Goryachev top*, J. Phys. A **22** (1989), no. 3, L73–L79.
29. ———, *Infinite series of Lie algebras and boundary conditions for integrable systems*, Zap. Nauchn. Sem. Leningrad. Otdel. Mat. Inst. Steklov. (LOMI) **172** (1989), Differentsialnaya Geom. Gruppy Li i Mekh. vol. 10, pp. 88–98; Soviet transl. J. Soviet Math. **59** (1992), no. 5, 1085–1092.
30. ———, *Quantum relativistic Toda lattices*, Zap. Nauchn. Sem. S.-Peterburg. Otdel. Mat. Inst. Steklov. (POMI) **205**, (1993), Differentsialnaya Geom. Gruppy Li i Mekh. **13**, 71–84, 180 (Russian); English transl. J. Math. Sci. **80** (1996), no. 3, 1802–1810.
31. D. Markushevich, *Kowalevski top and genus-2 curves*, Kowalevski Workshop on Mathematical Methods of Regular Dynamics (Leeds, 2000), J. Phys. A **34** (2001), no. 11, 2125–2135.
32. P. van Moerbeke, *The spectrum of Jacobi matrices*, Invent. Math. **37** (1976), no. 1, 45–81.
33. D. Mumford, *Tata lectures on theta.* II, Jacobian Theta Functions and Differential Equations, vol. 43. Birkhäuser Boston, Inc., Boston, MA, 1984.
34. E. Previato, *Flows on r-gonal Jacobians*, The Legacy of Sonya Kovalevskaya (Cambridge and Amherst, 1985), Contemp. Math., vol. 64, Amer. Math. Soc., Providence, RI, 1987, pp. 153–180.
35. A. G. Reyman and M. A. Semenov-Tian-Shansky, *Lax representation with a spectral parameter for the Kowalewski top and its generalizations*, Lett. Math. Phys. **14** (1987), no. 1, 55–61.
36. E. K. Sklyanin, *The Goryachev–Chaplygin top and the method of the inverse scattering problem*, Differential geometry, Lie groups and mechanics, vol. 6, Zap. Nauchn. Sem. Leningrad. Otdel. Mat. Inst. Steklov. (LOMI) vol. 133, 1984, pp. 236–257 (Russian).
37. ———, *The quantum Toda chain*, Nonlinear Equations in Classical and Quantum Field Theory (Meudon/Paris, 1983/1984), Lecture Notes in Phys., vol. 226, Springer, Berlin, 1985, pp. 196–233.
38. ———, *Boundary conditions for integrable quantum systems*, J. Phys. A **21** (1988), no. 10, 2375–2389.
39. ———, *Separation of variables—new trends*, Quantum Field Theory, Integrable Models and Beyond (Kyoto, 1994), Progr. Theoret. Phys. Suppl., no. 118 , 1995, pp. 35–60.
40. P. Vanhaecke, *Integrable systems in the realm of algebraic geometry*, Lecture Notes in Mathematics, vol. 1638, Springer-Verlag, Berlin, 1996.
41. A. P. Veselov and S. P. Novikov, *Poisson brackets and complex tori*, Algebraic Geometry and Its Applications, Trudy Mat.Inst. Steklov. **165** (1984), 49–61 (Russian).

Department of Applied Mathematics, University of Leeds, LEEDS LS2 9JT, United Kingdom.
E-mail address: `vadim@maths.leeds.ac.uk`

Centre de Recherches Mathématiques
CRM Proceedings and Lecture Notes
Volume **32**, 2002

Some Algebro-Geometric Integrable Systems Versus Classical Ones

Dimitri Markushevich

ABSTRACT. Mukai's integrable system and its generalisation by Beauville represent Lagrangian families of Jacobians of curves from a linear series on a K3 surface. Several classical integrable systems are shown to be degenerate cases of these, related to singular K3 surfaces: Kowalevski top, Jacobi problem, periodic Toda with three particles. New integrable systems that are Lagrangian families of Prym varieties associated to Enriques and Del Pezzo surfaces are constructed. A question on the description of one integrable system by different families of small genus curves is discussed, and the restricted Kowalevski top (that is, the symplectic leaf on which the angular momentum of the top is orthogonal to the gravity vector) is worked out in detail.

Mukai defined in [**Muk**] an integrable system related to a linear system on a K3 double cover of the projective plane $\mathbb{P}^2$. The linear system is the one of inverse images of lines in $\mathbb{P}^2$, and the phase space of the integrable system is the 4-dimensional holomorphically symplectic variety formed by the Jacobians of the elements of the linear system. When the K3 surface is sufficiently general, all the curves of the linear system are reduced and irreducible, and in this case the family of Jacobians can be compactified, for example, in using the Altman–Kleiman compactification [**AK**]. The principal interest of Mukai's construction from the point of view of an algebraic geometer is that it provided a new example of a compact holomorphically symplectic 4-fold. In the present paper, we will show that several classical integrable systems which are linearised on families of Jacobians of genus-2 curves, namely, Kowalevski top, Jacobi problem, periodic Toda with three particles, are related to certain singular K3 double covers of the projective plane. Thus they are degenerate cases of the Mukai system.

This is a manifestation of the general phenomenon which can be formulated, roughly speaking, by saying that any integrable system which is integrated on a family of Jacobians of genus-2 curves is associated to a K3 surface, possibly degenerate. Under certain genericity assumptions, this is proved in [**Mar-2**].

2000 *Mathematics Subject Classification.* Primary: 14H70; Secondary: 37J35.

The author thanks M. Audin, I. V. Komarov, Vadim B. Kuznetsov, and E. Previato for discussions.

This is the final form of the paper.

Sometimes the family of genus-2 curves related to the problem is not unique. Richelot's transform [**R1, R2**] allows to construct very explicitly, starting from one genus-2 curve, another one with isogeneous Jacobian. It is determined by a partition of the 6 Weierstrass points of the curve into three pairs. This construction can be realised globally on a family of genus-2 curves if the monodromy of the family allows to choose a partition into pairs in all the fibres depending holomorphically on the point of the base. For Kowalevski top, this is possible only for the orbit $(l,g)=0$, where l is the angular momentum of the top, g the gravity vector. We show that in this case the Richelot transform gives another family of genus-2 curves, already known in the literature: it was introduced by Bobenko–Reyman–Semenov-Tian-Shansky [**BRS**] in the solution of the equations of the Kowalevski top by the method of spectral curves, applied to the Lax representation of Reyman–Semenov-Tian-Shansky [**R-STS**].

After Mukai's system and classical systems of this type, we describe a generalisation, due to Beauville [**Beau-2**], who showed that the family of Jacobians of the curves of a complete linear system of any genus on a K3 surface is an integrable system. Here the inverse statement stays a conjecture and only partial results in this direction are known.

Next we introduce new integrable systems of Prym varieties related to Enriques and Del Pezzo surfaces. Both types of surfaces can be obtained as quotients of a K3 surface by an involution. We work in a general context of a linear system on a K3 surface S with an involution η and show that the associated relative Prym variety, constructed from the subfamily of curves invariant under the involution, is an integrable system with symplectic form induced by the one of Beauville's integrable system if and only if the holomorphic 2-form of the K3 surface is anti-invariant under the involution. This is the case when the quotient S/η is either an Enriques surface (η is fixpoint free), or a rather special rational surface (η is not free). In particular, the construction works for any Del Pezzo surface. In the other case, when the 2-form is η-invariant, the Prym family is never an integrable system with respect to the 2-form induced by Beauville's: Instead of being nondegenerate, this 2-form is identically zero, and in fact the relative Prym is a Lagrangian subvariety in the phase manifold of Beauville's system.

We conclude the survey by describing Kuznetsov's transform [**K1, K2, KK**] of a genus-2 curve for the particular cases of Kowalevski's curve (case $(l,g)=0$) and the BRS curve. Kuznetsov used this transform in his solution to the problem of separation of variables for the Kowalevski top. Remark, that it exists for arbitrary (l,g), but we use it only in the case $(l,g)=0$. According to [**Mar-3**], the application of Kuznetsov's transform to Kowalevski's curve (case $(l,g)=0$) gives a double cover $\widetilde{E}\to E$ from a genus-3 curve onto an elliptic curve whose Prymian is isogeneous to the Jacobian of the original curve. In fact, one can think of Kuznetsov's transform as a technique to pass from the Jacobian of a curve of genus 2 to an isogeneous Prym variety. As such, it represents a particular case of Audin's construction, described in [**Au-2**, Section 5.3].

Another interesting observation is that the result of Kuznetsov's transform applied to the BRS curve is also a well known curve: It is the spectral curve of the Kuznetsov–Tsiganov Lax representation [**KT**] of the Kowalevski top. We show also that the Prym families of the two Kuzntesov transforms are of Del Pezzo type, associated to some singular Del Pezzo surface.

1. Mukai's Integrable System

Let $P_6(x,y)$ be a degree 6 polynomial in two variables with complex coefficients, defining a degree-6 curve Δ in the projective plane $\mathbb{P}^2$. Assume Δ nonsingular. Then the surface

$$X\colon \{z^2 - P_6(x,y) = 0\}$$

has a nonsingular compactification over the infinite line of $\mathbb{P}^2$, defined similarly to that of the genus-2 curve $z^2 - P_6(x) = 0$ for a degree-6 polynomial in one variable. Let us denote the compactified surface by the same letter X, and let

$$\rho\colon X \to \mathbb{P}^2, \quad (x,y,z) \mapsto (x,y)$$

be the natural two-sheeted covering, ramified at Δ. The 2-form $\alpha_X = (dx \wedge dy)/z$ has no zeros or poles on X and defines a symplectic structure on it.

Let $\mathbb{P}^{2\vee}$ be the dual projective plane parametrising lines $l \subset \mathbb{P}^2$. For any $l \in \mathbb{P}^{2\vee}$, the curve $C_l = \rho^{-1}(l)$ is a double cover of $l = \mathbb{P}^1$ ramified at 6 points $l \cap \Delta$. This curve is nonsingular if and only if the 6 points are distinct, that is, if l is not tangent to Δ. Thus we have a family

$$\varphi\colon \mathcal{C} \to \mathbb{P}^{2\vee}$$

of genus-2 curves over $\mathbb{P}^{2\vee}$ degenerating over the curve $\Delta^\vee \subset \mathbb{P}^{2\vee}$ of tangent lines of Δ. The Jacobian JC_l is well defined even when C_l is singular as the algebraic group of isomorphism classes of line bundles on C_l. The Jacobians C_l over all lines l fit into a family

$$f_0 = J\varphi\colon J\mathcal{C} \to \mathbb{P}^{2\vee},$$

called the relative Jacobian of $\mathcal{C}$.

If $P_6(x,y)$ is sufficiently generic, so that Δ has neither tritangent, nor degenerate flexional tangent lines, then the $f_0\colon J\mathcal{C} \to \mathbb{P}^{2\vee}$ has a natural compactification $f\colon M = \bar{J}\mathcal{C} \to \mathbb{P}^{2\vee}$. Mukai shows that it possesses a holomorphic symplectic structure α_M, unique up to a constant factor, and the fibres $f^{-1}([l])$, $[l] \in \mathbb{P}^{2\vee}$ are Lagrangian. Here we denote by $[l]$ the point of $\mathbb{P}^{2\vee}$ representing a line l. Thus, if we choose any affine coordinate of $\mathbb{P}^{2\vee}$ as Hamiltonian, we will obtain a complex algebraically completely integrable system of dimension 4.

To give an idea of what the compactification is, say that there are several standard ways of compactifying $J\mathcal{C}$. The first one is Mukai's: the line bundles on C_l can be considered as torsion sheaves on X with support C_l, and M is the closure of the set of all these sheaves in the moduli space of simple sheaves on X. This definition does not give always a compact result, so in general this is a partial compactification. To get always a compact result, one should take the closure inside the Simpson moduli space of semistable equidimensional sheaves [**Sim**], the notion which extends "Gieseker semistable" to the case of torsion sheaves, but this compactification may be singular. Another one is the Altman–Kleiman compactified Jacobian, parametrising isomorphism classes of fractional ideal sheaves of degree 0 on C_l. It is defined only when all the curves C_l are reduced and irreducible, and in case Mukai's and the Altman–Kleiman compactifications give the same nonsingular projective variety $M \simeq \bar{J}\mathcal{C}$. This condition is equivalent to the above formulated genericity assumption that Δ has neither tritangent, nor degenerate flexional tangent lines.

A compact simply connected holomorphically symplectic manifold is called *irreducible symplectic* if its symplectic structure is unique up to proportionality. Thus, Mukai's system provides an example of an irreducible symplectic fourfold.

It is easy to describe both the compactification and the 2-form α_M in this example. First of all, notice that $f_0\colon JC \to \mathbb{P}^{2\vee}$ is birationally equivalent to the family of symmetric squares $\varphi^{(2)}\colon \mathcal{C}^{(2)} \to \mathbb{P}^{2\vee}$ of the curves $C_l = \varphi^{-1}(l)$, whose fibers are the surfaces $C_l^{(2)}$ of unordered pairs of points P, Q of C_l. Choose an affine chart with coordinates u, v in $\mathbb{P}^{2\vee}$, say the one representing lines $y = ux + v$. Then the family $\mathcal{C}$ is given by the equation

$$\mathcal{C}\colon \{z^2 - P_6(x, ux+v) = 0\}.$$

For any pair $(P,Q) = ((x_1, z_1), (x_2, z_2)) \in C_l^{(2)}$ such that $x_1 \neq x_2$ we can identify locally $\bar{J}\mathcal{C}$ with $\mathcal{C}^{(2)}$ near (P,Q) and write

$$\alpha_{\bar{J}\mathcal{C}} = \left(x_1\frac{dx_1}{y_1} + x_2\frac{dx_2}{y_2}\right) \wedge du + \left(\frac{dx_1}{y_1} + \frac{dx_2}{y_2}\right) \wedge dv.$$

It is interesting to note that the above expression for $\alpha_{\bar{J}\mathcal{C}}$ is a priori nonclosed and nonholomorphic, and it is really so for an arbitrary family of genus-2 curves $z^2 - P_6(x, u, v) = 0$, but the 2-form becomes both closed and holomorphic as soon as the dependence on u, v acquires the peculiar form $P_6(x, u, v) = P_6(x, ux + v)$.

The following assertion is verified [**Mar-1, Mar-2**]:
Let $\varphi\colon \mathcal{C} \to B$ be a family of curves of genus 2 over a nonsingular surface locally of the form $z^2 - P_6(x, u, v) = 0$, where u, v are local analytic coordinates on B and P_6 is a degree 6 polynomial in x whose coefficients depend on u, v. Assume that the Altman–Kleiman compactified Jacobian $\bar{J}\varphi\colon \bar{J}\mathcal{C} \to B$ is an irreducible symplectic fourfold such that $\bar{J}\varphi$ is a Lagrangian fibration. Then $B \simeq \mathbb{P}^2$ and $\mathcal{C}$ is identified with a linear system $\{\rho^{-1}(l)\}$ of inverse images of lines on a degree-2 K3 surface $\rho\colon X \to \mathbb{P}^2$.

The paper [**Mar-2**] gives also a criterion for the nonsingularity of $\bar{J}\mathcal{C}$ in terms of the local equations of $\mathcal{C}$ for any genus, which we will not reproduce here.

Mukai's transform. Let $\rho\colon X \to \mathbb{P}^2$ be a degree-2 K3 surface, and $\bar{J}\varphi\colon \bar{J}\mathcal{C} \to \mathbb{P}^{2\vee}$ the associated Lagrangian fibration. Then one can see that $\bar{J}\mathcal{C}$ is birational to the symmetric square $X^{(2)}$. Indeed, for two general points P, Q, there exists a unique curve $\rho^{-1}(l)$ from our family passing through P, Q, namely, the one which corresponds to the line $l = \overline{\rho(P)\rho(Q)}$ joining $\rho(P)$, $\rho(Q)$. Hence the map that sends a pair of points $(P,Q) \in X^{(2)}$ to the same pair (P,Q) but considered as an element of $C_l^{(2)}$, is generically one-to-one and thus defines a birational equivalence between $X^{(2)}$ and $\mathcal{C}^{(2)}$.

The $X^{(2)}$ has a 2-form α_2, induced by the 2-form $\mathrm{pr}_1^*\alpha_X + \mathrm{pr}_2^*\alpha_X$ on $X \times X$ via the quotient map $X \times X \to X^{(2)} = X \times X/\iota$, where $\iota(P,Q) = (Q,P)$. The symmetric square is singular along the diagonal $D = \{(P,Q) \mid P = Q\}$ and its blow up $V = X^{[2]}$ along D is a resolution of singularities. Let $\sigma\colon V \to X^{(2)}$ be the blow up map, then $\alpha_V = \sigma^*\alpha_2$ is a symplectic form on V, unique up to proportionality. Thus we have two symplectic fourfolds M and V, both birational to the symmetric square $X^{(2)}$. V is known as Fujiki's symplectic fourfold [**Fu**].

The composition $\mu\colon X^{[2]} \to \bar{J}\mathcal{C}$ is a birational equivalence, known as Mukai's transform. It blows up the plane $\Pi \subset X^{[2]}$ of pairs $\rho^{-1}(t)$ $(t \in \mathbb{P}^2)$ and blows down the exceptional divisor onto another plane $\Pi' \subset \bar{J}\mathcal{C}$, the image of the zero section $\mathbb{P}^{2\vee} \hookrightarrow J\mathcal{C}$. Thus Mukai's transform is a biregular isomorphism of the complements of the planes, $X^{[2]} \setminus \Pi \xrightarrow{\sim} \bar{J}\mathcal{C} \setminus \Pi'$, which does not extend continuously to the points of Π (neither does its inverse to the points of Π').

It is well known that any birational map between two compact manifolds transforms bijectively holomorphic forms on one of them to those on the other [**GH**]. Moreover, if we use μ to pull back a symplectic form from $\bar{J}\mathcal{C}$ to $X^{[2]}$, we will obtain a symplectic form on $X^{[2]} \setminus \Pi$. But codimension of Π is 2, so it extends to all of $X^{[2]}$ (Riemann's Extension Theorem, ibid.). The extended 2-form has no poles or points of degeneracy, because they should fill in divisors, but this is impossible, since the singularities of the 2-form might occur only at points of the plane Π, which is of codimension 2. This argument allows to deduce the irreducibility of $\bar{J}\mathcal{C}$ from that of $X^{[2]}$, the latter being an easy exercise.

2. Integrable Systems on Degenerate K3 Surfaces: Kowalevski Top, Jacobi, Periodic Toda

Kowalevski top. We will use the notations and the normalisation of dimensional parameters from [**BRS**]. The top represents a solid with a fixed point O in a constant gravity field. There are 6 dynamical variables: 3 components of the angular momentum $l = (l_1, l_2, l_3)$ and 3 components of the gravity vector $g = (g_1, g_2, g_3)$, everything with respect to a moving orthonormal frame, attached to the solid. The motion of the top can be described by the following system:

$$\begin{cases} \frac{dl}{dt} = [l, \omega] + [c, g], \\ \frac{dg}{dt} = [g, \omega], \\ l = I\omega \end{cases} \tag{2.1}$$

where c is the constant vector of the center of mass, I the inertia tensor.

In Kowalevski's integrable case, the principal inertia moments have the ratio $2m : 2m : m$ and c lies in the plane, perpendicular to the symmetry axis. One can normalise the parameters so that $I = \mathrm{diag}(1, 1, 1/2)$, $c = (1, 0, 0)$ and $|g| = 1$. The problem possesses a trivial integral of motion, the scalar product (l, g), which we will consider as a parameter. Fixing its value $(l, g) = \delta$, we obtain a 4-dimensional phase space M_δ, which is a symplectic manifold. The Hamiltonian

$$H = \frac{1}{2}(l_1^2 + l_2^2 + 2l_3^2) - g_1,$$

and Kowalevski's first integral

$$K = ({l_1}^2 - {l_2}^2 + 2g_1)^2 + 4(l_1 l_2 + g_2)^2$$

are in involution and define the moment map $\Phi_\delta \to \mathbb{R}^2$ whose fibres are the Liouville tori in our symplectic 4-fold.

Kowalevski's solution [**Kow**] of (2.1) includes two steps: the complexification and her famous change of variables, which we will not specify here. It brings the integrable system $\Phi_\delta \to \mathbb{R}^2$ to the form $J\mathcal{C} \to \mathbb{C}^2$, where $\mathcal{C} = \mathcal{C}_\delta \to \mathbb{C}^2$ is the family of Kowalevski curves

$$z^2 = \left[x\left((x-H)^2 + 1 - \frac{1}{4}K\right) - \delta^2\right]\left((x-H)^2 - \frac{1}{4}K\right)$$

parametrised by the points $(H, K) \in \mathbb{C}^2$. It is brought to the canonical form $z^2 - P_6(x, ux+v) = 0$ by choosing new coordinates in the plane $\mathbb{C}^2$:

$$u = -2H, \quad v = H^2 - \frac{1}{4}K.$$

Thus we obtain a degenerate Mukai's system on the singular K3 surface

$$X_1\colon \{z^2 = P_6(x,y)\}, \quad P_6(x,y) = \big(x(x^2+y+1) - \delta^2\big)(x^2+y)\}.$$

The fact that the degree of P_6 is 5 means that the infinity line is in the ramification curve Δ. In fact, the homogeneous equation of Δ in $\mathbb{P}^2$ is

$$\Delta\colon \{x_0\big(x_1(x_1^2 + x_0x_2 - x_0^2) - \delta^2 x_0^3\big)(x_1^2 + x_0x_2) = 0\}.$$

So, Δ has three components of degrees respectively 1, 2, and 3 (the last one can decompose further into a line and a conic when $\delta = 0$). The points of intersection of the components yield singularities of X_1, and there is one rather nasty singularity over the point $(0:0:1)$. Because of this singularity, X_1 has no resolution which would be a K3 surface. In fact, the surface singularities admitting a resolution that carries a nondegenerate 2-form are known in algebraic geometry under the name of Du Val singularities, see also [**Dur**] for several other characterisations. They can be given by the list of ADE normal forms modulo formal or local analytic equivalence:

$$\begin{aligned} f_{A_n} &= z^2 + x^2 + y^{n+1}, \quad (n \geq 1),\\ f_{D_n} &= z^2 + x^2y + y^{n-1}, \quad (n \geq 4),\\ f_{E_6} &= z^2 + x^3 + y^4,\\ f_{E_7} &= z^2 + x^3 + xy^3,\\ f_{E_8} &= z^2 + x^3 + y^5. \end{aligned}$$

The singularity ξ of X_1 over $(0:0:1)$ can be written in the form $z^2 = g(x,y)$ where the lowest degree of monomials contained in g is 4. So it cannot be brought by a formal change of variables to any of the ADE normal forms, because the latter contain quadratic or cubic terms, independent of z.

The Jacobian family JC has no compactification which would be a symplectic fourfold as well. This follows easily from the fact that JC has an open piece that is identified with an open piece of the symmetric square $X_1^{(2)}$, and the latter contains singular points locally analytically of type $\xi \times \mathbb{C}^2$, which cannot be resolved in a nice way because ξ does not. Moreover, remark that C_l is just the double line when l is the infinite line, so both Mukai's compactification and that of Altman–Kleiman are partial ones. We will see below a different proof for the nonexistence of a symplectic compactification of JC.

Another effect of the degeneracy of the Kowalevski system is the unirationality of JC. An algebraic variety M is called unirational if a Zariski open set of it can be parametrised by rational functions, or equivalently, if there exists a rational dominant map $\mathbb{C}^n \dashrightarrow M$. If moreover the parametrisation (or the ratioal map) is generically bijective, then M is rational. The complexified phase space $\Phi_\delta^{\mathbb{C}}$ of the Kowalevski top is rational, because it is a locally trivial fibre bundle over the quadric $g_1^2 + g_2^2 + g_3^2 = 1$, which is rational via the stereographic projection, with fibre $\{l \in \mathbb{C}^3 \mid (l,g) = \delta\} \simeq \mathbb{C}^2$.

It is well known [**Au-Si, HvM, AvM, BRS**] that the Liouville tori $\Phi^{\mathbb{C}}_{\delta,H,K}$ are isogeneous to the Jacobians $JC_{\delta,H,K}$ of Kowalevski's curves, and the isogeny can be defined simultaneously on all the fibres, so that we have a finite-to-one map $\Phi_\delta^{\mathbb{C}} \to JC$. This proves that JC is unirational. Nevertheless, none of the nonsingular compact symplectic varieties is unirational. This follows from the fact that the lift $f^*\alpha$ of a holomorphic p-form α on a compact variety V is well defined as a holomorphic form not only for a regular holomorphic map $f\colon W \to V$, but also for a meromorphic or rational map $f\colon W \dashrightarrow V$. If V is unirational, then there exists a rational parametrisation $\mathbb{C}^n \dashrightarrow V$, which can be thought of as a rational map $f\colon \mathbb{P}^n \dashrightarrow V$ with dense image. If there were a nonzero holomorphic 2-form α on V, then $f^*\alpha$ would be one on $\mathbb{P}^n$, which is nonsense: $\mathbb{P}^n$ has no holomorphic forms.

We can summarise as follows:

The Kowalevski system is a degenerate case of Mukai's integrable system, corresponding to the singular K3 *surface* X_1. *The nonsingular part of* X_1 *carries a symplectic* 2*-form* α_{X_1}, *but there is no resolution of singularities of* X_1 *having a symplectic* 2*-form.*

The relative Jacobian JC_δ *of the family of Kowalevski curves is unirational. Every compact symplectic manifold is not unirational. Hence* JC_δ *has no symplectic compactification.*

Jacobi system. This is the Hamiltonian system of the geodesic flow on the ellipsoid

$$Q_x\colon \left\{\frac{\xi_1^2}{\alpha_1 - x} + \frac{\xi_2^2}{\alpha_2 - x} + \frac{\xi_3^2}{\alpha_3 - x} = 0\right\}$$

where ξ_1, ξ_2, ξ_3 are coordinates in $\mathbb{R}^3$, x a parameter, and α_i are fixed constants (see a modern exposition in [**Au-1**], or the original Jacobi's solution in [**J**]). Its phase space is identified with the cotangent bundle T^*S^2 of the 2-sphere, and there are two integrals in involution F_1, F_2 defining the moment map

$$T^*S^2 \xrightarrow{F=(F_1,F_2)} \mathbb{R}^2.$$

This system possesses a Lax representation with spectral curve

$$C_{(h_1,h_2)}\colon \{z^2 = \big(h_1(\alpha_2 - x)(\alpha_3 - x) + h_2(\alpha_1 - x)(\alpha_3 - x) \\ + (1 - h_1 - h_2)(\alpha_1 - x)(\alpha_2 - x)\big)(\alpha_1 - x)(\alpha_2 - x)(\alpha_3 - x)\}$$

corresponding to the common level set $F_1 = h_1$, $F_2 = h_2$ of the first integrals. We can bring it to our canonical form $z^2 - P_6(x, ux+v) = 0$ in replacing the coordinates h_1, h_2 in $\mathbb{C}^2$ by the linear functions

$$u = h_1(\alpha_3 - \alpha_1) + h_2(\alpha_3 - \alpha_2) + \alpha_1 + \alpha_2,$$
$$v = h_1\alpha_2\alpha_3 + h_2\alpha_1\alpha_3 + (1 - h_1 - h_2)\alpha_1\alpha_2.$$

The polynomial $P_6(x, y)$ is again of degree 5:

$$P_6(x, y) = -(x - \alpha_1)(x - \alpha_2)(x - \alpha_2)(x^2 + y),$$

so, the infinite line is a component of the ramification curve.

Thus, we arrive at the following statement:

The Jacobi system of geodesics on the quadric Q_x *is a degenerate case of Mukai's system corresponding to the singular* K3 *surface*

$$X_2\colon \{z^2 = -(x - \alpha_1)(x - \alpha_2)(x - \alpha_2)(x^2 + y)\}.$$

Periodic toda with 3 particles. This is a Lax system (see [**Au-2**] and further references therein)

$$\dot{A}_\lambda = [A_\lambda, B_\lambda]$$

where

$$A_\lambda = \begin{bmatrix} b_1 & a_1 & a_3\lambda^{-1} \\ a_1 & b_2 & a_2 \\ a_3\lambda & a_2 & b_3 \end{bmatrix}, \quad B_\lambda = \begin{bmatrix} 0 & a_1 & -a_3\lambda^{-1} \\ -a_1 & 0 & a_2 \\ a_3\lambda & -a_2 & 0 \end{bmatrix}.$$

It has two trivial first integrals

$$\Sigma a_i = \text{const}, \quad \Sigma b_i = \text{const},$$

which define 4-dimensional symplectic leaves in the phase space. The change of variables $x_k = a_k^2$ brings us to a new integrable system of Lax type, whose spectral curves are of genus 2:

$$C_{(h_1,h_2)}\colon \{\lambda + \lambda^{-1} = -\mu^3 + h_1\mu + h_2\}$$

where h_1, h_2 stand for the values of the first integrals in involution

$$H_1 = x_1 + x_2 + x_3 - b_1b_2 - b_2b_3 - b_3b_1, \quad H_2 = b_1b_2b_3 - b_1x_1 - b_2x_2 - b_3x_3.$$

It is reduced to the standard hyperelliptic form $z^2 = \mathcal{D}$, where $\mathcal{D}$ is the discriminant of the equation of $C_{(h_1,h_2)}$ as a quadratic one in λ. It is represented in our normal form $z^2 - P_6(x, ux + v) = 0$ with

$$P_6(x, y) = (-x^3 + y - 2)(-x^3 + y + 2), \quad u = h_1, \ v = h_2.$$

Thus we arrive at the following conclusion:
The periodic Toda system with three particles is a degenerate case of Mukai's integrable system corresponding to the singular degree-2 K3 *surface*

$$X_3\colon \{z^2 = (-x^3 + y - 2)(-x^3 + y + 2)\}.$$

3. Richelot Transform

Let C be a genus-2 curve. There is a beautiful construction, discovered by Richelot [**R1, R2**] in 1836, of another genus-2 curve C' such that the Jacobians JC, JC' are isogeneous. This construction was described by Humbert [**H**] in 1901 in geometric terms, and surveyed in a modern language in [**BM**]; see also the monograph [**CF**].

The isogeny between JC, JC' is a group epimorphism with kernel $L = (\mathbb{Z}/2\mathbb{Z})^{\oplus 2}$ inside the subgroup $JC_2 = (\mathbb{Z}/2\mathbb{Z})^{\oplus 4}$ of points of order 2 of JC. In terms of periods, one can say that JC' is obtained from C by doubling a half of its periods. The choice of the half to be doubled is not arbitrary: in order that the result of doubling might be a principally polarised Abelian surface, it is necessary and sufficient that L is a Lagrangian plane in JC_2 considered as a 4-dimensional vector space over the field $\mathbb{F}_2 = \mathbb{Z}/2\mathbb{Z}$ endowed with Weil pairing. See also [**DL**] on the extension of Richelot's transform to genus-3 curves. The authors also prove there that there is no analog of Richelot's transform for curves of genus $g \geq 4$, or more exactly: Let C be a general curve of genus $g \geq 4$, p a prime number and L a Lagrangian subspace of the vector space JC_p of points of order p. Then JC/L with induced principal polarisation is never the Jacobian of a curve.

Let C be a genus 2 curve. Choosing a Lagrangian plane $L \subset JC_2$ is equivalent to dividing the 6 branch points of the double cover $C \to \mathbb{P}^1$ into three pairs, or to writing the equation of C in the form

$$z^2 = P_1(x)P_2(x)P_3(x)$$

where $P_j(x)$ are quadratic polynomials without multiple or common roots in $\mathbb{P}^1(\mathbb{C})$. The fact that we are looking at the roots in $\mathbb{P}^1(\mathbb{C})$ rather than those in $\mathbb{C}$ means that we admit the case when one of the quadratic polynomials is in fact linear, and then ∞ is one of its roots. Let a_i, $a_i' \in \mathbb{P}^1$ be the roots of $P_i(x)$, $i = 1, 2, 3$. Then $L = \{0, [(a_1, 0) - (a_1', 0)], [(a_2, 0) - (a_2', 0)], [(a_3, 0) - (a_3', 0)]\}$, and $JC_2 = \{0\} \cup \{[(x_i, 0) - (x_j, 0)]\}_{1 \le i < j \le 6}$, where x_i are all the 6 roots of the degree 6 polynomial $P_1(x)P_2(x)P_3(x)$.

Define a curve C' by the equation

$$\delta w^2 = U_1(u)U_2(u)U_3(u)$$

where

$$U_{i+2}(u) = [P_i, P_{i+1}] = P_i'(u)P_{i+1}(u) - P_i(u)P_{i+1}'(u)$$

(addition of subscripts modulo 3) and $\delta = \det(P_1P_2P_3)$ is the determinant of the 3×3 matrix of coefficients of the P_i. C' is called the *Richelot transform* of C. If one reapplies the Richelot transform to C' with the natural partition into pairs, one will get again C up to scaling of z, so that the Richelot transform is an involution in this case. But if one changes the partition on each step, one can get an infinite sequence of nonisomorphic curves.

There are no nonconstant maps between C and C'. However, there is a nice substitute for a map, a $(2, 2)$-correspondence. Ona can think of it as a bivalued map in either direction between C and C': $\Phi\colon C \overset{2:2}{\longleftrightarrow} C'$. Its graph is a curve $Z \subset C \times C'$ with double projections to both factors. It is defined by the explicit equations

$$Z = \begin{cases} P_1(x)U_1(u) + P_2(x)U_2(u) = 0, \\ zw = P_1(x)U_1(u)(x - u). \end{cases}$$

The correspondence allows to define a map between the holomorphic differentials on C and C', and also an isogeny between their Jacobians. The induced map $\Phi^*\colon H^0(\Omega^1_{C'}) \to H^0(\Omega^1_C)$ between the 2-dimensional vector spaces of 1-forms is defined by $\Phi^* := p_{1*}p_2^*$, where p_1, p_2 are the natural projections from Z to C, resp. C', p_2^* is the usual pullback of exterior forms and p_{1*} the trace map for the double covering p_1, a particular case of the integration over the fibers of a flat map. The calculation of Φ^*, done effectively by Richelot, and represented as a sequence of exercises in [**BM**], gives the following result: $\Phi^*(S(u)du/w) = S(x)dx/z$ for S a polynomial of degree ≤ 1.

The isogeny $\Phi_*\colon J(C) \to J(C')$ between the Jacobians of the two curves is defined by

$$\Phi_*([\Sigma n_i P_i]) = [\Sigma n_i p_2 p_1^{-1} P_i]$$

where $P_i \in C$ are points of the curve, n_i integers such that $\Sigma n_i = 0$, and the brackets denote divisor classes. The space of holomorphic differentials on a curve is identified with the dual of the tangent space of its Jacobian at 0, and the relation between Φ_* and Φ^* is very natural: $\Phi^* = (d_0\Phi_*)^*$, the transpose of the differential at 0.

If we want to apply the Richelot transform to a family of genus-2 curves $\{C_b\}_{b\in B}$, we encounter the monodromy obstruction: In general, there is no family of Lagrangian planes $L_b \subset (JC_b)_2$ simultaneously over all B, or even over any Zariski open subset of B. For the families $\mathcal{C} \to B$ of Mukai type, associated to a degree-2 K3 surface $X\colon \{z^2 = P_6(x,y)\}$, the global Richelot transform exists if and only if $P_6(x,y)$ is the product of three quadratic polynomials, $P_6(x,y) = P_1(x,y)P_2(x,y)P_3(x,y)$, or equivalently, the ramification curve Δ of the double cover map $\rho\colon X \to \mathbb{P}^2$ is the union of three conics. One of the P_i $(i = 1,2,3)$ can be linear, which means that the corresponding conic is reducible and is the infinite line plus the line $P_i(x,y) = 0$. For generic conics, this is a slightly degenerate Mukai's system. We say "slightly", because the intersection points of the three conics give rise to ordinary double singular points of X which can be resolved in such a way that the resolution is a (nonsingular) K3 surface. The relative Richelot transform in this case can be represented by the following diagram:

$$\begin{array}{ccc} \mathcal{C} & \xleftrightarrow{\text{Richelot}} & \mathcal{C}' \\ & \searrow \quad \swarrow & \\ & \mathbb{P}^{2\vee} & \end{array}$$

Special case of Kowalevski top: $(\delta = (l,g) = 0)$. It is impossible to apply the relative Richelot transform to the family of Kowalevski curves $\mathcal{C}_\delta$ for any δ, because the branch curve contains a cubic component and thus is not a union of three conics. But if $\delta = 0$, the cubic decomposes in a conic and a line, and we obtain the Richelot transform $\mathcal{C}'$ of $\mathcal{C} = \mathcal{C}_0$. It was studied in [**LM**]. It turns out that the family $\mathcal{C}'$ arising in this case has been already described in literature: this is the Bobenko–Reyman–Semenov-Tian-Shansky (BRS) curve [**BRS**]. Here are the equations of both curves: Kowalevski's [**Kow**],

$$\mathcal{C}\colon z^2 = x\Big((x-H)^2 - \frac{1}{4}K\Big)\Big((x-H)^2 - \frac{1}{4}K + 1\Big), \tag{3.1}$$

and that of Bobenko–Reyman–Semenov-Tian-Shansky:

$$\mathcal{C}'\colon w^2 = u\Big(u^2 + 2Hu + \frac{1}{4}K\Big)\Big(u^2 + 2Hu + \frac{1}{4}K - 1\Big). \tag{3.2}$$

Both equations are written for the case when the angular momentum l is orthogonal to the gravity vector g; only in this case the procedure of Bobenko–Reyman–Semenov-Tian-Shansky leads to a genus 2 curve. We slightly modified the second equation by applying a trivial change of variables, $w \mapsto \sqrt{-2}w$, $u \mapsto u - H$. Thus the Richelot transform gives a partial answer to the problem of finding a natural approach to the construction of Kowalevski's curve that avoids her mysterious change of variables. See also [**HvM**], where Kowalevski's curve emerges as a degenerate member of a pencil of genus-3 curves on the Liouville torus of the Kowalevski system.

Remark that the curves of the family $\mathcal{C}'$ vary on the same K3 surface

$$X_0\colon \{z^2 = x(x^2+y)(x^2+y+1)\},$$

so that the relative Richelot transform can be also considered as a correspondence $\mathcal{C} \overset{2:2}{\longleftrightarrow} \mathcal{C}$ which is not fibrewise; in fact, it induces a birational automorphism of the base $\mathbb{P}^{2\vee}$ of $\mathcal{C}$.

4. K3 Surfaces of Higher Degree: Beauville's Systems

DEFINITION. A compact complex surface X is called K3 if it is holomorphically symplectic and simply connected.

Enriques–Kodaira classification of surfaces implies that any holomorphically symplectic compact complex surface is either K3 or a complex torus of dimension 2 (for theory of algebraic surfaces, see the survey [**I-Sha**]). The condition that X is holomorphically symplectic can be equivalently reformulated as follows: the canonical class of X is trivial, $K_X = 0$.

EXAMPLE 1. A quartic surface $X \subset \mathbb{P}^3$ is K3 . If we define it by an equation $f(x, y, z) = 0$ in an affine chart of $\mathbb{P}^3$, then its symplectic form is given by the Poincaré residue

$$\alpha_X = \mathrm{Res}_X\left(\frac{dx \wedge dy \wedge dz}{f}\right) = \frac{dx \wedge dy}{\partial f/\partial z}.$$

More generally, if we have any variety M of dimension $r+2$ and r divisors $H_1, \dots, H_r$ on it such that $H_1 + \cdots + H_r$ is linear equivalent to minus the canonical class $-K_M$ and the intersection $X = H_1 \cap \cdots \cap H_r$ is a nonsingular surface, then X is K3 by adjunction formula for the canonical class. The quartic surface is obtained for $M = \mathbb{P}^3$ and $r = 1$. The intersection of a quadric and a cubic in $\mathbb{P}^4$, or of three quadrics in $\mathbb{P}^5$, or a section of the Grassmannian $G(2,5) \subset \mathbb{P}^9$ by a quadric and three hyperplanes are K3 surfaces.

EXAMPLE 2. Let S be a surface, and D a positive divisor, linearly equivalent to $2K_S$. If there exists a double cover $\pi\colon X \to S$ ramified at D, then $K_X = 0$. If X is simply connected, then it is K3 . The double cover of $\mathbb{P}^2$ ramified in a sextic is an example of this kind. Here are some other examples: the double cover of a quadric surface $Q \subset \mathbb{P}^3$ ramified in the intersection of Q with a quartic, the double cover of a cubic surface $S \subset \mathbb{P}^3$ ramified in the intersection of S with a quadric surface. In fact, any Del Pezzo surfaces S gives rise to a K3 double cover; all the Del Pezzo surfaces, except for Q, are the results of blowing up k points in $\mathbb{P}^2$ in a sufficiently general position ($k = 0, 1, \dots, 8$) [**Man**].

EXAMPLE 3. The quotient of an Abelian surface A by a finite nontrivial subgroup G of $\mathrm{SL}(2, \mathbb{C})$ acting linearly admits a resolution of singularities which is a K3 surface. The simplest G acting on any Abelian surface is $\{\pm 1\}$. Here G has 16 fixed points, the half-periods of A, and the quotient has 16 simplest singular points. They are resolved by a single blow up, pasting in 16 exceptional curves $\mathbb{P}^1$. The quotient and its resolution are called Kummer surface.

Special Abelian surfaces can have additional automorphisms allowing to quotient out by a bigger group G. Say, the product of two elliptic curves $E_\tau \times E_\tau$ with period $\tau = e^{2\pi i/3}$ has an automorphism $a = \mathrm{diag}(e^{2\pi i/3},\ e^{-2\pi i/3}) \in \mathrm{SL}(2, \mathbb{C})$ which allows to construct two new generalised Kummer surfaces: One for $G = \langle a \rangle$ of order 3, the other for $G = \langle -1, a \rangle$ of order 6.

Let X be either a K3 surface or a complex 2-torus, $\alpha = \alpha_X$ a holomorphic symplectic form. Assume that X contains a curve ($=$ a compact Riemann surface)

C_0 of genus $g \geq 2$. Associated to C_0 is the line bundle $\mathcal{O}_X(C_0)$ on X with a section $s_0 \in H^0(X, \mathcal{O}_X(C_0))$ whose zero locus $(s_0)_0$ is C_0. The family of curves $C_s = (s)_0$ which are the zero loci of all the sections of $\mathcal{O}_X(C_0)$ is called the *complete linear system* of C_0 and is denoted $|C_0|$. It is parametrised by the projective space $\mathbb{P}^{N-1} = \mathbb{P}H^0(X, \mathcal{O}_X(C_0))$, the projectivisation of the vector space $H^0(X, \mathcal{O}_X(C_0))$. The Riemann–Roch formula $\chi(\mathcal{O}_X(C_0)) = C_0^2/2 + \chi(\mathcal{O}_X)$, where $C_0^2 = 2g-2$ is the self-intersection number of C_0 and $\chi(L) = h^0(L) - h^1(L) + h^2(L)$, together with the Nakai–Moishezon vanishing theorem, which implies $h^1(\mathcal{O}_X(C_0)) = h^2(\mathcal{O}_X(C_0)) = 0$, yield $N = h^0(\mathcal{O}_X(C_0)) = g+1$ if X is K3 and $N = g-1$ if X is a 2-torus.

Let $\varphi\colon \mathcal{C} \to B$ be the linear system of C_0 in the case if X is K3, so that $B = \mathbb{P}^g$, and the family of all the curves $t + C_s$ over $t \in X$, $s \in H^0(X, \mathcal{O}_X(C_0))$ in the case if X is a 2-torus, so that $B = X \times \mathbb{P}^{g-1}$. Let $J\varphi\colon J\mathcal{C} \to B$ be the family of Jacobians. If $C_b = \varphi^{-1}(b)$ is nonsibular, then the Jacobian JC_b is a p.p.a.v. of dimension g parametrising the line bundles of degree 0 on C_b. If C_b is singular, then JC_b is an extension, as an Abelian group, of an Abelian variety of dimension $d < g$ by $g - d$ groups $\mathbb{C}^*$ or $\mathbb{C}$. This is the so called generalised Jacobian of a curve (see [**Ro**] or [**Serre**]).

Mukai [**Muk**] showed that the moduli space of simple sheaves on X carries a nondegenerate 2-form, whose closedness was proved later in [**Bo**]; see also [**HL**]. Hence the closure M of $J\mathcal{C}$ in the moduli space of simple sheaves on X has a symplectic form α_M. Beauville [**Beau-2**] noticed that the map $J\mathcal{C} \to B$ is a Lagrangian fibration, so $(M \to B, \alpha_M)$ is an integrable system. M is always nonsingular, according to Mukai, but it is not always compact; it is in general only a partial compactification of $J\mathcal{C}$. In the case when all the members of the linear system $|C_0|$ are reduced and irreducible, M is compact, projective and is isomorphic to the Altman–Kleiman relative compactified Jacobian $\bar{J}\mathcal{C}$. This condition is verified for a generic X containing a curve of genus g. Mukai's system considered in Section 1 is a particular case corresponding to $g = 2$.

For X K3 and any $g \geq 2$, one can interprete the linear system C_0 as the family of the inverse images $\varphi^{-1}(H \cap X_0)$ of hyperplane sections of the image $X_0 = \varphi(X)$ of the map $\varphi = \varphi_{|C_0|}\colon X \to \mathbb{P}^g$, $x \mapsto (s_0(x) \colon \ldots \colon s_g(x))$, where $s_0, \ldots, s_g$ is a basis of $H^0(X, \mathcal{O}_X(C_0))$. In a basis free language, $\varphi_{|C_0|}(x)$ is the hyperplane in $\mathbb{P}H^0(X, \mathcal{O}_X(C_0))$ of sections s such that $s(x) = 0$. So $X_0 \subset \mathbb{P}H^0(X, \mathcal{O}_X(C_0))^\vee$. If X is a generic K3 surface containing a curve of genus $g > 2$, then $\varphi_{|C_0|}$ is an isomorphism onto its image, and one can identify X with its image in $\mathbb{P}^g$, called a K3 surface of degree $2g-2$. The curves of the linear system $|C_0|$ are then just hyperplane sections of X and are parametrised by $\mathbb{P}^{g\vee}$.

Let us now restrict ourselves to the case when X is K3 , $X = X_0 \subset \mathbb{P}^g$ and X is sufficiently generic. Then we have the Beauville system $(M = \bar{J}\mathcal{C} \to \mathbb{P}^{g\vee}, \alpha_M)$ on a nonsingular projective algebraic $2g$-dimensional manifold M, and exactly as in the case $g = 2$, one can establish a birational isomorphism of M with the relative g-th symmetric power $\mathcal{C}^{(g)}$ of the linear system $|C_0|$ of hyperplane sections, and the birational isomorphism of the latter variety with the g-th symmetric power $X^{(g)}$ of X. The birationality of the last map follows from the fact that g generic points $x_1, \ldots, x_g$ of X lie on exactly one hyperplane of $\mathbb{P}^g$, hence on exactly one curve $C \in |C_0|$. The symmetric power $X^{(g)}$ is singular along the generalised diagonal, where two or more points $x_1, \ldots, x_g$ collide. It has a natural "minimal" resolution of singularities $\sigma\colon X^{[g]} \to X^{(g)}$, where $X^{[g]}$ is the punctual

Hilbert scheme of degree g of X, parametrising the ideal sheaves $\mathcal{I} \subset \mathcal{O}_X$ such that $\operatorname{colength}(\mathcal{I}) = \dim_{\mathbb{C}}(\mathcal{O}_X/\mathcal{I}) = g$. $X^{[g]}$ carries a unique (up to proportionality) holomorphic symplectic form α_g, induced by $\mathrm{pr}_1^*\,\alpha_X + \cdots + \mathrm{pr}_g^*\,\alpha_X$ on $X^g = X \times \cdots \times X$. Beauville was the first to observe that the varieties $X^{[n]}$ are symplectic [**Beau-1**]; for $n = 2$ this observation belongs to Fujiki [**Fu**]. As we mentioned in Sect. 1, in the case $g = 2$ the map $\sigma\colon X^{[g]} \to X^{(g)}$ is a plain blowup.

By analogy with the case $g = 2$, we will state the following conjecture:

CONJECTURE. Any irreducible holomorphically symplectic $2g$-dimensional variety of the form $\bar{J}\mathcal{C}$ for some family of genus-g curves $\varphi\colon \mathcal{C} \to B$ such that the fibres of $\bar{J}\varphi\colon \bar{J}\mathcal{C} \to B$ are Lagrangian, can be identified with one of Beauville's systems associated to a K3 surface X, so that $B \simeq \mathbb{P}^g$ is the parameter space of a linear system $|C|$ of curves of genus g on X.

Hurtubise [**Hur**] proved under certain hypotheses of general position a local analytic version of this conjecture: for any fibre C_{b_0}, there is a germ of a holomorphically symplectic complex surface X_{b_0} containing C_b for all b from a small neighbourhood of b_0.

Matsushita [**Mat-1,Mat-2**] proved in a more general situation of a Lagrangian fibration $f\colon V \to B$ on an irreducible symplectic $2n$-dimensional variety V whose fibres are not necessarily Jacobians of curves, that B has at worst log-canonical singularities, and that if B is smooth, then its Hodge numbers $h^{pq}(B)$ are equal to those of the projective space $\mathbb{P}^n$.

As another partial result towards the above conjecture, we can state the following: if $g = 3$ and $B = \mathbb{P}^3$, then the conjecture is true, that is, $\varphi\colon \mathcal{C} \to B$ is a linear system on a K3 surface X. The proof is similar to that of the genus-2 case (unpublished).

5. From Jacobians to Prymians: K3 Surfaces with Involution

Prymians. Let C be a genus-g curve, $\pi\colon \widetilde{C} \to C$ a double cover, $\tilde{g}$ the genus of $\widetilde{C}$, $\tau\colon \widetilde{C} \to \widetilde{C}$ the involution permuting the two points of each fiber of π. There is a natural involution $\iota = \tau^*$ on the Jacobian $J\widetilde{C}$ sending the divisor class $[\sum n_i P_i]$ into $[\sum n_i \tau(P_i)]$. The Prym variety $P(\widetilde{C}/C)$ is defined by

$$P(\widetilde{C}/C) = \operatorname{im}(\iota - \mathrm{id}) = \ker(\iota + \mathrm{id})^\circ$$

where the superscript $\circ$ denotes the connected component of 0. The Prym variety is complementary to $\pi^* JC$, that is, $\#\big(P(\widetilde{C}/C) \cap \pi^* JC\big) < \infty$ and $P(\widetilde{C}/C) + \pi^* JC = J\widetilde{C}$. The cotangent spaces of JC, $J\widetilde{C}$ and $P(\widetilde{C}/C)$ are canonically identified with $H^0(C, \Omega^1_C)$, $H^0(\widetilde{C}, \Omega^1_{\widetilde{C}})$ and $H^0(\widetilde{C}, \Omega^1_{\widetilde{C}})/\pi^* H^0(C, \Omega^1_C)$ respectively.

In the case when π is nonramified (or equivalently, τ is free), it is well known [**Mum**] that the restriction of the principal polarisation of $J\widetilde{C}$ on $P(\widetilde{C}/C)$ can be divided by 2 and the result is a principal polarisation on $P(\widetilde{C}/C)$. In general, if τ has fixed points, the type of the induced polarisation on $P(\widetilde{C}/C)$ is $(1, \ldots, 1, 2, \ldots, 2)$.

The above definition is naturally extended to the case when $\widetilde{C}, C$ are possibly singular curves, then the Jacobians and the Prymian may be noncompact extensions of smaller-dimensional Abelian varieties by a number of copies of $\mathbb{C}$ or $\mathbb{C}^*$. There is a special case when the Prymian stays compact in spite of the singularities of $\widetilde{C}$, C, provided the following *Beauville's condition* is verified: the singular points of $\widetilde{C}$, C

are ordinary double points, π maps bijectively $\operatorname{Sing}\widetilde{C}$ onto $\operatorname{Sing}C$, and each one of the two branches of $\widetilde{C}$ at any singular point P is a double cover with ramification at $\pi(P)$ of a branch of C at $\pi(P)$ (that is, τ must not permute the two branches at singular points of $\widetilde{C}$).

Let now $\widetilde{\varphi}\colon \widetilde{\mathcal{C}} \to B$ be a family of curves of genus $\tilde{g}$, $\tau : \widetilde{\mathcal{C}} \to \widetilde{\mathcal{C}}$ the corresponding fibrewise involution, $\varphi\colon \widetilde{\mathcal{C}}/\tau = \mathcal{C} \to B$ the quotient family of genus-g curves, and $\pi\colon \widetilde{\mathcal{C}} \to \mathcal{C}$ the quotient map. We can associate to these data a family of Prym varieties $P\widetilde{\varphi} : P = P(\widetilde{\mathcal{C}}/\mathcal{C}) \to B$. When is $P \to B$ a Lagrangian fibration for some symplectic structure?

Hurtubise and Markman [**HM**] prove a local analytical result: If one imposes some conditions of general position on the symplectic form on P, then the curves $\widetilde{C}_b$ must vary locally on a holomorphic symplectic surface. More exactly: for any $b_0 \in B$ there exists a germ of holomorphically symplectic complex surface S_{b_0} containing $\widetilde{C}_{b_0}$ together with an involution $\eta\colon S_{b_0} \to S_{b_0}$, so that in a small neighbourhood U_{b_0} of b_0, the family $\widetilde{\mathcal{C}}$ can be identified with a family of curves in S_{b_0} and τ is induced by η. So it is quiet a natural question to ask, whether a holomorphically symplectic surface with an involution globally defines an integrable system.

K3 surfaces with involution. Let S be a K3 surface with an involution $\eta\colon S \to S$, and $\mathcal{D} \to \mathbb{P}^{\tilde{g}}$ a linear system of curves of genus $\tilde{g}$ on S. Then we have the Beauville integrable system $(F\colon \bar{J}\mathcal{D} \to \mathbb{P}^{\tilde{g}}, \alpha_{\bar{J}\mathcal{D}})$. Let $\widetilde{\varphi}\colon \widetilde{\mathcal{C}} \to B$ be the subfamily of the curves, invariant under η. The base B may be empty, if η sends the curves of $\mathcal{D}$ to curves of some other linear system, or the disjoint union of two projective spaces $\mathbb{P}^{n_1}$, $\mathbb{P}^{n_2}$ such that $n_1 + n_2 = \tilde{g} - 1$, if η acts on $\mathcal{D}$ and then it has two eigenspaces of complementary dimensions in $\mathbb{P}^{\tilde{g}}$.

If B is nonempty, η induces a fibrewise involution τ on $\widetilde{\mathcal{C}}$ and we can define the quotient family $\varphi\colon \widetilde{\mathcal{C}}/\tau = \mathcal{C} \to B$ of genus-g curves and the relative Prym variety $P\widetilde{\varphi}\colon P = P(\widetilde{\mathcal{C}}/\mathcal{C}) \to B$ together with a closed 2-form $\alpha_P = (\alpha_{\bar{J}\mathcal{D}})|_P$. Now there are two possible cases: α_S is invariant under η, that is, $\eta^*(\alpha_S) = \alpha_S$, and α_S is anti-invariant under η, that is, $\eta^*(\alpha_S) = -\alpha_S$. We will see that only the second case provides an integrable system of Prymians.

Consider the first case: α_S is invariant. Then η has a finite nonempty set of fixed points in S. Indeed, it cannot have fixed curves, because the tangent line of the curve at any of its points is the $+1$-eigenspace of the differential of η; the other eigenvalue has to be -1, so α_S is anti-invariant, and this contradicts our assumption. Furthermore, η cannot be free. Indeed, in this case α_S induces on the quotient surface $X = S/\eta$ a symplectic form α_X and $\pi_1(X) \simeq \mathbb{Z}/2\mathbb{Z}$. This contradicts the classification of surfaces: as we already remarked, a symplectic surface is either K3, and then it is simply connected, or a 2-torus, and then its fundamental group is infinite.

Thus $X = S/\eta$ is a singular K3 surface: it has finitely many singular points and it (or more exactly, its nonsingular part) carries a symplectic form α_X. According to one of the possible characterisations of Du Val points as the singularities that are analytically equivalent to quotients of $\mathbb{C}^2$ by a finite subgroup of $\mathrm{SL}(2,\mathbb{C})$ (see [**Dur**]), X has a resolution X_0 which is again a K3 surface. So the linear systems of curves on X can be treated, on the one hand, as those on X_0, and on the other hand, as the quotients of subsystems of curves on S, invariant under η.

Restrict further the family $\widetilde{\varphi}\colon \widetilde{\mathcal{C}} \to B$ to one of the components of B; denote it by the same character B. Let n be the number of fixed points of η contained in a generic curve $\widetilde{C}$ from this family. The Hurwitz formula for the degree of the canonical class gives the relation $2\tilde{g} - 2 = 4g - 4 + n$. Since $\widetilde{\mathcal{C}}$ comes from a linear system of curves of genus g on the K3 surface X_0, we have $B = \mathbb{P}^g$, $\dim P(\widetilde{C}/C) = \tilde{g} - g = g - 1 + n/2$. So the only case when $P(\widetilde{C}/C) \to B$ can be a Lagrangian fibration by purely dimensional reasons is $n = 2$, $\tilde{g} = 2g$. But even in this case the 2-form α_P is degenerate. In fact, we will show now that α_P is identically zero independently of n.

To see this, let us write down the isomorphism between the vertical cotangent distribution of fibers of Beauville's system $F\colon \bar{J}\mathcal{D} \to \mathbb{P}^{\tilde{g}}$ and the lift of the tangent bundle of the base induced by $\alpha_{\bar{J}\mathcal{D}}$:

$$\mathcal{A}_{\bar{J}\mathcal{D}}\colon F^*T_{\mathbb{P}^{\tilde{g}}} \xrightarrow{\sim} \Omega^1_{\bar{J}\mathcal{D}/\mathbb{P}^{\tilde{g}}}.$$

Let D be one of the curves of $\mathcal{D}$, and $V = H^0\big(S, \mathcal{O}_S(D)\big)$, so that the base $\mathbb{P}^{\tilde{g}}$ of the family is identified with the projectivisation $\mathbb{P}V$ of V. Let $s \in V$ be nonzero, $D_s = (s)_0$ the corresponding curve, represented by the point $[s] = [D_s] \in \mathbb{P}V$. Then the tangent space $T_{[s]}\mathbb{P}V$ is canonically identified with $V/\mathbb{C}s$. Let now t be a nonsingular point of the fibre of F over $[s]$; then the fibres of our two vector bundles can be identified as

$$F^*T_{\mathbb{P}^{\tilde{g}},t} = V/\mathbb{C}s, \quad \Omega^1_{\bar{J}\mathcal{D}/\mathbb{P}^{\tilde{g}},t} = T^*_t JD_s = H^0(D_s, \Omega^1_{D_s}),$$

and the isomorphism $\mathcal{A}_{\bar{J}\mathcal{D}}$ at t is given by

$$\mathcal{A}_{\bar{J}\mathcal{D},t} : v + \mathbb{C}s \mapsto \omega_{s,v} = \operatorname{Res}_{D_s}\left(\frac{\alpha_S v}{s}\right)$$

where Res_D denotes the Poincaré residue on a curve D of a 2-form with a simple pole along D. It follows that the divisor of the holomorphic differential $\omega_{s,v}$ on D_s is just the intersection $D_s \cap D_v$.

The 2-form α_P on $P(\widetilde{C}/C)$ is the restriction of $\alpha_{\bar{J}\mathcal{D}}$. So, the Prymians $P(\widetilde{C}/C)$ are isotropic, because they are contained in the Jacobians $J\widetilde{C}$, which are Lagrangian with respect to $\alpha_{\bar{J}\mathcal{D}}$. Hence to see that $\alpha_P = 0$, it is sufficient to verify that the map

$$\mathcal{A}_P\colon (P\widetilde{\varphi})^*T_B \xrightarrow{\sim} \Omega^1_{P/B}$$

induced by $\mathcal{A}_{\bar{J}\mathcal{D}}$ via restriction and the passage to the quotient $\Omega^1_{P/B} = (\Omega^1_{\bar{J}\mathcal{D}/\mathbb{P}^{\tilde{g}}})\big|_P / J\pi^*\Omega^1_{\bar{J}\mathcal{C}/\mathbb{P}^g}$, is zero.

Choose s in such a way that $\widetilde{C} = D_s$ is invariant under η, so that $[s] \in B = \mathbb{P}^g \subset \mathbb{P}V$ and restrict $\mathcal{A}_{\bar{J}\mathcal{D},t}$ to $T_{[s]}B$. As B parametrises the curves D_b invariant under η, the image of a tangent vector $v + \mathbb{C}s \in T_{[s]}B$ is a holomorphic differential $\omega_{s,v}$ whose divisor $D_s \cap D_v$ is the intersection of two η-invariant curves, hence it is the lift of a holomorphic differential on $C = \pi(\widetilde{C})\colon \omega_{s,v} = \pi^*\nu$, $\nu \in H^0(C, \Omega^1_C)$. Hence it is in the fiber of the subbundle $J\pi^*\Omega^1_{\bar{J}\mathcal{C}/\mathbb{P}^g}$, by which we quotient out in order to get the cotangent bundle of the relative Prymian. Hence $\mathcal{A}_P$ is identically zero. Since $\dim P = \tilde{g}$, P is a Lagrangian subvariety of $\bar{J}\mathcal{D}$.

EXAMPLE 4. Take the Fermat quartic $S\colon \{x_0^4 + x_1^4 + x_2^4 + x_3^4 = 0\}$ with the involution $\eta\colon (x_0, x_1, x_2, x_3) \mapsto (-x_0, -x_1, x_2, x_3)$. It has 8 fixed points: $P_i\colon \{x_0 = x_1 = 0, x_2 + \zeta_i x_3 = 0\}$ and $P_i'\colon \{x_2 = x_3 = 0, x_0 + \zeta_i x_1 = 0\}$, where $\zeta_1, \ldots, \zeta_4$ are the

4th roots of -1. In the linear system $\mathbb{P}^3$ of hyperplane sections of S, there are two pencils $\mathbb{P}^1$ of hyperplane sections invariant under η: the one is $\{\lambda_0 x_0 + \lambda_1 x_1 = 0\}$ the other $\{\lambda_2 x_2 + \lambda_3 x_3 = 0\}$. For both of them the number of fixed points in the generic member of the pencil is $n = 4$. So $g = 1$ and we have two families P of 2-dimensional Pryms over $\mathbb{P}^1$.

If we take in the same example the linear system $\mathbb{P}^9$ of sections of S by quadrics, then two components of B are of different dimensions: one is $\mathbb{P}^3$, the other $\mathbb{P}^5$, so that $n = 8$ and $n = 0$ respectively.

Thus we have proved:

Let S be a K3 *surface with an involution η such that the symplectic form α_S is invariant under η. Let $|D| = \{\mathcal{D}/\mathbb{P}^{\tilde{g}}\}$ be a complete linear system of curves of genus $\tilde{g} \geq 2$, $(\bar{J}\mathcal{D} \to \mathbb{P}^{\tilde{g}}, \alpha_{\bar{J}\mathcal{D}})$ Beauville's integrable system, and $\widetilde{\varphi} : \widetilde{\mathcal{C}} \to B$ any one of the the two subfamilies of the curves in $|D|$, invariant under η, $B = \mathbb{P}^g$. Let τ be the involution induced by η on $\widetilde{\mathcal{C}}$ and $\varphi\colon \widetilde{\mathcal{C}}/\tau = \mathcal{C} \to B$ the quotient family of genus-g curves. Let $P\widetilde{\varphi} : P = P(\widetilde{\mathcal{C}}/\mathcal{C}) \to B$ be the relative Prym variety. Then P is a Lagrangian subvariety of $\bar{J}\mathcal{D}$ with respect to $\alpha_{\bar{J}\mathcal{D}}$. In particular, Beauville's system can never induce on P a structure of a Lagrangian fibration over B.*

6. Integrable Systems Associated to Enriques and Del Pezzo Surfaces

Now we will consider the case of an involution on a K3 surface with an anti-invariant 2-form. In the notations of the previous section, assume that $\eta^*(\alpha_S) = -\alpha_S$. Then the fixpoint set of η is either empty, or is a union of curves.

CASE 1. The fixpoint set of $\boldsymbol{\eta}$ is empty.

The quotient $X = S/\eta$ is a compact nonsingular surface with invariants

$$2K_X - 0 (K_X \neq 0), \quad \pi_1(X) = H_1(X, \mathbb{Z}) = \mathbb{Z}/2\mathbb{Z}, \quad H_2(X, \mathbb{Z}) = \mathbb{Z}^{10} + \mathbb{Z}/2\mathbb{Z},$$
$$h^{1,0}(X) = h^0(X, \Omega^1_X) = 0, \quad h^{2,0}(X) = h^0(X, \Omega^2_X) = 0.$$

It is called an *Enriques surface.* The following conditions are equivalent: X contains a curve C of genus $g \geq 1$ and S contains a curve $\widetilde{C}$ of genus $\tilde{g} = 2g - 1$ invariant under η. An Enriques surface always contains a pencil of elliptic curves ($g = 1$) (see [**Sha, I-Sha**]). For $g \geq 2$, the dimension of the linear system $|C|$ is always $g - 1$. Along with $|C|$, X contains another linear system of curves of genus g, namely, $|C + K_X|$. Their lifts to S form two subfamilies of the complete linear system $|\widetilde{C}|$, $\widetilde{\varphi}_i \colon \widetilde{\mathcal{C}}_i \to B_i$ $(i = 1, 2)$, where $B_i \simeq \mathbb{P}^{g-1}$ are the two eigenspaces of the action of η on $|\widetilde{C}|$.

As before, consider the relative Prym variety $P\widetilde{\varphi}\colon P = P(\widetilde{\mathcal{C}}/\mathcal{C}) \to B$ of one of these two families, say, for $i = 1$ $(B = B_1)$ together with the closed 2-form α_P, induced by the symplectic form of Beauville's system. An argument very similar to the one applied in the previous section shows that $\mathcal{A}_P\colon (P\widetilde{\varphi})^* T_B \xrightarrow{\sim} \Omega^1_{P/B}$, is an isomorphism at all the nonsingular points t of all the fibres of P over B. In fact, if $B_i = \mathbb{P}W_i$, where $W_i \subset V$, $W_1 \oplus W_2 = V$, and $s \in W_1$, then the holomorphic differentials $\omega_{s,v} = \mathcal{A}_{\bar{J}\mathcal{D},t}(v + \mathbb{C}s)$ over all all $v \in W_2$ are exactly the lifts to $\widetilde{C} = D_s$ of all the holomorphic differentials on $C = D_s/\tau$. This is true because the divisors $D_v \cap D_s$ are the lifts of the canonical divisors $\pi D_v \cap \pi D_s \sim C \cdot (C + K_X) = K_C$ on C. Hence $\mathcal{A}_{\bar{J}\mathcal{D},t}$ maps $W_2 + \mathbb{C}s/\mathbb{C}s$ isomorphically onto $\pi^* H^0(C, \Omega^1_C)$. Hence none of the tangent vectors $v + \mathbb{C}s$ with $v \notin W_2$ is sent to an element of $\pi^* H^0(C, \Omega^1_C)$ by the *isomorphism* $\mathcal{A}_{\bar{J}\mathcal{D},t}$. In particular, $T_{[C]}B = W_1/\mathbb{C}s$ is mapped injectively into the quotient by $\pi^* H^0(C, \Omega^1_C)$. This quotient is nothing but the cotangent space of $P(\widetilde{C}/C)$, so we are done.

Explicit constructions of Enriques surfaces. Let S be a complete intersection of three quadrics in $\mathbb{P}^5$. By Example 1, Section 4, it is a K3 surface. Take the equations of quadrics with separated variables:

$$f_i(x_0, x_1, x_2) + g_i(x_3, x_4, x_5) = 0, \quad i = 1, 2, 3$$

where f_i, g_i are quadratic forms such that the three conics $\{f_i = 0\} \subset \mathbb{P}^2$, as well as $\{g_i = 0\} \subset \mathbb{P}^2$ $(i = 1, 2, 3)$ have no common points. Then the map

$$\eta\colon (x_0, x_1, x_2, x_3, x_4, x_5) \mapsto (-x_0, -x_1, -x_2, x_3, x_4, x_5)$$

is an involution without fixed points and S/η is an Enriques surface. Verra [**V**] proved that any Enriques surface can be obtained in this way. The linear system of hyperplane section of S is 5-dimensional, and it gives rise to two relative Prymians $P_i \to B_i \simeq \mathbb{P}^2$, which are 4-dimensional integrable systems. Contrary to the case of K3 surfaces, it is not clear if there are linear systems on X, all of whose members are reduced and irreducible, so the Altman–Kleiman and Mukai compactifications of the relative Jacobians may be only partial ones.

There are several other descriptions of Enriques surfaces. One of them makes explicit the existence of two pencils of elliptic curves on a generic Enriques surface. The construction goes as follows. Let $Q = \mathbb{P}^1 \times \mathbb{P}^1$ (a quadric in $\mathbb{P}^3$), and Δ a divisor of bidegree $(4, 4)$ (a section of the quadric by a surface of degree 4 in $\mathbb{P}^3$). Let $\rho\colon S \to Q$ be the double cover ramified at Δ. Then S is a K3 surface (see Example 2 in Section 4). Let η_Q be the involution on the quadric defined by $\big((x_0 : x_1), (y_0 : y_1)\big) \mapsto \big((-x_0 : x_1), (-y_0 : y_1)\big)$. It has 4 fixed points. If one chooses Δ η_Q-invariant and not passing through the fixed points, then η_Q lifts to S and gives an involution $\eta\colon Q \to Q$ such that either η, or $\eta\nu$ is fixpoint free, where ν is the involution associated to the double covering ρ. A generic Enriques surface can be represented in this way, and the elliptic pencils are the images of $\{\rho^{-1}(u \times \mathbb{P}^1)\}_{u \in \mathbb{P}^1}$ and $\{\rho^{-1}(\mathbb{P}^1 \times u)\}_{u \in \mathbb{P}^1}$. One can apply a similar construction to a quadratic cone in place of the nonsingular quadric Q, which will give the so called special Enriques surfaces having only one pencil of elliptic curves.

We have proved:

Let X be an Enriques surface, S its K3 *double cover,* $|C| = \{\mathcal{C} \to \mathbb{P}^{g-1}\}$ *a linear system of curves of genus $g \geq 2$ on X,* $\widetilde{\mathcal{C}} \to \mathbb{P}^{g-1}$ *the family of their inverse images on S. Then the relative Prymian* $P = P(\widetilde{\mathcal{C}}/\mathcal{C}) \to \mathbb{P}^{g-1}$ *is an integrable system with respect to the symplectic structure induced by Beauville's system via restriction.*

CASE 2. The fixpoint set of η is 1-dimensional.

Since the differential of η is nondegenerate at any fixed point, the fixed curve Γ should be nonsingular. So the quotient $X = S/\eta$ is a nonsingular surface. By the ramification formula for the canonical class, $K_S = \pi^* K_X + \Gamma$. Multiplying by 2 and using the fact that $\pi^*\Delta = 2\Gamma$, where Δ is the ramification curve of the quotient map $S \to X$, we see that $2K_X$ can be represented by the negative divisor $-\Delta$. Hence no positive multiple of K_X can be a positive divisor, that is the plurigenera $P_r = h^0\big(X, \mathcal{O}_X(rK_X)\big)$ vanish for all $r > 0$. We have also $h^0(X, \Omega^1_X) \leq h^0(S, \Omega^1_S) = 0$. By classification of surfaces, X is a rational surface.

Conversely, one can apply the construction of S as a double covering to some of the rational surfaces, in particular, to any Del Pezzo surface (see Example 2 in Section 4). In this case, for any complete linear system $|lh|$ on X, the above construction associates an integrable system of Prym varieties. As in the case of Enriques surfaces, these systems on X have always reducible fibres and the Altman–Kleiman or Mukai's compactification does not exist, so the question on the existence of nonsingular symplectic varieties compactifying the relative Prymians stays open.

Let us start by checking dimensions. Let X be a Del Pezzo surface, h the positive generator of the Picard group of X, $h^2 = k$ the degree of X, $K_X = -ih$, where $i = 3$ for $X = \mathbb{P}^3$, $i = 2$ for $X = Q$ a quadric and $i = 1$ for all the remaining del Pezzo surfaces, which are blow-ups of $\mathbb{P}^2$ at $9 - k$ points, $k = 1, \ldots 8$. By Riemann–Roch, $\dim |lh| = C^2 - CK_X/2 = kl(l+i)/2$. We have $\Delta \in |2ih|$, so that the number of the ramification points of the covering $\widetilde{C} \to C$ is $n = lh \cdot \Delta = 2lih^2 = 2kli$. By adjunction, $g = \deg K_C/2 + 1 = C^2 + CK_X/2 + 1 = kl(l+i)/2 + 1$. We obtain $\tilde{g} = 2g - 1 + n/2 = kl^2 + 1$, and $\dim P(\widetilde{C}/C) = \tilde{g} - g = kl(l+i)/2 = \dim |lh|$. Hence the dimension of the fibre equals that of the base, so we can hope that the family of Prymians is Lagrangian.

The verification that α_P is nondegenerate is very similar to that in the case of Enriques surfaces. We have in $V = H^0(\widetilde{C}, \Omega^1_{\widetilde{C}})$ two complementary eigenspaces of the involution, W_1 and W_2, and $\mathbb{P}W_1 = \pi^*|lh|$ parametrises the inverse images of the curves of the family $\mathcal{C}$, whilst $\mathbb{P}W_2 = \Gamma + \pi^*\big|(l-i)h\big|$ is the linear subsystem of reducible curves having Γ as a fixed component. Taking $s \in W_1$, representing a η-invariant curve $\widetilde{C}$, and $v \in W_2$, we obtain for the divisor of $\omega_{s,v}$ on the curve D_s:

$$(\omega_{s,v}) = D_s \cap D_v \sim \widetilde{C} \cdot (\Gamma + \pi^*\big((l-i)h)\big) = \Gamma \cap \widetilde{C} + \pi^* K_C = K_{\widetilde{C}}$$

by Hurwitz, hence $\omega_{s,v} \in \pi^* H^0(C, \Omega^1_C)$. As in the case of Enriques surfaces, this implies that α_P induces an isomorphism $W_1/\mathbb{C}s \simeq V/\pi^* H^0(C, \Omega^1_C)$ and that α_P is nondegenerate.

We have proved the following statement:
Let X be a Del Pezzo surface, S its K3 *double cover, depending on the choice of a ramification curve $\Delta \in |-2K_X|$. Let $|C| = \{\mathcal{C} \to \mathbb{P}^m\}$ be a linear system of curves of genus $g \geq 0$ on X, $\widetilde{\mathcal{C}} \to \mathbb{P}^m$ the family of their inverse images on S. Then the relative Prymian $P = P(\widetilde{\mathcal{C}}/\mathcal{C}) \to \mathbb{P}^m$ is an integrable system with respect to the symplectic structure induced by Beauville's via restriction.*

7. Kuznetsov's Transformation

We will come back in this section to the special case of Kowalevski top when $\delta = (l, g) = 0$ and show that two other genus 3 curves, one due to Kuznetsov, the other to Kuznetsov–Tsiganov, fit into the framework of the Del Pezzo integrable systems.

The first one was introduced by Kuznetsov in his solution to the problem of separation of variables [**K1, K2, KK**] for Kowalevski top; it is defined for any δ, but we will use it only in the case $\delta = 0$, when it is of genus 3. Let us call it the curve of separation of variables and denote by $\widetilde{E}$. It is a double cover of an elliptic curve E. In [**Mar-3**], it was shown that the Prym variety $P(\widetilde{E}/E)$ is isogeneous to the Jacobian of Kowalevski's curve C. One can think of the passage from C to $\widetilde{E}$ as a systematic way of replacing the Jacobian of a curve by an isogeneous

Prymian; we will call it Kuznetsov's transform. See also the Appendix of [**Au-2**] for a coordinate free approach to this question.

The three curves fit into the diagram

$$\begin{array}{ccc} C & \xleftarrow{f} & \widetilde{E} \\ & & \downarrow{\scriptstyle \pi} \\ & & E \end{array} \tag{7.1}$$

where $g(C) = 2$, $g(\widetilde{E}) = 3$, $g(E) = 1$, and f, π are double coverings; f is unramified, and π is ramified at 4 points. C is given by the equation (3.1), and $\widetilde{E}$ is defined by

$$\widetilde{E}\colon \left\{ w^2 = 4\left(u^4 - 2Hu^2 + H^2 - \frac{1}{4}K + 1\right)\left(u^4 - 2Hu^2 + H^2 - \frac{1}{4}K\right)\right\};$$

the equation of E is obtained by replacing u^2 by an independent variable, and f is defined by the formulas

$$\begin{cases} u = \sqrt{x} \\ w = \frac{2z}{\sqrt{x}} \end{cases}, \quad \begin{cases} x = u^2 \\ z = \frac{uw}{2} \end{cases}. \tag{7.2}$$

A similar transform can be applied to the BRS curve given by the equation (3.2) which we will rewrite in changing the notations for the variables:

$$C'\colon \left\{ z'^2 = x'\left(x'^2 + 2Hx' + \frac{1}{4}K\right)\left(x'^2 + 2Hx' + \frac{1}{4}K - 1\right)\right\}.$$

It turns out that the resulting curve $\widetilde{E}'$ is nothing but the spectral curve of the Kuznetsov-Tsiganov Lax representation [**KT**] of the Kowalevski top for $\delta = 0$ by 2×2-matrices. Similarly to the above, we have the diagram

$$\begin{array}{ccc} C' & \xleftarrow{f'} & \widetilde{E}' \\ & & \downarrow{\scriptstyle \pi'} \\ & & E' \end{array} \tag{7.3}$$

where

$$\widetilde{E}'\colon \left\{ w'^2 = 4\left(u'^4 + 2Hu'^2 + \frac{1}{4}K\right)\left(u'^4 + 2Hu'^2 + \frac{1}{4}K - 1\right)\right\};$$

$$\begin{cases} u' = \sqrt{x'} \\ w' = \frac{2z'}{\sqrt{x'}} \end{cases}, \quad \begin{cases} x' = u'^2 \\ z' = \frac{u'w'}{2} \end{cases}. \tag{7.4}$$

The following result is proved in [**Mar-3**]:
Kuznetsov's transforms (7.1), (7.3) *and Richelot's transform give the following sequence of isogenies*

$$P(\widetilde{E}/E) \xrightarrow{f_*} J(C) \xrightarrow{\text{Richelot}} J(C') \xrightarrow{f'^*} P(\widetilde{E}'/E')$$

of degrees respectively 8, 4, 2, *their kernels being direct sums of copies of* $\mathbb{Z}/2\mathbb{Z}$. *The following assertions are verified:*

(i) *The composition* $P(\widetilde{E}/E) \xrightarrow{f_*} J(C) \xrightarrow{\text{Richelot}} J(C')$ *is an isogeny of degree* 32 *whose kernel contains all the points of order* 2 *of* $P(\widetilde{E}/E)$. *Hence it factors through the multiplication by* 2 *on* $P(\widetilde{E}/E)$, *and there exists an isogeny* $P(\widetilde{E}/E) \to J(C')$ *of degree* 2.

(ii) *The composition of all the three isogenies is an isogeny of degree* 64 *which factors through the multiplication by* 2, *hence there exists also an isogeny of order* 4 $P(\widetilde{E}/E) \to P(\widetilde{E}'/E')$ *with kernel* $\mathbb{Z}/2\mathbb{Z} \oplus \mathbb{Z}/2\mathbb{Z}$.

The same assertions hold for the maps in the opposite direction.

(iii) *The elliptic curves* E, E' *are nonisogeneous for generic constants of motion* K *and* H.

Let us identify now the algebraic surfaces related to the new families of curves that we have constructed in this section. First of all remind that both Kowalevski and BRS families vary on the singular K3 surface

$$X_0 \colon \{z^2 = x(x^2+y)(x^2+y+1)\},$$

to which one can apply the change of coordinates (7.2), (7.4) to obtain the surface $w^2 = 4(U^4+y+1)(U^4+y) = 0$. The curves $\widetilde{E}$ or $\widetilde{E}'$ both are given by $y = uU^2 + v$ for different choices of u, v. The things being similar for the two families, let us restrict ourselves to $\widetilde{E}$. The involution η is given by $U \mapsto -U$, so that our curves form the invariant part of the family of conics $y = u_1U^2 + u_2U + u_3$, where $u = u_1$, $v = u_3$. This looks quite unnatural: The family of conics is not a complete linear system (the complete system of conics in the plane of variables U, y is of dimension 5), and the curves live on the double cover of $\mathbb{P}^2$ ramified in an optic, which is neither K3, nor Del Pezzo. The solution is simple: replace the plane by a quadric in the 3-space, then the family of conics, which are just hyperplane sections of the quadric, is 3-dimensional. This is done by the following rational map:

$$\mathbb{P}^2 \dashrightarrow Q_0 = \{x_2^2 = x_1x_3\} \subset \mathbb{P}^3, \quad (U, y) \mapsto (x_0, x_1, x_2, x_3) = (y, U^2, U, 1).$$

It is a singular quadric, a quadratic cone, and the curves $\widetilde{E}$ vary on the singular K3 double cover S of Q_0, ramified in a section of Q_0 by a quartic $\Delta_0 \in |-2K_{Q_0}| = |4h|$:

$$S \colon \begin{cases} w^2 = 4(x_1^2 + x_0x_3 + x_3^2)(x_1^2 + x_0x_3), \\ x_2^2 = x_1x_3. \end{cases}$$

They form the η-invariant part of the *complete* linear system on S of the inverse images of conic sections on Q_0, where $\eta \colon S \to S$ is defined by $x_2 \mapsto -x_2$, and the quotient S/η is a singular Del Pezzo surface.

References

[AvM] M. Adler and P. van Moerbeke, *The Kowalewski and HénonHeiles motions as Manakov geodesic flows on* SO(4)*—A two-dimensional family of Lax pairs*, Comm. Math. Phys. **113** (1988), 659–700.

[AK] A. B. Altman and S. L. Kleiman, *The presentation functor and the compactified Jacobian*, The Grothendieck Festschrift, vol. I, Progr. Math. **86**, Birkhäuser Boston, Boston, MA, 1990, pp. 15–32.

[Au-1] M. Audin, *Courbes algébriques et systèmes intégrables: Géodésiques des quadriques*, Exposition. Math. **12** (1994), 193–226.

[Au-2] ———, *Spinning tops, a course on integrable systems*, Cambridge Stud. Adv. Math., vol. 51, Cambridge Univ. Press, Cambridge, 1996.

[Au-Si] M. Audin and R. Silhol, *Variétés abéliennes réelles et toupie de Kowalevski*, Compositio Math. **87** (1993), 153–229.

[Beau-1] A. Beauville, *Variétés Kähleriennes dont la première classe de Chern est nulle*, J. Differential Geom. **18** (1983), 755–782.

[Beau-2] ———, *Systèmes hamiltoniens complètement intégrables associés aux surfaces* K3, Problems in the Theory of Surfaces and Their Classification (Cortona, 1988), Sympos. Math., vol. 32, Academic Press, London, 1991, pp. 25–31.

[BRS] A. I. Bobenko, A. G. Reyman, and M. A. Semenov-Tian-Shansky, *The Kowalewski top 99 years later—A Lax pair, generalisations and explicit solutions*, Comm. Math. Phys, **122** (1989), 321–354.

[BM] J.-B. Bost and J.-F. Mestre, *Moyenne arithmético-géométrique et périodes des courbes de genre* 1 *et* 2, Gaz. Math. Soc. France **38** (1988), 36–64.

[Bo] F. Bottacin, *Poisson structures on moduli spaces of sheaves over Poisson surfaces*, Invent. Math. **121** (1995), 421–436.

[CF] J. W. S. Cassels and E. V. Flynn, *Prolegomena to a middlebrow arithmetic of curves of genus* 2, London Mathematical Society Lecture Note Series, vol. 230, Cambridge Univ. Press, Cambridge, 1996.

[DL] R. Donagi and R. Livné, *The arithmetic-geometric mean and isogenies for curves of higher genus*, Ann. Scuola Norm. Sup. Pisa Cl. Sci. (4) **28** (1999), 323–339.

[Dur] A. Durfée, *Fifteen characterisations of rational double points and simple critical points*, Enseign. Math. (2) **25** (1979), 131–163.

[Fu] A. Fujiki, *On primitively symplectic compact Kähler V-manifolds of dimension four*, Classification of Algebraic and Analytic Manifolds, Katata Symp. Proc. (Katata, 1982), Progr. Math., vol. 39, Birkhäuser Boston, Boston, MA, 1983, pp. 71–250.

[GH] Ph. Griffiths and J. Harris, *Principles of algebraic geometry*, John Wiley & Sons, Inc., New York, 1994.

[HvM] E. Horozov and P. van Moerbeke, *The full geometry of Kowalewski's top and* $(1,2)$*-Abelian surfaces*, Commun. Pure Appl. Math. **42** (1989), 357–407.

[H] G. Humbert, *Sur la transformation ordinaire des fonctions abéliennes*, J. de Math. (5) **7** (1901), 395–417.

[Hur] J. C. Hurtubise, *Integrable systems and algebraic surfaces*, Duke Math. J. **83** (1996), 19–50; *Erratum*, **84** (1996), 815.

[HM] J. C. Hurtubise and E. Markman, *Rank 2-integrable systems of Prym varieties*, Adv. Theor. Math. Phys. **2** (1998), 633–695.

[HL] D. Huybrechts and M. Lehn, *The geometry of moduli spaces of sheaves*, Aspects of Mathematics, vol. E31, Vieweg & Sohn, Braunschweig, 1997.

[I-Sha] V. A. Iskovskikh and I. R. Shafarevich, *Algebraic surfaces*, Algebraic geometry, II, Encyclopaedia Math. Sci., vol. 35, Springer, Berlin, 1996, pp. 127–262.

[J] C. G. J. Jacobi, *Vorlesungen über Dynamik*, Gesammelte Werke, Bd. VIII, Chelsea Publishing Co., New York, 1969.

[KK] I. V. Komarov and V. B. Kuznetsov, *Quasiclassical quantization of the Kowalewski top*, Theoret. and Math. Phys. **73** (1987), 1255–1263.

[Kow] S. Kowalevski, *Sur le problème de la rotation d'un corps solide autour d'un point fixe*, Acta Math. **14** (1889), 177–232.

[K1] V. B. Kuznetsov, *Quasiclassical quantization of the Kowalevski top*, M.Sc. Thesis, Leningrad University, 1985.

[K2] ———, *Applications of the inverse scattering method to* 2*-dimensional classical and quantum integrable systems*, PhD Thesis, Leningrad University, 1989.

[KT] V. B. Kuznetsov and A. V. Tsiganov, *A special case of Neumann's system and the Kowalewski-Goryachev-Chaplygin top*, J. Phys. A **22** (1989), L73–L79.

[LM] F. Leprévost and D. Markushevich, *A tower of genus two curves related to the Kowalewski top*, J. Reine Angew. Math. **514** (1999), 103–111.

[Man] Yu. I. Manin, *Cubic forms. Algebra, geometry, arithmetic*, North-Holland Publishing Co., Amsterdam, 1986.

[Mar-1] D. Markushevich, *Completely integrable projective symplectic* 4*-dimensional varieties*, Izv. Ross. Akad. Nauk Ser. Mat. **59** (1995), no. 1, 157–184; Russian transl., Izv. Math. **59** (1995), no. 1, 159–187.

[Mar-2] ———, *Lagrangian families of Jacobians of genus* 2 *curves*, Algebraic geometry, 5, J. Math. Sci. **82** (1996), 3268–3284.

[Mar-3] ———, *Kowalevski top and genus-*2 *curves*, J. Phys. A **34** (2001), 2125–2135.

[Mat-1] ———, *On fibre space structures of a projective irreducible symplectic manifold*, Topology **38** (1999), 79–83; *Addendum*, **40** (2001), 431–432.

[Mat-2] ———, *Equidimensionality of Lagrangian fibrations on holomorphic symplectic manifolds*, Math. Res. Lett. **7** (2000), 389–391.

[Muk] S. Mukai, *Symplectic structure of the moduli of sheaves on an Abelian or* K3 *surface*, Invent. Math. **77** (1984), 101–116.

[Mum] D. Mumford, *Prym varieties.* I, Contributions to Analysis, Academic Press, New York, 1974, pp. 325–350.

[R-STS] A. G. Reyman and M. A. Semenov-Tian-Shansky, *Lax representation with a spectral parameter for the Kowalewski top and its generalizations*, Lett. Math. Phys. **14** (1987), 55–61.

[R1] F. Richelot, *Essai sur une méthode générale pour déterminer la valeur des intégrales ultra-elliptiques, fondée sur des transformations remarquables de ces transcendantes*, C. R. Acad. Sci. Paris **2** (1836), 622–627.

[R2] ———, *De transformatione Integralium Abelianorum primiordinis commentation*, J. Reine Angew. Math. **16** (1837), 221–341.

[Ro] M. Rosenlicht, *Generalized Jacobian varieties*, Ann. of Math. (2) **59** (1954), 505–530.

[Serre] J.-P. Serre, *Algebraic groups and class fields*, Grad. Texts in Math., vol. 117, Springer-Verlag, New York, 1988.

[Sha] I. R. Safarevic, B. G. Averbuh, Ju. R. Vainberg, A. B. Zizcenko, Ju. I. Manin, B. G. Moisezon, G. N. Tjurina, and A. N. Tjurin, *Algebraic surfaces*, Trudy Mat. Inst. Steklov. **75** (1965), 1–215 (Russian); English Transl. Proc. Steklov Inst. Math., No. 75, Amer. Math. Soc., Providence, R.I. 1965.

[Sim] C. T. Simpson, *Moduli of representations of the fundamental group of a smooth projective variety.* I, Inst. Hautes Études Sci. Publ. Math. **79** (1994), 47–129; II, **80** (1994), 5–79.

[V] A. Verra, *On the Enriques' surface as a fourfold cover of* P^2, Math. Ann. **266** (1983), 241–250.

Département de Mathématiques Pures et Appliquées, Cité Scientifique—Bât. M2, Université Lille 1, F-59655 Villeneuve d'Ascq Cedex, France.

E-mail address: markushe@agat.univ-lille1.fr

Centre de Recherches Mathématiques
CRM Proceedings and Lecture Notes
Volume **32**, 2002

Painlevé Sixth Equation as Isomonodromic Deformations Equation of an Irregular System

Marta Mazzocco

1. Introduction

In this paper we represent the Painlevé Sixth equation

$$\text{(PVI)}\quad y_{xx} = \frac{1}{2}\left(\frac{1}{y} + \frac{1}{y-1} + \frac{1}{y-x}\right)y_x^2 - \left(\frac{1}{x} + \frac{1}{x-1} + \frac{1}{y-x}\right)y_x + \frac{y(y-1)(y-x)}{x^2(x-1)^2}\left[\alpha + \beta\frac{x}{y^2} + \gamma\frac{x-1}{(y-1)^2} + \delta\frac{x(x-1)}{(y-x)^2}\right]$$

where $x \in \mathbb{C}$ and α, β, γ, δ are arbitrary complex parameters, as isomonodromic deformations equation of a 3×3 linear system with a simple pole at 0 and a double one at ∞, i.e.

$$\frac{d}{d\lambda}Y = \left(\begin{pmatrix} 0 & & \\ & x & \\ & & 1\end{pmatrix} + \frac{V - \mathbb{1}}{\lambda}\right)Y$$

where V is a 3×3 matrix with diagonal part Θ of entries $(-\theta_1, -\theta_2, -\theta_3)$ and eigenvalues $\mu_1 = (\sum\theta_k + \vartheta_\infty)/2$, $\mu_2 = (\sum\theta_k - \vartheta_\infty)/2$, and $\mu_3 = 0$. The parameters α, β, γ, δ of the PVI equation are related to the parameters $(\theta_1, \theta_2, \theta_3, \vartheta_\infty)$ by the following:

$$\alpha = \frac{(\vartheta_\infty - 1)^2}{2}, \quad \beta = -\frac{\vartheta_1^2}{2}, \quad \gamma = \frac{\vartheta_3^2}{2}, \quad \delta = \frac{1-\vartheta_2^2}{2}.$$

We recall that the general solution $y(x; c_1, c_2)$ of PVI $(\alpha, \beta, \gamma, \delta)$ satisfies the following two important properties (see [**19**]):

1. The solution $y(x; c_1, c_2)$ can be analytically continued to a meromorphic function on the universal covering of $\overline{\mathbb{C}}\backslash\{0, 1, \infty\}$ (the so-called *Painlevé property*).
2. For generic values of the integration constants c_1, c_2 and of the parameters $\alpha, \beta, \gamma, \delta$, the solution $y(x; c_1, c_2)$ can not be expressed via elementary or classical transcendental functions.

2000 *Mathematics Subject Classification.* Primary: 34M55; Secondary: 34M50.

The author is grateful to B. Dubrovin and N. Hitchin who inspired this paper and is indebted to D. Guzzetti, for many fruitful conversations. The author was supported by an EPSRC research assistantship and by MSRI, Mathematical Sciences Research Institute, Berkeley.

This is the final form of the paper.

The second property means that the generic solutions to the PVI define some new functions, the so-called *Painlevé sixth transcendents*. Nevertheless, for some special values of the parameters $(\alpha, \beta, \gamma, \delta)$, special solutions may be expressed via classical functions. For example, particular classical solutions to PVI, expressed via hypergeometric functions, were first constructed by Lukashevich [**13**]. A general approach to study the classical solutions of PVI was proposed by Okamoto [**17, 18**]. One of the main tools of this approach is the symmetry group of PVI: the particular solutions are those being invariant with respect to some symmetry of PVI. The symmetries act in a non trivial way on the space of the parameters $(\alpha, \beta, \gamma, \delta)$. Okamoto described the fundamental region of the action of this symmetry group and showed that all the classical solutions known at that moment, fit into the boundary of this fundamental region. Recently some other classical solutions (algebraic) have been found that do not fit into such a boundary ([**6, 10, 15, 21**]). This fact leads to the natural question whether all symmetries of PVI have been found or there are some hidden ones that act in a special way on these new algebraic solutions. In [**7, 8**] a method to find new symmetries of the Painlevé equations via mappings preserving the locations of movable poles has been proposed, but did not yet work out for the case of PVI. We are not going to answer this question here, but we believe that our representation of PVI as isomonodromic deformations equation of the system (PVI) might lead to the discovery of new symmetries.

A second advantage of this representation appears in connection with the applications of PVI. In fact a particular case of the isomonodromic problem for the system (PVI) appears naturally in the theory of Frobenius manifolds (see [**4**]) and also in the problem of construction of self-dual Bianchi-type IX Einstein metrics [**3, 23**].

Our approach consists in constructing a map between the irregular system of first order differential equations and a Fuchsian system of the form

$$\frac{\mathrm{d}}{\mathrm{d}\lambda}\Phi = \left(\frac{\mathcal{A}_1}{\lambda} + \frac{\mathcal{A}_2}{\lambda - x} + \frac{\mathcal{A}_3}{\lambda - 1}\right)\Phi$$

where $\mathcal{A}_j$ are 2×2 matrices independent on λ satisfying the following conditions:

$$\operatorname{eigen}(\mathcal{A}_k) = \pm\frac{\theta_k}{2} \quad \text{and} \quad -\sum_{k=1}^{3}\mathcal{A}_k = \mathcal{A}_\infty = \frac{1}{2}\begin{pmatrix} \theta_\infty & 0 \\ 0 & -\theta_\infty \end{pmatrix}.$$

The isomonodromic deformations equations of such a Fuchsian system reduce to the PVI equation [**11, 12**].

This map was already discovered in [**9**] in terms of parameter-dependent pairs of moment maps from a symplectic vector space to the dual spaces of two different loop-algebras. We present here another (explicit) construction via the Laplace transform, already proposed in [**2, 5, 20**], that allows us to give an explicit expression of the matrix V of the system in terms of the solutions $y(x)$ of the Painlevé sixth equation:

$$V_{11} = -\theta_1, \quad V_{22} = -\theta_2, \quad V_{33} = -\theta_3,$$

$$V_{12} = \frac{k}{l}\left(\frac{x((x-1)y' + \theta_1)}{2y} + \frac{1 - \theta_1 - (x-1)\theta_2 + (y-x)\vartheta_\infty - y}{2}\right),$$

$$V_{13} = k\left(\frac{(x-1)y' + \theta_1}{2y} + \frac{1 + (x-1)\theta_3 - x\theta_1 + (y-1)\vartheta_\infty - y}{2x}\right),$$

$$V_{21} = \frac{l}{k}\left(\frac{-y\theta_2 - xy'}{2(y-x)} - \frac{\theta_1 + y\vartheta_\infty}{2(x-1)} + \frac{y(y-1)}{2(x-1)(y-x)}\right),$$

$$V_{23} = l\left(\frac{-(x-1)y' - (y-1)\theta_2}{2(y-x)} + \frac{\theta_3 - (y-1)\vartheta_\infty}{2x} + \frac{y(y-1)}{2x(y-x)}\right),$$

$$V_{31} = \frac{1}{k}\left(\frac{xy' - y\theta_3}{2(y-1)} + \frac{x\theta_1 + y\vartheta_\infty - y}{2(x-1)}\right),$$

$$V_{32} = \frac{1}{l}\left(\frac{-x(x-1)y' - (y-x)\theta_3}{2(y-1)} + \frac{x\theta_2 - (y-x)\vartheta_\infty + y}{2}\right)$$

where l, k are functions of x that can be determined by quadratures.

In Section 2 we describe the monodromy data and the isomonodromic deformation equations of the irregular 3×3 system (PVI). In Section 3, we deal with the reduction of the isomonodromic deformation equations of the 2×2 Fuchsian system to the Painlevé VI equation. In Section 4, we build our map via the Laplace transform.

2. The Irregular System

In this section, We describe the monodromy data of the system

$$\frac{\mathrm{d}}{\mathrm{d}z}Y = \left(U + \frac{W}{z}\right)Y \tag{2.1}$$

where U is a diagonal matrix of elements (u_1, u_2, u_3) and W is any constant matrix of diagonal part Ω and with distinct eigenvalues μ_1, μ_2, μ_3.

Observe that the system (2.1) has two poles, a simple one at 0 and a double one at ∞. The solutions $Y(z)$ are well defined analytic multi-valued functions outside 0 and ∞. We can analytically continue the solutions $Y(z)$ along any path in the punctured plane. When arriving back at the original point, one obtains a linear transformation on $Y(z)$. Such a linear transformation is called the *monodromy transformation*. In this section, we are going to compute this transformation in terms of matrices.

2.1. Local theory at zero. The system (2.1) has a simple pole at $z = 0$. We assume that a branch-cut between zero and infinity has been chosen along a fixed line l and a branch of $\log z$ has been selected.

PROPOSITION 2.1. *There exists a gauge transformation $Y = \mathcal{G}(z)\widetilde{Y}$ where $G(z) = \sum_{k=0}^{\infty} \mathcal{G}_k z^k$ is convergent near 0, with principal term $\mathcal{G}_0$ defined by $V = \mathcal{G}_0 \mu \mathcal{G}_0^{-1}$, where μ is the diagonal matrix of elements μ_1, μ_2, μ_3, that maps the system* (2.1) *into the* Birkhoff Normal form:

$$\frac{\mathrm{d}}{\mathrm{d}z}\widetilde{Y} = \left(\frac{\mu}{z} + \sum_{k\geq 1} \mathcal{R}_k z^{k-1}\right)\widetilde{Y} \tag{2.2}$$

where $\mathcal{R} = \sum_{k\geq 1} \mathcal{R}_k$, with

$$\mathcal{R}_{k_{ij}} \neq 0 \quad \Longleftrightarrow \quad \mu_i - \mu_j = k.$$

As a consequence there exists a fundamental matrix of the system (2.1) *of the form*

$$Y_0(z) = \mathcal{G}(z) z^{\mathcal{R}} z^{\mu}, \quad as \quad z \to 0. \tag{2.3}$$

The fundamental matrix $Y_0(z)$ is uniquely determined up to the ambiguity:

$$Y(z) \mapsto Y(z)L_0 \tag{2.4}$$

where L_0 is any constant invertible matrix such that

$$z^{\mu} z^{\mathcal{R}} L_0 z^{-\mathcal{R}} z^{-\mu} = \sum_{j=0}^{N} L_0^{(j)} z^j, \tag{2.5}$$

for $L_0^{(0)} = \mathcal{G}_0$ and for some $L_0^{(1)}, \dots, L_0^{(N)}$ constant matrices.

Observe that being the number of possible *resonances*, (i.e. possible integer differences between eigenvalues of W) finite, there are only a finite number of non zero matrices $\mathcal{R}_k$. In the *nonresonant case* , i.e. for $\mu_i - \mu_j \notin \mathbb{Z}$ for all $i, j = 1, 2, 3$, all matrices $\mathcal{R}_k$ are zero. So in the nonresonant case the fundamental matrix is diagonal, while in the resonant case there are logarithmic corrections.

DEFINITION 2.2. Two systems of the form

$$\frac{\mathrm{d}}{\mathrm{d}z} y^{(1),(2)} = \sum_{k \geq 0} A_k^{(1),(2)} z^{k-1} \, y^{(1),(2)} \tag{2.6}$$

where $\sum_{k\geq 0} A_k^{(1),(2)} z^{k-1}$ is convergent in a neighbourhood of $z = 0$, are *analytically equivalent* if there exists a power series

$$G(z) = G_0 + G_1 z + \cdots$$

convergent in a neighbourhood of $z = 0$, such that the transformation

$$y^{(2)} = G(z) y^{(1)}$$

maps the first system into the second one.

LEMMA 2.3. *Two Birkhoff normal forms*

$$\frac{\mathrm{d}}{\mathrm{d}z} \widetilde{Y} = \left(\frac{\mu^{(1),(2)}}{z} + \sum_{k \geq 1} \mathcal{R}_k^{(1),(2)} z^{k-1} \right) \widetilde{Y}$$

of the same system (2.1) *are gauge equivalent iff* $\mu^{(1)} = \mu^{(2)}$ *and* $R^{(1)} = \Delta^{-1} R^{(2)} \Delta$, *where* $\Delta = \mathbb{1} + \Delta_1 + \Delta_2 + \cdots$ *with*

$$z^{\mu} \Delta_k z^{-\mu} = z^k \Delta_k. \tag{2.7}$$

Observe that Δ has mostly 0 entries. Explicitly $\widetilde{R}_1 = R_1$, $\widetilde{R}_2 = R_2 + [R_1, \Delta_1]$, etc.

PROOF. Assume that we have two normal forms $(\mu^{(1)}, R^{(1)})$ and $(\mu^{(2)}, R^{(2)})$ of the same system (2.1). Then $\mu^{(1)} = \mu^{(2)} = \mu$ because we can always rearrange the order of the eigenvalues of W by a constant gauge transformation. Thus the two systems

$$\frac{\mathrm{d}}{\mathrm{d}z} y^{(1)} = \left(\frac{\mu}{z} + \sum_{k \geq 1} R_k^{(1)} z^{k-1} \right) y^{(1)}, \quad \frac{\mathrm{d}}{\mathrm{d}z} y^{(2)} = \left(\frac{\mu}{z} + \sum_{k \geq 1} R_k^{(2)} z^{k-1} \right) y^{(2)}$$

are gauge-equivalent, i.e. there exists a gauge transformation $G(z) = \mathbb{1} + \sum_{k \geq 1} G_k z^k$ that maps the first system in the second one. In particular this means that multiplying every solution of the first system by $G(z)$ we arrive to a solution of the second system, i.e.

$$G(z) y(z)^{(1)} = G(z) z^{\mu} z^{R^{(1)}} = z^{\mu} z^{R^{(2)}} \Delta = y(z)^{(2)} \Delta \tag{2.8}$$

where Δ is a constant invertible matrix. In particular it follows that

$$G_{ij} \neq 0 \quad \Longleftrightarrow \quad \mu_i - \mu_j = k, \ k \in \mathbb{Z}_+.$$

Now we multiply (2.8) by $z^{-\mu}$ on the left and z^{-R} on the right, we obtain

$$z^{-\mu} G(z) z^{\mu} = z^{\widetilde{R}} \Delta z^{-R}.$$

The left hand side contains only powers of z, the right hand side is a polynomial in $\log z$. Since the order of $\log z$ is always lower than any power of z, we have

$$z^{-\mu} G(z) z^{\mu} = z^{\widetilde{R}} \Delta z^{-R} = \text{const.},$$

and we can compute it by looking only at the first terms of the expansion. Since

$$\left(z^{-\mu} G(z) z^{\mu}\right)_{ij} = G(z)_{ij} z^{\mu_j - \mu_i},$$

and

$$z^{\widetilde{R}} \Delta z^{-R} = \left(\mathbb{1} + \sum \frac{\widetilde{R}^k}{k!} \log z^k\right) \Delta \left(\mathbb{1} + \sum \frac{-R^k}{k!} \log z^k\right) \sim \Delta + \mathcal{O}(\log z),$$

we obtain

$$\mathbb{1} + G_1 + G_2 + \cdots = \Delta$$

thus $\Delta_k = G_k$ and we obtain the needed transformation between the two systems. □

Given the fundamental matrix $Y_0(z)$ it is easy to compute its analytic continuation along loops around zero. The monodromy $\mathcal{M}_0$ of the system (2.1) with respect to the normalised fundamental matrix (2.3) generated by a simple closed loop around the origin is

$$\mathcal{M}_0 = \exp(2\pi i \mu) \exp(2\pi i \mathcal{R}).$$

2.2. Local theory at infinity. The local theory near infinity is rather different from the one near 0. In fact the system (2.1) has a pole of order two at infinity. Generically this leads to two main facts: the solutions are not defined in discs around infinity but only in sectors and they have exponential growth, instead of polynomial.

PROPOSITION 2.4. *For the system* (2.1) *there exists a unique formal power series*

$$P(z) = \sum_0^{\infty} P_k z^{-k} \tag{2.9}$$

with $P_0 = \mathbb{1}$*, such that the formal gauge transformation* $Y = P(z)\widetilde{Y}$ *changes the system* (2.1) *into the system in normal form:*

$$\frac{\mathrm{d}}{\mathrm{d}z} \widetilde{Y} = \left(U + \frac{\Omega}{z}\right) \widetilde{Y}. \tag{2.10}$$

As a consequence, there is a unique formal fundamental solution Y_∞ to the system (2.1) *at* ∞. *It has the following form*[1]

$$Y_\infty = P(z) z^{\Omega} e^{zU}. \tag{2.11}$$

The above statement is a particular case of a more general theorem proved in [**24**].

The above result establishes only the existence of *formal solutions*. Regarding the true solutions, we need to set up some more machinery.

DEFINITION 2.5. The half-line

$$R_{ij} = \{z \mid \mathrm{Re}[z(u_i - u_j)] = 0, \mathrm{Im}[z(u_i - u_j)] < 0\} \tag{2.12}$$

oriented from zero to infinity is called *Stokes-ray.*

DEFINITION 2.6. An oriented line l in the complex plane is called *admissible with respect to the points* (u_1, u_2, u_3) if it is such that all the Stokes rays R_{ij} with $i < j$ lie on the left of l. Vice-versa, fixed any oriented line l in the complex plane, the triple of points (u_1, u_2, u_3) is called *admissible with respect to the line* l, if all the Stokes rays R_{ij} with $i < j$ lie on the left of l.

DEFINITION 2.7. (Valid for the particular case we are working with). Let Σ be some sector near ∞ and $f(z)$ analytic for $z \in \Sigma \cap \{|z| > N\}$, $N \in \mathbb{R}$. We say that $\sum_{k=0}^{n} a_k/z^k$ is the *asymptotic series expansion* of $f(z)$ in $\Sigma \cap \{|z| > N\}$,

$$f(z) \sim \sum_{k=0}^{n} \frac{a_k}{z^k},$$

if for all $n \in \mathbb{Z}_+$, and for $z \to \infty$ inside Σ

$$\lim_{z\to\infty} z^n \left(f(z) - \sum_{k=0}^{n} \frac{a_k}{z^k} \right) = 0.$$

LEMMA 2.8. *Fix an admissible oriented line l and consider the system* (2.1) *and its unique formal fundamental solution* (2.11). *Then there exists $\varepsilon > 0$ small enough, two sectors Π_L and Π_R defined as*

$$\begin{aligned} \Pi_R &= \{z\colon \arg(l) - \pi - \varepsilon < \arg(z) < \arg(l) + \varepsilon\}, \\ \Pi_L &= \{z\colon \arg(l) - \varepsilon < \arg(z) < \arg(l) + \pi + \varepsilon, \}, \end{aligned} \tag{2.13}$$

and two fundamental solutions $Y_L(z)$ in Π_L and $Y_R(z)$ in Π_R such that

$$Y_{L,R} \sim \left(\mathbb{1} + \mathcal{O}\left(\frac{1}{z}\right)\right) z^{\Omega} e^{zU}, \quad \text{as } z \to \infty,\ z \in \Pi_{L,R}. \tag{2.14}$$

The fundamental solutions $Y_{L,R}(z)$ are uniquely determined by (2.14).

The proof can be found in [**1**].

We are now going to define one of the main objects in the monodromy theory of our system at infinity: the Stokes matrices. In both of the narrow sectors

$$\begin{aligned} \Pi_+ &:= \{z \mid \varphi - \varepsilon < \arg z < \varphi + \varepsilon\}, \\ \Pi_- &:= \{z \mid \varphi - \pi - \varepsilon < \arg z < \varphi - \pi + \varepsilon\} \end{aligned}$$

[1] We fix the branch of the logarithm required in the definition of z^{Ω} as in the previous section.

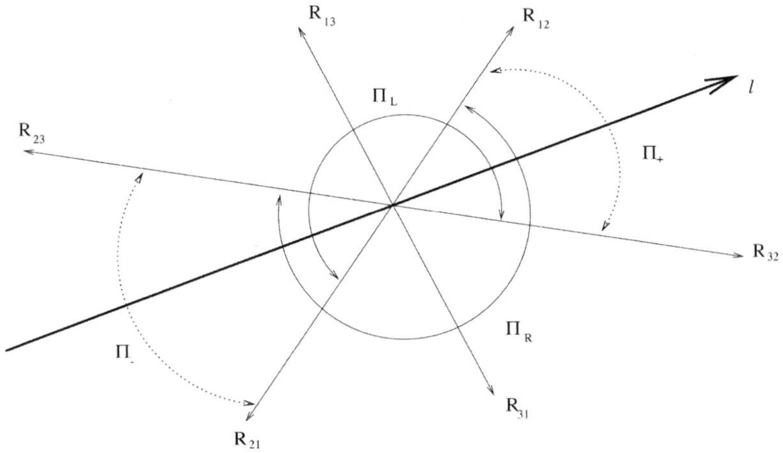

FIGURE 1. The admissible line l, the Stokes rays and the sectors.

obtained by the intersection of Π_l and Π_R, we have two fundamental matrices. They must be related by multiplication by a constant invertible matrix

$$Y_L(z) = Y_R(z)S_+,\ z \in \Pi_+. \qquad Y_L(z) = Y_R(z)S_-,\ z \in \Pi_-.$$

DEFINITION 2.9. The matrices S_+, S_- are called *Stokes matrices of the system* (2.1) with respect to the admissible line ℓ.

LEMMA 2.10. *Two systems of the form*

$$\frac{\mathrm{d}}{\mathrm{d}z}Y = \left(U + \frac{U_1}{z} + \frac{U_2}{z^2} + \cdots\right)Y, \quad \frac{\mathrm{d}}{\mathrm{d}z}\widetilde{Y} = \left(U + \frac{U_1}{z} + \frac{U_2}{z^2} + \cdots\right)\widetilde{Y}$$

are analytically equivalent near $z = \infty$[2] *if and only if they have the same Stokes matrices with respect to the same admissible oriented line* ℓ.

PROOF. Let us assume that the systems are analytically equivalent. Then

$$G(z)Y_{L,R}(z) = \widetilde{Y}_{L,R}(z), \quad z \in \Pi_{L,R}$$

where $G(z)$ is a convergent series in $1/z$. The same transformation holds true for the formal solutions

$$G(z)Y_f = \widetilde{Y}_f(z)$$

in the same sectors. Then

$$\widetilde{Y}_L = G(z)Y_L = G(z)Y_RS_+ = \widetilde{Y}_RS_+.$$

Vice-versa we want to prove that if two systems have the same Stokes matrices with respect to the same admissible oriented line ℓ, then they are analytically equivalent. Let $Y_L(z)$, $Y_R(z)$ be the solutions of the correspondent systems with the needed asymptotic developments in the sectors $\Pi_{L,R}$ respectively. Let us consider the following piecewise analytic matrix-valued function $G(z)$ defined for sufficiently large $|z|$:

$$G(z) = \begin{cases} \widetilde{Y}_R(z)Y_R^{-1}(z), & z \in \Pi_R, \\ \widetilde{Y}_L(z)Y_L^{-1}(z), & z \in \Pi_L. \end{cases}$$

[2]The notion of analytic equivalence at infinity is completely analogous to Definition 2.2.

This function is well defined and analytic near infinity in two the sectors $\Pi_{L,R}$. But it could have jumps on the intersections $\Pi_{+,-}$. In the sectors Π_+, Π_- we have

$$Y_L^{(1,2)}(z) = Y_R^{(1,2)}(z)S_+, \ z \in \Pi_+, \quad Y_L^{(1,2)}(z) = Y_R^{(1,2)}(z)S_-, \ z \in \Pi_-.$$

So, thanks to the fact that the Stokes matrices are the same, there are no jumps. So the function $G(z)$ is a single-valued analytic function for $|z| > M$ for some big constant M. In the sectors $\Pi_{R,L}$

$$G(z) \sim 1 + O\left(\frac{1}{z}\right).$$

Hence $z = \infty$ is a removable singularity for this function, and

$$G(\infty) = 1.$$

This function $G(z)$ establishes the needed gauge transformation between the systems. □

REMARK 2.11. Our system (2.1) is analytically equivalent to

$$\frac{\mathrm{d}}{\mathrm{d}z} y = U y$$

iff $S_+ = S_- = \mathbb{1}$. This is the obstruction to analytic reduction to the normal form.

2.3. Monodromy data. Resuming, for our system (2.1)

$$\frac{\mathrm{d}}{\mathrm{d}z} Y = \left(U + \frac{W}{z}\right) Y$$

where U is a diagonal matrix with pairwise distinct entries u_1, u_2, u_3 and W is diagonalisable, we have built three distinguished bases in the space of solutions, i.e. $Y_0(z)$ near 0, and $Y_{L,R}(z)$ near ∞ depending on the choice of the admissible line ℓ. As proved in Lemma 2.10, the Stokes matrices uniquely specify the class of analytic equivalence of our system (2.1) near ∞. Analogously (see Lemma 2.3) at the origin the analytic equivalence of our system (2.1) is specified by μ and $\mathcal{R}$.

To complete the list of the monodromy data we define the central connection matrix between 0 and ∞

$$Y_0(z) = Y_{R,L}(z) C_{R,L}, \quad z \in \Pi_{R,L}.$$

Recall that to define Y_0 we had to fix the branch cut of $\log(z)$. We can choose it along the negative part l_- of the same line l.

We can reduce the list of the monodromy data $(\mu, R, S_+, S_-, C_R, C_L)$, by noticing the following *cyclic relation*

$$C_R^{-1} S_-^T S_+^{-1} C_R = C_L^{-1} S_+^T S_-^{-1} C_L M_0 = \exp(2\pi i \mu) \exp(2\pi \mathcal{R}).$$

This expresses a simple topological fact: on the punctured plane $\mathbb{C} \setminus \{0\}$ a loop around infinity is homotopic to a loop around the origin. for a similar reason $C_L = S_+^{-1} C_R$.

DEFINITION 2.12. The *monodromy data* of the system (2.1) are a collection of constant matrices

$$(\mu, R, S_+, C_R).$$

THEOREM 2.13. *Two systems of the same form* (2.1) *coincide if and only of they have the same monodromy data.*

The proof can be found in [**5**].

2.4. Isomonodromic deformations equations. In this section we deal with the deformations of the systems (2.1) that preserve the monodromy data. We are only interested in the deformations described in the following

THEOREM 2.14. *Let $W(\mathbf{u})$, $\mathbf{u} = (u_1, u_2, u_3)$, be a diagonalisable matrix function with eigenvalues μ_1, μ_2, μ_3. Suppose that for every $i = 1, 2, 3$ the matrix function $W(\mathbf{u})$ satisfies*

$$\frac{\partial}{\partial u_i} W = [W_i, W] \tag{2.15}$$

where

$$W_i = \mathrm{ad}_{E_i} \mathrm{ad}_U^{-1}(W).$$

and $E_{i_{kl}} = \delta_{ik}\delta_{il}$. Then the monodromy data of the system

$$\frac{\mathrm{d}}{\mathrm{d}z} Y = \left(U + \frac{W(\mathbf{u})}{z}\right) Y \tag{2.16}$$

are constant. Moreover, for every $i = 1, 2, 3$, the solution $Y(z, u)$ of the system (2.16) satisfies

$$\frac{\partial}{\partial u_i} Y = (zE_i + W_i) Y. \tag{2.17}$$

The proof of this statement can be found in [**5, 11, 12**].

3. The Fuchsian System

Consider the following Fuchsian system with four pairwise distinct regular singularities at u_1, u_2, u_3, ∞:

$$\frac{\mathrm{d}}{\mathrm{d}\lambda}\Phi = \left(\frac{\mathcal{A}_1}{\lambda - u_1} + \frac{\mathcal{A}_2}{\lambda - u_2} + \frac{\mathcal{A}_3}{\lambda - u_3}\right)\Phi, \quad \lambda \in \overline{\mathbb{C}} \backslash \{u_1, u_2, u_3, \infty\}. \tag{3.1}$$

The $\mathcal{A}_j$ being 2×2 matrices independent of λ. Assume that the matrices $\mathcal{A}_j$ satisfy the following conditions:

$$\mathrm{eigen}(\mathcal{A}_j) = \pm\frac{\vartheta_j}{2} \quad \text{and} \quad -\mathcal{A}_1 - \mathcal{A}_2 - \mathcal{A}_3 = \mathcal{A}_\infty, \tag{3.2}$$

for some constants ϑ_j, $j = 1, 2, 3$ and

$$\mathcal{A}_\infty := \frac{1}{2}\begin{pmatrix} \vartheta_\infty & 0 \\ 0 & -\vartheta_\infty \end{pmatrix}, \qquad \text{for some constant } \vartheta_\infty \neq 0. \tag{3.3}$$

The solution $\Phi(\lambda)$ of the system (3.1) is a multi-valued analytic function on $\mathbb{C}\backslash\{u_1, u_2, u_3\}$, and its multivaluedness is described by the *monodromy matrices.* To define them, we fix a basis $\gamma_1, \gamma_2, \gamma_3$ of loops in $\pi_1\left(\overline{\mathbb{C}}\backslash\{u_1, u_2, u_3, \infty\}, \infty\right)$ as in Figure 2, and a fundamental matrix for the system (3.1).

PROPOSITION 3.1. *There exists a fundamental matrix of the system (3.1) of the form*

$$\Phi_\infty(\lambda) = \left(\mathbb{1} + \mathcal{O}\left(\frac{1}{\lambda}\right)\right)\lambda^{-\mathcal{A}_\infty}\lambda^{-\mathcal{R}_\infty}, \quad \textit{as} \quad \lambda \to \infty \tag{3.4}$$

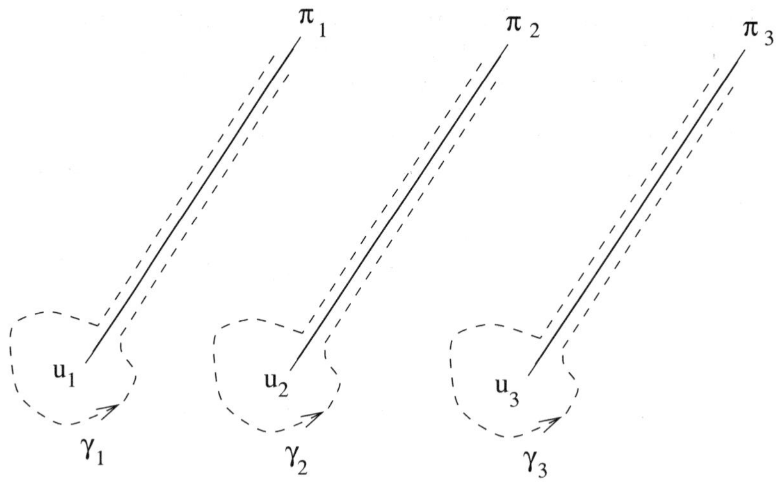

FIGURE 2. The branch-cuts π_j between the ordered singularities u_j and the oriented loops γ_j in the λ-plane.

where $\lambda^{-\mathcal{R}_\infty} := e^{-\mathcal{R}_\infty \log \lambda}$, *with the choice of the principal branch of the logarithm with the branch-cut along the common direction* η *of the cuts* π_1, π_2, π_3 *and the matrix entries of* $\mathcal{R}_\infty$ *given by* $\mathcal{R}_{\infty 11} = \mathcal{R}_{\infty 22} = 0$ *and*

$$\begin{array}{lll} \text{for } \vartheta_\infty \notin \mathbb{Z}, & \mathcal{R}_{\infty 12} = \mathcal{R}_{\infty 21} = 0, & \\ \text{for } \vartheta_\infty = n \in \mathbb{Z}_+, & \mathcal{R}_{\infty 12} = \mathcal{R}_{\infty 12}(\mathcal{A}_{1,2,3}, u_{1,2,3}), & \mathcal{R}_{\infty 21} = 0, \\ \text{for } \vartheta_\infty = -n,\ n \in \mathbb{Z}_+, & \mathcal{R}_{\infty 21} = \mathcal{R}_{\infty 21}(\mathcal{A}_{1,2,3}, u_{1,2,3}), & \mathcal{R}_{\infty 12} = 0 \end{array} \tag{3.5}$$

where the functions $\mathcal{R}_{\infty 12,21}(\mathcal{A}_{1,2,3}, u_{1,2,3})$ *are uniquely determined by* $(\mathcal{A}_{1,2,3}, u_{1,2,3})$. *Such a fundamental matrix* $\Phi_\infty(\lambda)$ *is uniquely determined up to*

$$\Phi_\infty(\lambda) \to \Phi_\infty(\lambda) L_\infty \tag{3.6}$$

where L_∞ *is any constant invertible matrix such that*

$$\lambda^{-\mathcal{A}_\infty} \lambda^{-\mathcal{R}_\infty} L_\infty \lambda^{\mathcal{A}_\infty} \lambda^{\mathcal{R}_\infty} = \mathbb{1} + \sum_{k=1}^{N} \frac{L_\infty^{(k)}}{\lambda^k}, \tag{3.7}$$

for some $L_\infty^{(1)}, \dots, L_\infty^{(N)}$ *constant matrices.*

The proof can be found in [**5**].

The fundamental matrix Φ_∞ can be analytically continued to an analytic function on the universal covering of $\overline{\mathbb{C}} \backslash \{u_1, u_2, u_3, \infty\}$. For any element $\gamma \in \pi_1(\overline{\mathbb{C}} \backslash \{u_1, u_2, u_3, \infty\}, \infty)$ denote the result of the analytic continuation of $\Phi_\infty(\lambda)$ along the loop γ by $\gamma[\Phi_\infty(\lambda)]$. Since $\gamma[\Phi_\infty(\lambda)]$ and $\Phi_\infty(\lambda)$ are two fundamental matrices in the neighbourhood of infinity, they are related by the following relation:

$$\gamma[\Phi_\infty(\lambda)] = \Phi_\infty(\lambda) \mathcal{M}_\gamma, \tag{3.8}$$

for some constant invertible 2×2 matrix $\mathcal{M}_\gamma$ depending only on the homotopy class of γ. Particularly, the matrix $\mathcal{M}_\infty := \mathcal{M}_{\gamma_\infty}$, γ_∞ being a simple loop around infinity in the clock-wise direction, is given by:

$$\mathcal{M}_\infty = \exp(2\pi i \mathcal{A}_\infty) \exp(2\pi i \mathcal{R}_\infty). \tag{3.9}$$

The resulting *monodromy representation* is an anti-homomorphism:

$$\begin{array}{ccc} \pi_1\left(\overline{\mathbb{C}}\backslash\{u_1,u_2,u_3,\infty\},\infty\right) & \to & \mathrm{SL}_2(\mathbb{C}) \\ \gamma & \mapsto & \mathcal{M}_\gamma \end{array} \tag{3.10}$$

$$\mathcal{M}_{\gamma\tilde{\gamma}} = \mathcal{M}_{\tilde{\gamma}}\mathcal{M}_\gamma. \tag{3.11}$$

The images $\mathcal{M}_j := \mathcal{M}_{\gamma_j}$ of the generators γ_j, $j = 1,2,3$ of the fundamental group, are called *the monodromy matrices* of the Fuchsian system (3.1). They generate the *monodromy group of the system,* i.e. the image of the representation (3.10). Since the loop $(\gamma_1\gamma_2\gamma_3)^{-1}$ is homotopic to γ_∞, the following relation between the generators holds:

$$\mathcal{M}_\infty\mathcal{M}_3\mathcal{M}_2\mathcal{M}_1 = \mathbb{1}. \tag{3.12}$$

Let us recall the definition of the *connection matrices.* Near the poles u_j, the fundamental matrices $\Phi_j(\lambda)$ of the system (3.1), are given by the following

PROPOSITION 3.2. *There exists a fundamental matrix of the system* (3.1) *of the form*

$$\Phi_j(\lambda) = G_j\left(\mathbb{1} + \mathcal{O}(\lambda-u_j)\right)(\lambda-u_j)^{J_j}(\lambda-u_j)^{\mathcal{R}_j}, \quad \textit{as} \quad \lambda\to u_j \tag{3.13}$$

where

$$\begin{aligned} &\textit{for } \vartheta_j \neq 0, \quad J_j = \frac{1}{2}\begin{pmatrix} \vartheta_j & 0 \\ 0 & -\vartheta_j \end{pmatrix}, \\ &\textit{for } \vartheta_j = 0, \quad J_j = J \equiv \begin{pmatrix} 0 & 1 \\ 0 & 0 \end{pmatrix}. \end{aligned} \tag{3.14}$$

The invertible matrix G_j is defined by $\mathcal{A}_j = G_jJ_jG_j^{-1}$, the diagonal elements of the matrix $\mathcal{R}_j$ are zero and the off-diagonal ones are defined as follows,

$$\begin{array}{lll} \textit{for } \vartheta_j \notin \mathbb{Z}, \textit{ and for }, \vartheta_j = 0, & \mathcal{R}_{j12} = \mathcal{R}_{j21} = 0, & \\ \textit{for } \vartheta_j = n \in \mathbb{Z}_+, & \mathcal{R}_{j12} = \mathcal{R}_{j12}(\mathcal{A}_{1,2,3},u_{1,2,3}), & \mathcal{R}_{j21} = 0, \\ \textit{for } \vartheta_j = -n, \ n\in\mathbb{Z}_+, & \mathcal{R}_{j21} = \mathcal{R}_{j21}(\mathcal{A}_{1,2,3},u_{1,2,3}), & \mathcal{R}_{j12} = 0 \end{array} \tag{3.15}$$

where the functions $\mathcal{R}_{j21,12}(\mathcal{A}_{1,2,3},u_{1,2,3})$ are uniquely determined by $\mathcal{A}_{1,2,3}$ and $u_{1,2,3}$. The choice of the branch of $\log(z-u_j)$ is along η as above.

The proof can be found in [**5**].

Continuing the solution $\Phi_\infty(\lambda)$ to a neighbourhood of u_j, along, say, the right-hand-side of the branch-cut π_j, one obtains another fundamental matrix around u_j, that must be related to $\Phi_j(\lambda)$ by:

$$\Phi_\infty(\lambda) = \Phi_j(\lambda)\mathcal{C}_j, \tag{3.16}$$

for some invertible matrix $\mathcal{C}_j$. The matrices $\mathcal{C}_1$, $\mathcal{C}_2$, $\mathcal{C}_3$ are called *connection matrices,* and they are defined by (3.16) up to the ambiguity $\mathcal{C}_j \to \mathcal{C}_jL_\infty$ due to (3.6). The connection matrices are related to the monodromy matrices as follows:

$$\mathcal{M}_j = \mathcal{C}_j^{-1}\exp(2\pi iJ_j)\exp(\mathcal{R}_j)\mathcal{C}_j, \quad j=1,2,3. \tag{3.17}$$

Thanks to the above relation it follows that

$$\mathrm{eigen}(\mathcal{M}_j) = \exp(\pm\pi i\vartheta_j). \tag{3.18}$$

DEFINITION 3.3. The *monodromy data* of the Fuchsian system (3.1) are

$$\{\mathcal{M}_1,\mathcal{M}_2,\mathcal{M}_3,\mathcal{R}_1,\mathcal{R}_2,\mathcal{R}_3\}. \tag{3.19}$$

REMARK 3.4. For noninteger ϑ_j, the correspondent $\mathcal{R}_j$ matrix is zero by definition and we drop it from the set of the monodromy data.

The theory of the deformations the poles of the Fuchsian system keeping the monodromy fixed is described by the following result:

THEOREM 3.5. *Let* $\{\mathcal{M}_1, \mathcal{M}_2, \mathcal{M}_3, \mathcal{R}_1, \mathcal{R}_2, \mathcal{R}_3\}$, *be the monodromy data of the Fuchsian system:*

$$\frac{\mathrm{d}}{\mathrm{d}\lambda}\Phi^0 = \left(\frac{\mathcal{A}_1^0}{\lambda - u_1^0} + \frac{\mathcal{A}_2^0}{\lambda - u_2^0} + \frac{\mathcal{A}_3^0}{\lambda - u_3^0}\right)\Phi^0, \tag{3.20}$$

of the above form (3.2), *with pairwise distinct poles* u_j^0, *and with respect to some basis* γ_1, γ_2, γ_3 *of the loops in* $\pi_1\left(\overline{\mathbb{C}}\backslash\{u_1^0, u_2^0, u_3^0, \infty\}, \infty\right)$. *If* $\mathcal{M}_k \neq \pm\mathbb{1}$ *for all* $k = 1, 2, 3, \infty$, *there exists a neighbourhood* $U \subset \mathbb{C}^3$ *of the point* $u^0 = (u_1^0, u_2^0, u_3^0)$ *such that, for any* $u = (u_1, u_2, u_3) \in U$, *there exists a unique triple* $\mathcal{A}_1(u)$, $\mathcal{A}_2(u)$, $\mathcal{A}_3(u)$ *of analytic matrix valued functions such that:*

$$\mathcal{A}_j(u^0) = \mathcal{A}_j^0, \quad i = 1, 2, 3, \tag{3.21}$$

and the monodromy matrices of the Fuchsian system

$$\frac{\mathrm{d}}{\mathrm{d}\lambda}\Phi = \mathcal{A}(\lambda; u)\Phi = \left(\frac{\mathcal{A}_1(u)}{\lambda - u_1} + \frac{\mathcal{A}_2(u)}{\lambda - u_2} + \frac{\mathcal{A}_3(u)}{\lambda - u_3}\right)\Phi \tag{3.22}$$

with respect to the same basis[3] γ_1, γ_2, γ_3 *of the loops, can be chosen to coincide with the given* $\mathcal{M}_1$, $\mathcal{M}_2$, $\mathcal{M}_3$. *The matrices* $\mathcal{A}_j(u)$ *are the solutions of the Cauchy problem with the initial data* $\mathcal{A}_j^0$ *for the following Schlesinger equations:*

$$\frac{\partial}{\partial u_j}\mathcal{A}_i = \frac{[\mathcal{A}_i, \mathcal{A}_j]}{u_i - u_j}, \quad \frac{\partial}{\partial u_i}\mathcal{A}_i = -\sum_{j\neq i}\frac{[\mathcal{A}_i, \mathcal{A}_j]}{u_i - u_j}. \tag{3.23}$$

The solution $\Phi_\infty^0(\lambda)$ *of* (3.20) *of the form* (3.4) *can be uniquely continued, for* $\lambda \neq u_i$ $i = 1, 2, 3$, *to an analytic function* $\Phi_\infty(\lambda, u)$, $u \in U$, *such that*

$$\Phi_\infty(\lambda, u^0) = \Phi_\infty^0(\lambda). \tag{3.24}$$

This continuation is the local solution of the Cauchy problem with the initial data Φ_∞^0 *for the following system that is compatible with the system* (3.22):

$$\frac{\partial}{\partial u_i}\Phi = -\frac{\mathcal{A}_i(u)}{\lambda - u_i}\Phi. \tag{3.25}$$

Moreover the functions $\mathcal{A}_i(u)$ *and* $\Phi_\infty(\lambda, u)$ *can be continued analytically to global meromorphic functions on the universal coverings of*

$$\mathbb{C}^3\backslash\{\mathrm{diags}\} := \{(u_1, u_2, u_3) \in \mathbb{C}^3 \mid u_i \neq u_j \ \textit{for}\ i \neq j\}, \tag{3.26}$$

and

$$\{(\lambda, u_1, u_2, u_3) \in \mathbb{C}^4 \mid u_i \neq u_j \ \textit{for}\ i \neq j \ \textit{and}\ \lambda \neq u_i, i = 1, 2, 3\}, \tag{3.27}$$

respectively.

The proof of this theorem can be found, for example, in [**14, 16, 22**].

REMARK 3.6. Observe that the isomonodromic deformations equations preserve the connection matrices $\mathcal{C}_i$ too.

[3] Observe that the basis γ_1, γ_2, γ_3 of $\pi_1\left(\overline{\mathbb{C}}\backslash\{u_1, u_2, u_3, \infty\}, \infty\right)$ varies continuously with small variations of u_1, u_2, u_3. This new basis is homotopic to the initial one, so one can identify them.

Let us now explain, following [**11, 12**], how to rewrite the Schlesinger equations (3.23) in terms of the PVI equation. We can assume $\vartheta_\infty \neq 0$ without loss of generality (see Remark 3.8 below). Schlesinger equations (3.23) with fixed $\mathcal{A}_\infty$ are invariant with respect to the gauge transformations of the form:

$$\mathcal{A}_i \mapsto D^{-1}\mathcal{A}_i D, \quad i=1,2,3, \qquad \text{for any } D \text{ diagonal matrix.} \tag{3.28}$$

We introduce two coordinates (p,q) on the quotient of the space of matrices satisfying (3.23) with respect to the equivalence relation (3.28) and a coordinate k that contains the above gauge freedom:

$$\begin{gathered}\left[\mathcal{A}(q;u_1,u_2,u_3)\right]_{12}=0,\\ p=\sum_{k=1}^{3}\frac{\mathcal{A}_{k_{11}}+\vartheta_k/2}{q-u_k},\quad k=\frac{2P(\lambda)[\mathcal{A}(\lambda;u_1,u_2,u_3)]_{12}}{\vartheta_\infty(q-\lambda)}\end{gathered} \tag{3.29}$$

where $\mathcal{A}(z;u_1,u_2,u_3)$ is given in (3.22) and $P(\lambda)=(\lambda-u_1)(\lambda-u_2)(\lambda-u_3)$. The matrices $\mathcal{A}_i$ are uniquely determined by the coordinates (p,q), and k and expressed rationally in terms of them:

$$\begin{aligned}\mathcal{A}_{i11}=\frac{1}{\vartheta_\infty P'(u_i)}\Bigg\{&P(q)(q-u_i)p^2+P(q)(q-u_i)p\left(\frac{\vartheta_\infty}{q-u_i}-\sum_{k=1}^{3}\frac{\vartheta_k}{q-u_k}\right)\\ &+(q-u_i)\left[\frac{\vartheta_\infty^2}{4}\left(q+2u_i-\sum_{k=1}^{3}u_k\right)m+\sum_{k=1}^{3}\frac{\vartheta_k^2}{4}\left(q+2u_k-\sum_{j=1}^{3}u_j\right)\right]\\ &+\frac{q-u_i}{2}\left(\vartheta_1\vartheta_2(q-u_3)+\vartheta_1\vartheta_3(q-u_2)+\vartheta_2\vartheta_3(q-u_1)\right)\\ &-\frac{\vartheta_\infty}{2}\sum_{k=1}^{3}\frac{\vartheta_k}{q-u_k}\Bigg\}\end{aligned} \tag{3.30}$$

$$\mathcal{A}_{i12}=-\vartheta_\infty k\frac{q-u_i}{2P'(u_i)},\quad \mathcal{A}_{i21}=\frac{1}{\mathcal{A}_{i12}}\left(\frac{\vartheta_i^2}{4}-\mathcal{A}_{i11}^2\right),\quad \mathcal{A}_{i22}=-\mathcal{A}_{i11},$$

for $i=1,2,3$, where $P'(z)=\mathrm{d}P/\mathrm{d}z$. The Schlesinger equations (3.23) in these variables are:

$$\begin{cases}\frac{\partial q}{\partial u_i}=\frac{P(q)}{P'(u_i)}\left[2p+\frac{1}{q-u_i}-\sum_{k=1}^{3}\frac{\vartheta_k}{q-u_k}\right]\\ \frac{\partial p}{\partial u_i}=-\left\{P'(q)p^2+\left[2q+u_i-\sum_j u_j-\sum_{k=1}^{3}\vartheta_k(2q+u_k-\sum_j u_j)\right]p\right.\\ \qquad\left.+\frac14\left(\sum_{k=1}^{3}\vartheta_k-\vartheta_\infty\right)\left(\sum_{k=1}^{3}\vartheta_k+\vartheta_\infty-2\right)\right\}\frac{1}{P'(u_i)},\end{cases} \tag{3.31}$$

and

$$\frac{\partial\log(k)}{\partial u_i}=(\vartheta_\infty-1)\frac{q-u_i}{P'(u_i)}, \tag{3.32}$$

for $i=1,2,3$. The system of the *reduced Schlesinger equations* (3.31) is invariant under the transformations of the form

$$u_i\mapsto au_i+b,\quad q\mapsto aq+b,\quad p\mapsto\frac{p}{a},\qquad \forall a,b\in\mathbb{C},\ a\neq0. \tag{3.33}$$

Introducing the following new invariant variables:

$$x=\frac{u_2-u_1}{u_3-u_1},\quad y=\frac{q-u_1}{u_3-u_1}; \tag{3.34}$$

the system (3.31), expressed in the these new variables, gives the PVI equation for $y(x)$ with parameters

$$\alpha = \frac{(\vartheta_\infty - 1)^2}{2}, \quad \beta = -\frac{\vartheta_1^2}{2}, \quad \gamma = \frac{\vartheta_3^2}{2}, \quad \delta = \frac{1-\vartheta_2^2}{2}. \tag{3.35}$$

REMARK 3.7. Observe that permutations of the poles u_i and of the values ϑ_i, $i = 1, 2, 3, \infty$ induce transformations of (y, x) of the type $x \to 1 - x$ and $y \to 1 - y$, $x \to 1/x$ and $y \to 1/y$, $x \to 1/(1-x)$ and $y \to (y-x)/(1-x)$ and their compositions. These transformations are the *symmetries* of the Painlevè VI equation.

REMARK 3.8. It is clear from (3.35) that changes of the signs of the parameters ϑ_k, $k = 1, 2, 3$ and transformations on ϑ_∞ of type $\vartheta_\infty \to 2 - \vartheta_\infty$ give rise to the same PVI equation.

4. Laplace Transform

In this section we build a map between the following two systems of first order differential equations

(i) irregular systems of the form

$$\frac{\mathrm{d}}{\mathrm{d}\lambda} Y = \left(U + \frac{V - \mathbb{1}}{\lambda} \right) Y \tag{4.1}$$

where U is a diagonal matrix of pairwise distinct elements (u_1, u_2, u_3) and V is any constant matrix with diagonal part Θ of entries $(-\theta_1, -\theta_2, -\theta_3)$ and eigenvalues $\mu_1 = (-\sum \theta_k + \vartheta_\infty)/2$, $\mu_2 = (-\sum \theta_k - \vartheta_\infty)/2$, and $\mu_3 = 0$.

(ii) Fuchsian systems of the form (3.1)

$$\frac{\mathrm{d}}{\mathrm{d}\lambda} \Phi = \left(\sum_{k=1}^{3} \frac{\mathcal{A}_k}{\lambda - u_k} \right) \Phi,$$

u_1, u_2, u_3 being pairwise distinct complex numbers and $\mathcal{A}_j$ are 2×2 matrices independent on λ satisfying the conditions 3.2 and 3.3.

More precisely we give the following

DEFINITION 4.1. Two 3×3 systems of the form (4.1) are *equivalent up to diagonal conjugation* if they have the same matrix U and the same matrix V up to conjugation $V \to D_3 V D_3^{-1}$, D_3 being any 3×3 diagonal matrix. Analogously, two 2×2 Fuchsian systems of the form (3.1) are *equivalent up to diagonal conjugation* if they have the same matrices $\mathcal{A}_k$, $k = 1, 2, 3$, up to conjugation $\mathcal{A}_k \to D_2^{-1} \mathcal{A}_k D_2$, D_2 being any diagonal matrix.

We prove the following

THEOREM 4.2. *There is a one to one correspondence between classes of equivalence of* 3×3 *systems* (4.1)

$$\frac{\mathrm{d}}{\mathrm{d}\lambda} Y = \left(U + \frac{V - \mathbb{1}}{\lambda} \right) Y$$

where U *is a diagonal matrix of elements* (u_1, u_2, u_3) *and* V *is any constant matrix of diagonal part* Θ *of entries* $(-\theta_1, -\theta_2, -\theta_3)$ *and with eigenvalues* $\mu_1 = (-\sum \theta_k +$

$\vartheta_\infty)/2$, $\mu_2 = (-\sum\theta_k - \vartheta_\infty)/2$, $\mu_3 = 0$, and classes of equivalence of 2×2 systems of the form (3.1)

$$\frac{\mathrm{d}}{\mathrm{d}\lambda}\Phi = \sum_{k=1}^{3}\frac{\mathcal{A}_k}{\lambda - u_k}\Phi$$

where $\mathcal{A}_k$ satisfy (3.2) *and* (3.3). *This one-to-one correspondence is realized by means of the following formulae:*

$$\mathcal{A}_k = \frac{1}{2}\begin{pmatrix} a_k b_k & -a_k^2 \\ b_k^2 - \frac{\theta_k^2}{a_k^2} & -a_k b_k \end{pmatrix}, \tag{4.2}$$

$$V = -\frac{1}{2}\begin{pmatrix} 2\theta_1 & a_2b_1 - a_1b_2 + \theta_1\frac{a_2}{a_1} + \theta_2\frac{a_1}{a_2} & a_1b_3 + a_3b_1 - \theta_3\frac{a_1}{a_3} + \theta_1\frac{a_3}{a_1} \\ a_1b_2 - a_2b_1 + \theta_1\frac{a_2}{a_1} + \theta_2\frac{a_1}{a_2} & 2\theta_2 & a_2b_3 + a_3b_2 - \theta_3\frac{a_2}{a_3} + \theta_2\frac{a_3}{a_2} \\ a_3b_1 + a_1b_3 - \theta_1\frac{a_3}{a_1} + \theta_3\frac{a_1}{a_2} & a_2b_3 + a_3b_2 + \theta_3\frac{a_2}{a_3} - \theta_2\frac{a_3}{a_2} & 2\theta_3 \end{pmatrix} \tag{4.3}$$

where a_k and b_k, $k = 1, 2, 3$ are some constants such that

$$\sum_{k=1}^{3} a_k b_k = -\vartheta_\infty, \quad \sum_{k=1}^{3} a_k^2 = 0, \quad \sum_{k=1}^{3}\frac{\theta_k^2}{a_k^2} - b_k^2 = 0. \tag{4.4}$$

The formulae (4.2), (4.3) were first obtained in [**9**] in terms of parameter-dependent pairs of moment maps from a symplectic vector space to the dual spaces of two different loop-algebras. We present here another derivation via the Laplace transform. This allows us to give an expression of the matrix V of the system (4.1) in terms of the coordinates (p, q) defined in (3.29), as in the following

COROLLARY 4.3. *The matrix V of the system* (4.1) *can be expressed in terms of the coordinates (p, q) defined in* (3.29) *as*

$$\begin{aligned}
&V_{11} = -\theta_1, \quad V_{22} = -\theta_2, \quad V_{33} = -\theta_3,\\
&V_{12} = \frac{k}{l(u_1 - u_3)}\Bigg(-(q-u_2)(q-u_3)p + \frac{\sum_{k=1}^3\theta_k - \vartheta_\infty}{2}q \\
&\qquad\qquad + \frac{\theta_2 - \theta_1 - \theta_3 + \vartheta_\infty}{2}u_2 - \theta_2 u_3\Bigg),\\
&V_{13} = \frac{k}{(u_1 - u_2)}\Bigg(-(q-u_2)(q-u_3)p + \frac{\sum_{k=1}^3\theta_k - \vartheta_\infty}{2}q \\
&\qquad\qquad + \frac{\theta_3 - \theta_1 - \theta_2 + \vartheta_\infty}{2}u_3 - \theta_3 u_2\Bigg),\\
&V_{21} = \frac{l}{k(u_2 - u_3)}\Bigg(-(q-u_1)(q-u_3)p + \frac{\sum_{k=1}^3\theta_k - \vartheta_\infty}{2}q \\
&\qquad\qquad + \frac{\theta_1 - \theta_2 - \theta_3 + \vartheta_\infty}{2}u_1 - \theta_1 u_3\Bigg),\\
&V_{23} = \frac{l}{(u_2 - u_1)}\Bigg(-(q-u_1)(q-u_3)p + \frac{\sum_{k=1}^3\theta_k - \vartheta_\infty}{2}q \\
&\qquad\qquad + \frac{\theta_3 - \theta_1 - \theta_2 + \vartheta_\infty}{2}u_3 - \theta_3 u_1\Bigg),
\end{aligned} \tag{4.5}$$

$$V_{31}=\frac{1}{k(u_3-u_2)}\Bigg(-(q-u_1)(q-u_2)p+\frac{\sum_{k=1}^3\theta_k-\vartheta_\infty}{2}q \\ +\frac{\theta_1-\theta_2-\theta_3+\vartheta_\infty}{2}u_1-\theta_1u_2\Bigg),$$

$$V_{32}=\frac{1}{l(u_3-u_1)}\Bigg(-(q-u_1)(q-u_2)p+\frac{\sum_{k=1}^3\theta_k-\vartheta_\infty}{2}q \\ +\frac{\theta_2-\theta_1-\theta_3+\vartheta_\infty}{2}u_2-\theta_2u_1\Bigg)$$

where $k(\mathbf{u})$ and $l(\mathbf{u})$ satisfy

$$\frac{\partial}{\partial u_1}\log k=\frac{q-u_2}{u_1-u_2}p+\frac{\theta_3-\theta_1}{u_3-u_1}+\frac{\theta_2-1}{u_2-u_1},$$
$$\frac{\partial}{\partial u_2}\log k=\frac{u_3-u_1}{(u_2-u_1)(u_2-u_3)}(-p(q-u_2)+\theta_2-1),$$
$$\frac{\partial}{\partial u_3}\log k=-\frac{q-u_2}{u_3-u_2}p-\frac{\theta_3-\theta_1}{u_3-u_1}-\frac{\theta_2-1}{u_2-u_3},$$
$$\frac{\partial}{\partial u_1}\log l=\frac{u_2-u_3}{(u_1-u_2)(u_1-u_3)}(p(q-u_1)+1-\theta_1),$$
$$\frac{\partial}{\partial u_2}\log l=\frac{q-u_1}{u_2-u_1}p+\frac{\theta_3-\theta_2}{u_3-u_2}+\frac{\theta_1-1}{u_1-u_2},$$
$$\frac{\partial}{\partial u_3}\log l=-\frac{q-u_1}{u_3-u_1}p-\frac{\theta_3-\theta_2}{u_3-u_2}-\frac{\theta_1-1}{u_1-u_3}.$$

The matrix $V(p,q)$ given by (4.5) satisfies the isomonodromic deformations equations

$$\frac{\partial}{\partial u_i}V=[V_i,V]\ \textit{with}\ V_i=\mathrm{ad}_{E_i}\mathrm{ad}_U^{-1}(V),$$

and $E_{i_{kl}}=\delta_{ik}\delta_{il}$ if and only if (p,q) satisfy the reduced Schlesinger equations (3.32).

PROOF. It is a direct computation. □

By (3.28), (3.29) we obtain the formulae for $V_{ij}(y,y',x)$ given in the introduction.

REMARK 4.4. In the case $\theta_1=\theta_2=\theta_3=0$ we can perform a diagonal conjugation on $V(p,q)$ in order to make it antisymmetric with off-diagonal elements such that $V_{ij}^2=-\mathrm{Tr}(\mathcal{A}_i\mathcal{A}_j)$ as in [**4**].

PROOF OF THEOREM 4.2. The main tool is the Laplace transform (see [**1, 2, 5**]). Consider the following 3×3 linear differential system

$$\frac{\mathrm{d}}{\mathrm{d}\lambda}\chi=(U-\lambda\mathbb{1})^{-1}V\chi \tag{4.6}$$

where U is a diagonal matrix of elements (u_1,u_2,u_3) and V is a constant matrix as in Theorem 4.2. The system (4.6) is a 3×3 Fuchsian system with four regular singularities at u_1, u_2, u_3, ∞. In fact it can be written in the form

$$\frac{\mathrm{d}}{\mathrm{d}\lambda}\chi=\sum_{k=1}^{3}\frac{B_k}{\lambda-u_k}\chi \tag{4.7}$$

where

$$B_k = -E_k V, \quad \text{i.e. } (B_k)_{ij} = -V_{ij},\delta_{ik} \tag{4.8}$$

and $B_\infty \equiv -\sum_{k=1}^3 B_k = V$. Observe that B_k has all the elements equal to zero, but the ones on the kth row that coincide with the kth row of the matrix $-V$. For $\theta_k \neq 0$, the matrix B_k is diagonalisable with eigenvalues $0, 0, \theta_k$. For $\theta_k = 0$, the matrix B_k has Jordan normal form $\begin{pmatrix} 0&1&0\\ 0&0&0\\ 0&0&0 \end{pmatrix}$.

Let us fix the branch cuts π_k between ∞ and each pole u_k as in Figure 2. Fix a fundamental solution χ_∞ of the system (4.7) as in Proposition 3.1 (we omit the analogous statement here)

$$\chi_\infty(\lambda) = \left(\mathbb{1} + \mathcal{O}\left(\frac{1}{\lambda}\right)\right)\lambda^{-\mu}\lambda^{-\mathcal{R}_\infty}.$$

Given a loop γ in the λ-plane avoiding the branch cuts, the inverse Laplace transform

$$Y(z) = \int_\gamma \chi_\infty(\lambda) e^{z\lambda} \mathrm{d}\lambda \tag{4.9}$$

converges in a nonempty sector $\mathcal{S}_\gamma$ as $z \to \infty$. This is due to the fact that $\chi_\infty(\lambda)$ grows at infinity not faster that some power of $|\lambda|$. It is easy to prove that by changing the loop γ one can obtain enough sectors $\mathcal{S}_\gamma$ to cover a disc around ∞. The inverse Laplace transform (4.9) maps solutions to (4.6) to solutions of the linear system (4.1) (this can be proved by direct substitution). We are interested in the relation between the systems (4.1) and (4.6), or its two-dimensional reduction discussed below.

Assume that V is diagonalisable, that is choose the parameter $\vartheta_\infty \neq 0$, $\pm\sum_{k=1}^3 \theta_k$.[4] Let $\mathcal{G}_0$ be the diagonalising matrix of V, i.e. $\mathcal{G}_0^{-1} V \mathcal{G}_0 = \mu$, and $\chi = \mathcal{G}_0\tilde{\chi}$. Then $\tilde{\chi}$ is such that the new Fuchsian system is

$$\frac{\mathrm{d}}{\mathrm{d}\lambda}\tilde{\chi} = \widetilde{B}(\lambda)\tilde{\chi} = \sum_{k=1}^3 \frac{\widetilde{B}_k}{\lambda - u_k}\tilde{\chi} \tag{4.10}$$

where $\widetilde{B}_k = \mathcal{G}_0^{-1} B_k \mathcal{G}_0 = -\mathcal{G}_0^{-1} E_k \mathcal{G}_0 \mu$. The matrices $\widetilde{B}_k$ have the form

$$\begin{aligned}
\widetilde{B}_1 &= \frac{1}{\det(\mathcal{G}_0)} \begin{pmatrix} (G_{23}G_{32}-G_{22}G_{33})G_{11}\mu_1 & (G_{23}G_{32}-G_{22}G_{33})G_{12}\mu_2 & 0 \\ (G_{21}G_{33}-G_{23}G_{31})G_{11}\mu_1 & (G_{21}G_{33}-G_{23}G_{31})G_{12}\mu_2 & 0 \\ (G_{22}G_{31}-G_{21}G_{32})G_{11}\mu_1 & (G_{22}G_{31}-G_{21}G_{32})G_{12}\mu_2 & 0 \end{pmatrix},\\
\widetilde{B}_2 &= \frac{1}{\det(G)} \begin{pmatrix} (G_{12}G_{33}-G_{13}G_{32})G_{21}\mu_1 & (G_{12}G_{33}-G_{13}G_{32})G_{22}\mu_2 & 0 \\ (G_{13}G_{31}-G_{11}G_{33})G_{21}\mu_1 & (G_{13}G_{31}-G_{11}G_{33})G_{22}\mu_2 & 0 \\ (G_{11}G_{32}-G_{12}G_{31})G_{21}\mu_1 & (G_{11}G_{32}-G_{12}G_{31})G_{22}\mu_2 & 0 \end{pmatrix},\\
\widetilde{B}_3 &= \frac{1}{\det(G)} \begin{pmatrix} (G_{13}G_{22}-G_{12}G_{23})G_{31}\mu_1 & (G_{13}G_{22}-G_{12}G_{23})G_{32}\mu_2 & 0 \\ (G_{11}G_{23}-G_{13}G_{21})G_{31}\mu_1 & (G_{11}G_{23}-G_{13}G_{21})G_{32}\mu_2 & 0 \\ (G_{12}G_{21}-G_{11}G_{22})G_{31}\mu_1 & (G_{12}G_{21}-G_{11}G_{22})G_{32}\mu_2 & 0 \end{pmatrix}
\end{aligned} \tag{4.11}$$

where G_{ij} are the entries of $\mathcal{G}_0$. Since the third column of each $\widetilde{B}_k$ is zero and the system for $\tilde{\chi}$ reduces to a system for the first two components of χ, $\begin{pmatrix} \tilde{\chi}_1 \\ \tilde{\chi}_2 \end{pmatrix} = \Psi$

$$\frac{\mathrm{d}}{\mathrm{d}\lambda}\Psi = \sum_{k=1}^3 \frac{A_k}{\lambda - u_k}\Psi \tag{4.12}$$

[4]We can impose this without loss of generality because we are interested in the reductions to the Painlevé sixth equation, for which the cases $\vartheta_\infty = 0, \pm\sum_{k=1}^3 \theta_k$ and $\vartheta_\infty = 2, 2 \mp \sum_{k=1}^3 \theta_k$ are equivalent.

where

$$A_k = \begin{pmatrix} \widetilde{B}_{k_{11}} & \widetilde{B}_{k_{12}} \\ \widetilde{B}_{k_{21}} & \widetilde{B}_{k_{22}} \end{pmatrix}, \tag{4.13}$$

and a quadrature for the third component

$$\frac{\mathrm{d}}{\mathrm{d}\lambda}\chi_3 = \sum_{k=1}^{3}\left(\frac{\widetilde{B}_k}{\lambda - u_k}\right)_{31}\chi_1 + \sum_{k=1}^{3}\left(\frac{\widetilde{B}_k}{\lambda - u_k}\right)_{32}\chi_2. \tag{4.14}$$

The eigenvalues of the matrices A_k are $(\theta_k, 0)$ and $A_\infty \equiv -\sum_{k=1}^{3} A_k = \begin{pmatrix} \mu_1 & \\ & \mu_2 \end{pmatrix}$, with $\mu_{1,2}$ given as in Theorem 4.2. □

LEMMA 4.5. *There is a one to one correspondence between classes of equivalence of* 3×3 *systems* (4.1)

$$\frac{\mathrm{d}}{\mathrm{d}\lambda}Y = \left(U + \frac{V - \mathbb{1}}{\lambda}\right)Y \tag{4.15}$$

where U *is a diagonal matrix of elements* (u_1, u_2, u_3) *and* V *is a constant matrix as in Theorem* (4.2), *and equivalence classes of* 2×2 *systems* (4.12)

$$\frac{\mathrm{d}}{\mathrm{d}\lambda}\Psi = \sum_{k=1}^{3}\frac{A_k}{\lambda - u_k}\Psi \tag{4.16}$$

where A_k *have eigenvalues* $\theta_k, 0$, $k = 1,2,3$, *and are such that* $-\sum_{k=1}^{3} A_k = \begin{pmatrix} \mu_1 & \\ & \mu_2 \end{pmatrix}$, *with* $\mu_{1,2}$ *given as in Theorem* (4.2).

PROOF. Let us fix a system (4.1) with a given diagonalisable V as in Theorem 4.2. The matrices A_k are uniquely determined by the diagonalising matrix $\mathcal{G}_0$, that is uniquely determined by V up to $\mathcal{G}_0 \to \mathcal{G}_0 D_3$ where D_3 is any 3×3 diagonal matrix of entries d_1, d_2, d_3. So, for a given V, the matrices A_k are determined up to diagonal conjugation $A_k \to D_2^{-1} A_k D_2$, D_2 being a diagonal matrix of entries $d_2/d_1, 1$.

Vice-versa, given a Fuchsian system of the form (4.12) with any matrices A_k, $k = 1,2,3$, having eigenvalues $\theta_k, 0$ and being such that $-\sum A_k = \begin{pmatrix} \mu_1 & \\ & \mu_2 \end{pmatrix}$, with $\mu_{1,2}$ given as in Theorem 4.2, we can determine a matrix $\mathcal{G}_0$ such that (4.11) and (4.13) hold. It is easy to check that $\mathcal{G}_0$ is determined up to $\mathcal{G}_0 \to D_3 \mathcal{G}_0$ where D_3 is a diagonal matrix of entries d_1, d_2, 1. Correspondingly V is uniquely determined up to $V \to D_3 V D_3^{-1}$. □

To conclude the proof of Theorem 4.2, it is enough to observe that the systems (4.12) and (3.1) are related trough the following Schlesinger transformation

$$\Psi = (z - u_1)^{\theta_1/2}(z - u_2)^{\theta_2/2}(z - u_3)^{\theta_3/2}\Phi. \tag{4.17}$$

The formulae (4.2), (4.3) can be obtained by comparison between (4.11) and the formula of V in terms of the diagonalising matrix $\mathcal{G}_0$:

$$\begin{aligned}
V_{11} =& G_{11}(G_{22}G_{33} - G_{23}G_{32})\mu_1 + G_{12}(G_{23}G_{31} - G_{21}G_{33})\mu_2,\\
V_{12} =& - G_{12}(G_{13}G_{31} - G_{11}G_{33})\mu_2 + G_{11}(G_{13}G_{32} - G_{12}G_{33})\mu_1,\\
V_{13} =& G_{12}(G_{13}G_{21} - G_{11}G_{23})\mu_2 + G_{11}(G_{12}G_{23} - G_{13}G_{22})\mu_1,\\
V_{21} =& G_{22}(G_{23}G_{31} - G_{21}G_{33})\mu_2 + G_{21}(G_{22}G_{33} - G_{23}G_{32})\mu_1,\\
V_{22} =& - G_{22}(G_{13}G_{31} - G_{11}G_{33})\mu_2 + G_{21}(G_{13}G_{32} - G_{12}G_{33})\mu_1,
\end{aligned} \tag{4.18}$$

$$\begin{aligned}
V_{23} =& G_{22}(G_{13}G_{21} - G_{11}G_{23})\mu_2 + G_{21}(G_{12}G_{23} - G_{13}G_{22})\mu_1,\\
V_{31} =& G_{32}(G_{23}G_{31} - G_{21}G_{33})\mu_2 + G_{31}(G_{22}G_{33} - G_{23}G_{32})\mu_1,\\
V_{32} =& - G_{32}(G_{13}G_{31} - G_{11}G_{33})\mu_2 + G_{31}(G_{13}G_{32} - G_{12}G_{33})\mu_1,\\
V_{33} =& G_{32}(G_{13}G_{21} - G_{11}G_{23})\mu_2 + G_{31}(G_{12}G_{23} - G_{13}G_{22})\mu_1
\end{aligned}$$

where G_{ij} are the entries of $\mathcal{G}_0$.

Observe that in the above proof the order of the eigenvalues of V is fixed. If we perform a Schlesinger transformation of the form

$$Y \to z^{-\mu_i} Y,$$

for some $i = 1, 2$, we obtain a new system

$$\frac{\mathrm{d}}{\mathrm{d}z}\widetilde{Y} = \left(U + \frac{\widetilde{V} - \mathbb{1}}{z}\right)\widetilde{Y},$$

with eigenvalues $\mu_1 - \mu_i, \mu_2 - \mu_i, -\mu_i$ and diagonal elements $\theta_1 + \mu_i, \theta_2 + \mu_i, \theta_3 + \mu_i$. This gives rise to a symmetry of the Painlevé VI equation listed by Okamoto (see [**17**]). For example for $i = 2$ we obtain the symmetry $w_1\ w_2\ w_1\ w_4\ w_1\ w_2\ w_1\ w_0\ w_1\ w_2\ w_3\ w_2\ w_1$ in Okamoto notation.

References

1. W. Balser, W. B. Jurkat, and D. A. Lutz, *Birkhoff invariants and stokes multipliers for meromorphic linear differential equations*, J. Math. Anal. Appl. **71** (1979), 48–94.
2. ———, *On the reduction of connection problems for differential equations with an irregular singular point to ones with only regular singularities*, SIAM J. Math. Anal. **12** (1981), 691–721.
3. S. Chakravarty, *A class of integrable conformally self-dual metrics*, Classical Quantum Gravity **11** (1994), no. 1, L1–L6.
4. B. Dubrovin, *Geometry of 2d topological field theories, integrable systems and quantum groups*, Springer Lecture Notes in Math., vol. 1620, Springer, Berlin, 1996.
5. ———, *Painlevé transcendents in two-dimensional topological field theory*, The Painlevé Property, CRM Ser. Math. Phys., Springer, New York, 1999, pp. 287–412.
6. B. Dubrovin and M. Mazzocco, *Monodromy of certain Painlevé–VI transcendents and reflection groups*, Invent. Math. **141** (2000), no. 1, 55–147.
7. P. R. Gordoa, N. Joshi, and A. Pickering, *Mappings preserving locations of movable poles: A new extension of the truncation method to ordinary differential equations*, Nonlinearity **12** (1999), no. 4, 955–968.
8. ———, *Mappings preserving locations of movable poles.* II. *The third and fifth Painlevé equations*, Nonlinearity **14** (2001), no. 3, 567–582.
9. J. Harnad, *Dual isomonodromic deformations and moment maps to loop algebras*, Comm. Math. Phys. **166** (1994), 337–365.
10. N. Hitchin, *Algebraic solutions of the Painlevé* VI *equation*, talk at Edinburgh, work in progress (1998).
11. M. Jimbo, T. Miwa, and K. Ueno, *Holonomic quantum fields.* I, Publ. Res. Inst. Math. Sci. **14** (1978), no. 1, 223–267.
12. ———, *Holonomic quantum fields.* II. *The Riemann–Hilbert problem*, Publ. Res. Inst. Math. Sci. **15** (1979), no. 1, 201–278.
13. N. A. Lukashevich, *On the theory of the third Painlevé equation*, Differ. Uravn. **3** (1967), no. 11, 1913–1923.
14. B. Malgrange, *Differential equations with polynomial coefficients*, Progress in Mathematics, vol. 96, Birkhäuser Boston, Inc., Boston, MA, 1991.
15. M. Mazzocco, *Picard and Chazy solutions to* PVI *equation*, Math. Ann. **321** (2001), no. 1, 131–169.
16. T. Miwa, *Painlevé property of monodromy preserving equations and the analyticity of τ-functions*, Publ. Res. Inst. Math. Sci. **17** (1981), 703–721.

17. K. Okamoto, *Studies on the Painlevé equations* I. *Sixth Painlevé equation* P_{VI}, Ann. Mat. Pura Appl. (4) **146** (1987), 337–381.
18. ———, *The Painlevé equations and the Dynkin diagrams*, Painlevé Transcendents (Sainte-Adèle, 1990), NATO Adv. Sci. Inst. Ser. B Phys., vol. 278, Plenum, New York, 1992, pp. 299–313.
19. P. Painlevé, *Sur les équations différentielles du second ordre et d'ordre supérieur, dont l'intégrable générale est uniforme*, Acta Math. **25** (1902), 1–86.
20. R. Schafke, *On the reduction of the Poincaré rank of singular systems of ordinary differential equations*, J. Reine Angew. Math. **365** (1986), 80–96.
21. J. Segert, *Frobenius manifolds from Yang–Mills instantons*, Math. Res. Lett. **5** (1998), no. 3, 327–344.
22. Y. Sibuya, *Linear differential equations in the complex domain: Problems of analytic continuation*, Mathematical Monographs, vol. 82, American Mathematical Society, Providence, RI, 1990.
23. K. P. Tod, *Self-dual Einstein metrics from the Painlevé* VI *equation*, Phys. Lett. A **190** (1994), 221–224.
24. W. Wasow, *Asymptotic expansions for ordinary differential equations*, John Wiley & Sons, Inc., New York, 1965.

MATHEMATICAL INSTITUTE, 24-29 ST GILES, OXFORD OX1 3LB, ENGLAND.
E-mail address: `mazzocco@maths.ox.ac.uk`

Centre de Recherches Mathématiques
CRM Proceedings and Lecture Notes
Volume **32**, 2002

Euler Characteristics of Theta Divisors of Jacobians for Spectral Curves

Atsushi Nakayashiki and Fedor A. Smirnov

ABSTRACT. We show how to calculate the Euler characteristic of an affine Jacobi variety of a spectral curve from its defining equations.

1. Introduction

Consider an algebraic curve X, its Jacobian $J(X)$ and the theta divisor $\Theta \subset J(X)$. The noncompact variety $J(X) - \Theta$ is known to be an affine algebraic variety, which we call the affine Jacobian variety of X.

In the paper [**6**] we consider, following [**4**], the algebraic model of an affine hyperelliptic Jacobian variety. Under certain strong assumptions we calculated the topological Euler characteristic and dimensions of singular cohomology groups of it. More precisely we made a conjecture on the explicit forms of cohomologies as the irreducible representations of the symplectic group. Later one of the authors [**5**] was able to prove these conjectures. In this proof certain very particular properties of hyperelliptic curves were used.

The algebraic model is available for the affine Jacobians of more general algebraic curves of the form:

$$w^N + t_1(z)w^{N-1} + \cdots + t_N(z) = 0 \tag{1.1}$$

where $\deg(t_j) \leq nj - 1$ for some n and the leading coefficient of t_N does not vanish (the corresponding construction is explained in detail in [**1,9**]). However, the cohomologies of the affine Jacobians of these curves seem much more complicated than in the hyperelliptic case. Up to now we were not able to conjecture their explicit form. Thus even a partial information about those cohomologies is important.

In this paper we shall show that under a certain assumption (Assumption 1 in Section 3) the topological Euler characteristic $\chi\big(J(X) - \Theta\big)$ of the affine Jacobian of a curve X from (1.1) can be calculated. It then gives the Euler characteristic of

---

2000 *Mathematics Subject Classification.* Primary: 14H40; Secondary: 14H70, 81-XX..

The second author is member of the CNRS.
One of the authors (FS) was partly supported by INTAS grant INTAS OPEN 97-01312.
We would like to thank Fu Baohua for comments on the appendix.
This is the final form of the paper.

the theta divisor Θ by the relation:

$$\chi(\Theta) = -\chi\big(J(X) - \Theta\big).$$

The paper is organized as follows. In the second section we present a rather abstract scheme into which the algebraic models of affine Jacobians can be put. We consider a certain family of affine algebraic sets parametrized by points of a complex affine space. In the third section we consider the special case when the parameter takes the origin. It is the only case when the affine ring becomes graded. In this case the Euler characteristic can be easily determined in terms of the character of the affine ring. We define a double complex which permits to identify the dimensions of cohomologies in question with the dimensions of the spaces of cochains in the minimal free resolution of the affine ring. In the fourth section we show that the Euler characteristic of the variety at a generic point in the parameter space coincides with the Euler characteristic of the variety at the origin by considering certain deformation of the previous construction. This is the main result of this paper. We apply this result to the family of the affine Jacobians of spectral curves (1.1) studied in [**9**] and get the formula for $\chi(\Theta)$. In the last section we shall show that our formula for $\chi(\Theta)$ produces correct values in the case of genus 3 and 4. The appendix contains a proof of the proposition given in Section 2.

2. General Setting

Consider the affine space $\mathbb{C}^m$ with coordinates $(a_1, \dots, a_m)$. Consider $m-g$ polynomials $f_j(a_1, \dots, a_m)$ and the affine algebraic set $\mathcal{J}_{f^0}$ defined by the equations

$$f_1(a_1, \dots, a_m) = f_1^0,$$
$$\vdots$$
$$f_{n-g}(a_1, \dots, a_m) = f_{m-g}^0$$

where f_j^0 are complex numbers. We assume that $\mathcal{J}_0 = \{p \in \mathbb{C}^m \mid f_j(p) = 0 \ \forall j\}$ is g-dimensional, that is, the maximum of the dimensions of irreducible components is g. In the case of the family of the affine Jacobians of spectral curves (1.1) this assumption is verified by using the explicit parametrization of $\mathcal{J}_0$ [**1, 9**]. We assign a positive degree $\deg(a_j)$ to each a_j and assume that the polynomial f_j is homogeneous of degree $\deg(f_j)$ for all j. We shall consider several rings:

$$A = \mathbb{C}[a_1, \cdots, a_m], \qquad F = \mathbb{C}[F_1, \cdots, F_{m-g}],$$
$$A_0 = A/\textstyle\sum(f_j A), \qquad A_{f^0} = A/\textstyle\sum((f_j - f_j^0)A).$$

The rings F and A are polynomial rings of m and $m-g$ variables respectively. The ring F acts on A by the multiplication of $F_j = f_j$. We define $\deg(F_j) = \deg(f_j)$. Then A, F and A_0 become graded rings. But this is not the case for A_{f^0} if some f_j^0 is not zero. The character of a graded vector space is defined as the generating function of the dimensions of homogeneous subspaces. It is obvious that the characters of A and F are given by

$$\mathrm{ch}(A) = \prod_{j=1}^{m} \frac{1}{1 - q^{\deg(a_j)}}, \qquad \mathrm{ch}(F) = \prod_{j=1}^{m-g} \frac{1}{1 - q^{\deg(f_j)}}.$$

The determination of the character of A_0 is more involved. But the result is simple.

PROPOSITION 1. *The character of A_0 is given by*

$$\mathrm{ch}(A_0) = \frac{\mathrm{ch}(A)}{\mathrm{ch}(F)}. \tag{2.1}$$

The proof of this proposition is given in Appendix A. For the case of the family of affine hyperelliptic Jacobians, this proposition were proved in [**6**] by determining the $\mathbb{C}$-basis of A_0 explicitly. In a similar way to Proposition 2 in [**6**] we have the following corollary of this proposition.

COROLLARY 1. *As a graded F module, A is a free module*: $A \simeq F \otimes_{\mathbb{C}} A_0$.

An important feature, which is special for the case related with the family of affine Jacobians, is that on A one can define the action of g commuting vector fields $D_1, \ldots, D_g$. Thus we assume that there exist algebraically independent g commuting vector fields $D_1, \ldots, D_g$ acting on A such that

$$D_j f_k = 0 \quad \forall j, k. \tag{2.2}$$

Moreover, we assume that, to every D_j a positive degree is prescribed such that

$$\deg(D_j x) = \deg(D_j) + \deg(x)$$

for every homogeneous $x \in A$. Then the ring $\mathcal{D} = \mathbb{C}[D_1, \ldots, D_g]$ becomes a graded ring and its character is given by

$$\mathrm{ch}(\mathcal{D}) = \prod_{j=1}^{g} \frac{1}{1 - q^{\deg(D_j)}}.$$

By the condition (2.2) the action of $\mathcal{D}$ on A descends to A_0 and A_{f^0}.

Let us introduce differentials dz_j dual to the vector fields D_j. We set $V = \bigoplus_{j=1}^{g} \mathbb{C} dz_j$. We define the spaces of cochains by

$$C^k = A \otimes \wedge^k V, \qquad C_0^k = A_0 \otimes \wedge^k V, \qquad C_{f^0}^k = A_{f^0} \otimes \wedge^k V.$$

The differential $d = \sum D_j \otimes dz_j$ defines complexes $(C^{\cdot}, d)$, $(C_0^{\cdot}, d)$, $(C_{f^0}^{\cdot}, d)$ which we shall study. We set $\deg(dz_j) = -\deg(D_j)$ in order that $\deg(d) = 0$. Then C^k and C_0^k become graded vector spaces.

3. Cohomologies at the Origin

Consider the complex:

$$0 \to C_0^0 \xrightarrow{d} C_0^1 \xrightarrow{d} \cdots \xrightarrow{d} C_0^{g-1} \xrightarrow{d} C_0^g \xrightarrow{d} 0. \tag{3.1}$$

This complex is graded. One easily finds its q-Euler characteristic [**6**]:

$$\chi_q(C_0^{\cdot}) = \sum_{k=0}^{g} (-1)^k \,\mathrm{ch}(C_0^k) = (-1)^g q^{-\sum \deg(D_j)} \frac{\mathrm{ch}(A_0)}{\mathrm{ch}(\mathcal{D})}.$$

Taking the limit $q \to 1$ we find the Euler characteristic as a number:

$$\chi(C_0^{\cdot}) = (-1)^g \frac{\prod_{j=1}^{m-g} \deg(f_j) \prod_{j=1}^{g} \deg(D_j)}{\prod_{j=1}^{m} \deg(a_j)}. \tag{3.2}$$

The fact that the Euler characteristic is finite does not mean that the cohomologies are finite dimensional. However we shall adopt the following assumption:

ASSUMPTION 1. $\dim H^g(C_0^{\cdot}) < \infty$.

Unfortunately, we do not know how to prove this assumption even in the case of the family of affine Jacobians of (1.1).

The Assumption 1 together with simple grading arguments guarantee that A_0 is a finitely generated $\mathcal{D}$-module. Then it has the minimal $\mathcal{D}$ free resolution (cf. [**2**, §19]) of the form

$$0\longrightarrow \mathcal{D}^{b_0}\xrightarrow{\varphi_0^0(D)}\cdots\longrightarrow \mathcal{D}^{b_{g-2}}\xrightarrow{\varphi_0^{g-2}(D)}\mathcal{D}^{b_{g-1}}\xrightarrow{\varphi_0^{g-1}(D)}\mathcal{D}^{b_g}\longrightarrow A_0\longrightarrow 0. \tag{3.3}$$

Here minimality condition is $\varphi_0^k(0)=0$ for all k. Consider another complex, the Koszul cpmplex of D_1, ..., D_g:

$$0\to\mathcal{D}\xrightarrow{d}\cdots\xrightarrow{d}\mathcal{D}\otimes\wedge^{g-2}V\xrightarrow{d}\mathcal{D}\otimes\wedge^{g-1}V\xrightarrow{d}\mathcal{D}\otimes\wedge^{g}V\to\mathbb{C}\to 0 \tag{3.4}$$

where d is given by the same formula as before $d=\sum D_j\otimes dz_j$. This complex is exact. For any nonnegative integer b, tensoring $\mathcal{D}^b$ (over $\mathcal{D}$) to (3.4), we get the exact sequence

$$0\to\mathcal{D}^b\xrightarrow{d}\cdots\xrightarrow{d}\mathcal{D}^b\otimes\wedge^{g-2}V\xrightarrow{d}\mathcal{D}^b\otimes\wedge^{g-1}V\xrightarrow{d}\mathcal{D}^b\otimes\wedge^{g}V\to\mathbb{C}^b\to 0.$$

Now we can consider the double complex

$$\begin{array}{ccccccccccc}
 & 0 & & 0 & & 0 & & 0 & & 0 & \\
 & \uparrow & & \uparrow & & \uparrow & & \uparrow & & \uparrow & \\
0\to & A_0 & \xrightarrow{d} & C_0^1 & \xrightarrow{d}\cdots\xrightarrow{d} & C_0^{g-1} & \xrightarrow{d} & C_0^g & \longrightarrow & 0 & \to 0\\
 & \uparrow & & \uparrow & & \uparrow & & \uparrow & & \uparrow & \\
0\to & \mathcal{D}^{b_g} & \xrightarrow{d} & \mathcal{D}^{b_g}\otimes V & \xrightarrow{d}\cdots\xrightarrow{d} & \mathcal{D}^{b_g}\otimes\wedge^{g-1}V & \xrightarrow{d} & \mathcal{D}^{b_g}\otimes\wedge^{g}V & \xrightarrow{d} & \mathbb{C}^{b_g} & \to 0\\
 & \uparrow{\scriptstyle \varphi_0^{g-1}(D)\otimes 1} & & \uparrow{\scriptstyle \varphi_0^{g-1}(D)\otimes 1} & & \uparrow{\scriptstyle \varphi_0^{g-1}(D)\otimes 1} & & \uparrow{\scriptstyle \varphi_0^{g-1}(D)\otimes 1} & & \uparrow{\scriptstyle \varphi_0^{g-1}(0)\otimes 1} & \\
0\to & \mathcal{D}^{b_{g-1}} & \xrightarrow{d} & \mathcal{D}^{b_{g-1}}\otimes V & \xrightarrow{d}\cdots\xrightarrow{d} & \mathcal{D}^{b_{g-1}}\otimes\wedge^{g-1}V & \xrightarrow{d} & \mathcal{D}^{b_{g-1}}\otimes\wedge^{g}V & \xrightarrow{d} & \mathbb{C}^{b_{g-1}} & \to 0\\
 & \uparrow & & \uparrow & & \uparrow & & \uparrow & & \uparrow & \\
 & \vdots & & \vdots & & \vdots & & \vdots & & \vdots & \\
 & \uparrow{\scriptstyle \varphi_0^{0}(D)\otimes 1} & & \uparrow{\scriptstyle \varphi_0^{0}(D)\otimes 1} & & \uparrow{\scriptstyle \varphi_0^{0}(D)\otimes 1} & & \uparrow{\scriptstyle \varphi_0^{0}(D)\otimes 1} & & \uparrow{\scriptstyle \varphi_0^{0}(0)\otimes 1} & \\
0\to & \mathcal{D}^{b_0} & \xrightarrow{d} & \mathcal{D}^{b_0}\otimes V & \xrightarrow{d}\cdots\xrightarrow{d} & \mathcal{D}^{b_0}\otimes\wedge^{g-1}V & \xrightarrow{d} & \mathcal{D}^{b_0}\otimes\wedge^{g}V & \xrightarrow{d} & \mathbb{C}^{b_0} & \to 0\\
 & \uparrow & & \uparrow & & \uparrow & & \uparrow & & \uparrow & \\
 & 0 & & 0 & & 0 & & 0 & & 0. &
\end{array}$$

In this double complex each column except the last one is exact, each row except the first one is exact. The minimality condition means that maps of the last column $\mathbb{C}^{b_k}\to\mathbb{C}^{b_{k+1}}$, $k\le g-1$ are all zero maps. Using this diagram one can easily prove that

$$H^k(C_0^{\cdot})\simeq\mathbb{C}^{b_k}.$$

4. Cohomologies at a Generic Point

We set $\mathcal{D}_F=F\otimes_{\mathbb{C}}\mathcal{D}$. Under Assumption 1, using Proposition 1, we can prove that A is a finitely generated $\mathcal{D}_F$ module in a similar manner to Proposition 4 in [**6**]. By the uniqueness of the minimal resolution (cf. [**2**, the graded version of

Theorem 20.2]), (3.3) implies the existence of the minimal $\mathcal{D}_F$-free resolution of A of the form

$$0 \longrightarrow \mathcal{D}_F^{b_0} \xrightarrow{\varphi^0(D)} \cdots \longrightarrow \mathcal{D}_F^{b_{g-2}} \xrightarrow{\varphi^{g-2}(D)} \mathcal{D}_F^{b_{g-1}} \xrightarrow{\varphi^{g-1}(D)} \mathcal{D}_F^{b_g} \longrightarrow A \longrightarrow 0,$$

such that $\varphi^k(D)|_{f_j=0} = \varphi_0^k(D)$. The minimality condition is that $\varphi^k(D)|_{f_j=0,D_j=0} = 0$. Again consider the double complex:

$$\begin{array}{ccccccccccccc}
 & & 0 & & 0 & & & & 0 & & 0 & & 0 \\
 & & \uparrow & & \uparrow & & & & \uparrow & & \uparrow & & \uparrow \\
0 & \to & A & \xrightarrow{d} & C^1 & \xrightarrow{d} & \cdots & \xrightarrow{d} & C^{g-1} & \xrightarrow{d} & C^g & \to & 0 \to 0 \\
 & & \uparrow & & \uparrow & & & & \uparrow & & \uparrow & & \uparrow \\
0 & \to & \mathcal{D}_F^{b_g} & \xrightarrow{d} & \mathcal{D}_F^{b_g}\otimes V & \xrightarrow{d} & \cdots & \xrightarrow{d} & \mathcal{D}_F^{b_g}\otimes \wedge^{g-1}V & \xrightarrow{d} & \mathcal{D}_F^{b_g}\otimes \wedge^{g}V & \to & F^{b_g} \to 0 \\
 & & \uparrow{\scriptstyle \varphi^{g-1}(D)\otimes 1} & & \uparrow{\scriptstyle \varphi^{g-1}(D)\otimes 1} & & & & \uparrow{\scriptstyle \varphi^{g-1}(D)\otimes 1} & & \uparrow{\scriptstyle \varphi^{g-1}(D)\otimes 1} & & \uparrow{\scriptstyle \varphi^{g-1}(0)\otimes 1} \\
0 & \to & \mathcal{D}_F^{b_{g-1}} & \xrightarrow{d} & \mathcal{D}_F^{b_{g-1}}\otimes V & \xrightarrow{d} & \cdots & \xrightarrow{d} & \mathcal{D}_F^{b_{g-1}}\otimes \wedge^{g-1}V & \xrightarrow{d} & \mathcal{D}_F^{b_{g-1}}\otimes \wedge^{g}V & \to & F^{b_{g-1}} \to 0 \\
 & & \uparrow & & \uparrow & & & & \uparrow & & \uparrow & & \uparrow \\
 & & \vdots & & \vdots & & & & \vdots & & \vdots & & \vdots \\
 & & \uparrow{\scriptstyle \varphi^{0}(D)\otimes 1} & & \uparrow{\scriptstyle \varphi^{0}(D)\otimes 1} & & & & \uparrow{\scriptstyle \varphi^{0}(D)\otimes 1} & & \uparrow{\scriptstyle \varphi^{0}(D)\otimes 1} & & \uparrow{\scriptstyle \varphi^{0}(0)\otimes 1} \\
0 & \to & \mathcal{D}_F^{b_0} & \xrightarrow{d} & \mathcal{D}_F^{b_0}\otimes V & \xrightarrow{d} & \cdots & \xrightarrow{d} & \mathcal{D}_F^{b_0}\otimes \wedge^{g-1}V & \xrightarrow{d} & \mathcal{D}_F^{b_0}\otimes \wedge^{g}V & \to & F^{b_0} \to 0 \\
 & & \uparrow & & \uparrow & & & & \uparrow & & \uparrow & & \uparrow \\
 & & 0 & & 0 & & & & 0 & & 0 & & 0.
\end{array}$$

In the diagram each column except the last one and each row except the first one are exact. Since every map in the diagram is F-linear and every object in the diagram is an F-free module, by tensoring $\mathbb{C} = F/\sum(f_j - f_j^0)F$ over F, we get the double complex:

$$\begin{array}{ccccccccccccc}
 & & 0 & & 0 & & & & 0 & & 0 & & 0 \\
 & & \uparrow & & \uparrow & & & & \uparrow & & \uparrow & & \uparrow \\
0 & \to & A_{f^0} & \xrightarrow{d} & C^1_{f^0} & \xrightarrow{d} & \cdots & \xrightarrow{d} & C^{g-1}_{f^0} & \xrightarrow{d} & C^g_{f^0} & \to & 0 \to 0 \\
 & & \uparrow & & \uparrow & & & & \uparrow & & \uparrow & & \uparrow \\
0 & \to & \mathcal{D}^{b_g} & \xrightarrow{d} & \mathcal{D}^{b_g}\otimes V & \xrightarrow{d} & \cdots & \xrightarrow{d} & \mathcal{D}^{b_g}\otimes \wedge^{g-1}V & \xrightarrow{d} & \mathcal{D}^{b_g}\otimes \wedge^{g}V & \to & \mathbb{C}^{b_g} \to 0 \\
 & & \uparrow{\scriptstyle \varphi_{f^0}^{g-1}(D)\otimes 1} & & \uparrow{\scriptstyle \varphi_{f^0}^{g-1}(D)\otimes 1} & & & & \uparrow{\scriptstyle \varphi_{f^0}^{g-1}(D)\otimes 1} & & \uparrow{\scriptstyle \varphi_{f^0}^{g-1}(D)\otimes 1} & & \uparrow{\scriptstyle \varphi_{f^0}^{g-1}(0)\otimes 1} \\
0 & \to & \mathcal{D}^{b_{g-1}} & \xrightarrow{d} & \mathcal{D}^{b_{g-1}}\otimes V & \xrightarrow{d} & \cdots & \xrightarrow{d} & \mathcal{D}^{b_{g-1}}\otimes \wedge^{g-1}V & \xrightarrow{d} & \mathcal{D}^{b_{g-1}}\otimes \wedge^{g}V & \to & \mathbb{C}^{b_{g-1}} \to 0 \\
 & & \uparrow & & \uparrow & & & & \uparrow & & \uparrow & & \uparrow \\
 & & \vdots & & \vdots & & & & \vdots & & \vdots & & \vdots \\
 & & \uparrow{\scriptstyle \varphi_{f^0}^{0}(D)\otimes 1} & & \uparrow{\scriptstyle \varphi_{f^0}^{0}(D)\otimes 1} & & & & \uparrow{\scriptstyle \varphi_{f^0}^{0}(D)\otimes 1} & & \uparrow{\scriptstyle \varphi_{f^0}^{0}(D)\otimes 1} & & \uparrow{\scriptstyle \varphi_{f^0}^{0}(0)\otimes 1} \\
0 & \to & \mathcal{D}^{b_0} & \xrightarrow{d} & \mathcal{D}^{b_0}\otimes V & \xrightarrow{d} & \cdots & \xrightarrow{d} & \mathcal{D}^{b_0}\otimes \wedge^{g-1}V & \xrightarrow{d} & \mathcal{D}^{b_0}\otimes \wedge^{g}V & \to & \mathbb{C}^{b_0} \to 0 \\
 & & \uparrow & & \uparrow & & & & \uparrow & & \uparrow & & \uparrow \\
 & & 0 & & 0 & & & & 0 & & 0 & & 0,
\end{array}$$

where

$$\varphi_{f^0}^k(D) = \varphi^k(D)|_{f_j=f_j^0}.$$

In the diagram each column except the last one and each row except the first one are exact. One can easily prove the following isomorphism from this diagram:

$$H^k(C^{\cdot}_{f^0}) \simeq \operatorname{Ker}(\varphi^k_{f^0}(0)\otimes 1)/\operatorname{Im}(\varphi^{k-1}_{f^0}(0)\otimes 1), \quad 1\le k\le g$$

where we set $\varphi^g_{f^0}(0)=0$. It follows that

$$\begin{aligned}\chi(C^{\cdot}_{f^0}) &= \sum_{k=0}^{g}(-1)^k\Big(\dim\big(\operatorname{Ker}(\varphi^k_{f^0}(0)\otimes 1)\big) - \dim\big(\operatorname{Im}(\varphi^{k-1}_{f^0}(0)\otimes 1)\big)\Big)\\ &= \sum_{k=0}^{g}(-1)^k\Big(\dim\big(\operatorname{Ker}(\varphi^k_{f^0}(0)\otimes 1)\big) + \dim\big(\operatorname{Im}(\varphi^{k}_{f^0}(0)\otimes 1)\big)\Big)\\ &= \sum_{k=0}^{g}(-1)^k b_k = \chi(C^{\cdot}_0).\end{aligned}$$

Thus the cohomologies at generic point can differ from the cohomologies at the origin. But the Euler characteristic is the same. Summarizing the results we have

THEOREM. *For any f^0, $\chi(C^{\cdot}_{f^0})=\chi(C^{\cdot}_0)$ and it is given by* (3.2).

Now we apply this theorem to the case studied in [**9**]. Then for the curve (1.1), which we assume nonsingular, of genus $g=\frac{1}{2}(N-1)(Nn-2)$, using the results of [**9**] one finds:

$$\chi(\Theta) = (-1)^{g-1}N^{N^2n-2N+1}(Nn-1)^{N-1}\big(\Gamma(N)\big)^2 \times\prod_{j=1}^{N-1}\frac{\Gamma(j)}{\Gamma(nN+j)}\left(\frac{\Gamma(j(nN-1)/N)}{\Gamma(j/N)}\right)^2. \tag{4.1}$$

This formula is quite different from what we have for a generic Abelian variety, $\chi(\Theta)=(-1)^{g-1}g!$ (see [**6**]).

The formula (4.1) is still a conjecture since Assumption 1 is not proved in this special case.

5. Examples

For the hyperelliptic case $N=2$, the formula (4.1) was proved in [**5**]. In this section we shall give other examples for which (4.1) is verified.

EXAMPLE 1 ($g=3$ case). Consider the case $N=4$, $n=1$ and X is given by $w^4+t_4(z)=0$, where $t_4(z)$ is a polynomial of degree 3. We assume that $t_4(z)$ does not have multiple zeros. In this case genus g of X is 3. The curve X is not hyperelliptic since w has third order poles at infinity and is holomorphic at all other points. Thus Θ is nonsingular and $\chi(\Theta)=6$. Our formula (4.1) recovers this.

EXAMPLE 2 ($g=4$ case). Consider the case when $N=3$, $n=2$ and the curve X is given by $w^3+t_3(z)=0$. The polynomial $t_3(z)$ is of degree 5, we suppose that it does not have multiple zeros. The genus g of X is 4. In this case the singularity of the theta divisor Θ consists of one point, say e (cf. [**3**]). We have the Abel–Jacobi map π from the symmetric products of X to $J(X)$. By Riemann's theorem the restriction of π to $S^3(X)$ gives the surjective map

$$\pi\colon S^3(X)\to\Theta.$$

Moreover the isomorphism

$$S^3(X) - \pi^{-1}(e) \simeq \Theta - \{e\}$$

holds. Since $\pi^{-1}(e) \simeq \mathbb{P}^1$, the Euler characteristic of Θ is given by

$$\chi(\Theta) = \chi\big(S^3(X)\big) - \chi\big(\pi^{-1}(e)\big) + 1 = -21.$$

Our formula (4.1) precisely recovers this. Notice that the curve X is not hyperelliptic and is not generic among genus 4 nonhyperelliptic curves.

Those examples support the validity of our assumptions.

For Jacobians of curves satisfying certain genericity conditions, the formula of $\chi(\Theta)$ is given in [**7**]. For genus 4 case, their formula produces $\chi(\Theta) = -22$. This corresponds to curves whose Θ has exactly two singular points.

Appendix A

We shall give a proof of Proposition 1.

It is obvious that (2.1) holds if and only if $(f_1, \dots, f_{m-g})$ is a regular sequence. We set $I_j = Af_1 + \cdots + Af_j$ and, for a subset $T \subset A$, $Z(T) = \{p \in \mathbb{C}^m \mid f(p) = 0\ \forall f \in T\}$.

We first notice that $f_{j+1} \notin I_j$ for any j. Because, otherwise $\dim(\mathcal{J}_0) > g$. Suppose that $(f_1, \dots, f_{m-g})$ is not a regular sequence. Then there exist i and $g \in A$ such that

$$f_{i+1}g \in I_i, \quad g \notin I_i.$$

Write I_i as an intersection of homogeneous primary ideals $\mathcal{Q}_j$ in an irredundant way (cf. [**8**, Ch. IV, §4])

$$I_i = \mathcal{Q}_1 \cap \cdots \cap \mathcal{Q}_r.$$

This is possible since f_j's are all homogeneous. Then $f_{i+1} \notin \sqrt{\mathcal{Q}_j}$ for any j. In fact suppose that $f_{i+1} \in \sqrt{\mathcal{Q}_{j_0}}$ for some j_0. Then

$$Z(f_{i+1}) \supset Z(\sqrt{\mathcal{Q}_{j_0}}) = Z(\mathcal{Q}_{j_0}). \tag{A.1}$$

Notice that

$$Z(I_i) = Z(\mathcal{Q}_1) \cup \cdots \cup Z(\mathcal{Q}_r)$$

is the irreducible decomposition of $Z(I_i)$. Since $Z(I_i)$ is the zero set of i polynomials, $\dim\big(Z(\mathcal{Q}_j)\big) \geq m-i$ for any j. Then the relation (A.1) implies that $\dim\big(Z(I_{i+1})\big) \geq \dim\big(Z(\mathcal{Q}_{j_0})\big) \geq m - i$. Since $\mathcal{Q}_j$'s and f_j's are homogeneous, the intersection $Z(\mathcal{Q}_{j_0}) \cap Z(f_{i+2}) \cap \cdots \cap Z(f_{m-g})$ is nonempty. Thus $\dim\big(Z(I_{m-g})\big) = \dim(\mathcal{J}_0) > g$, which contradicts the assumption.

On the other hand $g \notin \mathcal{Q}_{j_0}$ for some j_0, since $g \notin I_i$. This means that $f_{i+1}g \in \mathcal{Q}_{j_0}$ and $g \notin \mathcal{Q}_{j_0}$. It follows that $f_{i+1}^k \in \mathcal{Q}_{j_0}$ for some $k \geq 2$, which is impossible.

References

1. A. Beauville, *Jacobiennes des courbes spectrales et systèmes hamiltoniens complétement intégrables*, Acta Math. **164** (1990), 211–235.
2. D. Eisenbud , *Commutative algebra. With a view toward algebraic geometry*, Grad. Texts in Math., vol. 150, Springer-Verlag, New York, 1995.
3. H. M. Farkas and I. Kra , *Riemann surfaces*, 2nd ed., Grad. Texts in Math., vol. 71, Springer-Verlag, New York-Berlin, 1992.
4. D. Mumford, *Tata lectures on theta.* I, Progr. Math., vol. 28, Birkhäuser, Boston, 1983; II, vol. 43, 1984.

5. A. Nakayashiki, *On the cohomology of theta divisors of hyperelliptic Jacobians*, 2000, math.AG/001000, Proc. of 9th MSJ-IRI Tokyo 2000 (M. Guest et al., ed.) (to appear).
6. A. Nakayashiki and F. A. Smirnov, *Cohomologies of affine hyperelliptic Jacobi varieties and integrable systems*, Comm. Math. Phys. **217** (2001), 623–652.
7. A. Parusiński and P. Pragacz, *Chern-Schwartz-MacPherson classes and the Euler characteristic of degeneracy loci and special divisors*, J. Amer. Math. Soc. **8** (1995), 793–817.
8. O. Zariski and P. Samuel, *Commutative algebra*, Vol. I, corrected reprinting, Grad. Texts in Math., vol. 28, Springer-Verlag, New York-Heidelberg-Berlin, 1975.
9. V. Zeitlin, *Affine Jacobians for spectral curves* (to be published).

FACULTY OF MATHEMATICS, KYUSHU UNIVERSITY, ROPPONMATSU 4-2-1, FUKUOKA 810-8560, JAPAN.

E-mail address: `atsushi@rc.kyushi-u.ac.jp`

LABORATOIRE DE PHYSIQUE THÉORIQUE ET HAUTES ÉNERGIES, UNIVERSITÉ PIERRE ET MARIE CURIE, TOUR 16 1ER ÉTAGE, 4 PLACE JUSSIEU, 75252 PARIS CEDEX 05, FRANCE.

E-mail address: `smirnov@lpthe.jussieu.fr`

Centre de Recherches Mathématiques
CRM Proceedings and Lecture Notes
Volume **32**, 2002

Reduction Theory, Elliptic Solitons and Integrable Systems

Emma Previato

ABSTRACT. One goal of this paper is to connect two of the thesis topics of Sophie Kowalevski, namely reduction theory for Abelian functions and integrability of a dynamical system. Part I is a digest of classical and current research on splittable Jacobians and curves with automorphisms. Part II is a collection of applications of the splittable situation to the explicit integration of Hamiltonian systems of finite and infinite dimension. Both fields are full of open questions, and the last section in each part is a collection of such.

Introduction

This article is divided into two parts, each having to do with the work of S. Kowalevski's. Kowalevski devoted part of her thesis to the study of Abelian surfaces whose function theory reduces to elliptic functions. She also spent time pursuing the question of integrability of a rotating body, which had been proposed by her mentor, K. Weierstrass, but didn't obtain her striking results until later, when they earned her the Bordin prize. In the light of more recent discoveries in soliton equations and completely integrable systems, these two apparently unrelated areas of her work have come in close connection, as was remarked already in [**BBME**]. We give a survey of classical and recent questions of splittability for Abelian varieties and attempt to connect them with related questions arising for differential equations. Indeed, it is an established feature of this area of mathematics that connections prompt unexpected progress. In the theory of the KP equation, the question of reduction becomes that of expressing KP solutions algebraically by means of elliptic functions. The question of characterizing solutions to the KP equation that are elliptic in the first variable is surprisingly connected with an important Hamiltonian system. It was solved by Krichever, who defined an algebraically completely integrable Hamiltonian system parametrizing them all. This system was identified as a rich algebro-geometric object by Treibich and Verdier, and it was also identified as a special case of a (generalized) Hitchin system by Donagi and Markman. We describe the system and several generalizations of it, along with its salient features, including the question of reduction, where the word

2000 *Mathematics Subject Classification.* Primary: 14H40, 14H52, 14H70, 14K20, 14K25, 33E05, 34L05, 35Q53, 37J35.
NSF support under Grant DMS-9971966 is gratefully acknowledged.
This is the final form of the paper.

has a different meaning than in Part I. The KP solution reduces to, e.g., a KdV or Boussinesq solution according to certain flows being stationary, and the question of nonemptiness of these loci is quite difficult.

I am deeply indebted to the organizers of the Kowalevski meeting, for an opportunity of mathematical excitement and also for the moving exposure to details of her life in Russia. It was a special event, and it is the more special for me to have been asked to write a survey of problems that stemmed from her work. I am also thankful to my collaborators, J. C. Eilbeck and V. Z. Enolskii, who asked me to present on their behalf our joint work on splittability of certain genus-3 varieties of elliptic solitons (Section II-3), despite being among the participants: it was an honor and a pleasure for me.

Part I. Reduction for Abelian Integrals

1. Weierstrass, Kowalevski, Poincaré

Sophie Kowalevski spelled her name thus, when she published her paper on reduction of Abelian integrals [**Ko1**]. She found a necessary and sufficient condition for a curve of genus 3 to cover an elliptic curve. This was one of the three results each of which Weierstrass "without any hesitation (...) would have accepted (...) as a doctoral dissertation" [**Ke,** Appendix C]. Part of the depth of the work is by reason of the interplay between algebra and analysis; indeed, the condition can be stated in several different ways, and several variants of the question could be posed. Some of these variants are still actively pursued in genus 3, besides being wide open in higher genus, and the applications of reduction theory to another area of Kowalevski's work, integrable systems, are also the object of current research.

Weierstrass phrased the problem of reduction of Abelian integrals as follows [**Ke,** Appendix C]:

Give a condition that an algebraic function $f(x)$ must satisfy if among the integrals

$$\int F\big(x, f(x)\big)\, dx$$

where F is a rational function of x and $f(x)$, there exist some that can be transformed into elliptic integrals. This means that the curve X whose function field is $\mathbb{C}\big(x, f(x)\big)$ has a holomorphic differential which is elliptic, or equivalently $\operatorname{Jac} X$ contains an elliptic curve. Weierstrass gave the following answer: the required condition is the existence of a standard homology basis for X (namely, cycles $\alpha_1, \dots, \alpha_g$, $\beta_1, \dots, \beta_g$ whose intersection matrix is $\left[\begin{smallmatrix} 0 & I \\ -I & 0 \end{smallmatrix}\right]$) such that the corresponding basis of holomorphic differentials has period matrix τ_{ij}, $1 \le i, j \le g$, with $\tau_{12} = m/n$, $\tau_{13} = \cdots = \tau_{1g} = 0$ for some integers m, n. The algebraic side of this condition is then the determination of the moduli of the curves that satisfy it; L. Könisberger treated the case $g = 2$, $n = 2$ and Kowalevski the case $g = 3$, $n = 2$. In her case, the algebraic condition is that the canonical model of the curve, namely a plane smooth quartic, should have bitangents meeting at one point [**B,** §76].

In order to formulate a more general problem, we briefly recall some terminology; more details and proofs can be found in a standard text like [**LB**].

Definition. 1. An Abelian variety A of dimension g is a complex torus $\mathbb{C}/\Lambda$, where Λ is a lattice of rank $2g$, which is also an algebraic variety. If L is a period matrix for the torus, namely a $g \times 2g$ complex matrix whose columns span Λ, then a principal polarization of A is determined by a skew-symmetric matrix P in $\mathrm{SL}(2g, \mathbb{Z})$ such that $LP\,{}^tL = 0$ and the Hermitian matrix $iLP\,{}^t\bar{L}$ is positive definite. The $g \times 2g$ complex matrices arising this way are called Riemann matrices.

2. The algebraic counterpart of (1) is the following: a principal polarization is an ample line bundle $\mathcal{L}$ over A, such that the homomorphism

$$\phi_{\mathcal{L}}: A \to \mathrm{Pic}^0 A, \quad a \mapsto T_a^* \mathcal{L} \otimes \mathcal{L}^{-1}$$

is an isomorphism.

The link between (1) and (2) is given by Riemann's theta function ϑ: if A has a principal polarization, then L can be normalized to $[I\Omega]$ with Ω a $g \times g$ symmetric matrix and $\mathrm{Im}\,\Omega$ positive definite, so that for $z \in \mathbb{C}^g$

$$\vartheta(z, \Omega) = \sum_{n \in \mathbb{Z}^g} \exp(\pi i\,{}^t n \Omega n + 2\pi i\,{}^t z n)$$

is an analytic function. $\mathcal{L}$ corresponds to Riemann's theta divisor $\Theta = \{z \mid \vartheta(z, \Omega) = 0\}$.

3. Two complex tori A and A' with period matrices L and L' are said to be isomorphic if and only if there are matrices $C \in \mathrm{GL}(n, \mathbb{C})$ and $T \in \mathrm{GL}(2n, \mathbb{Z})$ such that $CL = L'T$. Notice that T can be taken to be in $\mathrm{SL}(2n, \mathbb{Z})$ by changing $C \mapsto C \cdot \det T$. Then, if $\phi: A' \to A$ is the isomorphism, the pull-back of a principal polarization P gives a principal polarization $P' = TP\,{}^tT$ for A'.

REMARK. By the previous definition then, an Abelian variety of dimension one (also called an elliptic curve) has a unique principal polarization, whereas for $g > 1$ it may have none—infinitely many examples of 2-dimensional Abelian varieties that do not are given in [**E2**]—or several, as the following theorem illustrates.

THEOREM [**L2,** Th. 1.5]. *If the Abelian variety A has a principal polarization, then the set of isomorphism classes of principal polarizations is in one-to-one correspondence with the set of automorphisms of A, which are symmetric with respect to any fixed principal polarization and have positive eigenvalues, modulo the natural action of* $\mathrm{Aut}\,A$.

In particular, the set of nonisomorphic principal polarizations is finite, taking care of a question posed by Martens [**Mar1**].

Before returning to Weierstrass' question let's recall two classical facts:

THEOREM. *A Jacobian $A = \mathrm{Jac}(X)$ has a canonical principal polarization given by the injective homomorphism $H_1(X, \mathbb{Z}) \cong \mathbb{Z}^{2g} \hookrightarrow H^0(X, \Omega^1)^*$ (integration over a cycle). The corresponding theta divisor is only determined up to translation, giving isomorphic principally polarized Abelian varieties, and is irreducible, i.e., the Jacobian with its canonical principal polarization is indecomposable.*

THEOREM. (Poincaré's Complete Reducibility) *If X is an Abelian variety and Y an Abelian subvariety of X, there exists an Abelian subvariety Z of X such that $Y \cap Z$ is finite and $Y + Z = X$.*

REMARK. 1. The statement of this theorem is equivalent to saying that X is isogenous to the product $Y \times Z$, where an isogeny is a surjective group homomorphism with finite kernel.

2. This statement is the algebraic version. Analytically, it could be stated as follows:

THEOREM (Weierstrass-Poincaré, [**Mar2**]). *Let* $[E \quad Z]$ *be a* $g \times 2g$ *Riemann matrix admitting reduction:* $H \cdot [E \quad Z] = \Pi \cdot M$, *where* H *is a maximal-rank* $p \times g$ *complex matrix,* Π *is a* $p \times 2p$ *complex matrix and* M *is a maximal-rank* $2p \times 2g$ *integral matrix, with* $1 \leq p \leq g$. *Then there exist matrices* $A \in \mathrm{GL}(g, \mathbb{C})$, $T \in \mathrm{Sp}(2g, \mathbb{C})$ *such that*

$$[E \quad Z] \cdot T = A \cdot \begin{bmatrix} E_1 & 0 & Z_1 & Q \\ 0 & E_2 & {}^tQ & Z_2 \end{bmatrix}$$

where Z_1, Z_2 *are* $p \times p$ *and* $(g-p) \times (g-p)$ *Riemann matrices, and* Q *is a* $p \times (g-p)$ *rational matrix, whose only possible nonzero entries occur at the beginning of the main diagonal.*

3. The analytic version of the result has the practical consequence that Riemann's theta function reduces to theta functions of lower dimension. Indeed (cf. [**Mar2**]), for a period matrix $[IZ]$ with

$$Z = \begin{bmatrix} Z_1 & Q \\ {}^tQ & Z_2 \end{bmatrix}$$

as above,

$$\vartheta(z; Z) = \sum_{m_2} \vartheta(z_1 + Qm_2; Z_1) \vartheta \begin{bmatrix} D^{-1}m_2 \\ 0 \end{bmatrix} (Dz_2; DZ_2D)$$

where $z = \left[\begin{smallmatrix} z_1 \\ z_2 \end{smallmatrix}\right]$, $D = \mathrm{diag}(d^1, \ldots, d^{g-p})$ is an integral matrix such that QD is integral, we rewrite any integral vector of length $g - p$ as $n_2 = m_2 + Dk_2$ with $0 \leq m_2^j < d^j$, and $\left[\begin{smallmatrix} D^{-1}m_2 \\ 0 \end{smallmatrix}\right] (Dz_2; DZ_2D)$ is the theta function with characteristics, defined more generally for real g-vectors r and s, in dimension g,

$$\vartheta \begin{bmatrix} r \\ s \end{bmatrix} (z; Z) = \sum_m \exp\big(\pi i^t(m + r)\big(Z(m + r) + 2(z + s)\big)\big).$$

It is now apparent that we can ask several important types of reduction questions, with practical applications to solving integrable systems (of finite and infinite dimension!); that there are advantages in formulating both the algebraic and the analytic versions of the questions; and that, even if we decide to restrict our attention to Jacobians, more general Abelian varieties may come into play, as we will see next.

2. Automorphisms

By Poincaré's complete reducibility, if the Abelian variety A contains an Abelian subvariety B, it is isogenous to a product $B \times C$ (with C isogenous to A/B). One question is, when is A actually isomorphic to a product of lower-dimensional Abelian varieties? A second question is whether A is isomorphic to a product of lower-dimensional Abelian varieties that admit a principal polarization; and if so, whether the product polarization induces a given principal polarization on A. These three questions are actually inequivalent. Let's follow Ries [**R**] and call a principally polarized Abelian variety "splittable" if it is isogenous or isomorphic to a product of lower-dimensional Abelian varieties. One way, though not the only one, to find splittable Jacobians is to consider curves with automorphisms.

Hurwitz [**Hu2**] proved that a curve of genus g can have at most $84(g-1)$ automorphisms. Klein's quartic, in homogeneous coordinates $X_0X_1^3 + X_1X_2^3 + X_0^3X_2 = 0$, is the only curve of genus 3 for which the Hurwitz bound is attained. As a consequence, it is actually possible to calculate the period matrix, a rare situation for a curve given algebraically [**BT, LB** §11.7, **RL**], because it is preserved by many symmetries:

$$Z = \begin{bmatrix} -\frac{1}{8} + \frac{3\sqrt{7}i}{8} & -\frac{1}{4} - \frac{\sqrt{7}i}{4} & -\frac{3}{8} + \frac{\sqrt{7}i}{8} \\ -\frac{1}{4} - \frac{\sqrt{7}i}{4} & \frac{1}{2} + \frac{\sqrt{7}i}{2} & -\frac{1}{4} - \frac{\sqrt{7}i}{4} \\ -\frac{3}{8} + \frac{\sqrt{7}i}{8} & -\frac{1}{4} - \frac{\sqrt{7}i}{4} & \frac{7}{8} + \frac{3\sqrt{7}i}{8} \end{bmatrix}.$$

As observed in [**RL**], all entries lie in the field generated (over the field k of definition of the curve, $k = \mathbb{Q}$, e.g.) by the character of the representation induced on the differentials of the first kind by the automorphism group of the curve. But another interesting phenomenon occurs: $\mathrm{Jac}(X) = \mathbb{C}^3/\Lambda$, where Λ is the lattice corresponding to $[\,I \quad Z\,]$, is actually isomorphic to the product of 3 elliptic curves.

Baker [**B,** §76] shows that

$$\begin{bmatrix} 1 & 0 & 0 & \frac{1+i\sqrt{7}}{4} & 0 & 0 \\ 0 & 1 & 0 & 0 & 2\frac{1+i\sqrt{7}}{4} & 0 \\ 0 & 0 & 1 & 0 & 0 & 2\frac{1+i\sqrt{7}}{4} \end{bmatrix}$$

is a matrix of Abelian integrals that gives an isomorphic Abelian variety. Since Jacobians are indecomposable, however, this choice of homology basis cannot be canonical, or the matrices would determine isomorphic polarizations, when a polarization is viewed as a bilinear form: $H = H\big((0,x_2),(0,y_2)\big) = {}^t x_2(\operatorname{Im} Z)^{-1} y_2$ on the complex structure $x = \Omega x_1 + x_2$ on $\mathbb{R}^{2n} = \big\{(x_1,x_2) \in \mathbb{R}^n \times \mathbb{R}^n\big\}$, according to the dictionary given in [**Mu,** Ch.II §4].

The point of this example is to exhibit a Jacobian that is not only isogenous to a product, as predicted by Poincaré's Reducibility, but actually isomorphic to one. Earle [**E1**] constructs 1-dimensional families of Jacobians $\mathrm{Jac}(X)$ of any even genus $2k$ which are both isogenous and isomorphic to $\mathrm{Jac}(Y_1) \times \mathrm{Jac}(Y_2)$, with Y_i quotients of X by an automorphism; X is the hyperelliptic curve given by

$$y^2 = (x^n - 1)(x^n - t), \quad n = 2k+1, \ t \in \mathbb{C}\backslash\{0,1\}.$$

It is then a natural question whether, for a Jacobian isomorphic to a product of lower-dimensional tori, these tori "will necessarily be Jacobians themselves, or at least will admit principal polarization" [**E2**]. Earle replies "emphatically no" by constructing perturbations $\mathrm{Jac}(Y)$ of the Klein Jacobian and showing that $\mathrm{Jac}(Y)$ admits isomorphisms onto the product of two complex tori of dimensions 1 and 2, and for any such isomorphism the 2-dimensional torus admits no principal polarization.

Since $X \hookrightarrow \mathrm{Jac}(X)$ projects to each elliptic curve E_i, there exist finite maps $X \to E_i$; in fact each of them is obtained by moding out by an automorphism of order 2, as illustrated in [**E3**]. First, we connect the actual period matrix of $\mathrm{Jac}(X)$ with Baker's matrix:

- $C[\,I \quad Z\,] = [\,I \quad Z'\,]N$

where $Z' = \begin{bmatrix} \tau & 0 & 0 \\ 0 & 2\tau & 0 \\ 0 & 0 & 2\tau \end{bmatrix}$, $C = \begin{bmatrix} 1 & 0 & -1 \\ 2\tau-1 & -1 & 0 \\ -2\tau & 0 & 1 \end{bmatrix}$ and the unimodular matrix

$$N = \begin{bmatrix} 1 & 0 & -1 & 0 & 0 & -1 \\ -1 & -1 & 0 & -1 & 1 & 0 \\ 0 & 0 & 1 & 1 & -1 & 1 \\ 0 & 0 & 0 & 1 & 0 & -1 \\ 1 & 0 & 0 & 0 & -1 & 0 \\ -1 & 0 & 0 & 0 & 0 & 1 \end{bmatrix}$$

is not symplectic, since the polarization of $\mathrm{Jac}(X)$ is indecomposable. But then the three resulting equations

$$[1 \quad 0 \quad -1]\,[\,I \quad Z\,] = [\,1 \quad \tau\,] \begin{bmatrix} 1 & 0 & -1 & 0 & 0 & -1 \\ 0 & 0 & 0 & 1 & 0 & -1 \end{bmatrix},$$

$$[\,2\tau - 1 \quad -1 \quad 0\,]\,[\,I \quad Z\,] = [\,1 \quad 2\tau\,] \begin{bmatrix} -1 & -1 & 0 & -1 & 1 & 0 \\ 1 & 0 & 0 & 0 & -1 & 0 \end{bmatrix},$$

$$[\,-2\tau \quad 0 \quad 1\,]\,[\,I \quad Z\,] = [\,1 \quad 2\tau\,] \begin{bmatrix} 0 & 0 & 1 & 1 & -1 & 1 \\ -1 & 0 & 0 & 0 & 0 & 1 \end{bmatrix}$$

each describe a map of $\mathrm{Jac}(X)$ to the corresponding elliptic curve E_i; the corresponding maps $X \to E_i$ have degree $a \cdot d - b \cdot c$ ($= 2$, in our case) where $\left[\begin{smallmatrix} a & b \\ c & d \end{smallmatrix}\right]$ is the matrix of the induced map $H_1(X) \to H_1(E_i)$ with respect to a canonical homology basis. Since all elements of order 2 in $\mathrm{Aut}\,X$ are conjugate, the three elliptic curves must be isomorphic, as seen by the equation

$$\tau\,[\,1 \quad 2\tau\,] = [\,1 \quad \tau\,] \begin{bmatrix} 0 & -1 \\ 1 & 1 \end{bmatrix}.$$

We note that the elliptic curve $E = E_i$ does not have extra automorphisms. However, it does have complex multiplication, in fact Ries [**R**] proves that E is the only elliptic curve with complex multiplication by τ and that $\mathrm{End}\, E = \mathbb{Z}[\tau]$.

Again the interplay of analysis and algebra has a role in some of the many, still wide-open questions that have been asked about the automorphism groups of Riemann surfaces. Hurwitz besides finding the bound $84(g-1)$ proved that it is achieved if and only if the group can be generated by two elements, t, u such that $t^2 = u^3 = (tu)^7 = 1$ [**Hu2**]. Later it was determined that the maximum possible order of an automorphism is $2(2g+1)$ (and this is optimal). Since I have not read the original papers I refer instead to the survey [**M**] where they are cited. Call $N(g)$ the maximum order of $\mathrm{Aut}\,X$, X a curve of genus g. Accola [**A1**] proved that $8(g+1) \leq N(g)$ and that equality obtains infinitely often, and on hyperelliptic curves, however Maclachlan independently verified the bound on a disjoint family of curves [**Ma**]. What seems to be more elusive is an understanding of the subvariety of moduli space where a particular group acts (or even its dimension!), except for the fact that the bound $84(g-1)$ can only be achieved on isolated curves (but we know now, not a unique one as is the case for the first two genera, $g = 3$ and 7 [**M** §7]) despite the importance of the locus of curves with automorphisms, in view of the fact that it is the branch locus of the covering map $\mathcal{T}_g \to \mathcal{M}_g$ from Teichmüller space (under the action of the mapping class group). As noted in [**R**], we can view the question of characterizing the family of principally polarized Abelian varieties with the action of a given group G of polarization-preserving automorphisms as a

Schottky-type problem. For example, Ries proves that the Klein curve belongs to a 1-parameter family of principally polarized Abelian varieties which split completely and whose automorphism group contains $\mathrm{SL}(2, 2^3)$; he uses the automorphism group and the attendant group ring in $\mathrm{End}\, A$ to produce a splitting of A.

Finally Accola [**A2**], starting with an analytic property of theta functions, constructs a plane model of the curve whose properties allow him to prescribe certain automorphisms of the curve and to calculate the dimension of the locus of such curves in Teichmüller space.

The lessons we learned in this section are that it is rare for a curve to have automorphisms, and that even if it has many, the question of splitting its Jacobian is by no means settled, in particular as regards polarization.

Before leaving this section we mention another classical problem, having to do with ordinary differential equations, where the Klein curve stars again. We use it as motivation for a sketch of differential Galois theory which we take from [**SU**].

The third-order equation

$$\begin{aligned} H(y) &= x^2(x-1)^2 y''' + (7x-4)x(x-1)y'' \\ &\quad + \left(\frac{72}{7}(x^2-x) - \frac{20}{9}(x-1) + \frac{3}{4}x\right)y' + \left(\frac{792}{7^3}(x-1) + \frac{5}{8} + \frac{2}{63}\right)y \end{aligned}$$

has regular-singular points at $x = 0, 1, \infty$. Hurwitz [**Hu1**] proved that its solutions are algebraic functions defined on Klein's curve. But more is true. The field generated by a basis of solutions for a linear differential equation $L(y) = 0$ is a differential field E, called a Picard-Vessiot extension. The differential Galois group $\mathcal{G}(L)$ of $E/\mathbb{C}(x)$ consists of all automorphisms of E that fix $\mathbb{C}(x)$ elementwise and commute with differentiation. The monodromy group is always a subgroup of the Galois group. In fact, let $c_1, \dots, c_n$ be the singular points of $L(y) = 0$ (including infinity if it is a singular point) and let c_0 be an ordinary point of the equation. Consider these points as lying on the Riemann sphere S^2. Let $\{y_1, \dots, y_n\}$ be a fundamental set of solutions of $L(y) = 0$ analytic at c_0 and let γ be a closed path in $S^2 - \{c_1, \dots, c_n\}$ that begins and ends at c_0. One can analytically continue $\{y_1, \dots, y_n\}$ along γ and get new fundamental solutions $\{\overline{y}_1, \dots, \overline{y}_n\}$ analytic at c_0. These two sets must be related via ${}^t(\overline{y}_1, \dots, \overline{y}_n) = M_\gamma {}^t(y_1, \dots, y_n)^T$ where $M_\gamma \in \mathrm{GL}(n, \mathbb{C})$. One can show that M_γ depends only on the homotopy class of γ and that the map $\gamma \mapsto M_\gamma$ defines a group homomorphism from $\pi_1(S^2 - \{c_1, \dots, c_n\})$ to $\mathcal{G}(L)$. The image of this map depends on the choice of c_0 and $\{y_1, \dots, y_n\}$ but is unique up to conjugacy and is called the monodromy group of $L(y)$. In general the image of this group will be a proper subgroup of $\mathcal{G}(L)$ but when $L(y)$ is Fuchsian, the Zariski closure of this group will be the full Galois group $\mathcal{G}(L)$. Thus, the Galois group of $H(y)$ coincides with the automorphism group of the Klein curve and Singer and Ulmer [**SU**] check this computationally.

3. Poncelet

As we saw in the previous sections, the reduction problem can be approached in two different ways: by cover, or by symmetries. The question of whether a Jacobian is splittable is still more general, but these are sufficient conditions:

- if a curve Γ of genus h covers a curve X of genus g then $\mathrm{Jac}\,\Gamma$ is splittable, more precisely it is isogenus to $B \times C$ with $B = \mathrm{Jac}(X)$,
- if σ is an automorphism of a curve Γ then Γ covers $X = \Gamma/\sigma$.

However, the question of whether $\operatorname{Jac}\Gamma$ contains an elliptic curve is equivalent to the fact that Γ covers an elliptic curve. For $h(\Gamma) = 2$, several results are known and work is in progress. Lange [**L1**] proved that the set $\mathcal{M}_2(d)$ of curves of genus 2 that admit a morphism of degree d to an elliptic curve is a subvariety of the moduli space $\mathcal{M}_2$ that has dimension 2 for each d divisible by 2, 3 or 5, and dimension ≤ 2 for any $d \leq 2$. He obtains algebraic equations for this subvariety by defining the cover explicitly and using the fact that Γ covers an elliptic curve E if and only if the surface $\Gamma \times E$ has a nontrivial section. However, in [**Kuh**] Kuhn produces a more combinatorial construction which shows (by Riemann's existence theorem) that the dimension of $\mathcal{M}_2(d)$ is 2 for every d and from which one could presumably calculate the number of its connected components (as [**L1**] notices, $\mathcal{M}_2(2)$ is connected). But a difficult question, which has arithmetic significance, remains: what is an algebraic equation for a complementary elliptic curve $J(\Gamma)/E$? Even ignoring the fact that a complementary curve is not canonical (cf. [**Kuh**]) and eliminating all the interesting arithmetic features that occur in finite characteristic, this question remains challenging. Kuwata [**Kuw**] had the idea of interpreting it in the framework of Poncelet's theorem and we sketch his construction in view of the significance of this theorem in dynamics, where it can be generalized to hyperelliptic curves of any genus $g > 1$ in place of E [**P**].

First we describe Kuhn's combinatorial parametrization.

THEOREM [**Kuh**]. *Let Γ be a hyperelliptic curve of genus $g = 2$ covering a curve E of genus 1 by a map $f\colon \Gamma \to E$, ι the hyperelliptic involution on Γ, $C_1, \ldots, C_6$ the Weierstrass points of X and $D_1, \ldots, D_4$ the ramification points of π_E in the following commutative diagram (where superscript ι means modulo ι):*

$$\begin{array}{ccc} \Gamma & \xrightarrow{f} & E \\ \pi_\Gamma \downarrow & & \downarrow \pi_E \\ \Gamma^\iota & \xrightarrow{f^\iota} & E^\iota \end{array}$$

Let c_j and d_i be the images of C_j and D_i in Γ^ι and E^ι; then for all i, $f^{\iota *}(d_i)$ contains an odd number of c_j's (counted with multiplicity) and any other point of Γ with even multiplicity; every such ramification configuration $f^\iota\colon \mathbb{P}^1 \to \mathbb{P}^1$ lifts to a morphism $f\colon \Gamma \to E$. For a given hyperelliptic curve Γ, the existence of a lifting of $f^\iota\colon \Gamma^\iota \to \mathbb{P}^1$ to a map from Γ to an elliptic curve is an algebraic condition in terms of the points c_j, and it falls into two different cases:

(a) *(generic) If f is unramified over the D_i, the ramification of f^ι over the d_i is as follows (with one more point of Γ^ι where f^ι is doubly ramified), with the symbol — representing an unramified point of Γ^ι over one of the d_i, which is therefore one of the c_j, and $*$ representing a doubly ramified point:*

$$\text{even } d: \quad \underset{\text{case 1}}{\begin{array}{cccc} * & * & * & * \\ \vdots & \vdots & \vdots & \vdots \\ * & * & * & * \\ * & = & = & = \\ d_1 & d_2 & d_3 & d_4 \end{array}} \qquad \underset{\text{case 2}}{\begin{array}{cccc} * & * & * & * \\ \vdots & \vdots & \vdots & \vdots \\ * & * & * & * \\ * & = & * & * \\ = & = & * & * \\ d_1 & d_2 & d_3 & d_4 \end{array}} \qquad \underset{\text{case 3}}{\begin{array}{cccc} * & * & * & * \\ \vdots & \vdots & \vdots & \vdots \\ * & * & * & * \\ = & * & * & * \\ = & * & * & * \\ = & * & * & * \\ d_1 & d_2 & d_3 & d_4 \end{array}}$$

$$\text{odd } d: \quad \begin{array}{cccc} * & * & * & * \\ \vdots & \vdots & \vdots & \vdots \\ * & * & * & * \\ = & * & * & * \\ - & - & - & - \\ d_1 & d_2 & d_3 & d_4 \end{array}$$

(b) (*special*) *If f is ramified over some D_i, then the ramification of f^ι over the d_i is as shown in* (a) *except that either one of the c_j has ramification of degree* 3, *or there is a unique point, not one of the c_j, with ramification of degree* 4.

We assume from now on that $d = 3$. There are two possible cases for the map $f: \Gamma \to E$, after normalizing the choice of origin in E so that the induced involution ι is ± 1:

CASE 1. f ramifies at two distinct points R and $\iota(R)$.

CASE 2. f ramifies at a Weierstrass point C_1.

After relabeling, we have the following situation:

$$f^{\iota *}(d_i) \ni c_i, \quad e_i \qquad (i = 1, 2, 3); \qquad f^{\iota *}(d_4) = \{c_4, c_5, c_6\}$$

and the ramification points of f^ι are

CASE 1. $r = \pi_\Gamma(R)$ and the e_i, $i = 1, 2, 3$.

CASE 2. e_1, e_2 and $c_3 = e_3$ with multiplicity 2.

Now we recall Poncelet's result in the briefest possible terms, only for the case of triangles and for a generic situation (the conics in question are assumed to meet at four distinct points, e.g.)

THEOREM. *If there is a triangle inscribed in a conic Q_1 and circumscribed to a conic Q_2, then there is a 1-parameter family of such triangles, obtained by moving the divisor $D = P_1 + P_2 + P_3$ (where P_i are the tangency points of the sides of the triangle to Q_2) in a linear equivalence class. This family of triangles corresponds to a curve of genus 1, $I = \{(P, \ell) \mid P \in Q_1, \ell \in Q_2^*, P \in \ell\}$ whose Jacobian* $\mathrm{Pic}^0(I)$ *is therefore an elliptic curve; and to the point of order three $\mathcal{D} =$ the divisor class $[\tau(P, \ell) - (P, \ell)]$ where τ is the fixed-point free involution that sends a pair (P, ℓ) to the pair (Q, m), (P, Q) being the side of a triangle.*

We now return to the map $f: \Gamma \to E$, Case 1. We choose a conic $Q_2 \cong \mathbb{P}^1 \cong \Gamma^\iota$. This is then Kuwata's observation [**Kuw,** Lemma 3.4]

PROPOSITION. *Let $\mathcal{D}$ be the divisor class determined by $f^{\iota *}$. Then the corresponding elliptic curve $E' =$* $\mathrm{Pic}^0(Q_1, Q_2)$ *can be viewed as a double covering of Q_2 branched over c_1, c_2, c_3 and r', where rr' is the side of a triangle, while r is the image of $\mathcal{D}$ under the projection $E' \to Q_2$. If $\overline{E}'$ is the quotient of E' by the subgroup $\langle \mathcal{D} \rangle$ of order* 3, *then $\overline{E}'$ can be viewed as a double covering of E^ι ramified* *at* $f^\iota(r)$, q_1, q_2, q_3, *as in the following commutative diagram:*

$$\begin{array}{ccc} \Gamma & \xrightarrow{f} & E \\ \pi_\Gamma \downarrow & & \downarrow \pi_E \\ \Gamma^\iota & \xrightarrow{f^\iota} & E/\{\pm 1\} \\ \uparrow & & \uparrow \\ E' & \xrightarrow{\phi} & \overline{E}' \end{array}$$

With this motivation, Kuwata constructs explicitly a family of curves Γ, parametrized by an open set U of $\mathbb{P}^1$, that admit a morphism of degree 3 to a fixed elliptic curve E. Let E be given in Weierstrass form: $y^2 = x^3+ax+b$, $\delta = 4a^3+27b^2$ and for $u \in \mathbb{C}$ let

$$\begin{aligned} d(u) &= 27bu^3 - 18a^2u^2 - 27abu - 2a^3 - 27b^2, \\ c(u) &= 3au^4 + 18bu^3 - 6a^2u^2 - 6abu - a^3 - 9b^2, \\ f_u(t) &= 4c(u)t^3 - 3\delta(3u^2 + a)t - 3\delta(2au + 3b), \\ g_u(t) &= (u^3 + au + b)t^3 - (3au^2 + 9bu - a^2)t^2 + \delta. \end{aligned}$$

PROPOSITION. *The set of all curves* Γ *of genus* 2 *that admit a morphism* $\Gamma \to E$ *of degree* 3 *can be parametrized by the open set* $U = \{u \in \mathbb{P}^1 \colon c(u)d(u) \neq 0\}$, *with the curve* Γ_u *given by equation:* $-\delta d(u)s^2 = f_u(t)g_u(t)$. *The morphism* $f\colon \Gamma \to E$ *is given by the formula:*

$$\begin{aligned} f(t,s) = \Bigg(& \frac{\delta d(u)(3(u^3 + au + b)t + 3au^2 + 9bu - a^2)}{c(u)f_u(t)} \\ & + \frac{9bu^4 - 8a^2u^3 - 18abu^2 - 18b^2u + a^2b}{c(u)}, \ \frac{\delta d(u)g_u(-2t)}{f_u(t)^2}s \Bigg) \end{aligned}$$

and is in case 2 *if and only if* $u^3 + au + b = 0$.

The reason for such polynomial expressions in the parameter is for the result to hold over any field of definition k (except the formulas need to be modified in characteristic 3). So, the point of order 3 $\big(x(u) = (-2au + 3b)/(3u^2 + a), y(u)\big)$ on the curve $\overline{E}'$ will be defined over k, with $\overline{E}'$:

$$\begin{aligned} d(u)y^2 &= (x - e(u))(x^3 + ax + b), \\ e(u) &= \frac{9bu^4 - 8a^2u^3 - 18abu^2 - 18b^2u + a^2b}{c(u)} \end{aligned}$$

as well as an explicit isogeny of degree 3: $E' \to \overline{E}'$. Γ_u is then $E'/\{\pm 1\} \times_{\overline{E}'/\{\pm 1\}} E$.

Poncelet's theorem now gives a way of writing an algebraic equation for the second elliptic curve $F = \mathrm{Jac}(\Gamma_u)/E$, which in this case is canonical.

THEOREM. $F_u = \mathrm{Jac}(\Gamma_u)/E$ *can be given by the equation:*

$$\begin{aligned} 3\delta y^2 = x^3 &- 3\delta(3u^4 + 6au^2 + 12bu - a^2)x \\ &- 6\delta\big(9bu^6 - 12a^2u^5 - 45abu^4 - 90b^2u^3 + 15a^2bu^2 \\ &\qquad - (4a^4 + 18ab^2)u - 3a^2b - 18b^3\big). \end{aligned}$$

The morphism $\Gamma \to F$ and the j-invariant of F are also computed explicitly in [**Kuw**]. The fact that allows F to be identified is the following:

PROPOSITION [**Kuw,** 5.2]. *Let Q_2 be a conic in $\mathbb{P}^2$ and let $w_1, \ldots, w_6$ be mutually distinct points on Q_2. We denote by ℓ_i the tangent line to Q_2 at w_i and we set $p_{ij} = \ell_i \cap \ell_j$. Then the following two conditions are equivalent:*

1. There exists a conic Q' passing through p_{12}, p_{23} and p_{31}, and tangent to ℓ_4, ℓ_5 and ℓ_6.
2. There exists a conic Q'' passing through p_{45}, p_{56} and p_{64}, and tangent to ℓ_1, ℓ_2 and ℓ_3.

Let Γ be a curve of genus 2 that is a double covering of Q_2 ramifying at $w_1, \ldots, w_6$. If one of the above conditions is satisfied, then the Jacobian $\mathrm{Jac}(\Gamma)$ of Γ is isogenous to a product of two elliptic curves E and F.

The key to this "duality" is the following lemma due to Kuhn, which generalizes Jacobi's description of all the degree-2 morphisms from a curve of genus 2 to an elliptic curve: given an involution α of $\mathbb{P}^1$ and three points a_1, a_2, a_3 not fixed by the involution, the curve of genus 2 ramified over a_1, a_2, a_3, $\alpha(a_1)$, $\alpha(a_2)$, $\alpha(a_3)$ maps to the two elliptic curves ramified over $\mathbb{P}^1/\alpha$ at the images of the a_i and one of the two fixed points of the involution.

LEMMA [**Kuh,** p. 47]. *If $f\colon \Gamma \to E$ and $g\colon \Gamma \to F = \mathrm{Jac}(\Gamma)/E$ are a given morphism of degree 3 and the canonically associated one, there is a labeling of the branch points of Γ such that $f^{*\iota}(d_i) \ni c_i$, $i = 1, 2, 3$, $f^{\iota *}(d_4) = \{c_4, c_5, c_6\}$ and $g^{*\iota}(b_i) \ni c_{i+3}$, $i = 1, 2, 3$, $g^{*\iota}(b_4) = \{c_1, c_2, c_3\}$ where $b_1, \ldots, b_4$ correspond to the points of period 2 of F, and d_4, b_4 correspond to the origins of E, F:*

$$\begin{array}{ccccc} F & \xleftarrow{g} & \Gamma & \xrightarrow{f} & E \\ \downarrow & & \downarrow & & \downarrow \\ F^{\iota} & \leftarrow & \Gamma^{\iota} & \rightarrow & E^{\iota} \end{array}$$

REMARK. There are exactly two conics (up to projective equivalence) that are both 3-inscribed in and 3-circumscribed to a given conic Q', according to Barth and Michel [**BM**]. They correspond to "control points": $[1, \omega, \omega^2]$ and $[1, (1+\sqrt{5})/2, (1-\sqrt{5})/2] \in \mathbb{P}^2$, resp. Since Kuwata points out that Q', Q'' in the previous proposition are mutually 3-circumscribed and the corresponding Poncelet elliptic curve has an automorphism of order 3, it seems reasonable that it must correspond to the solution $[1, \omega, \omega^2]$ (checking this would be a straightforward calculation).

4. Problems

In this part we have seen that many beautiful classical questions having to do with special Abelian varieties are still quite open, and in fact despite the progress of algebraic geometry no particularly powerful techniques have been introduced, as opposed to a classical area such as enumerative geometry, which has received so much impetus lately from the theory of Gromov–Witten invariants. While it is often the case that in group theory the methods are *ad hoc* (cf. [**A1**]), still I would pose some questions that seem both tractable and useful, because they have arisen in several contexts, one of which will be the theme of the next part.

4.1 Genus 2, splittable case. We refer to the notation of Section 3. Use the invariant theory of 6 points on a line, and the topology of finite covers of this moduli space, to investigate the number of connected components of $\mathcal{M}_2(d)$.

4.2 Genus 2, splittable case. Pursue Kuwata's idea and apply Poncelet's theorem to describe $\mathcal{M}_2(d)$, especially the complementary elliptic curve, which is canonical for an "optimal" cover [**Kuh**], namely one that doesn't factor nontrivially through another elliptic curve. In view of Kuhn's statement, to the effect that if $\Gamma \to E$ is optimal of degree d, there exists a unique E' and an optimal morphism $\Gamma \to E'$ also of degree d, with $\mathrm{Jac}(\Gamma) \sim E \times E'$, these pairs (E, E') should again give examples of the inscribed and circumscribed pairs that Barth and Michel suggest are worth understanding [**BM,** §6].

4.3 Higher genus hyperelliptic Γ and Poncelet's theorem. There is a higher-dimensional generalization of Poncelet's theorem, together with its interpretation as one of the aci's of II.2 below [**P**], and here closure holds only in the rare situation that the curve has a point P with $\mathcal{O}(P - \iota P)$ of finite order. But this might very well be related to the fact that Γ admits a morphism to an elliptic curve, with some restrictions on the inverse image of the origin. This is only speculative at this stage, but concrete calculations in genus 2 and 3 should provide a good indication.

4.4 Genus 3. Kowalevski's curve:

$$(z^2 - f_2)^2 = 4xy(ax + by)(cx + dy)$$

in projective coordinates, with f_2 a form of degree 2 in x and y, has genus 3 and admits a 2:1 morphism to an elliptic curve, yet is not hyperelliptic. She proved that a canonical quartic has splittable Jacobian if and only if it has two concurrent bitangents, as recalled above. This point of view seems to me worth pursuing in two directions. One question is the dimension of the subvariety $\mathcal{M}_g(d)$ of curves of genus g that admits a degree d morphism to an elliptic curve, then its topology and defining equations. Even the dimension question was nontrivial in $g = 2$ [**L1, Kuh**] and I am advocating finding geometric criteria on the canonical curve equivalent to the property of being in $\mathcal{M}_g(d)$. The technique goes back to Riemann and has been very fruitful, namely translating analytic properties of the theta function (with characteristics) into properties of linear series on the curve; as an example, again recalled above, Accola infers from the vanishing of ϑ with 1/2-period characteristics the existence of an automorphism of order 2 on the Riemann surface; then on the geometric side he uses the same vanishing to produce a plane model of the curve with given properties; by putting the two together, he derives the dimension of moduli of curves with an automorphism of order 2. In the second direction, geometric properties of (nonhyperelliptic) canonical curves reflect of course on the moduli space of rank 2-bundles with trivial determinant $\mathcal{M}_\Gamma(2, \mathcal{O}) \hookrightarrow |2\Theta_\Gamma|^* = \mathbb{P}^{2^g-1}$; for genus 3, Coble's analysis [**C**] highlights important features of this variety, with many questions remaining open. The special case corresponding to curves with splittable Jacobians seems fertile ground to plow. Lastly, it seems to be unknown whether the Jacobian of Kowalevski's curve is actually isomorphic to a product of three elliptic curves; number-theoretic applications of such a feature would include current techniques of coding theory and cryptography.

4.5 Differential Galois theory. L. Greenberg [**Gr**] proved that for any finite group G and nonnegative integer g there is a compact Riemann surface X with $\mathrm{Aut}\, X \cong G$ and X/G of genus g (cf. [**TT**]; as those authors note, the genus of X is not controlled by this statement). Given the recent advances in differential Galois theory, particularly in computation, it would seem worth pursuing the question: when X/G is $\mathbb{P}^1$, G can be viewed as the monodromy group of a differential equation with coefficients in $\mathbb{C}(z)$; find an equation explicitly and use the coefficients to gain insight on the moduli space of curves with $\mathrm{Aut}(X) = G$.

I won't mention here any other of the many fascinating problems on automorphisms because I can't offer any suggestion of a method (that's why I included some at the end of Section 2); it should be said that automorphisms of the curve X have also been used in the construction and investigation of Prym varieties with applications to integrable systems (cf. [**K, D**], and the recent [**LR**] for progress on the geometric side). In concluding, I will stress again the 'interdisciplinary' nature of the reduction problem; to wit, (4.1–4.3) are geometric, 4.4 links analysis and algebra (group theory), 4.5 is differential calculus; and since I am unable to touch at all on number theory I will cite a question in [**TT**] as an example: are the entries of the period matrix for a curve with $84(g-1)$ automorphisms algebraic numbers?

PART II: ELLIPTIC SOLITONS AND INTEGRABLE SYSTEMS

1. Calogero-Moser-Krichever: Aspects

Ever since Dubrovin and Novikov [**DN**] obtained the superposition of 'periodic solitons',

$$2\wp\big(x - x_1(t)\big) + 2\wp\big(x - x_2(t)\big) + 2\wp\big(x - x_3(t)\big)$$

the question of elliptic solutions of the KdV or KP equation has been with us. There are two (inequivalent) properties that one may seek in this regard; consider to begin with an 'algebro-geometric solution of the KP hierarchy' by which what is meant is a (complex-valued) function of g (complex) variables $\underline{t} = (t_1 = x, t_2 = y, t_3 = t, \dots, t_g)$ that is produced from algebro-geometric data (roughly speaking, a curve X of genus g with a choice of standard homology basis, a (smooth) point $\infty \in X$ and a divisor $D = P_1 + \cdots + P_g$ on X with $h^0(D - \infty) = 0$)

$$u(\underline{t}) = 2\frac{d^2}{dx^2}\log\vartheta\left(\sum_{j=1}^{g}\int_{P_0}^{P_j}\vec{\omega} + \Sigma t_i U_i - \Delta\right) + \text{const.}$$

where $U_j \in \mathbb{C}^g$ and Δ are constants that depend on the curve (as well as the additive const.).

This function could have one of two properties:

(a) $u(\underline{t})$ is elliptic (i.e., doubly periodic) in the first variable x;

(b) $u(\underline{t})$ is an algebraic expression in g elliptic functions $\wp_1, \dots, \wp_g$, $\wp_1', \dots, \wp_g'$ where each $(\wp_i, \wp_i')$ corresponds to a given lattice Λ_i.

The first question, where knowledge is much more advanced, also arises in an another way: suppose that one asks what is the dependence on time ($t = t_3$) of the poles in x of a solution of the KdV equation. It turns out [**AMcKM**] that they obey an "algebraically completely integrable Hamiltonian system" (aci for short) called the Calogero–Moser system, with Hamiltonian:

$$H = \frac{1}{2}\sum \dot{x}_i^2 - 2\sum_{i\neq j}\frac{1}{(x_i - x_j)^2}.$$

Inspired by this formula, Krichever [**Kr**] showed that the Hamiltonian system with

$$H = \frac{1}{2}\sum \dot{x}_i^2 - 2\sum_{i\neq j} \wp(x_i - x_j)$$

likewise is aci, and moreover that any algebro-geometric KP solution that is elliptic in x has the form: $u(\underline{t}) = \sum_{j=1}^{g} \wp\big(x - x_j(\underline{t})\big)$, with x_j depending on $t_2, t_3, \ldots, t_g$ according to this Hamiltonian, which is called the Calogero–Moser–Krichever system. To link these problems with the previous part, one needs only to observe that the formula given for $u(\underline{t})$ linearizes the KP flows on $\operatorname{Jac} X$, therefore periodicity in x is equivalent to requiring that an elliptic curve E, tangent at the origin of motion to U_1, be contained in $\operatorname{Jac} X$. Following this point of view, Treibich and Verdier in a series of papers gave a geometric description of what they named "elliptic solitons"; Markman interpreted the locus as a generalized Hitchin system (cf. [**DM**] for a survey).

As regards question (b), however, nothing is known apart from a few examples. First of all, since the functions $d^k/dx^k \log \vartheta(\sum_{j=1}^{g} \int_{P_0}^{P_j} \vec{\omega} + xU_1 - \Delta)$, $k \geq 2$, generate the function field of $\operatorname{Jac} X$ and the t-flows span the tangent space to $\operatorname{Jac} X$, in order for such a basic function to be elliptic $\operatorname{Jac} X$ would have to be splittable as an $E_1 \times \cdots \times E_g$, secondly, the time flows would have to be tangent to tori, which seems to be a more subtle question. We make this observation more precise in the following:

REMARK. The locus of elliptic solitons of genus g has dimension g; while the locus of Abelian varieties of dimension g isogenous to a product of g elliptic curves has the same dimension, it is very unlikely that the dimension of the Jacobian sublocus is the same. Thus, in principle the locus of elliptic solitons does not coincide with the locus of elliptic solutions.

We will give some concrete details on elliptic solitons in the next section. Then Section 3 is related to both problems in yet another way, and harks back to classical questions. We conclude this section with the examples of (b) that we know of.

EXAMPLE 1. In [**BBM1,2**] the authors implement a theorem of Appell and by suitable analytic conditions on the period matrix they produce elliptic solutions for integrable hierarchies; they give examples of hyperelliptic and trigonal curves where the condition is satisfied by choosing suitable homology bases. In [**BBME**], the more general reduction criterion due to Weierstrass is applied.

EXAMPLE 2. In [**BE1-3**], the authors work out explicit expressions for elliptic solutions in special cases, as well as equations for the locus of genus-2 curves which allow reduction ("Humbert surface").

EXAMPLE 3. In [**Le**] and in [**MS**] the authors calculate the expression of elliptic solutions for the KdV, Boussinesq equations, resp., coming from the curve: $y^2 = x(x^4 - \alpha^4)(x^4 - \beta^4)$, $y^4 = (x-\alpha)(x-\beta)(x-\gamma)(x-\delta)$, resp.

More examples will be given in Section 3.

2. Elliptic and Abelian Solitons

Perhaps the most remarkable fact about Krichever's aci is that for the moduli space of spectral curves Γ that gives the base of his system, $\operatorname{Jac}\Gamma$ are all splittable.

We provide a different interpretation of such moduli space, due to Treibich and Verdier, which allows us briefly to survey the Hitchin-Markman system, a 'non-Abelian' version of aci in the sense that the motion takes place on (the cotangent bundle to) a moduli space of vector bundles over a fixed curve X. To begin with, we take $X = E$, our fixed elliptic curve.

DEFINITION. Let S be the projectivization of the total space of the semistable rank-2 bundle with trivial determinant over E. S is the ruled surface $\pi_S: S \to E$ characterized by the existence of a unique section C_0 with zero self-intersection. Let's denote by S_q the fibre, $q \in E$.

DEFINITION. A finite pointed morphism $\pi: (\Gamma, p) \to (E, q)$ is called a tangential cover when $\pi^*(E)$ is tangent to $A_\Gamma(\Gamma)$, with $A_\Gamma: r \mapsto \mathcal{O}(r-p)$ the Abel map. It is called primitive if it does not factor nontrivially through another tangential cover.

THEOREM [**TV1**]. *Given a tangential cover $\pi: (\Gamma, p) \to (E, q)$ of degree n, there exists a morphism $\rho: \Gamma \to S$ of degree 1 that sends p to $p = C_0 \cap S_q$ and Γ to a member of the linear system $|nC_0 + S_q|$. Any two such morphisms ρ_1, ρ_2 are related by $\rho_1 = \sigma \circ \rho_2$, $\sigma \in \mathrm{Aut}(S/E) = \mathbb{C}$. Conversely, any irreducible member of $|nC_0 + S_q|$ with the natural projection to E is a tangential cover of degree n. The map $\Gamma \mapsto \rho(\Gamma)$ gives a $1:1$ correspondence between primitive tangential covers and the open affine set $V(n, E)$ of irreducible divisors in the linear system $|nC_0 + S_q|$, which has dimension n and only one fixed point, p;* the general member of $V(n, E)$ is smooth of genus n.

DEFINITION [**DM,** 4.6-4.9]. Let $\mathrm{Higgs}_X^{sm}(r, d, \omega(D))$ be the component of the moduli space parametrizing semistable Higgs pairs (E, ϕ) that have generically irreducible and reduced spectral curves. Here E is a rank-r, degree-d vector bundle on the curve X, ω is the canonical line bundle of X, D is a divisor on X (of sufficiently large degree) and $\phi: E \to E \otimes \omega(D)$ is a twisted endomorphism. Let $b_i = (-1)^i$ trace $(\Lambda^i \phi)$ and let the Hamiltonian be the characteristic polynomial map:

$$H: \mathrm{Higgs}_X^{sm}(r, d, \omega(D)) \to B_\omega: \bigoplus_{i=1}^{r} H^0(X, \omega(D)^{\otimes i}), \quad \phi \mapsto (b_i).$$

Let also $y \in H^0(X, \pi^*\omega(D))$ be the tautological section of the total space $\pi: \omega(D) \to X$. Then the spectral curve of ϕ is the inverse image Γ under $\mathrm{char}(\phi) = y^r - tr(\phi)y^{r-1} + \cdots + (-1)^r \det(\phi): \omega(D) \to \omega(D)^{\otimes r}$, in the total space of $\omega(D)$, of the zero section of $\omega(D)^{\otimes r}$.

THEOREM [**DM,** 4.9]. *The Hamiltonian map H is an algebraically completely integrable system. The generic (Lagrangian) fibre is the Jacobian of a smooth spectral curve of genus $r^2(g-1) + 1 + (\deg D)r(r-1)/2$.*

THEOREM [**DM,** 6.14]. *The variety of elliptic* KP *solitons of degree n corresponding to a fixed pointed elliptic curve (E, q) is birational to the divisor of traceless Higgs pairs in the ($2n$-dimensional symplectic leaf of* $\mathrm{Higgs}_E^{sm}(n, 0, \omega(q))$ *that have residue*

$$\begin{bmatrix} -1 & 0 & \dots & & 0 \\ 0 & -1 & \dots & & \\ \vdots & & & & \vdots \\ & & & -1 & 0 \\ 0 & \dots & & 0 & n-1 \end{bmatrix}$$

over q.

It is now a natural question whether there can be a curve X of genus $g > 1$ in this "tangential position" to a family of solitons, more specifically an integrable system equivalent to a Lax pair whose entries are rational functions over a curve of genus > 1. The answer must be negative for dimensional reasons, as noted in [**DKN,** 2.1] (comment on Lax pairs whose coefficients are sections of a line bundle over a curve), and indeed Treibich showed that if X with $g > 1$ and $\phi: (\Gamma, p) \to (X, q)$ are in tangential position, then Γ and X have to be hyperelliptic with q a Weierstrass point while p is not [**T1**]. Instead, in [**DP**] we ask the following question: are there families of KP "Abelian solitons," namely Jacobians $\mathrm{Jac}(\Gamma)$ containing an Abelian subvariety P of dimension $k > 1$ in such a way that the first k KP flows (which span a hyperosculating space to the Abel image of Γ at the image of ∞) are tangent to P? We found only a very special case, namely P is "coelliptic" with $\mathrm{Jac}(\Gamma)/P$ isomorphic to a fixed elliptic curve E, but we were able to show that they form a completely integrable system:

DEFINITION. Fix any two integers $n \geq g \geq 2$ and an elliptic curve E. We also fix the origin q of E, and define the $(g-1)$-dimensional family of coelliptic solitons to be the curves Γ with a map $\pi: \Gamma \to E$ of degree n, generically ramified simply over $g-1$ points, with a branch point of multiplicity $(g-1)$ at q.

THEOREM [**DP,** Th. 1]. *The family of degree n, genus g coelliptic solitons constructed above is birationally equivalent to an algebraically completely integrable system with $(g-1)$-dimensional* Prym *varieties P as Liouville tori.*

Note that by construction, the first $(g-1)$ KP flows are indeed tangent to the Prym variety $P = \mathrm{Prym}(\pi)$, namely the connected component through the origin of the kernel of the norm map: $\mathrm{Jac}\,\Gamma \to E$. Our integrable system is obtained by symplectic reduction from the Hitchin–Markman system of Higgs pairs $\mathrm{Higgs}_E(n, d, \omega(q + \max\{0, n-g\}q'))$ (for any $q \neq q' \in E$) over a subvariety of spectral curves cut out by Hamiltonians. Therefore, the Hamiltonian flows are compatible with the KP flows [**DM** §6].

In view of this result, for $g = 3$ e.g., one might still be able to write a Lax pair for the integrable systems, describing the motion of a divisor on P, a 2-dimensional Abelian variety ("ultraelliptic" case) and corresponding to genus-3 KP flows and their Baker functions. To that end, in [**EEP3, EP**] we investigate the hyperelliptic Kleinian functions and the differential equations that they satisfy. These analogs of the $\wp$ function were developed in [**BEL**] and in genus 2 also introduced by D. Grant [**G**] along with an analog of Jacobi's derivative formula, and extended to genus 3 by Y. Ônishi [**O**]. We plan to devise a proper Ansatz for the Baker function of an ultraelliptic soliton, and follow Krichever [**Kr**] in showing that the Lax pair for the system can be derived from a Frobenius–Stickelberger formula.

3. Lamé

We begin with a different point of view, which in truth is closely related to the theory of elliptic solitons. The classical spectral problem for differential operators with elliptic coefficients:

$$L = \partial^n + u_{n-2}(x)\partial^{n-2} + \cdots + u_0(x)$$

where $\partial = d/dx$ and $u_j(x) \in \mathbb{C}(\wp, \wp') = \mathbb{C}(E)$, with the normalization $u_{n-1}(x) = 0$, can be viewed in the context of isospectral deformations of commutative algebras of differential operators, namely the KP flows:

REMARK. If the ring of differential operators that commute with L is of (arithmetic) genus > 0, which is to say that L is finite-gap, then the corresponding KP solution is an elliptic soliton.

Indeed, the ring of operators that commute with L is given by the differential-operator part of the ring of formal pseudo-differential operators: $\{\sum_{j\leq N} c_j(L^{1/n})^j,\ c_j \in \mathbb{C}\}$ so all the coefficients are periodic in x.

The question of when L is finite-gap is surprisingly subtle:

THEOREM (cf. [**I**]). *$L = \partial^2 + c\wp(x)$ is finite gap if and only if the constant $c = -n(n+1)$, where n is a natural number.*

In fact, the first significant example of elliptic soliton cited in Section 1 arises this way for $n = 2$. A powerful and effective method was introduced recently by F. Gesztesy and his collaborators (cf. [**GW**] for a survey) by extending Picard's theorem to link the finite-gap property with the qualitative behavior of the eigenfunctions. One is then faced with the question of computing the equation for the isospectral curve Γ whose Jacobian splits along E. There are several recursive approaches, [**Al, DGU, GRT**], to finding a differential operator B of order n that commutes with L, so that the polynomial equation $B^2 = L^{2n+1}$ +(lower-order terms) that they satisfy gives the (possibly singular) curve, but a more classical one can be found in [**He**], namely an Ansatz for the expression of the eigenfunctions at the branched values of the spectral parameter. For example when n is odd and the branch point is of one of two possible types (the expressions for the other cases are analogous):

$$\psi(x) = \left((\wp(x) - e_1)^{1/2}(\wp(x) - e_2)\right)^{1/2} \sum_{j=0}^{(n-3)/2} b_j\left(\wp(x) - e_3\right)^{n/2-j-1}$$

(notice how this differs from the more general Ansatz for the KP elliptic solitons used by Krichever:

$$\psi = \sum_{j=1}^{n} a_i(t, k, \alpha)\phi(x - x_i, \alpha)e^{xk+tk^2}$$

where

$$\phi(x, \alpha) = -\frac{\sigma(x - \alpha)}{\sigma(\alpha)\sigma(x)}e^{\zeta(\alpha)x}$$

and σ, ζ are the classical Weierstrass functions).

By expanding the Lamé equation (using MAPLE, e.g.) and solving algebraic conditions of compatibility for the b_j, one can give examples of Lamé curves:

$n = 2$ $\quad w^2 = (z^2 - 3g_2)\prod_{i=1}^{3}(z + 3e_i)$,

$n = 3$ $\quad w^2 = z\prod_{i=1}^{3}(z^2 - 6ze_i + 45e_i^2 - 15g_2)$

and the general-n case can be calculated recursively. There are several questions to be addressed:

- Which Lamé Jacobians are isogenous to a product of elliptic curves (or even isomorphic to products?)

- When that happens, what is an explicit expression for a KdV solution in terms of elliptic functions?
- In view of recent work on elliptic solitons, more general elliptic potentials are known to be finite-gap, as reviewed in the next proposition: the two questions above should be asked for these too.

PROPOSITION ([**T1,2**]). *For any $n \geq 1$ and $(a_i) \in \mathbb{N}^4$ such that $\sum_{i=0}^{3} a_i(a_i + 1) = 2n$, there is a unique hyperelliptic tangential cover $\Gamma \to E$ and a half period $\omega \in \{\omega_0 = q, \omega_1, \omega_2, \omega_3\}$ such that the potential*

$$u(x) = \sum_{i=0}^{3} a_i(a_i + 1)\wp(x - \omega_i)$$

is the KdV *initial value associated to the divisor $\pi^*(\omega - q)$, and the other normalizing data. For any other point ρ on the elliptic curve, the potential*

$$u(x) = 2\wp(x - \rho) + 2\wp(x + \rho) + \sum_{i=0}^{3} a_i(a_i + 1)\wp(x - \omega_i)$$

is finite-gap if and only if $\sum_{i=0}^{3}(2a_i + 1)^2\wp'(\rho - \omega_i) = 0$.

Recall that a tangential cover is called primitive if it does not factor nontrivially through another one, and exceptional if the (geometric) genus of its canonical image on the surface S (defined in Section 2) is 0.

In [**T1**] Treibich proves some characterizing properties that allow him to compute the tangential polynomials, that is to say the equations as covers of the curve E, for the primitive exceptional hyperelliptic tangential covers of degree $n(= \deg \pi)$ up to 6, for example when:

$n = 3$ $\quad S_3(T) = T^3 - 3\wp T + \wp'$, $u(x) = 6\wp(x)$, $g = 2$,
$n = 4$ $\quad S_4(T) = T^4 + 3(e_i - 2\wp)T^2 + 4\wp' T - 3\prod_{j \neq i}(\wp - e_j)$, $u(x) = 2\big(3\wp(x) + \wp(x - \omega_i)\big)$, $g = 2$ (with $i = 1, 2, 3$),
$n = 5$ $\quad$ also corresponds to genus 2,;
$n = 6$ $\quad S_6(T) = T^6 - 15\wp T^4 + 20\wp' T^3 - 9/4(20\wp^2 - 3g_2)T^2 + 12\wp\wp' T - 5/4\wp'^2$, $u(x) = 12\wp(x)$, $g = 3$.

I posed these questions above rather than in the next section because this is work in progress, in collaboration with J. C. Eilbeck and V. Z. Enolskii [**EEP2**].

These are some (very) preliminary observations which we are pursuing in order to give a theory of Lamé tangential covers. The hyperelliptic curve Γ_n associated to the potential $-n(n+1)\wp$, for some n and some special elliptic curves will be singular, though the arithmetic genus is always n. In coordinates (z, w) as above, the "KP basis" of holomorphic differentials is $\{\omega_j\colon z^j dz/w\}$, $0 \leq j \leq g - 1$, in the sense that the KP flows on $\mathrm{Jac}(\Gamma_n)$ are parallel to the directions $(0, \dots, \int \omega_j, \dots 0)$.

OBSERVATION 1. Aside from the cover that defines the elliptic soliton, corresponding to the differential $z^{g-1}dz/w$ (cf. [**BC**]), the time direction t_3 also is elliptic for $n = 2, 3$ and, in the "equianharmonic" case (i.e. when $\wp$ is associated to the curve with an automorphism of order 3) also for $n = 4$ and 5.

OBSERVATION 2. $\mathrm{Jac}(\Gamma_3)$ splits completely in the equianharmonic case. For $n = 2$, $\mathrm{Jac}(\Gamma_2)$ splits into two non-isomorphic elliptic curves; for $n = 3$, the curves along the $x = t_1$, t_5 direction are both equianharmonic, while the $t = t_3$ curve is not.

The next insight into differential equations with periodic coefficients was provided by Halphen. He generalized Ince's result and proved [**H**]:

The operator $L = \partial^3 - (n^2-1)\wp\partial - (n^2-1)/2\wp' = 0$ is finite gap.

The resulting elliptic solitons will be solutions of the Boussinesq equation, because the third KP time evolution corresponds to the equation

$$\partial_{t_3} L = \left[(L^{3/3})_+, L\right] = 0.$$

In this situation, to find the equation of the spectral curve one uses the Halphen Ansatz for the eigenfunction:

$$\psi(x;\alpha) = e^{kx} \sum_{j=0}^{n-1} a_j(x,\mu,k) \frac{d^j}{dx^j} \phi(x;\alpha)$$

where α is a complex number viewed as a point of the elliptic curve $\nu^2 = 4\mu^3 - g_2\mu - g_3$; by expanding the eigenvalue problems $L\psi = z\psi$ and $B\psi = w\psi$, where B is the lowest-order differential operator that commutes with L (not a polynomial in L) and imposing compatibility, one obtains the equation $F(w,z) = 0$ of a (trigonal) $(3,N)$ curve.

Halphen also pointed out that the equation $\partial^3 - 4/3n^2\wp\partial + (2n(n-3)(4n+3))/27\wp' = 0$ has solutions that can be expressed in terms of elliptic functions, so it would be interesting to check whether the operator $L = \partial^3 - 4/3n^2\wp\partial + (2n(n-3)(4n+3))/27\wp'$ is finite-gap. A Gröbner-basis approach together with new examples of third-order finite-gap operators is worked out in [**Br**].

In [**EEP1**], we showed that the $(3,4)$ Halphen curve, which has genus 3, in the case of the equianharmonic $\wp$-function splits into three elliptic curves, and what's more, again the three elliptic curves are tangent to the three KP flows. In this case, all three curves are isomorphic.

The other task is to express explicitly the KP solution in terms of elliptic functions, which is achieved by the general formula once the evolution vectors U_j are computed. In [**EEP1**] we worked out the solution to the Krichever system, from which the KP flows are deduced. The method we used involves knowledge of the "Kleinian functions" $\wp_{ij}(\underline{z}) = -\partial^2/\partial z_i \partial z_j \log\sigma(\underline{z})$, and their use in giving an algebraic solution of Jacobi's inversion problem. Our final result is the following:

THEOREM. *The 5-particle Calogero–Moser–Krichever system for the equianharmonic elliptic curve, with trigonal spectral curve*

$$w^3 = \left(z^2 + \frac{25}{4}g_3\right)\left(z^2 - \frac{135}{4}g_3\right)$$

has evolution divisor $\sum_{i=1}^{3}(z_i, w_i)$ depending on the particles $x_i(t_1,t_2,t_4)$, $1 \le i \le 5$ according to the equations

$$\left(\frac{df}{dy}\right)^2 - f^6 - 4\left(\widetilde{\wp}\left(\frac{2}{3}y\right) + 2g(y)\right)f^2 = 0$$

where

$$f = \frac{dx_i(y)}{dy}, \quad g(y) = \left.\frac{\partial x_i(t,y)}{\partial t}\right|_{t=0},$$

denoting $y = t_1$, $t = t_2$, $\tilde{\wp}'^2 = 4\tilde{\wp}^3 + (40g_3)^2$ *(the curve tangent to the* t_4 *direction). In turn, the algebraic inversion problem is solved in terms of* f, g.

4. Problems

Kowalevski also gave the separation of variables for a dynamical problem, which is now known as the "Kowalevski top" [**Ko2**]. In this part, we looked at some links between her work on reduction and the theory of integrable systems; we have seen that it is quite exceptional that 'reduction persist' through an integrable system, and that this exceptional situation is related to the motion of the poles of rational/elliptic KP solutions. One area of KP theory which has so far eluded explicit integration is that of the higher-rank solutions, which will be parametrized by rank-r vector bundles over a curve as well as $(r-1)$ arbitrary functions, and I pose a problem in that area even if I don't have concrete suggestions, mainly because I believe Veselov's little-advertised example deserves further investigation.

4.1. As part of his thesis, Veselov [**V**] produced a Hamiltonian system for the poles of certain solutions of the KP equation which turned out to be strictly of higher rank [**La,LP**]. Can one use the poles of the elliptic KP solutions of higher rank [**KN, PW2**] to construct an integrable system? Also, Veselov's system does not have a spectral-curve or aci model yet, to the best of my knowledge; the data however may be truly infinite-dimensional, because of the presence of the arbitrary function [PW1].

4.2. As for the Abelian solitons of Section 2, any matter of explicit expression (Hamiltonians, Lax pairs, reduced KP solution) is open, and would give an application of the Kleinian (σ)-functions [**BEL**].

4.3. The two problems, reduction on one hand, and aci systems on the other, have come in contact in a fairly mysterious way, through elliptic/Abelian solitons mainly. I do not know of an example of elliptic soliton that can be proved to be non-reducible to elliptic functions, although probably this is the generic situation. More significantly, at least two questions could be asked for the moduli space of curves: firstly, from experimental evidence [**EEP2**], as mentioned in Section 3, when the elliptic soliton has lattice with extra symmetries there seem to be a better chance at reduction. Secondly, the question of finding equations for the subvarieties of $\mathcal{M}_g$ where reduction occurs has been asked already [**BE1**], and the answer in genus 2 degree N, is a Humbert surface $H_{N^2} \subset \mathcal{M}_2$, but one more nontrivial condition to detect would be whether the second KP flow is always tangent to the 'second elliptic curve' in the splitting of Jac(Γ). Again, this is unlikely, but the KP equation has a way of surprising us. Also, periodicity in t has been calculated for 2-dimensional elliptic solitons [**S**] but it might make sense to ask for periodicity of one of the higher flows t_k, $k > 1$, for any solution, and for a moduli space of curves with this property.

I would close by saying that the classical theory of elliptic operators with regular-singular points still needs to be brought out in the deeper context of algebraic geometry; one of Verdier's ideas was indeed to embed the elliptic solitons in a moduli space of local systems; this somewhat has been furthered by the theory of Higgs fields, and Hitchin systems, where finally a dictionary between completely integrable equations and monodromy problems may be found. Sophie Kowalevski's thesis was one early place where the two theories appeared together.

References

[A1] R. D. M. Accola, *On the number of automorphisms of a closed Riemann surface*, Trans. Amer. Math. Soc. **131** (1968), 398–408.

[A2] ———, *Plane models for Riemann surfaces admitting certain half-canonical linear series,* I, Riemann surfaces and related topics, Proc. of the 1978 Stony Brook Conference, vol. 97, Ann. of Math. Stud., Princeton, N.J., 1981, pp. 7–20.

[AMcKM] H. Airault, H. P. McKean, and J. Moser, *Rational and elliptic solutions of the Korteweg-de Vries equation and a related many-body problem*, Comm. Pure Appl. Math. **30** (1977), 95–148.

[Al] S. I. Al'ber, *Investigation of equations of Korteweg-de Vries type by the method of recurrence relations*, J. London Math. Soc. (2) **19** (1979), 467–480.

[BBM1] M. V. Babich, A. I. Bobenko, and V. B. Matveev, *Reductions of Riemann theta functions of genus g to theta functions of lesser genus, and symmetries of algebraic curves*, Dokl. Akad. Nauk SSSR **272** (1983), 13–17.

[BBM2] ———, *Solution of nonlinear equations, integrable by the inverse problem method, in Jacobi theta-functions and the symmetry of algebraic curves*, Izv. Akad. Nauk SSSR Ser. Mat. **49** (1985), 511–529.

[B] H. F. Baker, *An introduction to the theory of multiply periodic functions*, Cambridge Univ. Press, Cambridge, 1907.

[BM] W. Barth and J. Michel, *Modular curves and Poncelet polygons*, Math. Ann. **295** (1993), 25–49.

[BBME] E. D. Belokolos, A. I. Bobenko, V. B. Matveev, and V. Z. Ènol'skii, *Algebro-geometric principles of superposition of finite-zone solutions of integrable nonlinear equations*, Uspekhi Mat. Nauk. **41** (1986), 3–42.

[BE1] E. D. Belokolos and V. Z. Ènol'skii, *Algebraically integrable nonlinear equations and Humbert surfaces*, Plasma theory and nonlinear and turbulent processes in physics (Kiev, 1987), vol. 1, 2, World Sci. Publishing, Singapore, 1988, pp. 20–57.

[BE2] ———, *Verdier's elliptic solitons and the Weierstrass reduction theory*, Funktsional. Anal. i Prilozhen **23** (1989), 57–58.

[BE3] ———, *Reduction of theta functions and elliptic finite-gap potentials*, Acta Appl. Math. **36** (1994), 87–117.

[BT] K. Berry and M. Tretkoff, *The period matrix of Macbeath's curve of genus seven*, Curves, Jacobians, and Abelian Varieties (Amherst, 1990), Contemp. Math., vol. 136, Amer. Math. Soc., Providence, RI, 1992, pp. 31–40.

[Br] Yu. V. Brezhnev, *Elliptic solitons and Gröbner bases*, preprint nlin.SI/0007028.

[BEL] V. M. Buchstaber, V. Z. Enol'skii, and D. V. Leykin, *Kleinian functions, hyperelliptic Jacobians and applications*, Reviews in Mathematics and Mathematical Physics (S. P. Novikov and I. M. Krichever, eds.), vol. 10:2, Gordon and Breach, London, 1997, pp. 1–125.

[BC] J. L. Burchnall and T. W. Chaundy, *Commutative ordinary differential operators*, Proc. Roy. Soc. London Ser. A **118**, 557–583.

[C] A. B. Coble, *Algebraic geometry and theta functions*, American Mathematical Society Colloquium Publications, Providence, RI, 1982.

[DGU] R. Dickson, F. Gesztesy, and K. Unterkofler, *Algebro-geometric solutions of the Boussinesq hierarchy*, Rev. Math. Phys. **11** (1999), 823–879.

[D] R. Donagi, *Spectral covers*, Current topics in complex algebraic geometry (Berkeley, CA, 1992/93), Math. Sci. Res. Inst. Publ., vol. 28, Cambridge Univ. Press, Cambridge, 1995, pp. 65–86.

[DM] R. Donagi and E. Markman, *Spectral covers, algebraically completely integrable Hamiltonian systems, and moduli of bundles*, Integrable systems and quantum groups (Montecatini Terme, 1993), Lecture Notes in Math., vol. 1620, Springer, Berlin, 1996, pp. 1–119.

[DP] R. Y. Donagi and E. Previato, *Abelian solitons*, Nonlinear waves: Computation and theory (Athens, GA, 1999) Math. Comput. Simulation **55** (2001), 407–418.

[DKN] B. A. Dubrovin, I. M. Krichever, and S. P. Novikov, *Integrable systems.* I, Current problems in mathematics. Fundamental directions, vol. 4, Itogi Nauki i Tekhniki, Akad. Nauk SSSR, Vsesoyuz. Inst. Nauchn. i Tekhn. Inform., Moscow, 1985, pp. 179–284.

[DN] B. A. Dubrovin and S. P. Novikov, *Periodic and conditionally periodic analogs of the many-soliton solutions of the Korteweg-de Vries equation*, Soviet Physics JETP **40** (1974), 1058–1063.

[E1] C. J. Earle, *Some Jacobian varieties which split*, Complex Analysis Joensuu 1978 Lecture Notes in Math., vol. 747, Springer, Berlin, 1979, pp. 101–107.

[E3] ———, *Some Riemann surfaces whose Jacobians have strange product structures*, Curves, Jacobians, and Abelian varieties (Amherst, MA, 1990), Contemp. Math., vol. 136, Amer. Math. Soc., Providence, RI, 1992, pp. 127–137.

[E2] C. J. Earle and H. E. Rauch, *Function theorist*, Differential geometry and complex analysis, Springer, Berlin, 1985, pp. 15–31.

[EEP1] J. C. Eilbeck, V. Z. Enolskii, and E. Previato, *Varieties of elliptic solitons*, J. Phys. A **34** (2001), 2215–2227.

[EEP2] ———, *Splittable Ince and Halphen potentials*, in preparation.

[EEP3] ———, *Frobenius-Stickelberger addition formula for σ functions of higher genera*, in preparation.

[EP] V.Z. Enolskii and E. Previato, *Varieties of ultraelliptic solitons*, in preparation.

[GRT] F. Gesztesy, R. Ratnaseelan and G. Teschl, *The* KdV *hierarchy and associated trace formulas*, Recent Developments in Operator Theory and Its Applications (Winnipeg, MB, 1994) Oper. Theory. Adv. Appl., vol. 87, Birkhäuser, Basel, 1996, pp. 125–163.

[GW] F. Gesztesy and R. Weikard, *Elliptic algebro-geometric solutions of the* KdV *and* AKNS *hierarchies–an analytic approach*, Bull. Amer. Math. Soc. (N.S.) **35** (1998), 271–317.

[G] D. Grant, *A generalization of Jacobi's derivative formula to dimension two*, J. Reine Angew. Math. **392** (1988), 125–136.

[Gr] L. Greenberg, *Maximal groups and signatures*, Discontinuous groups and Riemann surfaces (Proc. Conf. Univ. Maryland, College Park, MD, 1973) Ann. of Math. Studies, vol. 79, Princeton Univ. Press, Princeton, NJ, 1974, pp. 207–226.

[H] M. Halphen, *Traité des fonctions elliptiques et des leurs applications*, Gauthier-Villars, Paris, 1888.

[He] C. Hermite, *Oeuvres de Charles Hermite,* III, Gauthier-Villars, Paris, 1912.

[Hu1] A. Hurwitz, *Über einige besondere homogene lineare Differentialgleichungen*, Math. Ann. **26** (1886), 117–126.

[Hu2] ———, *Über algebraische Gebilde mit eindeutingen Transformationen in sich*, Math. Ann. **41** (1893), 403–442.

[I] E. L. Ince, *Ordinary differential equations*, Dover Publications, New York, 1944.

[K] V. Kanev, *Spectral curves, simple Lie algebras, and Prym-Tjurin varieties*, Theta functions (Brunswick, ME, 1987), Proc. Sympos. Pure Math., vol. 49, Part I, Amer. Math. Soc., Providence, RI, 1989, pp. 627–645.

[Ke] D. H. Kennedy, *Little sparrow: A portrait of Sophia Kovalevsky*, Ohio University Press, Athens, OH, 1983.

[Ko1] S. Kowalevski, *Über die Reduction einer bestimmten Klasse abel'scher Integrale dritten Ranges auf elliptische Integrale*, Acta Math. **4** (1884), 393–414.

[Ko2] ———, *Mémoire sur un cas particulier de la rotation d'un corps solide autour d'un point fixe*, Acta Math. **12** (1889), 177–232.

[Kr] I. M. Krichever, *Elliptic solutions of the Kadomcev-Petviašvili equations, and integrable systems of particles*, Funktsional. Anal. i Prilozhen **14** (1980), 45–54.

[KN] I. M. Krichever and S. P. Novikov, *Holomorphic vector bundles over Riemann surfaces and the Kadomcev-Petviašvili equation,* I, Funktsional. Anal. i Prilozhen **12** (1978), 41–52.

[Kuh] R. Kuhn, *Curves of genus* 2 *with split Jacobian*, Trans. Amer. Math. Soc. **307** (1988), 41–49.

[Kuw] M. Kuwata, *Morphisms from curves of genus* 2 *to elliptic curves and Poncelet's closure theorem*, preprint 1999.

[L1] H. Lange, *Über die Modulvarietät der Kurven vom Geschlecht* 2, J. Reine Angew. Math. **281** (1976), 80–96.

[L2] ———, *Abelian varieties with several principal polarizations*, Duke Math. J. **55** (1987), 617–628.

[LB] H. Lange and C. Birkenhake, *Complex Abelian varieties*, Grundlehren der Mathematischen Wissenschaften, vol. 302, Springer-Verlag, Berlin, 1992.

[LR] H. Lange and S. Recillas, *Abelian varieties with group action*, math.AG/0106055, preprint.

[La] G. A. Latham, *Solutions of the* KP *equation associated to rank-three commuting differential operator over a singular elliptic curve*, Phys. D **41** (1990), 55–66.

[LP] G. A. Latham and E. Previato, KP *solutions generated from* KdV *by "rank 2" transference*, Phys. D **94** (1996), 95–102.

[Le] M. H. Lee, *Solutions of the Korteweg-de Vries equation associated to hyperelliptic curves*, Hadronic Mechanics and Nonpotential Interactions, Part 1 (Cedar Falls, IA, 1990), Nova Sci. Publ., Commack, NY, 1992, pp. 205–216.

[M] A. M. Macbeath, *Hurwitz groups and surfaces*, The Eightfold way, Math. Sci. Res. Inst. Publ., vol. 35, Cambridge Univ. Press, Cambridge, 1999, pp. 103–113.

[Ma] C. Maclachlan, *A bound for the number of automorphisms of a compact Riemann surface*, J. London Math. Soc. **44** (1969), 265–272.

[Mar1] H. H. Martens, *Riemann matrices with many polarizations*, Complex Analysis and its Applications (Lectures, Internat. Sem., Trieste, 1975), vol. III, Internat. Atomic Energy Agency, Vienna, 1976, pp. 35–48.

[Mar2] ______, *On the reduction of Abelian integrals and a problem of H. Hopf*, Curves, Jacobians and Abelian varieties (Amherst, MA, 1990), Contemp. Math., vol. 136, Amer. Math. Soc., Providence, RI, 1992, pp. 287–296.

[MS] V. B. Matveev and A. O. Smirnov, *On the Riemann theta function of a trigonal curve and solutions of the Boussinesq and* KP *equations*, Lett. Math. Phys. **14** (1987), 25–31.

[Mu] D. Mumford, *Tata lectures on theta*, Progress in Mathematics, vol. 28, Birkhäuser, Boston, MA, 1983, pp. 287–296.

[O] Y. Ônishi, *Complex multiplication formulae for hyperelliptic curves of genus three*, Tokyo J. Math. **21** (1998), 381–431.

[P] E. Previato, *Poncelet's theorem in space*, Proc. Amer. Math. Soc. **127** (1999), 2547–2556.

[PW1] E. Previato and G. Wilson, *Vector bundles over curves and solutions of the* KP *equations*, Proc. Sympos. Pure Math., vol. 49, 1989, pp. 553–569.

[PW2] ______, *Differential operators and rank two bundles over elliptic curves*, Compositio Math. **81** (1992), 107–119.

[RL] H. E. Rauch and J. Lewittes, *The Riemann surface of Klein with* 168 *automorphisms*, Problems in Analysis (papers dedicated to Salomon Bochner, 1969), Princeton Univ. Press, Princeton, NJ, 1970, pp. 297–308.

[R] J. F. X. Ries, *Splittable Jacobi varieties*, Curves, Jacobians, and Abelian varieties (Amherst, MA 1990), Contemp. Math., vol. 136, Amer. Math. Soc., Providence, RI, 1992, pp. 305-326.

[SU] M. F. Singer and F. Ulmer, *On a third order differential equation whose differential Galois group is the simple group of* 168 *elements*, Applied algebra, algebraic algorithms and error-correcting codes (San Juan, PR, 1993), Lecture Notes in Comput. Sci., vol. 673, Springer, Berlin, 1993, pp. 316–324.

[S] A. O. Smirnov, *Finite-gap elliptic solutions of the* KdV *equation*, Acta Appl. Math. **36** (1994), 125–166.

[T1] A. Treibich, *Tangential polynomials and elliptic solitons*, Duke Math. J. **59** (1989), 611-627.

[T2] ______, *New elliptic potentials*, Acta Appl. Math. **36** (1994), 27–48.

[TT] C. L. Tretkoff and M. D. Tretkoff, *Combinatorial group theory, Riemann surfaces and differential equations*, Contributions to group theory, Contemp. Math., vol. 33, Amer. Math. Soc., Providence, RI, 1984, pp. 467–519.

[TV1] A. Treibich, J.-L. Verdier, *Solitons elliptiques*, The Grothendieck Festschrift, III, Progr. Math., vol. 88, Birkhäuser, Boston, 1990, pp. 437–480.

[TV2] ______, *Variétés de Kritchever des solitons elliptiques de* KP, Proceedings of the Indo-French Conference on Geometry (Bombay, 1989), Hindustan Book Agency, Delhi, 1993, pp. 187–232.

[V] A. P. Veselov, *Rational solutions of the Kadomtsev Petviashvili equation and Hamilton systems*, Uspekhi Mat. Nauk. **35** (1980), 195–196.

Department of Mathematics and Statistics, Boston University, Boston, MA 02215-2411, USA.

E-mail address: ep@bu.edu

Centre de Recherches Mathématiques
CRM Proceedings and Lecture Notes
Volume **32**, 2002

Schwarzian Derivatives and Uniformization

Takeshi Sasaki and Masaaki Yoshida

In every textbook on elementary function theory you can find a definition of Schwarzian derivative

$$\{z; x\} = \frac{1}{2}\left(\frac{z''}{z'}\right)' - \frac{1}{4}\left(\frac{z''}{z'}\right)^2$$

where $' = d/dx$, of a nonconstant function $z = z(x)$ with respect to x. You would also find an exercise to show

(0.1) (PGL(2)-invariance). If a, b, c, and d are constants satisfying $ad - bc \neq 0$, then

$$\left\{\frac{az+b}{cz+d}; x\right\} = \{z; x\}.$$

(0.2) $\{z; x\} = 0$ if and only if $z(x) = (ax+b)/(cx+d)$ for some constants a, b, c, and d satisfying $ad - bc \neq 0$.

(0.3) (change of variable). If y is a nonconstant function of x, then

$$\{z; y\} = \{z; x\}\left(\frac{dx}{dy}\right)^2 + \{x; y\}.$$

(0.4) (local behavior). If $z = x^\alpha u$ $(\alpha \neq 0)$, where u is a holomorphic function of x nonvanishing at 0, then

$$\{z; x\} = \frac{1-\alpha^2}{4x^2} + \frac{\text{a function holomorphic at } 0}{x};$$

if $z = \log(xu)$, where u is as above, then

$$\{z; x\} = \frac{1}{4x^2} + \frac{\text{a function holomorphic at } 0}{x}.$$

In this paper we discuss various generalizations of the Schwarzian derivative. We start from recalling how it was found.

2000 *Mathematics Subject Classification.* Primary: 53A55, 32G13; Secondary: 53A20, 34A30.
This is the final form of the paper.

1. A Paper by H. A. Schwarz

Hermann Amandus Schwarz (1843–1921) is well known through Schwarz inequality, Schwarz's lemma in elementary function theory, reflection principle, etc. In his paper [**Sch**], he treated the hypergeometric differential equation

$$\frac{d^2u}{dx^2} + \frac{\gamma - (\alpha+\beta+1)x}{x(1-x)}\frac{du}{dx} - \frac{\alpha\beta}{x(1-x)}u = 0,$$

and studied conditions for the parameters (α, β, γ) that every solution is algebraic, and found explicit expressions of such solutions. Let us leave aside why he got interested in such a problem, and follow his line.

Consider in general a 2nd order linear differential equation

$$\frac{d^2u}{dx^2} + p\frac{du}{dx} + qu = 0. \tag{1.1}$$

Let u_1 and u_2 be two linearly independent solutions, and put $z = u_2/u_1$. One of his main discoveries is the relation

$$\{z; x\} = q - \frac{1}{4}p^2 - \frac{1}{2}\frac{dp}{dx} \tag{1.2}$$

between z and the coefficients of the equation. He proved this by a straightforward computation: We can assume $u_2 = zu_1$. Substitute $u_2' = zu_1' + z'u_1$ and $u_2'' = zu_1'' + 2z'u_1' + z''u_1$ into $u_2'' + pu_2' + qu_2 = 0$, and use the identity $u_1'' + pu_1' + qu_1 = 0$. We are led to $2z'u_1' + (z'' + pz')u_1 = 0$, that is,

$$p + \frac{z''}{z'} = -2\frac{u_1'}{u_1}.$$

Differentiate both sides and we have

$$\begin{aligned} p' + \left(\frac{z''}{z'}\right)' &= -2\frac{u_1u_1'' - (u_1')^2}{u_1^2} = -2\frac{u_1(-pu_1' - qu_1) - (u_1')^2}{u_1^2} \\ &= 2q + 2\left(p\frac{u_1'}{u_1} + \left(\frac{u_1'}{u_1}\right)^2\right). \end{aligned}$$

Since u_1'/u_1 is already expressed in terms of p and z, substituting this expression, we obtain the above relation.

For the hypergeometric equation, the relation above appears to be

$$\{z; x\} = \frac{1-\lambda^2}{4x^2} + \frac{1-\mu^2}{4(1-x)^2} + \frac{1+\nu^2-\lambda^2-\mu^2}{4x(1-x)} \tag{1.3}$$

where

$$\lambda = 1-\gamma, \quad \mu = \gamma-\alpha-\beta, \quad \nu = \alpha-\beta.$$

On the other hand, the Wronskian $u_2du_1/dx - u_1du_2/dx$ of the solutions u_1 and u_2 is equal to a constant times $\exp(-\int p\,dx)$, which, in the hypergeometric case, turns out to be a constant multiple of $x^{-\gamma}(1-x)^{\gamma-\alpha-\beta-1}$. Thus if both u_1 and u_2 are algebraic, then γ and $\alpha+\beta$ must be rational numbers. Applying this argument after the change of variable $x \to 1/x$, we know that all of α, β, and γ must be rational numbers.

He next considers a map

$$s\colon x \mapsto z(x) = u_2(x)/u_1(x),$$

which is nowadays called Schwarz's s-map, and the s-image of the upper x-plane, which turns out to be a triangle bounded by three arcs, is called Schwarz's triangle. The local behavior at $x = 0$ and $x = 1$ can be known by (1.3) and the property (0.4) of the Schwarzian derivative (the change of variable $x \to 1/x$ leads also the behavior at $x = \infty$); this tells you the three angles of the triangle. Let us sketch his strategy: Possible analytic continuation of s can be illustrated by repeated reflection images along the arcs of the triangle. Since u_1 and u_2 are algebraic, so is s; thus these reflection images cannot be chaotic. This would determine possible three angles of the triangle and so the values of the three parameters.

There are two essential points in his paper: The formula (1.3), which is the origin of the name Schwarzian derivative, and the map s. In the following we study these by considering their generalizations. A survey of Shwarzian derivatives can be found also in [**MSY2**].

2. Normal Forms of Differential Equations

Let us consider a system of linear differential equations in n variables x with m unknowns u with rank (dimension of the solution space at a generic point) r. Two such systems are said to be *strongly equivalent* if a change of unknowns $u_j \to \sum_i k_j^i u_i$ takes one to the other, where $\det k_j^i$ is a nonzero function of x. Two such systems are said to be *weekly equivalent* if a change of unknowns like above and a change of variables x take one to the other.

Problem is *to find a set of invariants for these equivalence relations.* Invariants should be functions of the coefficients of the equations.

Let us consider the simplest nontrivial case: $n = m = 1$ and $r = 2$, i.e., equations in the form

$$u'' + pu' + qu = 0. \tag{2.1}$$

If we change the unknown as $u \to ku$ (this means 'substitute $u = kv$ into the equation, derive an equation with unknown v, and then write u instead of v'), we obtain

$$u'' + \left(p + 2\frac{k'}{k}\right)u' + \left(q + p\frac{k'}{k} + \frac{k''}{k}\right)u = 0.$$

If we choose k as a nonzero solution of (2.1), then the coefficient of u vanishes, and moreover if we change the variable x into a nontrivial solution of this new equation, then eventually the equation changes into $u'' = 0$. Therefore, any equation (2.1) is weakly equivalent to $u'' = 0$; so there is no invariant in this case. Thus we consider the strong equivalence class represented by the equation (2.1). If we choose k as above, the survived coefficient $p + 2k'/k$ would serve as an invariant; but unfortunately this quantity is not determined uniquely by the coefficient of the equation (2.1). If we choose k so that the coefficient of u' vanishes, then the survived coefficient can be luckily expressed in terms of p, q, and their derivatives; indeed, using $k'/k = -p/2$, we have

$$q + p\frac{k'}{k} + \frac{k''}{k} = q - \frac{1}{4}p^2 - \frac{1}{2}p',$$

which is exactly the Schwarzian derivative $\{z; x\}$ according to (1.2). This can be paraphrased as follows: In the (strong) equivalence class of the equation (2.1), there is a unique equation without the term u', the coefficient of u is given by (1.2), the

Schwarzian derivative. In other words, for a given function $q(x)$, the nonlinear equation

$$\{z; x\} = q(x)$$

can be solved as $z = u_2/u_1$, where u_1 and u_2 are linearly independent solution of the linear equation

$$u'' + qu = 0.$$

We present several lucky generalizations and a semi-lucky case in this note.

3. Geometrical Treatment

Schwarz's s-map suggests a geometrical counterpart of a system of linear differential equations in n variables x with m unknowns u with rank r: The map

$$s\colon x \longmapsto u_1(x) : \cdots : u_r(x) \in \mathrm{Gr}(m, r)$$

where $u_1, \ldots, u_r$ are column m-vectors giving linearly independent solutions, and $\mathrm{Gr}(m, r)$ denotes the Grassmannian variety

$$\mathrm{GL}(m)\backslash\{m \times r - \text{matrix of rank } m\};$$

in particular $\mathrm{Gr}(1, r) = \mathbb{P}^{r-1}$, the $(r-1)$-dimensional projective space. Note that the group $\mathrm{GL}(r)$ acts naturally on $\mathrm{Gr}(m, r)$ (from the right); in other words, the group $\mathrm{GL}(r)$ is the group of motions of $\mathrm{Gr}(m, r)$.

Note that two strongly equivalent systems define the same s-maps, up to the group of motions, and that two weekly equivalent systems give rise to the same images under the s-maps, up to the group of motions.

The problem in §2 can be geometrized as follows: For two maps from the x-space to $\mathrm{Gr}(m, r)$, decide when they are the same up to the group of motions. The weak version of this problem is given by replacing 'they' by 'their images'.

The solution in case $m = n = 1$ and $r = 2$ is given as follows: For given maps s_1 and s_2 both from x-space to $\mathbb{P}^1$, the maps s_1 and s_2 are related linear fractionally if and only if $\{s_1; x\} = \{s_2; x\}$.

In general the study concerning the problem above is called *geometry of submanifolds* or *projective differential geometry.*

We only have quite restricted results, some of which we present in this note.

4. $m = n = 1$: Projective Curves

We consider maps of 1-dimensional source space to the target $\mathbb{P}^{r-1}$. Since such a map is called a curve, the problem is to find invariants for curves in the projective space up to the projective motion group $\mathrm{PGL}(r)$.

An analogous problem is often taught and solved in an undergraduate course: A set of invariants of curves, curvature and torsion, in the euclidean 3-space up to the group of rigid motions appear as coefficients of a normalized equation (called Frenet–Serret equation). Our problem is an variant of this problem; we utilize projective motions in place of rigid motions.

This is solved by Halphen, Laguerre and Forsyth; a modern treatment can be found in [**Sea**]. Let us summarize their results when $r = 3$, that is, when the target is the plane $\mathbb{P}^2$. A curve in the plane can be expressed by a system of homogeneous coordinates as $u(x) = u_0(x) : u_1(x) : u_2(x)$. The differential equation for the unknown u

$$\begin{vmatrix} u''' & u'' & u' & u \\ u_0''' & u_0'' & u_0' & u_0 \\ u_1''' & u_1'' & u_1' & u_1 \\ u_2''' & u_2'' & u_2' & u_2 \end{vmatrix} = 0$$

admits u_0, u_1, and u_2 as solutions. If

$$\begin{vmatrix} u_0'' & u_0' & u_0 \\ u_1'' & u_1' & u_1 \\ u_2'' & u_2' & u_2 \end{vmatrix} \neq 0,$$

that is, if the curve is nondegenerate, then the equation has the form

$$u''' + p_1 u'' + p_2 u' + p_3 u = 0.$$

Conversely, for a 3rd order linear ordinary differential equation, three linearly independent solutions give rise to a plane curve; other choice of solutions give another curve which is projectively equivalent to the old one. The equation above for a given curve u is not unique either, indeed, though ρu where ρ is a nonzero function gives the same curve as u, the corresponding differential equation changes. It is easy to see that a suitable choice of ρ makes the coefficient of u'' zero. Let us write the resulting equation as

$$u''' + P_2 u' + P_3 u = 0$$

where the coefficients can be expressed as rational functions in $p_1, \dots, p_3$ and their derivatives (parallel computation as in §2). Now take a solution $f(x)$ of the equation

$$\{f; x\} = \frac{1}{4} P_2 \tag{4.1}$$

and change the variable x into $y = f(x)$, and take a new unknown $w = f'u$, then the equation with unknown w and variable y is of the form

$$\frac{d^3 w}{dy^3} + Rw = 0, \tag{4.2}$$

actually $R = (P_3 - P_2'/2)/(f')^3$. In this way, a plane curve determines uniquely an equation of the form (4.2).

Moreover, since any solution of (4.1) is expressed linear fractionally by f, the parametrization of the curve determined by (4.2) has exactly a freedom of PGL(2). In this sense, a plane curve carries a natural projective structure.

Anyway in this case the quantity R (expressible as a rational function of $p_1, \dots, p_3$ and their derivatives) serves as a complete invariant of curves for the weak equivalence relation. Precisely speaking we should take $R\,dy^3$ rather than R itself. An immediate but important consequence is that R vanishes identically if and only if the curve is a conic. A generalization of this statement will be given in §7.

For general $r \geq 3$, a similar argument is known as Laguerre–Forsyth's theory.

5. $m = 1$ and $r = n + 1$: Projective Structures

In this section, we find invariants for nondegenerate maps

$$x = (x^1, \dots, x^n) \mapsto u(x) = u^0(x) : \cdots : u^n(x) \in \mathbb{P}^n$$

under the group $\mathrm{PGL}(n+1)$ of motions of $\mathbb{P}^n$.

From the middle of the 19th century, study of systems of linear differential equations admitting algebraic solutions became popular. Algebraic functions are multi-valued, not easy to handle. So they wanted to derive, from the coefficients of the equations, rational functions in the unknown, which are single-valued after substituting (multi-valued) algebraic solutions. Around 1870, several people such as J. Liouville, P. Pepin, L. Fuchs, F. Klein, F. Brioschi, C. Jordan, and E. Goursat tried to make a general theory. Schwarz's paper introduced in §1 solved this problem for the hypergeometric equation. P. Painlevé studied this problem for equations for which $m = 1$, $n = 2$, and $r = 3$. (cf. Notes by Painlevé on 1887.5.31, and by E. Goursat on 1887.5.16 in Comptes Rendus.) Historical background can be found in [**Bou**]. Let us follow their line. Consider a (nondegenerate) map

$$(x, y) \mapsto (z, w),$$

which is multi-valued in a linear fractional way:

$$(z, w) \mapsto (Z, W) = \left(\frac{az + bw + c}{a''z + b''w + c''}, \frac{a'z + b'w + c'}{a''z + b''w + c''} \right).$$

Find *simple* relations in z, w, Z, W, and their derivatives with respect to (x, y), which are independent of the coefficients a, b, c, a', ... Their results are as follows: Put

$$\begin{aligned} I(z, w) &= \frac{z_{xx}w_x - w_{xx}z_x}{z_x w_y - w_x z_y}, \qquad J(z, w) = \frac{w_{yy}z_y - z_{yy}w_y}{z_x w_y - w_x z_y}, \\ M(z, w) &= \frac{z_{xx}w_y - w_{xx}z_y + 2(z_{xy}w_x - w_{xy}z_x)}{3(z_x w_y - w_x z_y)}, \\ N(z, w) &= \frac{w_{yy}z_x - z_{yy}w_x + 2(w_{xy}z_x - z_{xy}w_x)}{3(z_x w_y - w_x z_y)} \end{aligned}$$

then

$$\begin{aligned} I(z, w) &= I(Z, W), \qquad J(z, w) = J(Z, W), \\ M(z, w) &= M(Z, W), \quad N(z, w) = N(Z, W). \end{aligned}$$

They called these expression the fundamental differential invariants. Note that the (original) Schwarzian derivative involves 3rd derivative, but these invariants involve only derivatives up to 2nd order.

Next, for the map $(x, y) \mapsto (z, w)$, define functions u^0, u^1, and u^2 as

$$u^0 = \rho = (z_x w_y - z_y w_x)^{-1/3}, \quad u^1 = z\rho, \; u^2 = w\rho.$$

Note that the map can be regarded as the map

$$(x, y) \longmapsto 1 : z : w = u^0 : u^1 : u^2 \in \mathbb{P}^2$$

to the target space $\mathbb{P}^2$. It is very important to realize that the factor ρ is so chosen that any linear fractional transformation of (z, w) causes (u^0, u^1, u^2) a linear transformation. Let us derive differential equations, with unknown u, satisfied by

u^0, u^1, and u^2:

$$\begin{vmatrix} u_{xx} & u_x & u_y & u \\ u^0_{xx} & u^0_x & u^0_y & u^0 \\ u^1_{xx} & u^1_x & u^1_y & u^1 \\ u^2_{xx} & u^2_x & u^2_y & u^2 \end{vmatrix} = 0, \quad \begin{vmatrix} u_{xy} & u_x & u_y & u \\ u^0_{xy} & u^0_x & u^0_y & u^0 \\ u^1_{xy} & u^1_x & u^1_y & u^1 \\ u^2_{xy} & u^2_x & u^2_y & u^2 \end{vmatrix} = 0,$$
$$\begin{vmatrix} u_{yy} & u_x & u_y & u \\ u^0_{yy} & u^0_x & u^0_y & u^0 \\ u^1_{yy} & u^1_x & u^1_y & u^1 \\ u^2_{yy} & u^2_x & u^2_y & u^2 \end{vmatrix} = 0.$$

The coefficients of u_{xx}, u_{xy}, and u_{yy} are

$$\begin{vmatrix} u^0_x & u^0_y & u^0 \\ u^1_x & u^1_y & u^1 \\ u^2_x & u^2_y & u^2 \end{vmatrix} = \begin{vmatrix} \rho_x & \rho_y & \rho \\ u_x\rho & u_y\rho & 0 \\ v_x\rho & v_y\rho & 0 \end{vmatrix} = (\rho)^3(u_xv_y - u_yv_x) = 1.$$

The other coefficients turn out to be as

$$\begin{aligned} u_{xx} &= Mu_x - Iu_y + Au, \\ u_{xy} &= -Nu_x - Mu_y + Bu, \\ u_{yy} &= -Ju_x + Nu_y + Cu \end{aligned}$$

where

$$A = 2(M^2 + IN) - M_x + I_y, \quad B = IJ - MN + M_y + N_x,$$
$$C = 2(N^2 + JM) - N_y + J_x.$$

The computation above can be summarized as follows: For a given map

$$s\colon (x,y) \mapsto v^0 : v^1 : v^2 \in \mathbb{P}^2,$$

one can find a nonzero function ρ such that $u^0 = \rho v^0$, $u^1 = \rho v^1$, and $u^2 = \rho v^2$ solve a system

$$\begin{aligned} u_{xx} &= p_1u_x + q_1u_y + r_1u, \\ u_{xy} &= p_2u_x + q_2u_y + r_2u, \\ u_{yy} &= p_3u_x + q_3u_y + r_3u \end{aligned}$$

satisfying

$$p_1 + q_2 = 0, \quad q_3 + p_2 = 0;$$

the coefficients p_i, q_i, and r_i are uniquely determined by s, and can be expressed in terms of $z = v^1/v^0$ and $w = v^2/v^0$; these give the *relation between the solutions and the coefficients.*

Let us formulate this argument in general for $n \geq 2$; formulae will be shorter. We consider a map

$$s\colon x = (x^1, \dots, x^n) \mapsto z = (z^1, \dots, z^n).$$

Denote the Jacobi matrix by

$$j(z,x) = (j^k_i), \quad j^k_i = \frac{\partial z^k}{\partial x^i},$$

and set $J_i^k(z,x)=\partial x^k/\partial z^i$. We assume that s is nondegenerate, that is, $\det j(z,x)\neq 0$. Putting

$$\sigma(z,x)=\frac{1}{n+1}\log\det j(z,x),\quad \sigma_i(z,x)=\frac{\partial\sigma}{\partial x^i},$$
$$\gamma_{ij}^k(z,x)=\sum_\ell \frac{\partial^2 z^\ell}{\partial x^i\partial x^j}J_\ell^k(z,x),$$

we define the Schwarzian derivatives of z as

$$S_{ij}^k(z;x)=\gamma_{ij}^k(z,x)-\delta_i^k\sigma_j(z,x)-\delta_j^k\sigma_i(z,x).$$

Note that if $n=1$ this expression is just zero, and that if $n=2$ they coincide the fundamental differential invariants above. Easy to check that

$$S_{ij}^k(z;x)=S_{ji}^k(z;x),\quad \sum_k S_{ik}^k=0.$$

Let us compare the properties of these derivatives with those of the original Schwarzian derivative (cf. (0.1),...,(0.4)). We have (cf. [**Y1**])

(5.1) (PGL$(n+1)$-invariance).

$$S_{ij}^k(Az;x)=S_{ij}^k(z;x),\quad A\in \mathrm{PGL}_{n+1}.$$

(5.2)
$$S_{ij}^k(z;x)=0\longleftrightarrow z=Ax.$$

(5.3) (change of variables). If y is another set of variables,

$$S_{ij}^k(z;y)-\sum_{p,q,r}S_{pq}^r(z;x)j_i^p(x;y)j_j^q(x;y)J_r^k(x,y)=S_{ij}^k(x;y).$$

(5.4) (local behavior). If a map $x=(x^1,\dots,x^n)\mapsto z=(z^1,\dots,z^n)$, $n\geq 2$ ramifies along $x^1=0$ as

$$z^1(x)=(x^1)^\alpha v^1,\quad z^2(x)=v^2,\dots,z^n(x)=v^n,\quad \left|\frac{\partial z}{\partial x}\right|=(x^1)^{\alpha-1}u$$

where $v^j(1\leq j\leq n)$ and u are holomorphic not divisible by x^1. Then for $2\leq i,j,\ k\leq n$,

$$S_{ij}^k\{z;x\},\quad S_{1j}^k\{z;x\}+\delta_j^k\frac{1}{n+1}\frac{\alpha-1}{x^1},\quad \frac{1}{x^1}S_{ij}^1\{z;x\},$$
$$S_{1j}^1\{z;x\},\quad x^1S_{11}^k\{z;x\},\quad S_{11}^1\{z;x\}-\frac{n-1}{n+1}\frac{\alpha-1}{x^1}$$

are holomorphic. Logarithmic ramification implies $\alpha=0$.

Let us state the conclusion obtained:

CONCLUSION. Two maps z_1 and z_2 from the $x=(x^1,\dots,x^n)$-space to $\mathbb{P}^n$ are projectively equivalent if and only if

$$S_{ij}^k(z_1;x)=S_{ij}^k(z_2;x),\quad i,j,k=1,\dots,n.$$

If we put $\rho=\left(\det j(z,x)\right)^{-1/(n+1)}$, then $u^0=\rho$, $u^1=\rho z^1,\dots,u^n=\rho z^n$ solve the system

$$u_{ij}=\sum_k S_{ij}^k u_k+S_{ij}^0 u$$

where

$$S^0_{ij} = \frac{1}{n-1}\Big(\sum_{\ell,k} S^\ell_{ik} S^k_{\ell j} - \sum_k \frac{\partial}{\partial x^k} S^k_{ij}\Big);$$

this is the *relation between the solutions and the coefficients.*

6. The Uniformizing Equation of the Moduli Space of the Marked Cubic Surfaces

As an application of the relation between the solutions and the coefficients given in the previous section, we present in this section a system of differential equations which gives a uniformization of the moduli space M of the marked cubic surfaces.

Let us recall just an essence of cubic surfaces. Any nonsingular cubic surface in $\mathbb{P}^3$ can be obtained from $\mathbb{P}^2$ by blowing up six points, such that no three points are collinear and no conic passes through the six points. The marking means, in this setup, just the numbering of these six points. Let us represent these six points by a 3×6-matrix, where the jth column gives homogeneous coordinates of the jth point. Since no three points are collinear, we can assume that the first four points have the coordinates $1:0:0$, $0:1:0$, $0:0:1$, and $1:1:1$; so the matrix is of the form

$$\begin{pmatrix} 1 & 0 & 0 & 1 & 1 & 1 \\ 0 & 1 & 0 & 1 & x^1 & x^2 \\ 0 & 0 & 1 & 1 & x^3 & x^4 \end{pmatrix}. \tag{6.1}$$

The assumption on the six points can be now stated as the nonvanishing of

$$\begin{aligned} D(x) := \prod_{j=1}^{4} x^j(x^j-1)\cdot(x^1-x^2)(x^1-x^3)(x^2-x^4)(x^3-x^4) \\ \times(x^1x^4-x^2x^3)\{(x^1-1)(x^4-1)-(x^2-1)(x^3-1)\} \\ \times\{x^1(x^2-1)(x^3-1)x^4-(x^1-1)x^2x^3(x^4-1)\}. \end{aligned}$$

The moduli space M of the marked (nonsingular) cubic surfaces can be identified with

$$\{x = (x^1, \dots, x^4) \in \mathbb{C}^4 \mid D(x) \neq 0\}.$$

In [**ACT**], for each cubic surface $S(x)$ determined by $x \in M$, they considers five *periods* $u^0, \dots, u^4$ of $S(x)$ (precisely speaking, periods of the triple cyclic cover of $\mathbb{P}^3 - S(x)$), and shows that the (multi-valued) map

$$s\colon M \ni x \longmapsto u^0(x) : \cdots : u^4(x) \in \mathbb{P}^4$$

has its image in the 4-ball

$$B_4 = \{1 : z^1 : \cdots : z^4 \in \mathbb{P}^4 \mid |z^1|^2 + \cdots + |z^4|^2 < 1\},$$

and that the inverse map is defined on B_4 and is single-valued. This can be paraphrased as 'the moduli space M is isomorphic to the quotient of B_4 under a discontinuous group acting on it' or simply 'the moduli space admits a complex hyperbolic structure'.

Now we are ready to apply our Schwarzian derivatives. The period map s above should be a Schwarz map of a system, say E, of differential equations defined on M. Analogous stories such as 'elliptic curves and the hypergeometric equation', 'certain curves admitting a cyclic automorphism group and Appell–Lauricella's hypergeometric equation' (see [**Y2**]), and 'a 4-parameter family of K3 surfaces and the hypergeometric equation of type (3, 6)' (see §8) have been known and loved, and these fed many mathematicians. So if we can explicitly know the system E, that would be a lot of fun.

Let us denote the Schwarzian derivatives $S^k_{ij}(s;x)$ of the period map with respect to $x = (x^1, x^2, x^3, x^4)$ by S^k_{ij}, then the system E should be expressed as

$$u_{ij} = \sum_k S^k_{ij} u_k + S^0_{ij} u$$

where the coefficients S^0_{ij} can be determined by S^k_{ij} as is explained in §5. Important is that since the Schwarzian derivatives are invariant under PGL(4), the coefficients S^k_{ij} must be rational functions in x with poles only along the divisor $D(x) = 0$. Moreover the poles along this divisor are fairly restricted by the local properties (5.4), since it is known that the period map s ramifies along this divisor with index 3.

This information is sufficient to determine E, however it needs a complicated computation. If we recall a classical fact that the moduli space M admits a regular action of the Weyl group of type E_6, and require E to be invariant under this group action, then the computation becomes much simpler. We do not dare tabulate all the coefficients of E, but we show just one of them:

$$S^1_{23} = -\frac{1}{3}\prod_{j=1}^{4} x^j(x^j - 1)\cdot(x^1 - x^2)(x^1 - x^3)(x^2 - x^4)(x^3 - x^4) \\ \times x^1(x^1 - 1)\cdot x^4(x^4 - 1)(x^1 - x^3)(x^1 - x^2)/D(x).$$

On the other hand, as is suggested in [**MT**], this system E has a relation with Appell–Lauricella's hypergeometric differential equation. Indeed, E can be obtained from Appell–Lauricella's system of type D in nine variables with special parameters by restricting this system on a 4-dimensional subvariety, extracting from this restricted one a subsystem of rank 5, and finally performing an algebraic coordinate change on this subsystem; for more detail, see [**SY3**] and [**SY4**].

7. $m = 1$ and $r = n + 2$: Conformal Structure

We consider systems in n variables $x = (x^1, \dots, x^n)$ with an unknown u of rank $n + 2$ (cf. [**SY2**]). When $n = 1$, these are equations for plane curves studied in §4. Let us regard one variable say, x^n, special and write the system as

$$\frac{\partial^2 u}{\partial x^i \partial x^j} = g_{ij}\frac{\partial^2 u}{\partial x^1 \partial x^n} + \sum_{k=1}^{n} A^k_{ij}\frac{\partial u}{\partial x^k} + A^0_{ij}u, \quad (1 \le i, j \le n) \tag{7.1}$$

where

$$g_{ij} = g_{ji}, \quad g_{1n} = 1, \quad A^k_{ij} = A^k_{ji}, \quad A^0_{ij} = A^0_{ji}, \quad A^k_{1n} = A^0_{1n} = 0,$$

and assume $\det g_{ij} \ne 0$. The *normalization factor* is defined as

$$e^\theta = \det\left(\mathfrak{u}, \frac{\partial \mathfrak{u}}{\partial x^1}, \dots, \frac{\partial \mathfrak{u}}{\partial x^n}, \frac{\partial^2 \mathfrak{u}}{\partial x^1 \partial x^n}\right)$$

where $\mathfrak{u}$ is a column vector consisting of $(n + 2)$ linearly independent solutions; this quantity is independent of the choice of $\mathfrak{u}$ up to multiplicative constants. By multiplying a suitable function to the unknown u, we can assume that the condition

$$\det(e^\theta g_{ij}) = 1$$

holds.

This system is said to satisfy *quadric condition* if the image of the Schwarz map

$$x \mapsto \mathfrak{u}(x) \in \mathbb{P}^{n+1}$$

is contained in a certain quadratic hypersurface.

We assume $n \geq 3$; we studied already the case $n = 1$ and, when $n = 2$, the situation is a bit different (see [**SY1**]). Then the main result is as follows: If the system satisfies the quadric condition, then the coefficients are expressed as rational functions in g_{ij} and their derivatives:

$$A^j_{ik} = \Gamma^j_{ik} - g_{ik}\Gamma^j_{1n}, \quad A^0_{ik} = -S_{ik} + g_{ik}S_{1n}$$

where Γ^j_{ik} and S_{ik} are the Christoffel symbols and the Schouten tensor of $e^\theta g_{ij} =: h_{ij}$. They are defined as follows:

$$\Gamma^j_{ik} = \frac{1}{2}\sum_l h^{jl}(h_{il,k} + h_{kl,i} - h_{ik,l}), \quad dh_{il} = \sum_k h_{il,k}\, dx^k.$$

Let $R^j{}_{ikl}$ be the Riemannian curvature tensor:

$$d\pi^j_i - \sum_k \pi^k_i \wedge \pi^j_k = \frac{1}{2}\sum_{k,l} R^j{}_{ikl}\, dx^k \wedge dx^l, \quad \pi^j_i = \sum_k \Gamma^j_{ik}\, dx^k.$$

The Ricci and the scalar curvatures are defined by

$$R_{ij} = \sum_l R^l{}_{ilj}, \quad R = \sum_{i,j} h^{ij}R_{ij},$$

respectively, and finally the Schouten tensor (relative to h_{ij}) is defined as

$$S_{ik} - \frac{1}{n-2}\left(R_{ik} - \frac{R}{2(n-1)}h_{ik}\right).$$

8. The Uniformizing Equation of the Moduli Space of a 4-Dimensional Family of K3 Surfaces

We present a system of differential equations which gives a uniformization of the moduli space X of the K3 surfaces obtained as double covers of $\mathbb{P}^2$ branching along six lines, that no three lines meet (cf. [**MSY1, Y2**]). Let us represent six lines in the plane by a 3×6-matrix, where the jth column gives coefficients of the linear equation of the jth line. Since we can assume that the first four lines are defined by $t^1 = 0$, $t^2 = 0$, $t^3 = 0$, $t^1 + t^2 + t^3 = 0$, these matrices are of the form (6.1). The assumption on the six lines can now be stated as the nonvanishing of

$$\begin{aligned} D_1(x) := \prod_{j=1}^4 x^j(x^j - 1) \cdot (x^1 - x^2)(x^1 - x^3)(x^2 - x^4)(x^3 - x^4) \\ \times (x^1x^4 - x^2x^3)\{(x^1 - 1)(x^4 - 1) - (x^2 - 1)(x^3 - 1)\}, \end{aligned}$$

and the moduli space X can be identified with

$$\{x = (x^1, x^2, x^3, x^4) \in \mathbb{C}^4 \mid D_1(x) \neq 0\}.$$

Let K3(x) be the surface obtained by desingularizing the double cover of the plane branching along the six lines parameterized by $x \in X$. On each surface K3(x), there are exactly six linearly independent 2-cycles, say $\gamma_0(x), \dots, \gamma_5(x)$, and a unique holomorphic 2-form, say $\phi(x)$ up to multiplicative constant. The period map

$$x \longmapsto u^0(x) : \cdots : u^5(x) \in \mathbb{P}^5, \quad \text{where} \quad u^j(x) = \int_{\gamma_j(x)} \phi(x)$$

happens to have its image in a hyperquadric. The system of differential equations is of the form (7.1) in the previous section; so all the coefficients can be expressed in terms of the coefficients g_{ij} of $\partial^2 u/\partial x^1 \partial x^4$. They are given as

$$\begin{aligned}
g_{12} &= \frac{x^4 - x^3}{x^1 - x^2}, g_{13} = \frac{x^4 - x^3}{x^1 - x^2}, \quad g_{14} = 1, \\
g_{23} &= 1, \quad g_{23} = \frac{x^3 - x^1}{x^2 - x^4}, \quad g_{34} = \frac{x^2 - x^1}{x^3 - x^4}, \\
g_{11} &= \frac{x^2 x^3 - x^4}{x^1(1 - x^1)} - \frac{x^3(x^4 - x^2)}{x^1(x^1 - x^3)} - \frac{x^2(x^4 - x^3)}{x^1(x^1 - x^2)}, \\
g_{22} &= \text{make exchanges } x^1 \leftrightarrow x^2 \text{ and } x^3 \leftrightarrow x^4 \text{ in } g_{11}, \\
g_{33} &= \text{make an exchange } x^2 \leftrightarrow x^3 \text{ in } g_{22}, \\
g_{44} &= \text{make an exchange } x^1 \leftrightarrow x^4 \text{ in } g_{11}.
\end{aligned}$$

On the other hand, it is known that this system is equivalent to the hypergeometric system $E(3, 6; \alpha_1, \dots, \alpha_6)$ of type $(3, 6)$ with parameters $\alpha_j = 1/3$.

9. $n = 1$ and $r = md$: Systems of Ordinary Differential Equations of Order d

Let us consider a system of ordinary differential equations with m unknowns $\mathbb{u} = {}^t(u_1, \dots, u_m)$ of order d

$$\mathbb{u}^{(d)} = \sum_{k=0}^{d-1} P_k \mathbb{u}^{(k)}. \tag{9.1}$$

Under a transformation of unknowns

$$\mathbb{v} = K\mathbb{u}, \quad K = \left(k_i^j(x)\right)_{1 \le i,j \le m}, \quad \det K \neq 0, \tag{9.2}$$

the system (9.1) is transformed into

$$\mathbb{v}^{(d)} = (KP_1 + dK')K^{-1}\mathbb{v}^{(d-1)} + \cdots .$$

Choosing K so that $KP_1 + dK' = 0$, we can assume

$$P_1 = 0. \tag{9.3}$$

Change of variable $y = f(x)$ together with the transformation (9.2), then the system (9.1) changes into

$$\begin{aligned}
(f')^d \mathbb{v}^{(d)} + a_d (f')^{d-2} f'' \mathbb{v}^{(n-1)} + (b_d (f')^{d-3} f''' + c_d (f')^{d-4} (f'')^2) \mathbb{v}^{(d-2)} + \cdots \\
= K\mathbb{u}^{(d)} + dK'\mathbb{u}^{(d-1)} + \frac{d(d-1)}{2} K'' \mathbb{u}^{(n-2)} + \cdots
\end{aligned}$$

where

$$a_d = \frac{d(d-1)}{2}, \quad b_d = \frac{d(d-1)(d-2)}{6}, \quad c_d = \frac{d(d-1)(d-2)(d-3)}{8}.$$

From this, we get

$$(f')^d \mathbb{v}^{(d)} = \big(d(f')^{d-1}K'K^{-1} - a_d(f')^{d-2}f''\big)\mathbb{v}^{(d-1)} + (f')^{d-2}\widetilde{P}_2\mathbb{v}^{(d-2)} + \cdots$$

where

$$(9.4)\quad \widetilde{P}_2 = KP_2K^{-1} - b_d\frac{f''}{f'} - c_d\left(\frac{f''}{f'}\right)^2 + da_{d-1}\frac{f''}{f'}K'K^{-1} - d(d-1)K'K^{-1}K'K^{-1} + \frac{d(d-1)}{2}K''K^{-1}.$$

Hence, to preserve the condition (9.3), K must be of the form

$$K = (f')^{(n-1)/2}B, \quad B\text{: a constant matrix.}$$

On the other hand, from (9.4), we have

$$\frac{\operatorname{tr}\widetilde{P}_2}{m} = \frac{d(d-1)(d+1)}{6}\{f;x\} + \frac{\operatorname{tr}P_2}{m}.$$

Thus, by choosing f satisfying $\operatorname{tr}\widetilde{P}_2 = 0$, we get the following conclusion:

1. The system (9.1) can be normalized as

$$P_1 = 0 \quad\text{and}\quad \operatorname{tr}P_2 = 0.$$

2. A transformation of unknown $\mathbb{u}$ and that of variable preserving this condition has the form

$$y = \frac{\alpha x + \beta}{\gamma x + \delta},$$
$$\mathbb{v} = (\gamma x + \delta)^{1-d}B\mathbb{u}, \quad (B\text{: a constant matrix}).$$

3. Under this transformation, the matrix-valued quadratic form $P_2\,dx^2$ changes only by a conjugate action of B.

We can moreover see that the i-differential form

$$R_i - \sum_{j=0}^{i-2} a_{i,j}\left(\frac{d}{dx}\right)^j P_{i-j}(dx)^i \quad (i \geq 2)$$

where

$$a_{i,j} = (-1)^j\frac{(2i-j-2)!(n-i+j)!}{j!(i-j-1)!}$$

are also invariants up to conjugation of B [**W, Sea**]. Note that when $m = 1$, $\operatorname{tr}P_2 = 0$ implies $P_2 = 0$, of course. When $m = 1$ and $d = 3$, the conclusion above reduces to that in §4, where projective curves are studied.

When $m = 2$ and $d = 2$, our argument can be translated into the geometry of ruled surfaces in $\mathbb{P}^3$. In fact, four $(= r = md)$ linearly independent solutions

$$\begin{pmatrix} u_1^0 \\ u_2^0 \end{pmatrix}, \ldots, \begin{pmatrix} u_1^3 \\ u_2^3 \end{pmatrix},$$

viewed as two curves

$$u_1 = (u_1^0, \ldots, u_1^3) \quad\text{and}\quad u_2 = (u_2^0, \ldots, u_2^3)$$

define a ruled surface swept by lines joining $u_1(x)$ and $u_2(x)$; these lines can be thought of points in $\mathrm{Gr}(2,4)$.

10. $n = m^2$ and $r = 2m$: Systems Modelled After $\mathrm{Gr}(m, m^2)$

The reader might think that for any m, n, and r, there would be some analogue of the Schwarzian derivative. Unfortunately things are not so simple. In this section, we consider a seemingly simple case: Systems of linear differential equations in m^2 variables x^{ij} $(1 \leq i, j \leq m)$ of rank $2m$ with m unknowns u^k $(1 \leq k \leq m)$ and the changes K of unknowns

$$(u^k) \to \left(\sum_l K_l^k u^l\right), \quad \det(K_l^k) \not\equiv 0;$$

two systems related under such changes are said to be equivalent ([**SY5**], cf. [**SYY**]). As the ratio of two linearly independent solutions of 2nd order ordinary differential equation defines a map (the Schwarz map) from the x-space to the projective line, any $2m$ linearly independent solutions of our system defines a map from the $x = (x^{ij})$-space to the $(m, 2m)$-Grassmannian variety $\mathrm{Gr}(m, 2m)$; two equivalent systems define the same Schwarz map. We assume that this map is nondegenerate. We shall show that we can follow the classical argument to some extent but not to the quite same.

The system in question is an innocent analogue of the classical model case $(n = 1, m = 1, r = 2m)$, we just replace scalar variable and scalar unknown by $m \times m$-matrices. Or, this can be seen as a generalization of the case treated in §9 of order $d = 2$ to a high dimensional source space. When $m = 2$, since $\mathrm{Gr}(2, 4)$ can be embedded in $\mathbb{P}^5$ as a quadratic hypersurface, the uniformizing equation of the K3 surfaces presented in §8 can be transformed into a system modelled after $\mathrm{Gr}(2, 4)$.

Let us regard one variable, say x^{11}, special and write down our system as

$$E = E_m(a, b, \alpha, \beta) \begin{cases} u^k_{:11:11} &= \sum_l \alpha_l^k u^l_{:11} + \sum_l \beta_l^k u^l, \\ u^k_{:ij} &= \sum_l a^k_{ijl} u^l_{:11} + \sum_l b^k_{ijl} u^l, \end{cases}$$

$1 \leq k, l, i, j \leq m$, where $f_{:ij}$ stands for $\partial f / \partial x^{ij}$, and

$$a^k_{11l} = \delta^k_l, \quad b^k_{11l} = 0.$$

The Schwarz map of the system E is nondegenerate if and only if $m^2 \times m^2$-determinant

$$W = \det(a^k_{ijl})_{(i,j),(k,l)}$$

does not vanish identically.

The transformation

$$u^k \to \sum_l K_l^k u^l, \quad \det K_l^k \neq 0$$

changes the coefficients a as

$$a^k_{ijl} \to \sum K^k_p a^p_{ijq} (K^{-1})^q_l,$$

in other words,

$$\mathbb{A} = (a^k_l) \to K \mathbb{A} K^{-1}, \quad a^k_l = \sum a^k_{ijl}\, dx^{ij},$$

and α as

$$\mathcal{A} = (\alpha^k_l) \to (2K_{:11} + K\mathcal{A})K^{-1}.$$

Thus we can normalize the system as $\mathcal{A} = 0$, but still remains a freedom of transformations K satisfying $K_{:11} = 0$; this implies that the survived coefficients of this normalized system can not be expressed by the coefficients of the given system

E. Nevertheless we can control the system: We can extract a set of essential coefficients as follows.

Under the assumptions $W \neq 0$ and $\mathbb{A} \wedge \mathbb{A} \neq 0$, where $\mathbb{A} = (a_l^k)$, the coefficients a determine the other coefficients b and β up to adding an exact 1-form $dk(x)$, where $k(x)$ is independent of x^{11}, to b_i^i $(i = 1, \dots, m)$. This ambiguity is caused by the scalar transformation $K = k(x)I_n$.

Hence we have the following conclusion: Two systems $E_m(a, b, \alpha, \beta)$ and $E_m(\bar{a}, \bar{b}, \bar{\alpha}, \bar{\beta})$ are equivalent if only if there is K such that

$$\bar{\mathbb{A}} = K^{-1}\mathbb{A}K, \quad K = (K_l^k), \quad \mathbb{A} = (a_l^k), \quad a_l^k = \sum a_{ijl}^k \, dx^{ij},$$

provided that $W \neq 0$ and $\mathbb{A} \wedge \mathbb{A} \neq 0$.

11. Generalizations in Different Context

The Schwarzian derivative is used in the theory of univalent functions and in the study of the Teichmüller spaces. A fundamental fact known as Nehari's theorem can be stated as
If a function z holomorphic in $|x| < 1$ is univalent, then

$$\left|\{z; x\}\right| \leq 6\left(1 - |z|^2\right)^{-2}.$$

Conversely if

$$\left|\{z; x\}\right| \leq 2\left(1 - |z|^2\right)^{-2},$$

then z is univalent in $|x| < 1$. This leads to the boundedness of the Bers embedding of the Teichmüller space. Kobayashi–Wada [**KY**] defined a variant of the Schwarzian derivative for nondegenerate maps between two Riemann manifolds, and established a generalization of Nehari's theorem.

On the other hand, Sato [**Sat**] studied obstructions for 2nd order ordinary differential equations $y'' = f(x, y, y')$ to be reduced to the equation $y'' = 0$ under transformations of (x, y)-space, and found that the obstructions can be expressed in terms of the fundamental differential invariants I, J, M, and N appeared in §5. He moreover derived obstructions for 3rd order ordinary differential equations $y''' = f(x, y, y', y'')$ to be reduced to the equation $y''' = 0$ under contact transformations of (x, y, y')-space, and called them contact Schwarzian derivatives.

References

[ACT] D. Allcock, J. Carlson, and D. Toledo, *A complex hyperbolic structure for moduli of cubic surfaces*, C. R. Acad. Sci. Paris Sér. I Math. **326** (1998), no. 1, 49–54.

[Bou] A. Boulanger, *Contribution à l'étude des équations différentielles linéaires homogènes intégrable algébriquement*, J. de l'École Polytechnique **4** (1898), 1–122.

[KY] O. Kobayashi and M. Wada, *Circular geometry and the Schwarzian*, Far East J. Math. Sci. (FJMS) 2000, Special Volume, Part III, pp. 335–363.

[MSY1] K. Matsumoto, T. Sasaki, and M. Yoshida, *The monodromy of the period map of a 4-parameter family of* K3 *surfaces and the Aomoto–Gel'fand hypergeometric function of type* (3, 6), Internat. J. Math. **3** (1992), 1–164.

[MSY2] ———, *Recent progress of Gauss–Schwarz theory and related geometric structures*, Kyushu J. Math. **47** (1993), 283–381.

[MT] K. Matsumoto and T. Terasoma, *Theta constants associated to cubic threefolds*, math.AG/0008024.

[Sas] T. Sasaki, *Projective differential geometry and linear homogeneous differential equations*, Rokko Lectures in Math **5** (1999).

[Sat] H. Sato, *Schwarzian derivatives of contact diffeomorphisms*, Lobachevskii J. Math. **4** (1999), 89–98.

[Sch] H. A. Schwarz, *Ueber diejenigen Fälle, inwelchen die Gaussische hypergeometrische Reihe eine algebraische Funktion ihres vierten Elementes darstellt*, J. Reine Angew. Math. **75** (1872), 292–335.

[Sea] Y. Se-ashi, *A geometric construction of Laguerre–Forsyth's canonical forms of linear ordinary differential equations*, Adv. Stud. Pure Math. **22** (1993), 265–297.

[SYY] T. Sasaki, K. Yamaguchi, and M. Yoshida, *On the rigidity of differential systems modelled on hermitian symmetric spaces and disproofs of a conjecture concerning modular interpretations of configuration spaces*, Adv. Stud. Pure Math. **25** (1997), 315–354.

[SY1] T. Sasaki and M. Yoshida, *Linear differential equations in two variables of rank four*. I, Math. Ann. **282** (1988), 69–93.

[SY2] ———, *Linear differential equations modeled after hyperquadrics*, Tôhoku Math. J. **41** (1989), 321–348.

[SY3] ———, *The uniformizing differential equation of the complex hyperbolic structure on the moduli space of marked cubic surfaces*, Proc. Japan Acad. Ser. A Math. Sci. **75** (1999), 129–133.

[SY4] ———, *The uniformizing differential equation of the complex hyperbolic structure on the moduli space of marked cubic surface*. II, Kowalevski Workshop on Mathematical Methods of Regular Dynamics (Leeds, 2000), J. Phys. A **34** (2001), no. 11, pp. 2319–2328.

[SY5] ———, *Invariant theory for linear differential systems modelled after the Grassmannian* $\mathrm{Gr}(n, 2n)$, Preprint, Kyushu Univ., 2001-1.

[W] E. J. Wilczynski, *Projective differential geometry and curves and ruled surfaces*, Chelsea Publishing Co., New York, 1962.

[Y1] M. Yoshida, *Fuchsian differential equations*, Aspects Math. **E11**, Vieweg, Wiesbaden 1987.

[Y2] ———, *Hypergeometric functions, my love*, Vieweg, Wiesbaden 1997.

Department of Mathematics, Faculty of Science, Kobe University Rokko, Kobe 657, Japan.

E-mail address: sasaki@math.kobe-u.ac.jp

Faculty of Mathematics, Ropponmatsu 4-2-1, Chuoku, Fukuoka 810, Japan.

E-mail address: myoshida@math.kyushu-u.ac.jp

Centre de Recherches Mathématiques
CRM Proceedings and Lecture Notes
Volume **32**, 2002

Elliptic Solitons and Heun's Equation

Alexander O. Smirnov

ABSTRACT. We find a new class of algebraic geometric solutions of Heun's equation with the accessory parameter belonging to a hyperelliptic curve. Dependence of these solutions from the accessory parameter as well as their relation to Heun's polynomials is studied. Methods of calculating the algebraic genus of the curve, and its branching points, are suggested. Monodromy group is considered. Numerous examples are given.

Introduction

The Heun equation (see, for instance, [**17, 24, 31, 32**])

$$\frac{d^2y}{dz^2} + \left(\frac{\gamma}{z} + \frac{\delta}{z-1} + \frac{\varepsilon}{z-a}\right)\frac{dy}{dz} + \frac{\alpha\beta z - q}{z(z-1)(z-a)}y = 0 \tag{0.1}$$

where

$$1 + \alpha + \beta - \gamma - \delta - \varepsilon = 0,$$

is a Fuchsian equation with four regular singularities at 0, 1, a and ∞. Recently, it started to attract much attention due to a large number of its applications in mathematics and physics [**1, 9, 25, 26, 28, 48**].

As a matter of fact, this equation has been known for a very long time. Already in 1882 Darboux mentioned that it is the next one beyond the hypergeometric equation and that "it has already been studied by many geometers" [**10**]. In the works of Darboux this equation appears when he investigated the minimal surfaces in [**11**] and when he studied harmonic equations in [**12**]. In the work [**10**] Darboux points out at a close connection between the equation (0.1) and the generalization of Lamé's equation

$$\psi_{xx} - u(x)\psi = E\psi, \tag{0.2}$$

$$u(x) = m_0(m_0+1)\wp(x) + \sum_{i=1}^{3} m_i(m_i+1)\wp(x-\omega_i), \tag{0.3}$$

$$\wp(\omega_i) = e_i, \qquad \wp(x - 2\omega_i) \equiv \wp(x).$$

2000 *Mathematics Subject Classification.* Primary: 33E30; Secondary: 33E10, 34L40, 34M35.

This paper was translated from Russian by V. B. Kuznetsov.

This is the final form of the paper.

Here $\wp(x)$ is the Weierstrass function [**2**],

$$[\wp'(x)]^2 = 4\wp^3(x) - g_2\wp(x) - g_3 = 4\prod_{j=1}^{3}(\wp(x) - e_j). \tag{0.4}$$

In the same work Darboux remarks that the equation (0.2), (0.3) appeared already in the works of Hermite on solution of a special case (integer N) of Lamé's equation:

$$\psi_{xx} - N(N+1)\wp(x)\psi = E\psi. \tag{0.5}$$

Using the methods "proposed by Hermite in 1872 in the course of lectures in École Polytechnique", Darboux solves the equation (0.2), (0.3) with integer m_i and points out that one can find solutions of the equation (0.2), (0.3) for any value of E in other special cases. In particular, "when all m_i are half-odd-integer and N is odd integer" [1].

Unfortunately, until recently the work [**10**] was not well-known[2], in connection with this the equation (0.2), (0.3) has the name of Treibich and Verdier who proved the finite-gapness of the potential (0.3) for any integer m_i [**39, 42, 43**].

In the present paper we study exact solutions of a special case of Heun's equation:

$$\frac{d^2y}{dz^2} + \frac{1}{2}\left(\frac{1-2m_1}{z} + \frac{1-2m_2}{z-1} + \frac{1-2m_3}{z-a}\right)\frac{dy}{dz} + \frac{N(N-2m_0-1)z+\lambda}{4z(z-1)(z-a)}y = 0 \tag{0.6}$$

where

$$N = m_0 + m_1 + m_2 + m_3, \qquad m_0, m_1, m_2, m_3 \in \mathbb{Z}, \quad \lambda, z \in \mathbb{C}. \tag{0.7}$$

It is well-known (see, e.g., [**17, 24, 31, 32**]) that by putting

$$m_1 = m_2 = m_3 = 0, \quad m_0 = N$$

and changing the independent variable

$$z = \frac{\wp(x) - e_1}{e_2 - e_1}, \quad a = \frac{e_3 - e_1}{e_2 - e_1}, \qquad y(z) = \psi(x), \quad \lambda = \frac{E - N(N+1)e_1}{e_1 - e_2}, \tag{0.8}$$

this equation is reduced to the familiar Lamé's equation (0.5).

When $N \in \mathbb{N}$ the solutions of the equation (0.5) are the eigenfunctions of the Schrödinger operator (0.2) with the N-gap elliptic potential

$$u(x) = N(N+1)\wp(x).$$

They can be found either in terms of solutions of a differential equation satisfied by products of these eigenfunctions [**10, 17, 24, 46**] or by using the Krichever–Hermite ansatz [**4–7**].

---

[1] so far we do not know by which method one can find solutions of this special case of Heun's equation.

[2] the author learned about the work [**10**] in 1999 from Prof. V. B. Matveev.

In general, i.e. when $m_i \in \mathbb{C}$, the change of variables (see, for instance, [**10–12**])

$$(0.8') \qquad \begin{aligned} z &= \frac{\wp(x) - e_1}{e_2 - e_1}, & a &= \frac{e_3 - e_1}{e_2 - e_1}, \\ y(z) &= \psi(x)\prod_{i=1}^{3}(\wp(x) - e_i)^{m_i/2}, & \lambda &= \frac{E}{e_1 - e_2} + \text{const.}, \end{aligned}$$

transforms Heun's equation into Schrödinger's equation (0.2) with the potential (0.3), which for integer m_i will be finite-gap [3].

Because the Treibich–Verdier equation (0.2), (0.3) with integer m_i has been actively studied over the years by many researchers [**4, 5, 39, 42, 43**], including the author [**33–35**], it became possible to apply the obtained results to the theory of solutions of Heun's equation.

In particular, in this paper it is proved that for any nonnegative integer m_i and for all $\lambda \neq \lambda_j$ $(j = 1, \dots, 2g+1)$ the solutions of the Heun equation (0.6) are the functions of the following form:

$$\begin{aligned} &Y_{1,2}(m_0, m_1, m_2, m_3; \lambda; z) \\ &\qquad = \sqrt{\Psi_{g,N}(\lambda, z)}\exp\left(\pm\frac{\mathrm{i}\nu(\lambda)}{2}\int\frac{z^{m_1}(z-1)^{m_2}(z-a)^{m_3}\,dz}{\Psi_{g,N}(\lambda, z)\sqrt{z(z-1)(z-a)}}\right). \end{aligned}$$

Here $\mathrm{i}^2 = -1$,

$$\Gamma\colon \nu^2 = \prod_{j=1}^{2g+1}(\lambda - \lambda_j), \quad \lambda_j = \lambda(E_j),$$

E_j are the gap edges of the finite-gap elliptic potential $u(x)$ (0.3) and $\Psi_{g,N}(\lambda, z)$ is some polynomial of the degree N in z and of the degree g in λ. We develop the methods of calculating the polynomial $\Psi_{g,N}(\lambda, z)$, the branching points λ_j of the algebraic curve $\Gamma = \{(\nu, \lambda)\}$ and, respectively, the gap edges E_j of the spectrum of the Treibich–Verdier potential. We find the dependence of the genus g of the curves Γ and $\widetilde{\Gamma} = \{(w, E)\}$,

$$(0.9) \qquad \widetilde{\Gamma}\colon w^2 = \prod_{j=1}^{2g+1}(E - E_j),$$

as a function of the characteristics (exponents) m_i. The monodromy group of these solutions is also considered.

We also prove that for any integer m_i's and for $\lambda = \lambda_j$ the solutions of Heun's equation are expressed in terms of Heun polynomials [**17, 31, 32**], i.e. Heun polynomials with half-odd-integer $\beta - \alpha$, γ, δ, ϵ are special cases of the found solutions.

The needed facts from the theory of finite-gap elliptic potentials for the Schrödinger operator are collected without proof in the first paragraph.

Because the exact solutions of the Heun equation are of much interest, in the Appendix we give several simplest solutions of the Heun equation belonging to this class.

---

[3] changes of variables (0.8), (0.8′) are easily generalized to the case of Schrödinger's equation with any even elliptic potential. Based on our results in [**33, 35**] we plan a further study of a special case of the Fuchsian equation with five singularities.

The author thanks S. Yu. Slavyanov, V. B. Matveev, V. B. Kuznetsov and A. Ya. Kazakov for useful discussions and the referees for helpful suggestions about improving the text.

1. Schrödinger Operator with Finite-Gap Elliptic Potential

PROPOSITION 1 ([**21, 41, 44**]). For any $m_i \in \mathbb{Z}$ the Treibich–Verdier potential (0.3) is a finite-gap elliptic potential for the Schrödinger operator (0.2). For any nonnegative integer m_i the following formula for the genus g of the spectral curve $\widetilde{\Gamma}$ (0.9) for the potential (0.3) is true:

$$g = \frac{1}{2}\max\left\{2\max_{0\leqslant i\leqslant 3} m_i, 1+N-(1+(-1)^N)\left(\min_{0\leqslant i\leqslant 3} m_i + \frac{1}{2}\right)\right\}. \tag{1.1}$$

PROPOSITION 2 ([**8, 29, 30, 47**]). Any g-gap potential $u(x)$ of the Schrödinger operator (0.2) satisfies the Novikov equation:

$$J_g + \sum_{m=1}^{g} c_m J_{g-m} = d, \tag{1.2}$$

or, which is the same, the stationary 'higher' Korteweg–de Vries (KdV) equation:

$$\partial_x\left(J_g + \sum_{m=1}^{g} c_m J_{g-m}\right) = 0.$$

Here c_m, d are constants and the functions J_m are the flows of the 'higher' KdV equations

$$\partial_{t_m} u = \partial_x J_m.$$

The expressions for the flows J_m are found from the relations

$$L\psi = \psi_{xx} - 4u\psi + 2u_x\int_x^\infty \psi(\tau)\,d\tau, \tag{1.3a}$$

$$(J_n)_x = L^n(u_x) \tag{1.3b}$$

where $u(x)$ is a potential decreasing fast at ∞. In particular,

$$J_0 = u, \quad J_1 = u_{xx} - 3u^2, \quad J_2 = u_{xxxx} - 10u_{xx}u - 5u_x^2 + 10u^3,$$

$$J_3 = u_{xxxxxx} - 14uu_{xxxx} - 28u_xu_{xxx} - 21u_{xx}^2 + 70u^2u_{xx} + 70uu_x^2 - 35u^4.$$

In the case of decreasing fast at ∞ potential the variables

$$C_j = \int_{-\infty}^{\infty} J_j(x,t)\,dx$$

constitute an infinite set of integrals of motion for the KdV equation

$$\partial_t u = \partial_x J_1.$$

PROPOSITION 3 ([**13, 29, 47**]). The function

$$\widehat{\Psi}(x,E) = E^g + \sum_{j=1}^{g} \gamma_j(x)E^{g-j} \tag{1.4}$$

where

$$\gamma_j(x) = -\frac{2}{4^j}\left(J_{j-1} + \sum_{m=1}^{j-1} c_m J_{j-m-1} - \frac{c_j}{2}\right), \tag{1.5}$$

obeys the equation

$$\widehat{\Psi}_{xxx} = 4(u+E)\widehat{\Psi}_x + 2u_x\widehat{\Psi},$$

the solutions of which are the products of any two solutions of the equation (0.2). Here c_j, J_j are the same as in (1.2) and $u(x)$ is a g-gap potential of the Schrödinger operator (0.2).

PROPOSITION 4 ([**4–7**]). The equation (0.2) with the potential of Treibich–Verdier (0.3) has a solution of the form (Krichever–Hermite ansatz)

$$\psi(x;k,\alpha) = \sum_{i=0}^{3}\sum_{j=0}^{m_i-1} a_{ij}(k,\alpha)e^{kx}\frac{d^j}{dx^j}\Phi(x-\omega_i,\alpha). \tag{1.6}$$

Here k, α are auxiliary spectral parameters, $\omega_0 \equiv 0$,

$$\Phi(x,\alpha) = \frac{\sigma(\alpha-x)}{\sigma(x)\sigma(\alpha)}\exp\{\zeta(\alpha)x\}, \tag{1.7}$$

and $\sigma(\alpha)$, $\zeta(\alpha)$ are Weierstrass' elliptic functions [**2**].

Substituting the ansatz (1.6) for the ψ-function into the equation (0.2) one can find the equation of the spectral curve Γ for the finite-gap potential $u(x)$

$$\Gamma = \{(k,\alpha)\}: R(k,\alpha) = 0,$$

as well as the dependence of the coefficients $a_{ij}(k,\alpha)$ and the main spectral parameter E

$$E(-k,-\alpha) = E(k,\alpha)$$

as functions of the auxiliary spectral parameters k and α.

The gap edges of the spectrum of the finite-gap potential $u(x)$ are the values of the spectral parameter E calculated at the points $\mathcal{P}$ of the curve Γ

$$E_j = E(\mathcal{P}_j), \quad \tau\mathcal{P}_j = \mathcal{P}_j, \qquad j = 1,\dots,2g+2, \; E_{2g+2} = \infty$$

which are invariant with respect to the involution τ

$$\tau\colon (k,\alpha) \to (-k,-\alpha).$$

Linear independent solutions of the equation (0.2) for all values of the spectral parameter E, except for the gap edges E_j, are the functions

$$\psi_1(x,E) = \psi(x;k,\alpha) \quad \text{and} \quad \psi_2(x,E) = \psi(x;-k,-\alpha). \tag{1.8}$$

A product of these functions is a doubly periodic function and, modulo the normalizing factor, it is equal to $\widehat{\Psi}(x,E)$ (1.4), (1.5).

2. Finite-Gap Solutions of Heun's Equation

Since the potential (0.3) is finite-gap, one can introduce the notion of 'finite-gap' solutions to the Heun equation which are obtained from the finite-gap solutions of the Treibich–Verdier equation by the inverse change of variables (see (0.8))

$$\wp(x) = e_1 + (e_2-e_1)z, \quad e_2 = \frac{a-2}{a+1}e_1, \quad e_3 = \frac{1-2a}{a+1}e_1, \tag{2.1a}$$

$$E = (e_1-e_2)\lambda + \text{const.}, \quad u(x) = 4(e_2-e_1)U(z) + \text{const.}_1. \tag{2.1b}$$

DEFINITION. By Novikov's equation for the Heun equation of order g we will call the following equality:

$$I_g + \sum_{j=1}^{g} \widetilde{c}_j I_{g-j} = \widetilde{d} \tag{2.2}$$

where $\widetilde{c}_j$, $\widetilde{d}$ are some constants,

$$I_0 = U(z), \quad I_{j+1} = \mathcal{L}(I_j), \tag{2.3a}$$

$$\mathcal{L}(f) = z(z-1)(z-a)\frac{d^2f}{dz^2} + \frac{3z^2 - 2(a+1)z + a}{2} \cdot \frac{df}{dz} - \int \left(4I_0 \frac{df}{dz} + 2f\frac{dI_0}{dz} \right) dz, \tag{2.3b}$$

$U(z)$ is the 'potential' (2.1b)[4]

$$U(z) = \frac{m_0(m_0+1)}{4} z + \frac{m_1(m_1+1)}{4} \cdot \frac{a}{z} + \frac{m_2(m_2+1)}{4} \cdot \frac{z-a}{z-1} + \frac{m_3(m_3+1)}{4} \cdot \frac{a(z-1)}{z-a}. \tag{2.4}$$

THEOREM 1. *The 'potential' $U(z)$ (2.4) satisfies Novikov's equation for the Heun equation of order g (2.2), (2.3) if and only if the potential $u(x)$ (0.3) satisfies Novikov's equation (1.2), (1.3) of the same order.*

DEFINITION. 'Potential' $U(z)$ (2.4) will be called g-gap potential if it satisfies Novikov's equation for the Heun equation of order g (2.2), (2.3).

REMARK 1. The flows J_m (1.3) in the Novikov equation (1.2) were originally defined for the potentials decreasing fast at ∞. Although for finite-gap as well as for any periodic potentials these functions can not be obtained from the equations (1.3) because of the divergent integral with an infinite upper limit, the expressions for J_m are still valid in the case of a finite-gap potential.

The constant of integration in the definition of the 'flows' I_n (2.3) can not be easily fixed too. Because of that, all the 'flows' I_n are defined modulo a linear combination of lower order 'flows'. However, it is easily seen that the property of 'finite-gapness' of the 'potential' $U(z)$ (2.4) does not depend from the concrete values of integration constants in the definition of I_n. The latter affect only the values of the constants $\widetilde{c}_m$ and $\widetilde{d}$ in the equation (2.2).

THEOREM 2. *For any $m_i \in \mathbb{Z}$ there exists a nonnegative integer g such that the 'potential' $U(z)$ (2.4) satisfies Novikov's equation for the Heun equation of order g* (2.2), (2.3).

PROOF. Assume that all characteristics m_i in the equation (2.4) are nonnegative. If anyone $m_i < 0$ then one can always make the transformation $m_i \to -m_i - 1$ which does not change the 'flows' I_n (2.3), (2.4).

From the properties of the flows J_n (1.3) (see, e.g., [**8**]), from the properties of elliptic functions [**2**] and from the equation of the change of variable (2.1) it follows

[4]the term "potential" attached to the function $U(z)$ is used here only in analogy with the function $u(x)$, and it does not carry any additional meaning.

that all 'flows' I_n (2.3) are rational functions of the variable z (i.e. these functions do not have logarithmic singularities).

It is also not very difficult to check that the order of poles of the functions $I_n(z)$ in the singularities of Heun's equation does not exceed the corresponding characteristic m_i.

Indeed, it is easy to verify that for $m_1 = 0$ neither $I_0(z) = U(z)$ (2.4) nor all other $I_n(z)$ have a pole at $z = 0$, but for $m_1 \neq 0$ the function $I_0(z)$ has at this point a pole of first order.

If we now assume that the function $I_n(z)$ has at the point $z = 0$ a pole of order $\alpha \leqslant m_1$ (which is undoubtedly true for $n = 0$ and $m_1 \neq 0$):

$$I_n(z) = \frac{A}{z^\alpha} + O(z^{1-\alpha}), \quad z \to 0,$$

then we obtain that the function $I_{n+1}(z)$,

$$I_{n+1}(z) = \mathcal{L}\big(I_n(z)\big) = \frac{(2\alpha+1)(\alpha - m_1)(\alpha + m_1 + 1)}{2(\alpha+1)} \cdot \frac{aA}{z^{\alpha+1}} + O(z^{-\alpha}),$$

has at the same point a pole of order $\alpha' = \min\{m_1, \alpha + 1\}$.

By similar means one can check the other poles of the function $I_n(z)$.

Hence, for any integer $n \geqslant 0$ the dimension of the linear span of rational functions $1, I_0, \dots, I_n$ does not exceed $N+1$ and therefore there exists a number g,

$$\max_{0 \leqslant i \leqslant 3} m_i \leqslant g \leqslant N, \tag{2.5}$$

such that the equality (2.2) is fulfilled. □

COROLLARY 1. *For any $m_i \in \mathbb{Z}$ the potentials $U(z)$ (2.4) and $u(x)$ (0.3) are finite-gap. The genus g of the spectral curve Γ, associated with the finite-gap potential $u(x)$ (0.3), satisfies the inequality (2.5).*

DEFINITION. Solutions of the Heun equation (0.6) with integer characteristics m_i will be called finite-gap solutions of Heun's equation.

REMARK 2. As it was said above (cf. Introduction) the proof of finite-gapness of the potential $u(x)$ (0.3) was first given in the works by Treibich and Verdier [**39, 43**]. Here we gave another proof which allowed us to specify the explicit form of the finite-gap solutions of Heun's equation.

In order to find solutions of Heun's equation (0.6) let us use the results of the theory of finite-gap elliptic potentials for the Schrödinger operator and consider the equation ($m_i \geqslant 0$, $i = 0, 1, 2, 3$)

$$\begin{aligned} \frac{d^3\Psi}{dz^3} + 3P(z)\frac{d^2\Psi}{dz^2} + \big(P'(z) + 4Q(z) + 2P^2(z)\big)\frac{d\Psi}{dz} \\ + \big(2Q'(z) + 4P(z)Q(z)\big)\Psi = 0, \end{aligned} \tag{2.6}$$

$$P(z) = \frac{1}{2}\left(\frac{1-2m_1}{z} + \frac{1-2m_2}{z-1} + \frac{1-2m_3}{z-a}\right),$$

$$Q(z) = \frac{N(N - 2m_0 - 1)z + \lambda}{4z(z-1)(z-a)},$$

solutions of which are the products of any two solutions of Heun's equation (0.6).

THEOREM 3. *Equation (2.6) with nonnegative integer characteristics m_i has as its solution the function $\Psi_{g,N}(\lambda, z)$, which is a polynomial in λ of the degree g and in z of the degree N (0.7). The leading coefficient of this function considered as a polynomial in λ is equal*

$$\widetilde{a}_0(z) = z^{m_1}(z-1)^{m_2}(z-a)^{m_3}. \tag{2.7}$$

In other words,

$$\begin{aligned} \Psi_{g,N}(\lambda, z) &= a_0(\lambda)z^N + a_1(\lambda)z^{N-1} + \dots + a_N(\lambda) \\ &= z^{m_1}(z-1)^{m_2}(z-a)^{m_3}\lambda^g + \widetilde{a}_1(z)\lambda^{g-1} + \dots + \widetilde{a}_g(z). \end{aligned} \tag{2.8}$$

PROOF. From (1.4)–(1.8) it follows that the product $\widehat{\Psi}(x, E)$ of eigenfunctions of the Schrödinger operator with the Treibich–Verdier potential is an elliptic meromorphic function in the variable x, and it is a polynomial of degree g in the spectral parameter E. As a function of the variable x the function $\widehat{\Psi}(x, E)$ has poles of multiplicity $2m_i$ at the points $x = \omega_j$ ($\omega_0 \equiv 0$).

From (1.4), (0.3) and from the properties of the Weierstrass $\wp$-function it follows that $\widehat{\Psi}(x, E)$ is a rational function in $\wp(x)$. Hence, the function

$$\Psi_{g,N}(\lambda, z) = \text{const.} \cdot \widehat{\Psi}(x, E) \prod_{j=1}^{3} \bigl(\wp(x) - e_j\bigr)^{m_i} \tag{2.9}$$

where λ and z are related with E and x by equalities (2.1), is a polynomial in λ of degree g and in z of degree N (i.e. it is a rational function in the variable z with the unique pole of order N at the point $z = \infty$). The constant in the equality (2.9) is chosen such that the leading coefficient of $\Psi_{g,N}(\lambda, z)$, considered as a polynomial in λ, to be equal (2.7). □

COROLLARY 2. *Coefficients $\widetilde{a}_j(z)$ of the polynomial $\Psi_{g,N}(\lambda, z)$ have the form:*

$$\widetilde{a}_j(z) = z^{m_1}(z-1)^{m_2}(z-a)^{m_3}\widetilde{I}_{j-1}, \quad j = 1, \dots, g \tag{2.10}$$

where $\widetilde{I}_j$ is a linear combination of the rational functions $I_j, \dots, I_0, 1$ having poles in the singularities of Heun's equation.

PROOF. The proof follows from the equalities (1.5), (2.9) and from the change of variable (2.1). □

Knowing the product of the solutions of Heun's equation it is not difficult to find the solutions themselves (see, for instance, [**46**, §19.53, §23.7, §23.71]).

THEOREM 4. *Finite-gap solutions of Heun's equation with nonnegative characteristics m_i have the form*

$$\begin{aligned} &Y_{1,2}(m_0, m_1, m_2, m_3; \lambda; z) \\ &\qquad = \sqrt{\Psi_{g,N}(\lambda, z)}\exp\left(\pm\frac{\mathrm{i}\nu(\lambda)}{2}\int\frac{z^{m_1}(z-1)^{m_2}(z-a)^{m_3}\,dz}{\Psi_{g,N}(\lambda, z)\sqrt{z(z-1)(z-a)}}\right). \end{aligned} \tag{2.11}$$

Here $\mathrm{i}^2 = -1$,

$$\Gamma\colon \nu^2 = \prod_{j=1}^{2g+1}(\lambda - \lambda_j), \quad \lambda_j = \lambda(E_j), \tag{2.12}$$

E_j are the gap edges of the finite-gap elliptic potential $u(x)$ (0.3).

PROOF. From the Liouville formula it follows that the Wronskian of two linearly independent solutions of the linear homogeneous differential equation,

$$y'' + P(z)y' + Q(z)y = 0,$$

has the following dependence from z:

$$W\big[y_1(\lambda, z), y_2(\lambda, z)\big] = W_0(\lambda) \exp\Big\{-\int_{z_0}^{z} P(t)\,dt\Big\}$$

where $W_0(\lambda)$ is Wronskian's value at z_0. Hence, the Wronskian of two linearly independent solutions of Heun's equation (0.6) is equal to

$$W\big[y_1(\lambda, z), y_2(\lambda, z)\big] = -\mathrm{i}\nu(\lambda) \cdot \frac{z^{m_1}(z-1)^{m_2}(z-a)^{m_3}}{\sqrt{z(z-1)(z-a)}} \tag{2.13}$$

where

$$\nu(\lambda) = \mathrm{i}W_0(\lambda) \cdot \frac{\sqrt{z_0(z_0-1)(z_0-a)}}{z_0^{m_1}(z_0-1)^{m_2}(z_0-a)^{m_3}}.$$

If we now divide the Wronskian of two solutions (2.13) by their product,

$$y_1(\lambda, z) \cdot y_2(\lambda, z) = \Psi_{g,N}(\lambda, z), \tag{2.14}$$

we obtain a simple differential equation of the first oder:

$$\frac{y_2'}{y_2} - \frac{y_1'}{y_1} = -\mathrm{i}\nu(\lambda) \cdot \frac{z^{m_1}(z-1)^{m_2}(z-a)^{m_3}}{\Psi_{g,N}(\lambda, z)\sqrt{z(z-1)(z-a)}}. \tag{2.15}$$

The equation (2.15) can be easily integrated:

$$\frac{y_2(\lambda, z)}{y_1(\lambda, z)} = C(\lambda) \cdot \exp\left[-\mathrm{i}\nu(\lambda)\int_{z_1}^{z} \frac{t^{m_1}(t-1)^{m_2}(t-a)^{m_3}\,dt}{\Psi_{g,N}(\lambda, t)\sqrt{t(t-1)(t-a)}}\right] \tag{2.16}$$

where $C(\lambda) = y_2(\lambda, z_1)/y_1(\lambda, z_1)$.

If we now consider solutions $Y_{1,2}(\lambda, z)$ with the same product (2.14) and Wronskian (2.13)

$$Y_1(\lambda, z) = \sqrt{C(\lambda)} \cdot y_1(\lambda, z), \quad Y_2(\lambda, z) = \frac{y_2(\lambda, z)}{\sqrt{C(\lambda)}}, \tag{2.17}$$

then from (2.14) and (2.16) we get (2.11).

Substituting the ansatz (2.11) into the Heun equation (0.6) we get

$$\nu^2(\lambda) = \frac{2\Psi\Psi'' - (\Psi')^2 + 2P(z)\Psi\Psi' + 4Q(z)\Psi^2}{z^{2m_1-1}(z-1)^{2m_2-1}(z-a)^{2m_3-1}} \tag{2.18}$$

where

$$\Psi = \Psi_{g,N}(\lambda, z), \quad \Psi' = \frac{d\Psi}{dz}, \quad \Psi'' = \frac{d^2\Psi}{dz^2}.$$

From (2.8), (2.18) it follows that $\nu^2(\lambda)$ is a polynomial in λ of the degree $2g+1$ with the leading coefficient equal to 1.

It is not difficult to show that under the change (0.8′) the solutions $Y_{1,2}(\lambda, z)$ of Heun's equation (0.6) turn into Floquet solutions of the Treibich–Verdier equation (0.2), (0.3). Therefore, zeros λ_j ($j = 1, \ldots, 2g+1$) of the polynomial $\nu(\lambda)$ or, which is the same, zeros of the Wronskian of the solutions $Y_{1,2}(\lambda, z)$ correspond to zeros of the Wronskian of the Floquet solutions of the Treibich–Veridier equation, i.e. they correspond to the gap edges E_j ($j = 1, \ldots, 2g+1$) of spectrum of the

Triebich–Verdier potential. Hence, the hyperelliptic curve Γ (2.12) is isomorphic to the spectral curve $\widetilde{\Gamma}$ (0.9) of the finite-gap elliptic potential $u(x)$ (0.3). □

REMARK 3. The fact that the equation (2.6) for any nonnegative integer m_i and for any λ has as its solution a polynomial in z of degree N was known to Darboux who in the work [**10**] pointed put at the method of finding some exact solutions of the generalized Lamé equation, which is now known as the 'Treibich–Verdier equation'. However, he did not study the dependence of this polynomial and, respectively, solutions of the Heun equation, from λ.

REMARK 4. In order to find the 'finite-gap' solutions of Heun's equation with negative characteristics one must use the equalities (see [**17, 24, 31, 32**]).

1. Negative m_1:

$$Y(m_0, -m_1 - 1, m_2, m_3; \lambda; z) = z^{-m_1-1/2} Y(m_0, m_1, m_2, m_3; \mu; z) \tag{2.19}$$

where

$$\mu = \lambda - (2m_1 + 1)(2m_2 - 1)a - (2m_1 + 1)(2m_3 - 1).$$

2. Negative m_2:

$$Y(m_0, m_1, -m_2 - 1, m_3; \lambda; z) = (z - 1)^{-m_2-1/2} Y(m_0, m_1, m_2, m_3; \mu; z) \tag{2.20}$$

where

$$\mu = \lambda - (2m_1 - 1)(2m_2 + 1)a.$$

3. Negative m_3:

$$Y(m_0, m_1, m_2, -m_3 - 1; \lambda; z) = (z - a)^{-m_3-1/2} Y(m_0, m_1, m_2, m_3; \mu; z) \tag{2.21}$$

where

$$\mu = \lambda - (2m_1 - 1)(2m_3 + 1).$$

4. Negative m_0:

$$Y(-m_0 - 1, m_1, m_2, m_3; \lambda; z) = Y(m_0, m_1, m_2, m_3; \lambda; z). \tag{2.22}$$

3. Finite-Gap Solutions and Heun Polynomials

Looking at the analytical properties of the solutions (2.11) it is not difficult to prove the following statements.

COROLLARY 3. *For $\lambda \neq \lambda_j$ ($j = 1, \ldots, 2g+1$), where λ_j are the branching points of the hyperelliptic curve Γ (2.12):*

1. *the polynomial $\Psi_{g,N}(\lambda, z)$ does not have zeros at the singularities of Heun's equation;*
2. *the leading coefficient $a_0(\lambda)$ of the polynomial $\Psi_{g,N}(\lambda, z)$ is not equal to zero;*
3. *all zeros z_k ($k = 1, \ldots, N$) of the polynomial $\Psi_{g,N}(\lambda, z)$ are simple zeros satisfying the relations*

$$\prod_{\substack{j=1 \\ j\neq k}}^{N} (z_k - z_j)^2 = -\frac{\nu^2(\lambda)}{a_0^2(\lambda)} z_k^{2m_1-1}(z_k - 1)^{2m_2-1}(z_k - a)^{2m_3-1}.$$

COROLLARY 4. *Zeros of the equations*

$$a_0(\lambda) = 0,\ \Psi_{g,N}(\lambda, 0) = 0,\ \Psi_{g,N}(\lambda, 1) = 0,\ \Psi_{g,N}(\lambda, a) = 0,\ \Delta(\lambda) = 0 \tag{3.1}$$

where $\Delta(\lambda)$ is the discriminant of the algebraic equation of Nth order in z:

$$\Psi_{g,N}(\lambda, z) = 0,$$

are among the branching points λ_j of the curve Γ (2.12).

LEMMA 1. *For $\lambda = \lambda_j$ the solutions (2.11) cease to be linearly independent. In this case the linear independent 'finite-gap' solutions of Heun's equation (0.6) are the functions*

$$Y_1(m_0, m_1, m_2, m_3; \lambda_j; z) = \sqrt{\Psi_{g,N}(\lambda_j, z)} \tag{3.2a}$$

and

$$Y_2(m_0, m_1, m_2, m_3; \lambda_j; z) = \sqrt{\Psi_{g,N}(\lambda_j, z)} \int \frac{z^{m_1}(z-1)^{m_2}(z-a)^{m_3}\, dz}{\Psi_{g,N}(\lambda_j, z)\sqrt{z(z-1)(z-a)}}. \tag{3.2b}$$

COROLLARY 5. *For $\lambda = \lambda_j$ $(j = 1, \dots, 2g+1)$, where λ_j are the branching points of the hyperelliptic curve Γ (2.12):*

1. *the polynomial $\Psi_{g,N}(\lambda_j, z)$ has the form*

$$\Psi_{g,N}(\lambda_j, z) = z^{M_{1j}}(z-1)^{M_{2j}}(z-a)^{M_{3j}} F_{nj}^2(z) \tag{3.3}$$

   where $M_{ij} \in \{0, 2m_i + 1\}$ and $F_{nj}(z)$ is a polynomial in z of order n;

2. *the polynomial $F_{nj}(z)$ is a solution of Heun's equation with the P-symbol*

$$P\begin{Bmatrix} 0 & 1 & a & \infty & ; z \\ 0 & 0 & 0 & -n & ; -\widetilde{\lambda}_j/4 \\ 1/2 + \widetilde{m}_{1j} & 1/2 + \widetilde{m}_{2j} & 1/2 + \widetilde{m}_{3j} & -n + 1/2 + \widetilde{m}_{0j} & \end{Bmatrix} \tag{3.4}$$

   where

$$\begin{gathered} M_{0j} \in \{0, 2m_0 + 1\}, \\ \widetilde{m}_{ij} = m_i - M_{ij}, \quad \widetilde{m}_{ij} \in \{m_i, -m_i - 1\}, \qquad i = 0,1,2,3, \\ n = (\widetilde{m}_{0j} + \widetilde{m}_{1j} + \widetilde{m}_{2j} + \widetilde{m}_{3j})/2, \\ \widetilde{\lambda}_j = \lambda_j + 2m_1 M_{3j} + 2m_3 M_{1j} - 2M_{1j}M_{3j} - M_{1j} - M_{3j} + \\ + (2m_1 M_{2j} + 2m_2 M_{1j} - 2M_{1j}M_{2j} - M_{1j} - M_{2j})a, \end{gathered} \tag{3.5}$$

   i.e. the polynomial $F_{nj}(z)$ modulo a constant factor is a Heun polynomial (see, e.g., **[31, 32]**);

3. *the order of the polynomial $\Psi_{g,N}(\lambda_j, z)$, as a polynomial in z, is equal to $(N - M_{0j})$;*

4. *all zeros z_k $(k = 1, \dots, n)$ of the polynomial $F_{nj}(z)$ are simple and satisfy the relations*

$$\sum_{\substack{i=1 \\ i \neq k}}^{n} \frac{1}{z_k - z_i} = \frac{1}{4}\left(\frac{2\widetilde{m}_{1j} - 1}{z_k} + \frac{2\widetilde{m}_{2j} - 1}{z_k - 1} + \frac{2\widetilde{m}_{3j} - 1}{z_k - a}\right).$$

As a consequence, to every branching point λ_j $(j = 1, \dots, 2g+1)$ there corresponds a set of numbers $\widetilde{m}_{ij}$ $(i = 0, 1, 2, 3)$ such that n (3.5) is a nonnegative integer. Solutions of Heun's equation for $\lambda = \lambda_j$ are expressed in terms of Heun's polynomials of order n.

COROLLARY 6. *For $N \neq 0$ every branching point λ_j of the hyperelliptic curve Γ (2.12) is a zero of one of the equations* (3.1).

PROOF. The proof is done by assuming the opposite. Let there exists a branching point λ_j which is not a zero of anyone of the equations (3.1). From the condition it follows that in this case the singularities of Heun's equation are not the zeros of the polynomial $\Psi_{g,N}(\lambda_j, z)$ (3.3). Therefore

$$M_{1j} = M_{2j} = M_{3j} = 0. \tag{3.6}$$

From the equalities (3.3), (3.6) and from the condition $\Delta(\lambda_j) \neq 0$ it follows that $\Psi_{g,N}(\lambda_j, z) = \text{const.}$. But this, for $N \neq 0$, is in contradiction with the condition $a_0(\lambda_j) \neq 0$. Hence, every branching point λ_j of the hyperelliptic curve Γ (2.12) for $N \neq 0$ is a zero of one of the equations (3.1). □

LEMMA 2. *For any set of integers m_i $(i = 0, 1, 2, 3)$ there exists not less than one and not more than four numbers n_k such that*

$$\begin{gathered} n_k = 1 + \frac{1}{2}\sum_{i=0}^{3} \widetilde{m}_{ik}, \quad n_k \in \mathbb{N}, \\ \widetilde{m}_{ik} \in \{m_i, -m_i - 1\}, \quad i = 0, 1, 2, 3. \end{gathered} \tag{3.7}$$

PROOF. Rewrite the equality (3.7) in the form

$$n_k = \frac{1}{2}\sum_{i=0}^{3}\left(\widetilde{m}_{ik} + \frac{1}{2}\right) - \frac{1}{2}\sum_{i=0}^{3}(-1)^{\varepsilon_{ik}}\left(m_i + \frac{1}{2}\right), \quad \varepsilon_{ik} \in \{0, 1\}. \tag{3.8}$$

It is easy to see that from sixteen values which the number n_k can take there are only eight which are integer, and that to every positive number n_k there corresponds a sign-opposite negative one. Therefore the number of positive integer n_k's can be not more than four.

On the other hand, all integer n_k can not be equal to zero because, as it is not difficult to show, this was possible only if all $m_i = -1/2$, $(i = 0, 1, 2, 3)$ which is in contradiction with the condition $m_i \in \mathbb{Z}$. Hence, among integer n_k there is at least one positive and one negative number. □

THEOREM 5. *Let m_i $(i = 0, 1, 2, 3)$ be an arbitrary set of integers satisfying the condition*:

$$n = 1 + \frac{1}{2}\sum_{i=0}^{3} m_i, \quad n \in \mathbb{N}. \tag{3.9}$$

Then:

1. *there are exactly n values λ_j such that for $\lambda = \lambda_j$ the Heun equation* (0.6) *has as its solution the Heun polynomial of order* $(n-1)$;
2. *the points $\lambda = \lambda_j$ $(j = 1, \dots, n)$ are the branching points of the hyperelliptic curve Γ* (2.12) *for the finite-gap solution of Heun's equation with characteristics m_i* (3.9).

PROOF. For

$$m_0 = m_1 = m_2 = m_3 = 0$$

the statement of the Theorem is easily checked by the straightforward substitution of the polynomial of zero order into the Heun equation.

Now, let at least one of the characteristics m_i be different from zero. The first part of the statement is then a consequence of the general theory of Heun's polynomials (see, e.g., [**31, 32**]).

The second part will be proved assuming the opposite. Let for some value $\lambda = \lambda_1$ a solution of Heun's equation with the characteristics m_i (3.9) be a polynomial $F(z)$ of order $(n-1)$. Also, assume that the point $\lambda = \lambda_1$ is not a branching point of the hyperelliptic curve Γ (2.12) for the finite-gap solution of Heun's equation with these characteristics. Then one obtains that the 3-rd order equation (2.6) with nonnegative integer characteristics $\widetilde{m}_i \geqslant 0$

$$\widetilde{m}_i \in \{m_i, -m_i - 1\}, \quad i = 0, 1, 2, 3,$$

has as its solution two polynomials of zero order:

$$\begin{gathered}\mathcal{P}_1(z) = \Psi(\widetilde{\lambda}_1, z),\\ \mathcal{P}_2(z) = z^{M_1}(z-1)^{M_2}(z-a)^{M_3}F^2(z),\\ M_i = \widetilde{m}_i - m_i, \qquad i = 1, 2, 3,\end{gathered}$$

$$\begin{aligned}\widetilde{\lambda}_1 = \lambda_1 - 2m_1M_3 - 2m_3M_1 + 2M_1M_3 + M_1 + M_3&\\ - (2m_1M_2 + 2m_2M_1 - 2M_1M_2 - M_1 - M_2)a.&\end{aligned}$$

Moreover, because the point $\lambda = \lambda_1$ of the curve Γ, as was assumed, is not a branching point, the polynomial $\mathcal{P}_1(z)$ does not have multiple zeros and it also does not have zeros at the singularities of Heun's equation (cf. Corollary 3). Hence the polynomials $\mathcal{P}_1(z)$ and $\mathcal{P}_2(z)$ are linearly independent. It is not difficult to show that the equation (2.6) can not simultaneously have as its solutions the polynomials $\mathcal{P}_1(z)$ and $\mathcal{P}_2(z)$. This means that the point $\lambda = \lambda_1$ is a branching point of the hyperelliptic curve Γ (2.12) with the characteristics m_i (3.9). □

Based on the statements of the Lemma 2, the Theorem 5 and Corollary 5 one can suggest one more method of finding the branching points of the curve Γ (2.12).

COROLLARY 7. *In order to find all branching points λ_j of the hyperelliptic curve Γ (2.12) associated with 'finite-gap' solution of Heun's equation with the characteristics m_i it is necessary and sufficient to find all Heun's polynomials being the solutions of Heun equations (from one to four) with the characteristics $\widetilde{m}_i$ (3.7).*

The algebraic genus g of the curve Γ (2.12) can be calculated by finding a sum of all $n_k \in \mathbb{N}$ (3.7), i.e. by counting all the branching points λ_j $(j = 1, \ldots, 2g+1)$.

COROLLARY 8. *Let g be an algebraic genus of the hyperelliptic curves:*

(a) *Γ (2.12), which is associated with the 'finite-gap' solution of Heun's equation, and*

(b) *$\widetilde{\Gamma}$ (0.9), which is a spectral curve of the Treibich–Verdier potential. If all characteristics m_i $(i = 0, 1, 2, 3)$ are nonnegative integer then:*

  (a) *in the case of even N (0.7)*

$$g = \max\left\{\max_{0 \leqslant i \leqslant 3} m_i, \frac{N}{2} - \min_{0 \leqslant i \leqslant 3} m_i\right\}; \tag{3.10}$$

(b) *in the case of odd N*

$$g = \max\left\{ \max_{0 \leqslant i \leqslant 3} m_i, \frac{N+1}{2} \right\}. \tag{3.11}$$

4. Monodromy Group of the Finite-Gap Solutions

Let $a > 1$, $\lambda \neq \lambda_j$, $\operatorname{Im} \nu(\lambda) = 0$, $m_i \geqslant 0$. In the vicinity of points $z = 0$ and $z = 1$ let us consider the following 'finite-gap' solutions of Heun's equation:

$$y_1(z) = \sqrt{\Psi_{g,N}(\lambda, z)} \cos\left(\frac{\nu(\lambda)}{2} \int_0^z \frac{t^{m_1}(t-1)^{m_2}(t-a)^{m_3}\, dt}{\Psi_{g,N}(\lambda, t)\sqrt{t(t-1)(t-a)}} \right), \tag{4.1a}$$

$$y_1(z) = y_{10} + O(z), \qquad z \to 0; \tag{4.1b}$$

$$y_2(z) = \sqrt{\Psi_{g,N}(\lambda, z)} \sin\left(\frac{\nu(\lambda)}{2} \int_0^z \frac{t^{m_1}(t-1)^{m_2}(t-a)^{m_3}\, dt}{\Psi_{g,N}(\lambda, t)\sqrt{t(t-1)(t-a)}} \right), \tag{4.1c}$$

$$y_2(z) = z^{m_1+1/2}(y_{20} + O(z)), \quad z \to 0; \tag{4.1d}$$

$$y_3(z) = \sqrt{\Psi_{g,N}(\lambda, z)} \cos\left(\frac{\nu(\lambda)}{2} \int_1^z \frac{t^{m_1}(t-1)^{m_2}(t-a)^{m_3}\, dt}{\Psi_{g,N}(\lambda, t)\sqrt{t(t-1)(t-a)}} \right), \tag{4.1e}$$

$$y_3(z) = y_{30} + O(1-z), \quad z \to 1; \tag{4.1f}$$

$$y_4(z) = \sqrt{\Psi_{g,N}(\lambda, z)} \sin\left(\frac{\nu(\lambda)}{2} \int_1^z \frac{t^{m_1}(t-1)^{m_2}(t-a)^{m_3}\, dt}{\Psi_{g,N}(\lambda, t)\sqrt{t(t-1)(t-a)}} \right), \tag{4.1g}$$

$$y_4(z) = (1-z)^{m_2+1/2}(y_{40} + O(1-z)), \quad z \to 1 \tag{4.1h}$$

where y_{10}, y_{20}, y_{30} and y_{40} are some functions of λ.

From the Corollary 3 it follows that the function $\Psi_{g,N}(\lambda, z)$ does not have zeros in the real interval $z \in [0, 1]$. Therefore, the connection matrix of the solutions $y_1(z)$, $y_2(z)$ and $y_3(z)$, $y_4(z)$,

$$\begin{pmatrix} y_1 \\ y_2 \end{pmatrix} = T_{01} \begin{pmatrix} y_3 \\ y_4 \end{pmatrix}$$

can be found relatively simple

$$T_{01} = \begin{pmatrix} \cos\varphi(\lambda) & -\sin\varphi(\lambda) \\ \sin\varphi(\lambda) & \cos\varphi(\lambda) \end{pmatrix} \tag{4.2}$$

where

$$\varphi(\lambda) = \frac{\nu(\lambda)}{2} \int_0^1 \frac{t^{m_1}(t-1)^{m_2}(t-a)^{m_3}\, dt}{\Psi_{g,N}(\lambda, t)\sqrt{t(t-1)(t-a)}}. \tag{4.3}$$

From (4.1)–(4.3) it is not difficult to obtain the expressions for two generators of the monodromy group for 'finite-gap' solutions of Heun's equation (0.6) which correspond to simple loops starting at the point $z = 0$ and positively encircling the points $z = 0$ and $z = 1$, respectively

$$M_0 = \begin{pmatrix} 1 & 0 \\ 0 & -1 \end{pmatrix}, \tag{4.4}$$

$$M_1 = T_{01} \begin{pmatrix} 1 & 0 \\ 0 & -1 \end{pmatrix} T_{01}^{-1} = \begin{pmatrix} \cos 2\varphi(\lambda) & \sin 2\varphi(\lambda) \\ \sin 2\varphi(\lambda) & -\cos 2\varphi(\lambda) \end{pmatrix}. \tag{4.5}$$

Analogously, one can find the third generator

$$M_2 = T_{02}\begin{pmatrix}1 & 0\\ 0 & -1\end{pmatrix}T_{02}^{-1} = \begin{pmatrix}\cos 2\psi(\lambda) & \sin 2\psi(\lambda)\\ \sin 2\psi(\lambda) & -\cos 2\psi(\lambda)\end{pmatrix}, \tag{4.6}$$

$$T_{02} = \begin{pmatrix}\cos\psi(\lambda) & -\sin\psi(\lambda)\\ \sin\psi(\lambda) & \cos\psi(\lambda)\end{pmatrix}, \tag{4.7}$$

$$\psi(\lambda) = \frac{\nu(\lambda)}{2}\int_{\mathcal{C}} \frac{t^{m_1}(t-1)^{m_2}(t-a)^{m_3}\,dt}{\Psi_{g,N}(\lambda,t)\sqrt{t(t-1)(t-a)}}. \tag{4.8}$$

Here $\mathcal{C}$ is a path in the complex plane connecting the points $z=0$ and $z=a$ and avoiding zeros of the polynomial $\Psi_{g,N}(\lambda,z)$.

By exactly the same way one can find the generators of the monodromy group for the 'finite-gap' solutions of Heun's equation with negative values of the characteristics m_i.

Concluding Remarks[5]

Recently, there have appeared papers [**36, 37**] in which, following [**17, 24, 46**], the solutions of Lamé (0.5) and Treibich–Verdier (0.2), (0.3) equations are studied by making use of periodic solution of a differential equation for the product of eigenfunctions of Schrödinger's operator with periodic potential [**17, 24, 46**]. As seen from these papers, the author is not familiar with the results of Darboux [**10–12**], Krichever [**27**], Treibich and Verdier [**39, 40, 42, 43, 45**], Belokolos and Enol'skii [**3–7, 14–16, 23**], Gesztesy [**18–22**] who investigated in detail the finite-gap solutions of these equations as well as their connection with the Calogero–Moser system with elliptic interaction [6] for 1994 is dedicated to the memory of Verdier and contains many works directly or indirectly related to the Treibich–Verdier equation.

Despite this, in the part of studying the properties of solutions of Lamè and Treibich–Verdier equations with integer characteristics m_i the author of those papers not only re-discovered already known facts, such as the finite number of eigenvalues for the periodic and anti-periodic problems, but for two special cases he calculated their number. In this part, the results of the papers [**36, 37**] coincide with those of the present paper (cf. Corollary 9), taking into account that the eigenvalues for periodic and anti-periodic problems are the gap edges of the spectrum of finite-gap periodic potential (see, for instance, [**38**]) or, which is the same, are the branching points of the hyperelliptic spectral curve Γ (2.12). Unfortunately, in [**36, 37**] in the general case the formula for the number of eigenvalues is an unproven hypothesis (although the right one as follows from the results of the present work), and apart from that there are no methods suggested for calculating those eigenvalues.

Moreover, from our point of view, the analysis of finite-gap solutions of Lamè and Treibich–Verdier equations, in contrast to the case of Heun's equation, should be carried out not by making use of the product of eigenfunctions of Schrödinger's operator but by using the method of Krichever–Hermite ansatz [**4–7**]. This is because, in the first case a solution is expressed in terms of elliptic functions in N

[5]the author learned about the work [**37**] from V. B. Kuznetsov.

[6]notice also that the whole volume 36 of Acta Appl.Math. for 1994 is dedicated to the memory of Verdier and contains many works directly or indirectly related to the Treibich-Verdier equation.

(0.7) unknown parameters solving N transcendental equations. In the second case, one needs to solve only an algebraic equation of Nth order.

Although, perhaps the formulas used in [**36, 37**] serve best the problem considered there, i.e. to apply the method of Bethe ansatz to the Treibich–Verdier equation aiming to study the one-particle Inozemtsev's model.

Appendix A. Simplest 'Finite-Gap' Solutions

At the end of this paper we would like to list polynomials $\Psi(\lambda, z)$ and canonical equations (2.12) of the hyperelliptic curves $\Gamma = \{(\nu, \lambda)\}$ for some simplest 'finite-gap' solutions (2.11) of Heun's equation with nonnegative characteristics m_i $(i = 0, 1, 2, 3)$. Our examples are indexed by the characteristics (m_0, m_1, m_2, m_2).

(0,0,0,0):

$$\Psi(\lambda, z) = 1,$$
$$\nu^2 = \lambda.$$

(1,0,0,0):

$$\Psi(\lambda, z) = \lambda + z - a - 1,$$
$$\nu^2 = (\lambda - 1)(\lambda - a)(\lambda - a - 1).$$

(0,1,0,0):

$$\Psi(\lambda, z) = z\lambda + a,$$
$$\nu^2 = \lambda(\lambda + a)(\lambda + 1).$$

(1,1,0,0):

$$\Psi(\lambda, z) = z\lambda + z^2 + a,$$
$$\nu^2 = (\lambda + a + 1)(\lambda^2 - 4a).$$

(0,0,1,1):

$$\Psi(\lambda, z) = (z - 1)(z - a)\lambda + (a + 1)z^2 - 4az + a(a + 1),$$
$$\nu^2 = (\lambda + a + 1)(\lambda^2 - 4a).$$

(1,1,1,0):

$$\Psi(\lambda, z) = z(z - 1)\lambda^2 + \big(z^3 + 3(a - 1)z^2 - 3(a - 1)z - a\big)\lambda - 3a(a - 1),$$
$$\nu^2 = \lambda(\lambda + 3a)\big(\lambda + 3(a - 1)\big)\big(\lambda^2 + 2(2a - 1)\lambda - 3\big).$$

(1,1,1,1):

$$\Psi(\lambda, z) = z(z - 1)(z - a)\lambda + (z^2 - a)^2,$$
$$\nu^2 = \lambda(\lambda + 4a)(\lambda + 4).$$

(2,0,0,0):

$$\Psi(\lambda, z) = \lambda^2 + \big(3z - 5(a + 1)\big)\lambda + 9z^2 - 12(a + 1)z + (a + 4)(4a + 1),$$
$$\nu^2 = (\lambda - a - 1)(\lambda - a - 4)(\lambda - 4a - 1)\big(\lambda^2 - 4(a + 1)\lambda + 12a\big).$$

(0,2,0,0):

$$\Psi(\lambda, z) = z\lambda^2 + \big(3z^2 - 3(a + 1)z + a\big)\lambda + 9z^3 - 9(a + 1)z^2 - 3a(a + 1),$$
$$\nu^2 = (\lambda + 3)(\lambda + 3a)\big(\lambda + 3(a + 1)\big)\big(\lambda^2 + 4(a + 1)\lambda + 12a\big).$$

(2,1,0,0):

$$\Psi(\lambda, z) = z^2\lambda^2 + 3\big((a+1)z + a\big)z\lambda + 9az^2 + 9a^2,$$
$$\nu^2 = \big(\lambda - 3(a+1)\big)\big(\lambda^2 - 2\lambda - 3(4a+1)\big)\big(\lambda^2 - 2a\lambda - 3a(a+4)\big).$$

(0,0,2,1):

$$\begin{aligned}\Psi(\lambda, z) = {}& (z-1)^2(z-a)\lambda^2 \\ & + \big(4az^2 - (a+3)(3a-1)z + 2a(3a-1)\big)(z-1)\lambda \\ & - 12az^3 + 9(a+1)^2z^2 - 36az - 3a(3a^2 - 6a - 1),\end{aligned}$$
$$\nu^2 = (\lambda - 3)(\lambda^2 + 4a\lambda - 12a)\big(\lambda^2 + 2(3a-1)\lambda + 3(3a^2 - 6a - 1)\big).$$

(2,1,1,0):

$$\begin{aligned}\Psi(\lambda, z) = {}& z(z-1)\lambda^2 + \big(3z^3 + (3a-10)z^2 - (3a-8)z - a\big)\lambda \\ & + 9z^4 - 24z^3 - 2(3a-8)z^2 - a(3a-8),\end{aligned}$$
$$\nu^2 = (\lambda + 3a - 8)(\lambda + 3a + 1)\big(\lambda^3 + 4(a-2)\lambda^2 - 16a\lambda + 64a\big).$$

(0,1,1,2):

$$\begin{aligned}\Psi(\lambda, z) = {}& z(z-1)(z-a)^2\lambda^2 \\ & + \big(3(2a+3)z^3 - (3a^2 + 18a + 8)z^2 + a(3a+10)z + a^2\big)(z-a)\lambda \\ & + 9a(a+3)z^4 - 24a(3a+1)z^3 + 18a^2(a+3)z^2 - 3a^3(a+3),\end{aligned}$$
$$\nu^2 = \big(\lambda + 3a)(\lambda + 3(a+3)\big)\big(\lambda^3 + 4(a+4)\lambda^2 + 16(3a+4)\lambda + 192a\big).$$

(2,1,1,1):

$$\begin{aligned}\Psi(\lambda, z) = {}& z(z-1)(z-a)\lambda^3 \\ & + \big(3z^4 - 4(a+1)z^3 + 2(a+1)^2z^2 - 2a(a+1)z + a^2\big)\lambda^2 \\ & + \big(9z^5 - 15(a+1)z^4 - 5(a^2 - 7a + 1)z^3 + 15(a-1)^2(a+1)z^2 \\ & - 15a(a-1)^2z - 2a^2(a+1)\big)\lambda - 15a^2(a-1)^2,\end{aligned}$$
$$\begin{aligned}\nu^2 = {}& \lambda\big(\lambda^2 - 2(a+1)\lambda - 15(a-1)^2\big) \\ & \times \big(\lambda^2 - 2(a-2)\lambda - 15a^2\big)\big(\lambda^2 + 2(2a-1)\lambda - 15\big).\end{aligned}$$

(3,0,0,0):

$$\begin{aligned}\Psi(\lambda, z) = {}& \lambda^3 + 2\big(3z - 7(a+1)\big)\lambda^2 \\ & + \big(45z^2 - 78(a+1)z + 49a^2 + 158a + 49\big)\lambda \\ & + 225z^3 - 405(a+1)z^2 + 9(3a+8)(8a+3)z - 12(a+1)(3a^2 + 26a + 3),\end{aligned}$$
$$\begin{aligned}\nu^2 = {}& \big(\lambda - 4(a+1)\big)\big(\lambda^2 - 10(a+1)\lambda + 3(3a^2 + 26a + 3)\big) \\ & \times \big(\lambda^2 - 2(2a+5)\lambda + 3(8a+3)\big)\big(\lambda^2 - 2(5a+2)\lambda + 3a(3a+8)\big).\end{aligned}$$

References

1. D. I. Abramov and I. V. Komarov, Integral equations and relations for the Coulomb spheroidal functions, Teoret. Mat. Fiz. **29** (1976), no. 2 235–243 (Russian).
2. N. I. Akhiezer, *Elements of the theory of elliptic functions*, Transl. Math. Monogr., vol. 79, Amer. Math. Soc., Providence, RI, 1990.
3. E. D. Belokolos and V. Z. Enol'skii, *Isospectral deformations of elliptic potentials*, Uspekhi Mat. Nauk **44** (1989), no. 5(269), 155–156.

4. ———, Funktsional. Anal. i Prilozhen. **23** (1989), no. 1, 57–58 (Russian); English transl. Funct. Anal. Appl. **23** (1989), no. 1, 46–47.
5. ———, *Reduction of theta function and elliptic finite-gap potentials*, Acta Appl. Math. **36** (1994), 87–117.
6. E. D. Belokolos, A. I. Bobenko, V. Z. Enol'skii, A. R. Its, and V. B. Matveev, *Algebro-geometrical approach to nonlinear evolution equations*, Springer Ser. Nonlinear Dynam., Springer-New York, 1994.
7. E. D. Belokolos, A. I. Bobenko, V. B. Matveev, and V. Z. Enol'skii, *Algebro-geometric principles of superposition of finite-zone solutions of integrable nonlinear equations*, Uspekhi Mat. Nauk **41** (1986), no. 2(248), 3–42.
8. F. Calogero and A. Degasperis, *Spectral transform and solitons.* I. *Tools to solve and investigate nonlinear evolution equations*, Stud. Math. Appl., vol. 13, Lecture Notes in Comput. Sci., vol. 144, North-Holland Publishing Co., Amsterdam-New York, 1982.
9. S. Chandrasekhar, *The mathematical theory of black holes*, Internat. Ser. Monogr. Phys., vol. 69, Oxford Sci. Publ., The Clarendon Press, Oxford Univ. Press, New York, 1992.
10. G. Darboux, *Sur une équation linéaire.* C. R. **94** (1882), no. 25, 1645–1648.
11. ———, *Leçons sur la théorie générale des surfaces.* I, 2nd ed., Paris, Gauthier-Villars, 1914.
12. ———, *Leçons sur la théorie générale des surfaces.* II, 2nd ed., Paris, Gauthier-Villars, 1915.
13. B. A. Dubrovin, V. B. Matveev, and S. P. Novikov, *Nonlinear equations of Korteweg-de Vries type, finite-band linear operators and Abelian varieties*, Uspehi Mat. Nauk **31** (1976), no. 1(187), 55–136.
14. J. C. Eilbeck and V. Z. Enol'skii, *Elliptic Baker–Akhiezer functions and application to anintegrable dynamical system*, J. Math. Phys. **35** (1994), no. 3, 1192–1201.
15. ———, *On the two-gap locus for the elliptic Calogero–Moser model*, J. Phys. A. **28** (1995), 1069–1088.
16. V. Z. Enol'skii and N. A. Kostov, *On the geometry of elliptic solitons*, Acta Appl. Math. **36** (1994), 57–86.
17. A. Erdélyi, W. Magnus, F. Oberhettinger, and F. G. Tricomi, *Higher transcendental functions.* III. *Based on notes left by Harry Bateman*, Robert E. Krieger Publ. Co., Inc. Melbourne, Fla., 1981.
18. F. Gesztesy and R. Weikard, *Lamé potentials and the stationary* (m)KdV *hierarchy*, Math. Nachr. **176** (1995), 73–91.
19. ———, *Picard potentials and Hill's equation on torus*, Acta Math. **176** (1996), 73–107.
20. ———, *On Picard potentials*, Differential Integral Equations **8** (1995), no. 6, 1453–1476.
21. ———, *Treibich–Verdier potentials and the stationary* (m)KdV *hierarchy*, Math. Z. **219** (1995), no. 3, 451–476.
22. ———, *A characterization of elliptic finite-gap potentials*, C. R. Acad. Sci. Paris, Sér. I, Math. **321** (1995), no. 7, 837-841.
23. A. R. Its and V. Z. Enol'skii, *The dynamics of the Calogero–Moser system and reduction of hyperelliptic integrals to elliptic integrals*, Funktsional. Anal. i Prilozhen. **20** (1986), no. 1, 73–74.
24. E. Kamke, *Differentialgleichungen*, Lösungsmethoden und Lösungen, vol. I, Gewöhnliche Differentialgleichungen, Neunte Auflage, Mit einem Vorwort von Detlef Kamke, B. G. Teubner, Stuttgart, 1977 (German).
25. A. Ya. Kazakov and S. Yu. Slavyanov, *Integral equations for special functions of Heun class*, Methods Appl. Anal. **3** (1996), no. 4, 447–456.
26. I. V. Komarov, L. I. Ponomarev, and S. Yu. Slavyanov, *Sferoidalnye i kulonovskie sferoidalnye funktsii*, Izdat. "Nauka", Moscow, 1975 (Russian).
27. I. M. Krichever, *Elliptic solutions of the Kadomcev–Petviašvili equations, and integrable systems of particles*, Funktsional. Anal. i Prilozhen. **14** (1980), no. 4, 45–54 (Russian); English transl. Functional Anal. Appl. **14** (1981), no. 4, 282–290.
28. W. Lay, A. Seeger, and S. Yu. Slavyanov, *A generalisation of the Riemann scheme for equations of hypergeometric and Heun class*, Internat. J. Math. Edu. Sci. Tech. **28** (1997), no. 5, 641–660.
29. V. B. Matveev, *Abelian functions and solitons*, Preprint no. 373, Univ. of Wrocław, 1976.
30. S. P. Novikov, *A periodic problem for the Korteweg-de Vries equation.* I, Funkcional. Anal. i Priložen. **8** (1974), no. 3 54–66.

31. A. Ronveaux (ed.), *Heun's differential equations*, Oxford Science Publ., The Clarendon Press, Oxford Univ. Press., New York, 1995.
32. S. Yu. Slavyanov and W. Lay, *Special functions*, Oxford Univ. Press., New York, 2000.
33. A. O. Smirnov, *Elliptic solutions of the Korteweg-de Vries equation*, Mat. Zametki **45** (1989), no. 6, 66–73; English transl. Math. Notes **45** (1989), 476–481.
34. ———, *Elliptic solutions of integrable nonlinear equations*, Mat. Zametki **46** (1989), no. 5, 100–102.
35. ———, *Finite-gap elliptic solutions of the* KdV *equation*, Acta Appl. Math. **36** (1994), 125–166.
36. K. Takemura, *On the eigenstates of the elliptic Calogero–Moser model*, Lett. Math. Phys. **53** (2000), no. 3, 181–194.
37. ———, *The Heun equation and the Calogero–Moser-Sutherland system.* I. *The Bethe Ansatz method*, (2001), `arXiv: math.CA/0103077`.
38. E. C. Titchmarsh, *Eigenfunction expansions associated with second-order differential equations*, The Clarendon Press, Oxford Univ. Press, New York, 1958.
39. A. Treibich, *Revêtements exceptionnels et sommes de 4 nombres triangulaires*, Duke Math. J. **68** (1992), no. 2, 217–236 (French).
40. ———, *Tangential polynomials and elliptic solitons*, Duke Math. J. **59** (1989), no. 3, 611–627.
41. ———, *Beyond the exceptional covers and their canonical hyper-elliptic potentials*, Preprint, Université d'Artois no. 99-3, 1999.
42. A. Treibich and J.-L. Verdier, *Elliptic solitons*, Progr. Math. **88**, The Grothendieck Festschrift, vol. III, Birkhäuser-Boston, Boston, MA, 1990, pp. 437–480 (French).
43. ———, *Revêtements tangentiels et sommes de 4 nombres triangulaires*, C. R. Acad. Sci. Paris **311** (1990), no. 1, 51–54.
44. ———, *Au delà des potentiels et revêtements tangentiels hyperelliptiques exceptionnels*, C. R. Acad. Sci. Paris Sér. I Math. **325** (1997), 1101–1106.
45. J.-L. Verdier, *New elliptic solitons*, Algebraic Analisys, vol. 2. Academic Press, Boston, MA., 1988, pp. 901–910.
46. E. T. Whittaker and G. N. Watson, *A course of modern analysis*, An Introduction to the General Theory of Infinite Processes and of Analytic Functions; With an Account of the Principal Transcendental Functions, Cambridge Mathematical Library, Cambridge Univ. Press, Cambridge, 1996.
47. V. E. Zaharov, S. V. Manakov, S. P. Novikov, and L. P. Pitaevskiĭ, *Theory of solitons*, "Nauka", Moscow, 1980 (Russian).
48. A. Zeeger, V. Laĭ, and S. Y. Slavyanov (1995), *Confluence of second-order Fuchsian differential equations*, Teoret. Mat. Fiz. **104** (1995), no. 2, 233–247.

DEPARTMENT OF MATHEMATICS, ST. PETERSBURG STATE UNIVERSITY OF AIRCRAFT ENGINEERING, BOLSHAYA MORSKAYA STR. 67, ST.-PETERSBURG 190000, RUSSIA.

E-mail address: `smirnov@as.stoic.spb.su`

Centre de Recherches Mathématiques
CRM Proceedings and Lecture Notes
Volume **32**, 2002

Generalized Kowalevski Top: New Integrable Cases on $e(3)$ and so(4)

Vladimir V. Sokolov

ABSTRACT. A three-parameter family of integrable quadratic Hamiltonians on $e(3)$ and so(4) is presented. When one of the parameters vanishes, the Hamiltonian coincides with the Kowalevski Hamiltonian with the girostatic term. If the other two parameters are equal to zero then the Hamiltonian reduces to a new integrable case of the Kirchhoff equations found recently by the author.

1. A New Integrable Case of the Kirchhoff Equations

The Kirchhoff equations (see for example [**7**]) for the motion of a rigid body in an ideal fluid read as follows:

$$M_t = M \times \frac{\partial H}{\partial M} + \Gamma \times \frac{\partial H}{\partial \Gamma}, \quad \Gamma_t = \Gamma \times \frac{\partial H}{\partial M} \tag{1.1}$$

where $M = (M_1, M_2, M_3)$ and $\Gamma = (\gamma_1, \gamma_2, \gamma_3)$ are three-dimensional vectors, and $\times$ stands for the vector product. Without loss of generality the Hamiltonian H can be taken in the form

$$H = \sum a_i M_i^2 + \sum b_{ij}(\gamma_i M_j + \gamma_j M_i) + \sum c_{ij}\gamma_i\gamma_j. \tag{1.2}$$

Apart from H, for arbitrary values of the parameters a_i, b_{ij}, c_{ij}, equation (1.1) has the following first integrals:

$$I_1 = \gamma_1^2 + \gamma_2^2 + \gamma_3^2, \quad I_2 = M_1\gamma_1 + M_2\gamma_2 + M_3\gamma_3. \tag{1.3}$$

These integrals are the Casimirs for the Poisson structure

$$\{M_i, M_j\} = \varepsilon_{ijk}M_k, \quad \{M_i, \gamma_j\} = \varepsilon_{ijk}\gamma_k, \quad \{\gamma_i, \gamma_j\} = 0 \tag{1.4}$$

on $e(3)$, which corresponds (see [**11**]) to equations (1.1). Therefore, for the integrability of (1.1) we need one additional first integral I_3, functionally independent of H, I_1, I_2.

2000 *Mathematics Subject Classification.* Primary: 58F07; Secondary: 70E15.

The author is grateful to T. Wolf for his help during the work with the package "Crack" and to A. V. Borisov, Yu. N. Fedorov, V. B. Kuznetsov, I. S. Mamaev, and A. V. Tsiganov for useful discussions. The research was partially supported by RFBR grant 99-01-00294, INTAS grant 99-1782, and EPSRC grant GR K99015.

This is the final form of the paper.

There are classical integrable cases found by Kirchhoff, Clebsch and Steklov–Lyapunov [**16**]. For all these cases the matrices $B = \{b_{ij}\}$ and $C = \{c_{ij}\}$ are diagonal and the Hamiltonian is of the form

$$H = a_1 M_1^2 + a_2 M_2^2 + a_3 M_3^2 + 2b_{11} M_1 \gamma_1 + 2b_{22} M_2 \gamma_2 + 2b_{33} M_3 \gamma_3 + c_{11}\gamma_1^2 + c_{22}\gamma_2^2 + c_{33}\gamma_3^2. \tag{1.5}$$

The Kirchhoff case is described by the relations

$$a_1 = a_2, \quad b_{11} = b_{22}, \quad c_{11} = c_{22}.$$

For the Clebsch and Steklov–Lyapunov cases the coefficients a_i are arbitrary and the remaining parameters satisfy the following conditions:

$$b_{11} = b_{22} = b_{33},$$

$$\frac{c_{11} - c_{22}}{a_3} + \frac{c_{33} - c_{11}}{a_2} + \frac{c_{22} - c_{33}}{a_1} = 0,$$

and

$$\frac{b_{11} - b_{22}}{a_3} + \frac{b_{33} - b_{11}}{a_2} + \frac{b_{22} - b_{33}}{a_1} = 0,$$

$$c_{11} - \frac{(b_{22} - b_{33})^2}{a_1} = c_{22} - \frac{(b_{33} - b_{11})^2}{a_2} = c_{33} - \frac{(b_{11} - b_{22})^2}{a_3},$$

respectively. For each of these three cases there exists an additional quadratic integral I_3.

In the paper [**14**] the following Hamiltonian

$$H = M_1^2 + M_2^2 + 2M_3^2 + 2M_3(c_1\gamma_1 + c_2\gamma_2) + 2(c_1 M_1 + c_2 M_2)\gamma_3 + 4(c_2\gamma_1 - c_1\gamma_2)^2 - 4(c_1^2 + c_2^2)\gamma_3^2 \tag{1.6}$$

was considered. It turns out that (1.6) is in involution with some polynomial I_3 of fourth degree with respect to the brackets (1.4) or, equivalently, the corresponding Kirchhoff equations (1.1) possess an additional fourth degree integral.

Obviously, I_3 is defined up to a quadratic form of I_1, I_2, and H. In [**14**] this form has been chosen in such a way that I_3 depends on a minimal number of new variables. These variables,

$$\begin{array}{lll} x_1 = M_3 - c_1\gamma_1 - c_2\gamma_2, & x_2 = M_1 + c_1\gamma_3, & x_3 = M_2 + c_2\gamma_3, \\ x_4 = 3(c_2\gamma_1 - c_1\gamma_2), & x_5 = M_3 + 2c_1\gamma_1 + 2c_2\gamma_2, & x_6 = \gamma_3 \end{array} \tag{1.7}$$

are very useful for any computations related to the Hamiltonian (1.6).

In terms of these variables the additional integral of the Kirchhoff equations with the Hamiltonian (1.6) has the form

$$I_3 = x_1^2 P + Q^2 \tag{1.8}$$

where

$$P = (c_1^2 + c_2^2)x_5^2 + (c_2 x_2 - c_1 x_3)^2, \quad Q = c_2 x_2 x_4 + c_1 x_2 x_5 - c_1 x_3 x_4 + c_2 x_3 x_5.$$

It follows from these formulas that I_3 does not depend on the variable x_6.

Notice that one of the quadratic integrals becomes especially simple in the variables (1.7):

$$H + 5(c_1^2 + c_2^2)I_1 = x_1^2 + x_2^2 + x_3^2 + x_4^2 + x_5^2.$$

Simultaneous linear transformations of M and Γ with the orthogonal matrix of the form

$$T = \begin{pmatrix} \cos\phi & \sin\phi & 0 \\ -\sin\phi & \cos\phi & 0 \\ 0 & 0 & 1 \end{pmatrix} \tag{1.9}$$

preserve the form (1.6) of H while changing the parameters as follows

$$\bar{c}_1 = c_1 \cos\phi - c_2 \sin\phi, \quad \bar{c}_2 = c_1 \sin\phi + c_2 \cos\phi.$$

Therefore the only invariant is $\delta = c_1^2 + c_2^2$.

The parameter δ can be normalized by rescaling the γ_i. If $\delta \neq 0$ we can take $c_1 = 1$, $c_2 = 0$. In addition, there exists a particular (complex) case $c_2 = ic_1$. Here one can put $c_1 = 1$.

According to the classification theorem (see [**14**]), the Kirchhoff equations with the Hamiltonian (1.6) is the only integrable case with $a_1 = a_2 \neq a_3$ and an additional first integral of fourth degree.

It was noticed by Borisov and Mamaev (see [**5**]) that the change of variables: $\bar{\gamma}_i = \gamma_i$,

$$\overline{M}_1 = M_1 + c_1\gamma_3, \quad \overline{M}_2 = M_2 + c_2\gamma_3, \quad \overline{M}_3 = M_3 - c_1\gamma_1 - c_2\gamma_2, \tag{1.10}$$

defined by the first three of the variables (1.7), preserves the brackets (1.4) and simplifies the Hamiltonian. Namely subtracting $4(c_1^2 + c_2^2)I_1$ from (1.6) and performing the transformation (1.10), we get a new form of the Hamiltonian

$$\overline{H} = M_1^2 + M_2^2 + 2M_3^2 + 2(a_1\gamma_1 + a_2\gamma_2)M_3 - (a_1^2 + a_2^2)\gamma_3^2 \tag{1.11}$$

where we have put $a_1 = 3c_1$, $a_2 = 3c_2$. In the sequel we shall use this more elegant form of the Hamiltonian.

Borisov and Mamaev [**5**] have also observed that the freedom in the definition of I_3 can be used to bring I_3 into a factorized form $I_3 = k_1 k_2$. In terms of the variables (1.10) the factors are given by $k_1 = M_3$ and

$$\begin{aligned} k_2 = {} & (M_1^2 + M_2^2 + M_3^2)M_3 + 2(a_1M_1 + a_2M_2)(M_1\gamma_1 + M_2\gamma_2) \\ & + 2(a_1\gamma_1 + a_2\gamma_2)M_3^2 + (a_1\gamma_1 + a_2\gamma_2)^2 M_3 \\ & - (a_1^2 + a_2^2)(2M_1\gamma_1 + 2M_2\gamma_2 + M_3\gamma_3)\gamma_3. \end{aligned} \tag{1.12}$$

It turns out that both $k_1 = 0$ and $k_2 = 0$ are invariant submanifolds. This fact follows from the following formulas:

$$\dot{k}_1 = 2(a_2\gamma_1 - a_1\gamma_2)k_1, \quad \dot{k}_2 = -2(a_2\gamma_1 - a_1\gamma_2)k_2.$$

The particular case $a_1^2 + a_2^2 = 0$ is simpler than the generic case. For this complex case a Lax representation has been found in [**14**]. For the case $a_1^2 + a_2^2 \neq 0$ a Lax representation recently has been obtained in [**15**].

2. The Corresponding Integrable Case on so(4)

Let us consider the standard deformation

$$\{M_i, M_j\} = \varepsilon_{ijk}M_k, \quad \{M_i, \gamma_j\} = \varepsilon_{ijk}\gamma_k, \quad \{\gamma_i, \gamma_j\} = x\varepsilon_{ijk}M_k. \tag{2.1}$$

of bracket (1.4), where x is an arbitrary parameter. It is well known that if $x > 0$ then (2.1) defines a Poisson bracket on so(4). The Casimirs of this bracket are given by

$$(2.2) \qquad I_1 = x(M_1^2 + M_2^2 + M_3^2) + \gamma_1^2 + \gamma_2^2 + \gamma_3^2, \quad I_2 = M_1\gamma_1 + M_2\gamma_2 + M_3\gamma_3.$$

The main result of the paper [**5**] consists of the following observation. The Hamiltonian (1.11) possesses a commuting integral of fourth degree $\bar{I}_3 = \bar{k}_1\bar{k}_2$, where

$$\bar{k}_1 = k_1, \quad \bar{k}_2 = k_2 + x(a_2M_1 - a_1M_2)^2M_3.$$

This integrable deformation of the integrable case from the Section 1 has been found independently by Sokolov and Borisov–Mamaev. Here we present the most elegant form of the deformed additional integral proposed by Borisov and Mamaev.

Note that the dynamical system with Hamiltonian (1.11) and brackets (2.1) has just two one-parameter families of solutions of the form $M_i = X_it^{-1}$, $\gamma_i = Y_it^{-1}$. First of them is defined by

$$X_3 = Y_3 = 0, \quad Y_1^2 + Y_2^2 = 0, \quad 2a_1Y_2 - 2a_2Y_1 = 1, \quad X_1 = 2X_2(a_1Y_1 + a_2Y_2).$$

For the second family we have

$$X_1 = a_1Y_3, \quad X_2 = a_2Y_3, \quad X_3 = -a_1Y_1 - a_2Y_2,$$
$$\big(1 + x(a_1^2 + a_2^2)\big)\big(Y_1^2 + Y_2^2 + Y_3^2\big) = \frac{x}{4}, \quad 2a_1Y_2 - 2a_2Y_1 = -1.$$

For both families the Kowalevski exponents are $\{-1, 0, 1, 2, 2, 2\}$.

3. Further Generalizations

This section contains new results concerning possible integrable generalizations of the Hamiltonian (1.11).

It turns out that extra linear terms can be added to the Hamiltonian (1.11). The most general form of such terms is given by

$$(3.1) \quad H = M_1^2 + M_2^2 + 2M_3^2 + 2(a_1\gamma_1 + a_2\gamma_2)M_3 - (a_1^2 + a_2^2)\gamma_3^2 + p_1(M_3 + a_1\gamma_1 + a_2\gamma_2) + p_2(a_1\gamma_2 - a_2\gamma_1)$$

where p_i are arbitrary constants. This Hamiltonian possesses an additional first integral of fourth degree on so(4). Historically, the author found first that this is true for the Hamiltonian (3.1) on $e(3)$ with $p_1 = 0$. After that, independently, V. B. Kuznetsov and A. V. Tsiganov informed the author that the same Hamiltonian is integrable on so(4). Finally, the term with nonzero p_1 has been added by the author very recently.

Since the explicit expression for the additional integral I_4 is rather long, we put[1] $a_1 = 0$ in the Hamiltonian (3.1) and rewrite it as follows

$$(3.2) \qquad \widetilde{H} = M_1^2 + M_2^2 + 2M_3^2 + 2\lambda_1\gamma_2M_3 - \lambda_1^2\gamma_3^2 + 2\lambda_2(M_3 + \lambda_1\gamma_2) + 2\lambda_3\gamma_1$$

where λ_i are arbitrary constants. If $\lambda_1 = \lambda_2 = 0$ then the Hamiltonian reduces to the famous Kowalevski Hamiltonian. The case $\lambda_1 = 0$ corresponds to the Kowalevski Hamiltonian with the additional gyrostatic term.

[1]without loss of generality.

For the shortened Hamiltonian (3.2) the additional integral has the form

$$\begin{aligned} I_4 &= M_3^2(M_1^2+M_2^2+M_3^2) \\ &+ \lambda_1 M_3(2M_1M_2\gamma_1 + 2M_2^2\gamma_2 + 2M_3^2\gamma_2 + \lambda_1 M_3\gamma_2^2 - 2\lambda_1 M_1\gamma_1\gamma_3 \\ &\qquad - 2\lambda_1 M_2\gamma_2\gamma_3 - \lambda_1 M_3\gamma_3^2) \\ &+ 2\lambda_2\big(M_3(M_1^2+M_2^2+M_3^2) + \lambda_1 M_1M_2\gamma_1 + \lambda_1 M_2^2\gamma_2 + 2\lambda_1 M_3^2\gamma_2 \\ &\qquad + \lambda_1^2 M_3\gamma_2^2 - \lambda_1 M_2M_3\gamma_3 - \lambda_1^2 M_1\gamma_1\gamma_3 - \lambda_1^2 M_2\gamma_2\gamma_3\big) \\ &+ 2\lambda_3(M_1^2\gamma_1 + M_3^2\gamma_1 + M_1M_2\gamma_2 + \lambda_1 M_3\gamma_1\gamma_2) \\ &- \lambda_2^2 M_3^2 + 2\lambda_2\lambda_3 M_3\gamma_1 + 2\lambda_1\lambda_2\lambda_3\gamma_1\gamma_2 + \lambda_1^2\lambda_2^2\gamma_2^2 - \lambda_3^2\gamma_2^2 \\ &- 2\lambda_2\lambda_3 M_1\gamma_3 - 2\lambda_1\lambda_2^2 M_2\gamma_3 + 2\lambda_1^2\lambda_2^2\gamma_3^2 - 2\lambda_2^2(\lambda_2 M_3 + \lambda_3\gamma_1 + \lambda_1\lambda_2\gamma_2) \\ &+ x(\lambda_1^2 M_1^2 M_3^2 + 2\lambda_1^2\lambda_2 M_1^2 M_3 - 2\lambda_1\lambda_3 M_1M_2M_3 \\ &\qquad + \lambda_1^2\lambda_2^2 M_1^2 - \lambda_3^2 M_1^2 - 2\lambda_1\lambda_2\lambda_3 M_1M_2 - \lambda_3^2 M_3^2). \end{aligned}$$

In order to get the integral I_4 on $e(3)$ we can simply put $x=0$ in this formula. The explicit form of the integral I_4 on so(4) for (3.1) can be easily reconstructed with the help of the transformation (1.9).

For the next generalization let us consider the following Hamiltonian

$$\widetilde{H} = M_1^2 + M_2^2 + 2M_3^2 + 2(a_1\gamma_1 + a_2\gamma_2)M_3 - (a_1^2+a_2^2)\gamma_3^2 + F(\gamma_1,\gamma_2) \tag{3.3}$$

on $e(3)$. It is not difficult to verify that the corresponding Kirchhoff equations possess an additional integral of fourth degree of the form $\bar{I}_3 = p_1p_2 + T$, where T is a cubic polynomial in M_1, M_2, M_3 with the coefficients depending on γ_1, γ_2 iff

$$F = 2\lambda_1(a_2\gamma_1 - a_1\gamma_2) + \frac{\lambda_2}{\sqrt{\gamma_1^2+\gamma_2^2}}$$

where λ_i are arbitrary constants. In this case the function T reads as follows

$$\begin{aligned} T &= 2a_2\lambda_1\gamma_1 M_1^2 + 2\lambda_1(a_2\gamma_2 - a_1\gamma_1)M_1M_2 - 2a_1\lambda_1\gamma_2 M_2^2 \\ &+ \left(2a_2\lambda_1\gamma_1 - 2a_1\lambda_1\gamma_2 + \frac{\lambda_2}{\sqrt{\gamma_1^2+\gamma_2^2}}\right)M_3^2 \\ &+ (a_1\gamma_1 + a_2\gamma_2)\left(2a_2\lambda_1\gamma_1 - 2a_1\lambda_1\gamma_2 + \frac{\lambda_2}{\sqrt{\gamma_1^2+\gamma_2^2}}\right)M_3 \\ &- \lambda_1\left(\lambda_1(a_1\gamma_1 + a_2\gamma_2)^2 + (a_1\gamma_2 - a_2\gamma_1)\frac{\lambda_2}{\sqrt{\gamma_1^2+\gamma_2^2}}\right). \end{aligned}$$

Performing the scaling $\gamma_i \to \varepsilon\gamma_i$ and redefining the constants λ_i, we get the following integrable Hamiltonian

$$\begin{aligned} \widetilde{H} = M_1^2 + M_2^2 + 2M_3^2 + 2\varepsilon(a_1\gamma_1 + a_2\gamma_2)M_3 - \varepsilon^2(a_1^2+a_2^2)\gamma_3^2 \\ + 2\lambda_1(a_2\gamma_1 - a_1\gamma_2) + \frac{\lambda_2}{\sqrt{\gamma_1^2+\gamma_2^2}}. \end{aligned} \tag{3.4}$$

The case $\varepsilon = 0$ corresponds to the known generalization of the Kowalevski Hamiltonian found by H. M. Yehia [**18**].

It would be interesting to investigate possible extensions of the Hamiltonian (1.11) on the fixed level $I_2 = 0$ of the Casimir I_2 to find a generalization of the Goryachev–Chaplygin case.

4. Historical Remarks

A fourth degree integral for the Kirchhoff equations on $e(3)$ was found by S. A. Chaplygin under the additional assumption that the scalar product (M, Γ) is equal to zero [**6, 9**]. The corresponding partially integrable case on so(4) was found by O. I. Bogoyavlenski [**3**].

An integrable case on so(4) with an integral of fourth degree was found by M. Adler and P. van Moerbeke [**1**]. A Lax pair for that case has been constructed by A. Reyman and M. Semenov–Tian–Shansky [**12**]. The fact that $p_1 = 0$ is an invariant relation has been used in [**14**] to prove that the Kirchhoff equations with the Hamiltonian (1.6) are not linearly equivalent to the Adler-van Moerbeke–Reyman–Semenov–Tian–Shansky case on so(4).

R. Liouville [**10**] has obtained necessary conditions for the existence of algebraic integrals for the Kirchhoff equations with nondiagonal matrix B.

In recent investigations of algebraic integrability [**13**] it was assumed that all matrices $\mathbf{A}$, $\mathbf{B}$, $\mathbf{C}$ are diagonal. In [**4, 8**] the matrix $\mathbf{A}$ was assumed to be defined by the inertial tensor $\mathbf{I}$ of a real rigid body: $\mathbf{A} = \mathbf{I}^{-1}$, and all entries a_i to be distinct. The latter restriction has been also imposed in the papers [**2, 17**] devoted to integrability conditions of the equations on so(4).

References

1. M. Adler and P. van Moerbeke, *A new geodesic flow on* SO(4), Probability, Statistical Mechanics, and Number Theory, Adv. Math. Suppl. Stud., vol. 9, Academic Press, Orlando, FL, 1986, pp. 81–96.
2. ———, *The Kowalevski and Henon–Heiles motions as Manakov geodesic flows on* so(4)—*A two dimensional family of Lax pairs*, Comm. Math. Phys. **113** (1988), 659–700.
3. O. I. Bogoyavlenski, *Breaking solitons, nonlinear integrable equations*, M. Nauka, 1991.
4. A. V. Borisov, *Necessary and sufficient integrability conditions for the Kirchhoff equations*, Regul. Khaoticheskaya Din. **1** (1996), no. 2, 61–73 (Russian).
5. A. V. Borisov, S. I. Mamaev, and V. V. Sokolov, *A new integrable case on* so(4), Dokl. Akad. Nauk **381** (2001), no. 5, 614–615.
6. S. A. Chaplygin, *A new partial solution of the problem of rotation of heavy rigid body around a fixed point*, Sobranie Sochineny, vol. 1, M.-L., GITTL, 1948, pp. 125–132.
7. V. V. Kozlov, *Symmetries, topology and resonances in Hamiltonian mechanics*, Izdatel′stvo Udmurt·skogo Universiteta, Izhevsk, 1995 (Russian).
8. V. V. Kozlov and D. A. Onischenko, *Nonintegrability of Kirchhoff's equations*, Dokl. Akad. Nauk SSSR **226** (1982), no. 6, 1298–1300 (Russian).
9. V. B. Kuznetsov and A. V. Tsiganov, *A special case of Neumann's system and the Kowalevski–Chaplygin–Goryachev top*, J. Phys. A **22** (1989), L73–L79.
10. R. Liouville, *Sur le mouvement d'un solide dans un liquide indéfini*, C. R. Acad. Sci. Paris **2** (1896), 874–876.
11. S. P. Novikov, *Hamiltonian formalism and multi-valued analog of Morse theory*, Uspekhi Mat. Nauk **37** (1982), no. 5, 3–49.
12. A. G. Reyman and M. A. Semenov-Tian-Shansky, *A new integrable case of the motion of the 4-dimensional rigid body*, Comm. Math. Phys. **105** (1986), 461–472.
13. S. T. Sadetov, *Conditions for the integrability of the Kirchhoff equations*, Vestnik Moskov. Univ. Ser. I Mat. Mekh. (1990), no. 3, 56–62 (Russian).
14. V. V. Sokolov, *A new integrable case for the Kirchhoff equations*, Theoret. and Math. Phys. **129** (2001), no. 1, pp. 31–37.
15. V. V. Sokolov and A. V. Tsiganov, *On Lax pairs for the generalized Kowalewski and Goryachev–Chaplygin tops*, Theoret. and Math. Phys, 2002 (to appear).
16. V. A. Steklov, *On the motion of a rigid body in a fluid*, Kharkov, 1893.
17. A. P. Veselov, *Conditions for the integrability of Euler equation on* so(4), Dokl. Akad. Nauk SSSR **270** (1983), no. 6, 1298–1300 (Russian).

18. H. M. Yehia, *New generalizations of the integrable problems in rigid body dynamics*, J. Phys. A **30** (1997), 7269–7275.

Landau Institute for Theoretical Physics, Kosygina 2, Moscow 117334, Russia.
E-mail address: sokolov@landau.ac.ru

Reprint of the original paper by *Sophie Kowalevski*

published in
Acta Mathematica **12** (1889), 177–232

SUR LE PROBLÈME DE LA ROTATION D'UN CORPS SOLIDE AUTOUR D'UN POINT FIXE[1]

PAR

SOPHIE KOWALEVSKI

À STOCKHOLM.

§ 1.

Le problème de la rotation d'un corps solide pesant autour d'un point fixe peut se ramener, comme on sait, à l'intégration du système d'équations différentielles suivant:

$$A\frac{dp}{dt} = (B - C)qr + Mg(y_0\gamma'' - z_0\gamma'), \qquad \frac{d\gamma}{dt} = r\gamma' - q\gamma'',$$

$$(1) \qquad B\frac{dq}{dt} = (C - A)rp + Mg(z_0\gamma - x_0\gamma''), \qquad \frac{d\gamma'}{dt} = p\gamma'' - r\gamma,$$

$$C\frac{dr}{dt} = (A - B)pq + Mg(x_0\gamma' - y_0\gamma), \qquad \frac{d\gamma''}{dt} = q\gamma - p\gamma'.$$

Les constantes A, B, C, Mg, x_0, y_0, z_0 qui figurent dans ces équations ont la signification suivante.

A, B, C sont les axes principaux de l'ellipsoïde d'inertie du corps considéré, relativement au point fixe.

M est la masse du corps;

g l'intensité de la force de gravité;

---

[1] Ce mémoire est le résumé d'un travail auquel l'Académie des Sciences de Paris, dans sa séance solennelle du 24 décembre 1888, a décerné le prix Bordin élevé de 3000 à 5000 francs.

This paper was originally published in Acta Mathematica **12** (1889), 177–232. Reprinted with permission.

x_0, y_0, z_0 les coordonnées du centre de gravité du corps considéré dans un système de coordonnées, dont l'origine est au point fixe et dont la direction coïncide avec celle des axes principaux de l'ellipsoïde d'inertie.

Jusqu'à présent on n'était parvenu à intégrer ces équations que dans deux cas particuliers:

1) Le cas de POISSON (ou d'EULER) où l'on a $x_0 = y_0 = z_0 = 0$,

2) Le cas de LAGRANGE où l'on a $A = B$, $x_0 = y_0 = 0$.

Dans ces deux cas l'intégration s'opère à l'aide des fonctions $\vartheta(u)$ dont l'argument est une fonction entière linéaire du temps.

Les six quantités $p, q, r, \gamma, \gamma', \gamma''$ sont dans ces deux cas des fonctions uniformes du temps, n'ayant d'autres singularités que des pôles pour toutes les valeurs finies de la variable.

Les intégrales des équations différentielles considérées conservent-elles cette propriété dans le cas général?

Si tel était le cas il faudrait pouvoir intégrer ces équations différentielles à l'aide de séries de la forme

$$
\begin{aligned}
p &= t^{-n_1}(p_0 + p_1 t + p_2 t^2 + \ldots),\\
q &= t^{-n_2}(q_0 + q_1 t + q_2 t^2 + \ldots),\\
r &= t^{-n_3}(r_0 + r_1 t + r_2 t^2 + \ldots),\\
\gamma &= t^{-m_1}(f_0 + f_1 t + f_2 t^2 + \ldots),\\
\gamma' &= t^{-m_2}(g_0 + g_1 t + g_2 t^2 + \ldots),\\
\gamma'' &= t^{-m_3}(h_0 + h_1 t + h_2 t^2 + \ldots),
\end{aligned}
\tag{2}
$$

où $n_1, n_2, n_3, m_1, m_2, m_3$ désignent des nombres entiers positifs, et ces séries, pour pouvoir représenter le système général d'intégrales des équations différentielles considérées, devraient contenir *cinq* constantes arbitraires.

Il faut donc examiner si une pareille intégration est possible.

On s'assure facilement, en comparant les exposants des premiers termes dans les membres gauches et dans les membres droits des équations considérées que l'on doit avoir

$$n_1 = n_2 = n_3 = 1, \qquad m_1 = m_2 = m_3 = 2.$$

On trouve ensuite, en posant par brièveté

$$A_1 = B - C, \qquad B_1 = C - A, \qquad C_1 = A - B$$

et en supposant l'unité de longueur choisie de manière à ce que l'on ait $Mg = 1$, que les coefficients p_0, q_0, r_0 doivent satisfaire aux équations suivantes

$$\begin{aligned} &- Ap_0 = A_1 q_0 r_0 + y_0 h_0 - z_0 g_0, \qquad && - 2f_0 = r_0 g_0 - q_0 h_0, \\ (3) \qquad &- Bq_0 = B_1 r_0 p_0 + z_0 f_0 - x_0 h_0, && - 2g_0 = p_0 h_0 - r_0 f_0, \\ &- Cr_0 = C_1 p_0 q_0 + x_0 g_0 - y_0 f_0, && - 2h_0 = q_0 f_0 - p_0 g_0. \end{aligned}$$

Pour que les trois dernières de ces équations puissent être satisfaites par des valeurs de f_0, g_0, h_0 différentes de zéro, il faut que le déterminant

$$\begin{vmatrix} 2 & r_0 & -q_0 \\ -r_0 & 2 & p_0 \\ q_0 & -p_0 & 2 \end{vmatrix} = 2(4 + p_0^2 + q_0^2 + r_0^2)$$

soit nul.

Si l'on ajoute les trois premières des équations (3) après les avoir multipliées respectivement par Ap_0, Bq_0, Cr_0, on trouve

$$\begin{aligned} (4) \qquad & A^2 p_0^2 + B^2 q_0^2 + C^2 r_0^2 \\ &= x_0(Bq_0 h_0 - Cr_0 g_0) + y_0(Cr_0 f_0 - Ap_0 h_0) + z_0(Ap_0 g_0 - Aq_0 f_0); \end{aligned}$$

mais des trois dernières des équations (3) il résulte

$$\begin{aligned} &2(Bq_0 h_0 - Cr_0 g_0) = p_0(Ap_0 f_0 + Bq_0 g_0 + Cr_0 h_0) - f_0(Ap_0^2 + Bq_0^2 + Cr_0^2), \\ (5) \quad &2(Cr_0 f_0 - Ap_0 h_0) = q_0(Ap_0 f_0 + Bq_0 g_0 + Cr_0 h_0) - g_0(Ap_0^2 + Bq_0^2 + Cr_0^2), \\ &2(Ap_0 g_0 - Bq_0 f_0) = r_0(Ap_0 f_0 + Bq_0 g_0 + Cr_0 h_0) - h_0(Ap_0^2 + Bq_0^2 + Cr_0^2). \end{aligned}$$

Les six quantités $p, q, r, \gamma, \gamma', \gamma''$ devant satisfaire aux deux relations algébriques

$$Ap^2 + Bq^2 + Cr^2 = 2(x_0\gamma + y_0\gamma' + z_0\gamma'') + l_1,$$
$$Ax_0 p + By_0 q + Cz_0 r = l,$$

l et l_1 désignant des constantes d'intégration, il faut que l'on ait

$$Ap_0f_0 + Bq_0g_0 + Cr_0h_0 = 0,$$

$$x_0f_0 + y_0g_0 + z_0h_0 = \frac{1}{2}(Ap_0^2 + Bq_0^2 + Cr_0^2).$$

Il résulte donc des équations (4) et (5)

$$A^2p_0^2 + B^2q_0^2 + C^2r_0^2 = -\frac{1}{4}(Ap_0^2 + Bq_0^2 + Cr_0^2)^2.$$

Il faut maintenant distinguer deux cas.

1^r *cas.* Supposons qu'aucune des quantités

$$A_1 = B - C, \qquad B_1 = C - A, \qquad C_1 = A - B$$

ne soit nulle. Si l'on pose alors

$$\frac{1}{2}(p_0^2 + q_0^2 + r_0^2) = \lambda_0,$$

$$\frac{1}{2}(Ap_0^2 + Bq_0^2 + Cr_0^2) = \lambda,$$

$$\frac{1}{2}(A^2p_0^2 + B^2q_0^2 + C^2r_0^2) = \lambda_1,$$

on a

$$\frac{1}{2}p_0^2 = -\frac{BC\lambda_0 - (B + C)\lambda + \lambda_1}{B_1C_1},$$

$$\frac{1}{2}q_0^2 = -\frac{CA\lambda_0 - (C + A)\lambda + \lambda_1}{C_1A_1},$$

$$\frac{1}{2}r_0^2 = -\frac{AB\lambda_0 - (A + B)\lambda + \lambda_1}{A_1B_1}.$$

Or nous avons trouvé

$$\lambda_0 = -2, \qquad \lambda_1 = -\frac{1}{2}\lambda^2.$$

On a donc

$$
(6) \qquad
\begin{aligned}
p_0^2 &= \frac{4BC + 2(B + C)\lambda + \lambda^2}{B_1 C_1} = \frac{(2B + \lambda)(2C + \lambda)}{B_1 C_1}, \\
q_0^2 &= \frac{4CA + 2(C + A)\lambda + \lambda^2}{C_1 A_1} = \frac{(2C + \lambda)(2A + \lambda)}{C_1 A_1}, \\
r_0^2 &= \frac{4AB + 2(A + B)\lambda + \lambda^2}{A_1 B_1} = \frac{(2A + \lambda)(2B + \lambda)}{A_1 B_1}.
\end{aligned}
$$

Si l'on ajoute les trois premières des équations (3) après les avoir multipliées respectivement par x_0, y_0, z_0, on trouve

$$
(7) \qquad x_0(Ap_0 + A_1 q_0 r_0) + y_0(Bq_0 + B_1 r_0 p_0) + z_0(Cr_0 + C_1 p_0 q_0) = 0.
$$

Cette équation sert à déterminer la quantité λ.

Les trois quantités f_0, g_0, h_0 doivent satisfaire aux trois équations suivantes

$$
\begin{aligned}
p_0 f_0 + q_0 g_0 + r_0 h_0 &= 0, \\
Ap_0 f_0 + Bq_0 g_0 + Cr_0 h_0 &= 0, \\
x_0 f_0 + y_0 g_0 + z_0 h_0 &= \lambda,
\end{aligned}
$$

d'où l'on tire, en posant

$$
\mu = -(A_1 x_0 q_0 r_0 + B_1 y_0 r_0 p_0 + C_1 z_0 p_0 q_0) = Ax_0 p_0 + By_0 q_0 + Cz_0 r_0,
$$

$$
(8) \qquad
\begin{aligned}
f_0 &= -A_1 q_0 r_0 \frac{\lambda}{\mu}, \\
g_0 &= -B_1 r_0 p_0 \frac{\lambda}{\mu}, \\
h_0 &= -C_1 p_0 q_0 \frac{\lambda}{\mu}.
\end{aligned}
$$

Les équations (6) et (7) déterminent les quantités p_0, q_0, r_0 au signe près.

Si l'on pose

$$
a = \sqrt{\frac{2A + \lambda}{A_1}}, \qquad b = \sqrt{\frac{2B + \lambda}{B_1}}, \qquad c = \sqrt{\frac{2C + \lambda}{C_1}}
$$

en fixant le signe des quantités a, b, c d'une manière arbitraire, on trouve que pour satisfaire à toutes les équations (3) il faut poser

$$p_0 = bc,$$

$$q_0 = ca,$$

$$r_0 = -ab.$$

L'équation (7) peut donc être écrite de la manière suivante

$$x_0(A + \lambda)p_0 + y_0(B + \lambda)q_0 + z_0(C + \lambda)r_0 = 0$$

ou bien

$$x_0(A + \lambda)bc + y_0(B + \lambda)ca - z_0(C + \lambda)ab = 0.$$

2[d] *cas.* Supposons maintenant que l'une des trois quantités A_1, B_1, C_1, (p. ex. C_1) soit égale à zéro. (Dans ce cas on peut toujours aussi supposer $y_0 = 0$.) Pour satisfaire aux deux équations

$$A(p_0 f_0 + q_0 g_0) + C r_0 h_0 = 0,$$

$$p_0 f_0 + q_0 g_0 + r_0 h_0 = 0,$$

il faut nécessairement que l'on ait

$$r_0 h_0 = 0.$$

Les équations (3) admettent donc deux systèmes de solutions, qui peuvent s'écrire, en désignant par $\varepsilon \pm 1$,

I. $$p_0 = \varepsilon i \cdot \frac{2C}{A - 2C} \cdot \frac{z_0}{x_0}, \qquad f_0 = -\frac{2C}{x_0},$$

$$q_0 = \varepsilon i p_0, \qquad g_0 = -i\varepsilon \frac{2C}{x_0},$$

$$r_0 = 2\varepsilon i, \qquad h_0 = 0.$$

II. $$p_0 = 0; \qquad f_0 = -\frac{2A}{x_0 - i\varepsilon z_0},$$

$$q_0 = 2\varepsilon i, \qquad g_0 = 0,$$

$$r_0 = 0, \qquad h_0 = \varepsilon i \frac{2A}{x_0 - i\varepsilon z_0}.$$

Les coefficients

$$p_0, q_0, r_0, f_0, g_0, h_0$$

une fois determinés, on obtient les coefficients

$$p_m, q_m, r_m, f_m, g_m, h_m$$

en résolvant le système d'équations linéaires suivantes

$$\begin{aligned}
(m-1)Ap_m - A_1(q_0r_m + r_0q_m) - z_0q_m + y_0h_m &= P_m,\\
(m-1)Bq_m - B_1(r_0p_m + p_0r_m) - x_0h_m + z_0f_m &= Q_m,\\
(m-1)Cr_m - C_1(p_0q_m + q_0p_m) - y_0f_m + x_0g_m &= R_m,\\
(m-2)f_m - r_0g_m + q_0h_m - g_0r_m + h_0q_m &= F_m,\\
(m-2)g_m - p_0h_m + r_0f_m - h_0p_m + f_0r_m &= G_m,\\
(m-2)h_m - q_0f_m + p_0g_m - f_0q_m + g_0p_m &= H_m.
\end{aligned}$$

Les membres droits de ces équations sont des fonctions entières des coefficients $p_\mu, q_\mu, r_\mu, f_\mu, g_\mu, h_\mu$ dans lesquels l'index $\mu < m$.

Afin que les séries

$$\begin{aligned}
p &= t^{-1}\sum p_m t^m, & \gamma &= t^{-2}\sum f_m t^m,\\
q &= t^{-1}\sum q_m t^m, & \gamma' &= t^{-2}\sum g_m t^m,\\
r &= t^{-1}\sum r_m t^m, & \gamma'' &= t^{-2}\sum h_m t^m,
\end{aligned}$$

contiennent le nombre suffisant de constantes arbitraires il faut que le déterminant de ces équations linéaires, lequel est une fonction entière du 6me degré en m, s'évanouisse pour cinq valeurs différentes de m égales à des nombres entiers positifs. De plus il faut que les quantités P_m, Q_m, R_m, F_m, G_m, H_m soient telles que les équations précédentes soient compatibles entre elles.

En effectuant les calculs je me suis assurée que ces conditions ne sont pas remplies dans le cas général; mais que, outre les deux cas déjà connus, elles sont encore remplies dans un cas nouveau, où les constantes satisfont aux équations suivantes

$$A = B = 2C, \qquad z_0 = 0.$$

C'est ce cas-là que je me propose d'étudier dans les paragraphes suivants.

§ 2.

Dans le cas que nous allons considérer, on a

$$A = B = 2C, \qquad z_0 = 0.$$

Par une rotation des axes de coordonnées dans le plan de xy et par un choix convenable de l'unité de longueur, on peut toujours effectuer, que l'on ait de plus dans ce cas

$$y_0 = 0, \qquad C = 1.$$

Si je pose alors

$$c_0 = Mgx_0,$$

les équations différentielles que nous avons à considérer sont les suivantes:

$$(1) \qquad \begin{cases} 2\dfrac{dp}{dt} = qr, & \dfrac{d\gamma}{dt} = r\gamma' - q\gamma'', \\ 2\dfrac{dq}{dt} = -pr - c_0\gamma'', & \dfrac{d\gamma'}{dt} = p\gamma'' - r\gamma, \\ \dfrac{dr}{dt} = c_0\gamma', & \dfrac{d\gamma''}{dt} = q\gamma - p\gamma'. \end{cases}$$

Les trois intégrales algébriques du cas général, sont dans ce cas-ci:

$$(2) \qquad \begin{cases} 2(p^2 + q^2) + r^2 = 2c_0\gamma + 6l_1, \\ 2(p\gamma + q\gamma') + r\gamma'' = 2l, \\ \gamma^2 + \gamma'^2 + \gamma''^2 = 1. \end{cases}$$

Outre ces trois intégrales algébriques on en trouve facilement encore une quatrième.

On a en effet (en posant $i = \sqrt{-1}$),

$$2\frac{d}{dt}(p + qi) = -ri(p + qi) - c_0 i\gamma'',$$

$$\frac{d}{dt}(\gamma + \gamma' i) = -ri(\gamma + \gamma' i) + \gamma'' i(p + qi).$$

Par conséquent

$$\frac{d}{dt}\{(p + qi)^2 + c_0(\gamma + \gamma' i)\} = - ri\{(p + qi)^2 + c_0(\gamma + \gamma' i)\},$$

et de même

$$\frac{d}{dt}\{(p - qi)^2 + c_0(\gamma - \gamma' i)\} = \quad ri\{(p - qi)^2 + c_0(\gamma - \gamma' i)\},$$

d'où il suit

$$\frac{d}{dt} lg\{(p + qi)^2 + c_0(\gamma + i\gamma')\} + \frac{d}{dt} lg\{(p - qi)^2 + c_0(\gamma - i\gamma')\} = 0,$$

et, en intégrant,

$$(3) \qquad \{(p + qi)^2 + c_0(\gamma + i\gamma')\}\{(p - qi)^2 + c_0(\gamma - i\gamma')\} = k^2,$$

en désignant par k une constante réelle arbitraire.

Posons

$$x_1 = p + qi, \qquad y_1 = \gamma + \gamma' i,$$
$$x_2 = p - qi, \qquad y_2 = \gamma - \gamma' i,$$

$$\xi_1 = (p + qi)^2 + c_0(\gamma + i\gamma') = x_1^2 + c_0 y_1,$$
$$\xi_2 = (p - qi)^2 + c_0(\gamma - i\gamma') = x_2^2 + c_0 y_2.$$

L'équation (3) s'écrit alors

$$\xi_1 \xi_2 = k^2.$$

Des équations (2) on tire les valeurs de $r^2, r\gamma'', \gamma''^2$ exprimées en x_1, x_2, ξ_1, ξ_2;

$$(4) \qquad \begin{cases} r^2 = 6l_1 - (x_1 + x_2)^2 + \xi_1 + \xi_2, \\ c_0 r\gamma'' = 2lc_0 + x_1 x_2(x_1 + x_2) - x_2\xi_1 - x_1\xi_2, \\ c_0^2\gamma''^2 = c_0^2 - k^2 - x_1^2 x_2^2 + x_2^2\xi_1 + x_1^2\xi_2, \end{cases}$$

ou bien, en posant

$$\mathfrak{A} = 6l_1 - (x_1 + x_2)^2,$$

$$\mathfrak{B} = 2lc_0 + x_1 x_2(x_1 + x_2),$$

$$\mathfrak{C} = c_0^2 - k^2 - x_1^2 x_2^2,$$

$$(4') \quad \begin{cases} r^2 = \mathfrak{A} + \xi_1 + \xi_2, \\ c_0 r\gamma'' = \mathfrak{B} - x_2\xi_1 - x_1\xi_2, \\ c_0^2\gamma''^2 = \mathfrak{C} + x_2^2\xi_1 + x_1^2\xi_2. \end{cases}$$

Les quatre quantités x_1, x_2, ξ_1, ξ_2 satisfont donc aux deux relations algébriques suivantes

$$(5) \quad \begin{aligned} &\xi_1\xi_2 = k^2, \\ &(\mathfrak{A} + \xi_1 + \xi_2)(\mathfrak{C} + x_2^2\xi_1 + x_1^2\xi_2) - (\mathfrak{B} - x_2\xi_1 - x_1\xi_2)^2 = 0. \end{aligned}$$

Posons

$$R(x_1) = \mathfrak{A}x_1^2 + 2\mathfrak{B}x_1 + \mathfrak{C} = -x_1^4 + 6l_1x_1^2 + 4lc_0x_1 + c_0^2 - k^2,$$

$$R(x_2) = \mathfrak{A}x_2^2 + 2\mathfrak{B}x_2 + \mathfrak{C} = -x_2^4 + 6l_1x_2^2 + 4lc_0x_2 + c_0^2 - k^2,$$

$$\begin{aligned} R_1(x_1x_2) = \mathfrak{A}\mathfrak{C} - \mathfrak{B}^2 = &-6l_1x_1^2x_2^2 - (c_0^2 - k^2)(x_1 + x_2)^2 \\ &- 4lc_0(x_1 + x_2)x_1x_2 + 6l_1(c_0^2 - k^2) - 4l^2c_0^2. \end{aligned}$$

Les équations (5), développées, deviennent alors

$$(5') \quad \begin{aligned} &\xi_1\xi_2 = k^2, \\ &R(x_2)\xi_1 + R(x_1)\xi_2 + R_1(x_1x_2) + k^2(x_1 - x_2)^2 = 0. \end{aligned}$$

Posons de plus

$$\begin{aligned} R(x_1x_2) = \mathfrak{A}x_1x_2 + \mathfrak{B}(x_1 + x_2) + \mathfrak{C} = &-x_1^2x_2^2 + 6l_1x_1x_2 \\ &+ 2lc_0(x_1 + x_2) + c_0^2 - k^2; \end{aligned}$$

on a alors l'identité

$$(6) \quad R(x_1)R(x_2) - R(x_1x_2)^2 = (x_1 - x_2)^2R_1(x_1x_2).$$

Posons:

$$(7)\quad \begin{aligned} W^2 &= \{R_1(x_1x_2) + k^2(x_1 - x_2)^2\}^2 - 4k^2R(x_1)R(x_2) \\ &= \{R_1(x_1x_2) + k^2(x_1 - x_2)^2\}^2 - 4k^2\{R(x_1x_2)^2 + (x_1 - x_2)^2R_1(x_1x_2)\} \\ &= \{R_1(x_1x_2) - k^2(x_1 - x_2)^2\}^2 - 4k^2R(x_1x_2)^2 \end{aligned}$$

$$=\{R_1(x_1x_2)-k^2(x_1-x_2)^2-2kR(x_1x_2)\}\{R_1(x_1x_2)-k^2(x_1-x_2)^2+2kR(x_1x_2)\}.$$

Vu que d'après l'identité (6) on a

$$k^2(x_1 - x_2)^2 + 2kR(x_1x_2) - R_1(x_1x_2)$$

$$= \frac{1}{(x_1 - x_2)^2}\{k^2(x_1 - x_2)^4 + 2k(x_1 - x_2)^2R(x_1x_2) + R(x_1x_2)^2 - R(x_1)R(x_2)\}$$

$$= \frac{1}{(x_1 - x_2)^2}\{[k(x_1 - x_2)^2 + R(x_1x_2)]^2 - R(x_1)R(x_2)\}$$

$$= (x_1 - x_2)^2\left\{\frac{R(x_1x_2) - \sqrt{R(x_1)}\sqrt{R(x_2)}}{(x_1 - x_2)^2} + k\right\}\left\{\frac{R(x_1x_2) + \sqrt{R(x_1)}\sqrt{R(x_2)}}{(x_1 - x_2)^2} + k\right\},$$

et de même

$$k^2(x_1 - x_2)^2 - 2kR(x_1x_2) - R_1(x_1x_2)$$

$$= (x_1 - x_2)^2\left\{\frac{R(x_1x_2) - \sqrt{R(x_1)}\sqrt{R(x_2)}}{(x_1 - x_2)^2} - k\right\}\left\{\frac{R(x_1x_2) + \sqrt{R(x_1)}\sqrt{R(x_2)}}{(x_1 - x_2)^2} - k\right\},$$

on peut écrire

$$(7')\quad W^2 = (x_1 - x_2)^4\left\{\frac{R(x_1x_2) - \sqrt{R(x_1)}\sqrt{R(x_2)}}{(x_1 - x_2)^2} - k\right\}\left\{\frac{R(x_1x_2) - \sqrt{R(x_1)}\sqrt{R(x_2)}}{(x_1 - x_2)^2} + k\right\}$$

$$\times\left\{\frac{R(x_1x_2) + \sqrt{R(x_1)}\sqrt{R(x_2)}}{(x_1 - x_2)^2} + k\right\}\left\{\frac{R(x_1x_2) + \sqrt{R(x_1)}\sqrt{R(x_2)}}{(x_1 - x_2)^2} - k\right\},$$

et on a alors

$$(8)\qquad \left\{\begin{aligned} \xi_1 &= -\frac{R_1(x_1x_2) + k^2(x_1 - x_2)^2 - W}{2R(x_2)}, \\ \xi_2 &= -\frac{R_1(x_1x_2) + k^2(x_1 - x_2)^2 + W}{2R(x_1)}. \end{aligned}\right.$$

Au lieu des deux variables x_1 et x_2 introduisons maintenant deux variables nouvelles s_1 et s_2, définies par les équations

$$(9)\quad \begin{cases} s_1 = \dfrac{R(x_1x_2) - \sqrt{R(x_1)}\sqrt{R(x_2)}}{2(x_1 - x_2)^2} + \dfrac{1}{2}l_1. \\ s_2 = \dfrac{R(x_1x_2) + \sqrt{R(x_1)}\sqrt{R(x_2)}}{2(x_1 - x_2)^2} + \dfrac{1}{2}l_1; \end{cases}$$

s_1 et s_2 sont, comme on le voit, les deux racines de l'équation algébrique du second degré

$$(x_1 - x_2)^2\left(s - \frac{1}{2}l_1\right)^2 - R(x_1x_2)\left(s - \frac{1}{2}l_1\right) - \frac{1}{4}R_1(x_1x_2) = 0.$$

On a alors

$$W^2 = (x_1 - x_2)^4(2s_1 - l_1 - k)(2s_1 - l_1 + k)(2s_2 - l_1 - k)(2s_2 - l_1 + k);$$

ou bien, en posant

$$k_1 = \frac{l_1 + k}{2}, \qquad k_2 = \frac{l_1 - k}{2},$$

on a

$$(10)\quad W^2 = 16(x_1 - x_2)^4(s_1 - k_1)(s_2 - k_1)(s_1 - k_2)(s_2 - k_2),$$

$$(11)\quad \begin{cases} \xi_1 = \dfrac{(x_1 - x_2)^2}{R(x_2)}\{[\sqrt{(s_1 - k_1)(s_2 - k_1)} + \sqrt{(s_1 - k_2)(s_2 - k_2)}]^2 - k^2\}, \\ \xi_2 = \dfrac{(x_1 - x_2)^2}{R(x_1)}\{[\sqrt{(s_1 - k_1)(s_2 - k_1)} - \sqrt{(s_1 - k_2)(s_2 - k_2)}]^2 - k^2\}. \end{cases}$$

Etablissons maintenant les équations différentielles, qui définissent ces deux variables nouvelles s_1 et s_2 en fonction du temps.

En posant

$$(12)\quad \begin{aligned} g_2 &= k^2 - c_0^2 + 3l_1^2, \\ g_3 &= l_1(k^2 - c_0^2 - l_1^2) + l^2c_0^2, \end{aligned}$$

$$S_1 = 4s_1^3 - g_2s_1 - g_3, \qquad S_2 = 4s_2^3 - g_2s_2 - g_3,$$

on trouve

$$\sqrt{S_1} = \frac{R(x_1) - \frac{1}{4}(x_1 - x_2)R'(x_1)}{(x_1 - x_2)^2}\sqrt{R(x_2)} - \frac{R(x_2) + \frac{1}{4}(x_1 - x_2)R'(x_2)}{(x_1 - x_2)^2}\sqrt{R(x_1)},$$

$$\sqrt{S_2} = \frac{R(x_1) - \frac{1}{4}(x_1 - x_2)R'(x_1)}{(x_1 - x_2)^2}\sqrt{R(x_2)} + \frac{R(x_2) + \frac{1}{4}(x_1 - x_2)R'(x_2)}{(x_1 - x_2)^2}\sqrt{R(x_1)},$$

$$(13)\quad \begin{cases} \dfrac{ds_1}{\sqrt{S_1}} = \dfrac{dx_1}{\sqrt{R(x_1)}} + \dfrac{dx_2}{\sqrt{R(x_2)}}, \\[2ex] \dfrac{ds_2}{\sqrt{S_2}} = -\dfrac{dx_1}{\sqrt{R(x_1)}} + \dfrac{dx_2}{\sqrt{R(x_2)}}. \end{cases}$$

Des équations différentielles (1) il suit, vu que $x_1 = p + qi$, et $x_2 = p - qi$,

$$(14)\quad \begin{cases} 2i\dfrac{dx_1}{dt} = rx_1 + c_0\gamma'', \\[2ex] -2i\dfrac{dx_2}{dt} = rx_2 + c_0\gamma''. \end{cases}$$

Par conséquent

$$-4\left(\frac{dx_1}{dt}\right)^2 = r^2x_1^2 + 2rc_0\gamma''x_1 + c_0^2\gamma''^2$$

ou bien, en écrivant au lieu de r^2, $c_0r\gamma''$, $c_0^2\gamma''^2$ leurs valeurs, définies par les équations (4'),

$$-4\left(\frac{dx_1}{dt}\right)^2 = x_1^2(\mathfrak{A} + \xi_1 + \xi_2) + 2x_1(\mathfrak{B} - x_2\xi_1 - x_1\xi_2) + \mathfrak{C} + x_2^2\xi_1 + x_1^2\xi_2$$
$$= R(x_1) + (x_1 - x_2)^2\xi_1.$$

De la même manière on trouve

$$-4\left(\frac{dx_2}{dt}\right)^2 = R(x_2) + (x_1 - x_2)^2\xi_2$$

et

$$4\frac{dx_1}{dt}\cdot\frac{dx_2}{dt} = r^2x_1x_2 + rc_0\gamma''(x_1 + x_2) + c_0^2\gamma''^2$$

$$= x_1x_2(\mathfrak{A} + \xi_1 + \xi_2) + (x_1 + x_2)(\mathfrak{B} - x_2\xi_1 - x_1\xi_2) + \mathfrak{C} + x_2^2\xi_1 + x_1^2\xi_2$$
$$= R(x_1x_2).$$

En élevant au carré les deux membres de la première des équations (13), on trouve

$$-4\cdot\frac{1}{S_1}\left(\frac{ds_1}{dt}\right)^2=-\frac{4}{R(x_1)}\left(\frac{dx_1}{dt}\right)^2-\frac{4}{R(x_2)}\left(\frac{dx_2}{dt}\right)^2-\frac{8}{\sqrt{R(x_1)}\sqrt{R(x_2)}}\frac{dx_1}{dt}\frac{dx_2}{dt}$$

$$=2+(x_1-x_2)^2\left(\frac{\xi_1}{R(x_1)}+\frac{\xi_2}{R(x_2)}\right)-2\frac{R(x_1x_2)}{\sqrt{R(x_1)}\sqrt{R(x_2)}},$$

ou bien, par suite des équations (5'),

$$-4\left(\frac{1}{\sqrt{S_1}}\frac{ds_1}{dt}\right)^2$$

$$=2-\frac{(x_1-x_2)^2}{R(x_1).R(x_2)}\{R_1(x_1x_2)+k^2(x_1-x_2)^2\}-\frac{2R(x_1x_2)}{\sqrt{R(x_1)}.\sqrt{R(x_2)}}$$

$$=\frac{R(x_1).R(x_2)+\overline{R(x_1x_2)}^2-2R(x_1x_2)\sqrt{R(x_1)}\sqrt{R(x_2)}-k^2(x_1-x_2)^4}{R(x_1)R(x_2)}$$

$$=\frac{(x_1-x_2)^4}{R(x_1).R(x_2)}\left\{\frac{[R(x_1x_2)-\sqrt{R(x_1)}\sqrt{R(x_2)}]^2}{(x_1-x_2)^4}-k^2\right\}$$

$$=4\frac{(x_1-x_2)^4}{R(x_1)R(x_2)}\left(\frac{R(x_1x_2)-\sqrt{R(x_1)}\sqrt{R(x_2)}}{2(x_1-x_2)^2}-\frac{1}{2}k\right)\left(\frac{R(x_1x_2)-\sqrt{R(x_1)}\sqrt{R(x_2)}}{2(x_1-x_2)^2}+\frac{1}{2}k\right)$$

$$=4\frac{(x_1-x_2)^4}{R(x_1).R(x_2)}(s_1-k_1)(s_1-k_2).$$

Mais on a

$$s_2-s_1=\frac{\sqrt{R(x_1)}\sqrt{R(x_2)}}{(x_1-x_2)^2}.$$

Par conséquent

$$-\frac{1}{S_1}\left(\frac{ds_1}{dt}\right)^2=\frac{(s_1-k_1)(s_1-k_2)}{(s_1-s_2)^2}.$$

Posons

$$R_1(s)=-S(s-k_1)(s-k_2)=-4(s-e_1)(s-e_2)(s-e_3)(s-k_1)(s-k_2);$$

on a alors

$$\frac{ds_1}{\sqrt{R_1(s_1)}}=\frac{dt}{s_1-s_2}.$$

On trouve tout à fait de la même manière

$$\frac{ds_2}{\sqrt{R_1(s_2)}} = \frac{dt}{s_2 - s_1}.$$

D'où il suit

$$(15) \qquad \begin{aligned} 0 &= \frac{ds_1}{\sqrt{R_1(s_1)}} + \frac{ds_2}{\sqrt{R_1(s_2)}}, \\ dt &= \frac{s_1 ds_1}{\sqrt{R_1(s_1)}} + \frac{s_2 ds_2}{\sqrt{R_1(s_2)}}. \end{aligned}$$

$R_1(s)$ étant un polynôme du cinquième degré, et les racines de l'équation $R_1(s) = 0$ étant toutes différentes entre elles, les équations différentielles (15) nous conduisent aux fonctions ultraelliptiques, ou autrement dit, aux fonctions de M. Rosenhain.

Cherchons d'abord les expressions des six quantités $p, q, r, \gamma, \gamma', \gamma''$ exprimées à l'aide des deux quantités s_1 et s_2. Nous y arriverons à l'aide de formules, que j'emprunte à un cours inédit de M. Weierstrass sur les fonctions elliptiques, et dont une partie se trouvent aussi développées dans le *Traité des fonctions elliptiques* etc. de M. Halphen et dans les *Formeln und Lehrsätze* etc. de M. Schwartz.

§ 3.

Soit

$$R(x) = Ax^4 + 4Bx^3 + 6Cx^2 + 4B'x + A'$$

un polynôme du quatrième degré de la variable x. Les coefficients A, B, C, B', A' sont des constantes, assujetties à la condition que $R(x)$ n'ait point de diviseur quadratique. Soit u une seconde variable, liée à x par l'équation différentielle

$$(1) \qquad du = \frac{dx}{\sqrt{R(x)}}.$$

La relation la plus générale entre u et x peut être exprimée de la manière suivante.

Posons

$$
\begin{aligned}
&g_2 = AA' - 4BB' + 3C^2,\\
(2)\qquad &g_3 = ACA' + 2BCB' - AB'B' - A'BB - C^3,\\
&D = B^2 - AC, \qquad E = A^2B' - 3ABC + 2B^3;
\end{aligned}
$$

on a alors

$$
(3)\qquad 4\left(\frac{D}{A}\right)^3 - g_2\frac{D}{A} - g_3 = \frac{E^2}{A^3}.
$$

Nous désignerons par $\wp(u)$ la fonction $\wp(u, g_2, g_3)$, et par $\bar\wp(u)$ la fonction $\wp(u, g_2, -g_3)$.[1] En vertu de l'équation (3) on peut définir un argument w tel, que, le signe de $\sqrt{A}$ étant fixé arbitrairement, les deux équations

$$
(4)\qquad \wp(w) = \frac{D}{A}, \qquad \wp'(w) = -\frac{E}{A\sqrt{A}}
$$

aient lieu simultanément. Posons

$$
\begin{aligned}
\varphi_0(u) &= -\frac{B}{A} + \frac{1}{2\sqrt{A}}\frac{\wp'u + \wp'w}{\wp u - \wp w}\\
&= -\frac{B}{A} + \frac{1}{\sqrt{A}}\left\{\frac{\sigma'(u-w)}{\sigma(u-w)} - \frac{\sigma'(u)}{\sigma(u)} + \frac{\sigma'(w)}{\sigma(w)}\right\}.
\end{aligned}
$$

L'expression la plus générale de x en fonction de u est alors

$$
(6)\qquad x = \varphi_0(u - u_0)
$$

où u_0 désigne une constante arbitraire. (Pour $u = u_0$, $x = \infty$ et $\frac{1}{x^2}\frac{dx}{du} = \sqrt{A}$.) On a

$$
\varphi_0(-u + w) = \varphi_0(u), \qquad \varphi_0\left(u + \frac{w}{2}\right) = \varphi_0\left(-u + \frac{w}{2}\right)
$$

---

[1] Voir HALPHEN, *Traité des fonctions elliptiques.*
SCHWARTZ, *Formeln und Lehrsätze zum Gebrauche der elliptischen Functionen.*

et par conséquent, si l'on pose

$$(8)\qquad \varphi(u\,,w)=\varphi_0\left(u+\frac{w}{2}\right)=-\frac{B}{A}+\frac{1}{\sqrt{A}}\left\{\frac{\sigma'}{\sigma}\left(u-\frac{w}{2}\right)-\frac{\sigma'}{\sigma}\left(u+\frac{w}{2}\right)+\frac{\sigma'}{\sigma}(w)\right\}$$

$$=-\frac{B}{A}+\frac{1}{\sqrt{A}}\left\{\frac{\wp'\left(\frac{w}{2}\right)}{\wp u-\wp\left(\frac{w}{2}\right)}+\frac{1}{2}\frac{\wp''\left(\frac{w}{2}\right)}{\wp'\left(\frac{w}{2}\right)}\right\}.$$

$\varphi(u\,,w)$ est une fonction *paire* de u.

Les équations (4) subsistent sans changement, si l'on ajoute à w une période quelconque de la fonction $\wp(u)$. Si nous désignons par conséquent par $(2\tilde{\omega}\,,2\tilde{\omega}')$ une paire *quelconque* de périodes primitives de la fonction $\wp(u)$, et si de toutes les valeurs de w (en nombre infini) qui satisfont aux équations (4), nous fixons une arbitrairement les quatre expressions

$$\varphi(u\,,w)\,,\,\varphi(u\,,w+2\tilde{\omega})\,,\,\varphi(u\,,w+2\tilde{\omega}+2\tilde{\omega}')\,,\,\varphi(u\,,w+2\tilde{\omega})$$

représenteront *quatre fonctions paires, différentes* de u, qui toutes, mises à la place de x, satisfont à l'équation différentielle (1). En effet si 2Ω et $2\Omega'$ sont deux périodes quelconques de la fonction $\wp(u)$, il résulte de l'équation (8)

$$(9)\qquad \begin{aligned}\varphi(u\,,w+2\Omega)&=\varphi(u+\Omega\,,w)\\ \varphi(u\,,w+2\Omega')&=\varphi(u+\Omega',w).\end{aligned}$$

Pour que l'équation

$$\varphi(u\,,w+2\Omega)=\varphi(u\,,w+2\Omega')$$

puisse subsister pour toutes les valeurs de u, il faut donc que l'on ait

$$\varphi(u+\Omega\,,w)=\varphi(u+\Omega',w),$$

$$\varphi(u+\Omega'-\Omega\,,w)=\varphi(u\,,w),$$

c'est à dire $\Omega'-\Omega$ doit être une période de la fonction $\varphi(u)$ et par conséquent, (en vue de la dernière des équations (8)) $\Omega'-\Omega$ doit être aussi une période de $\wp u$.

Si tel n'est pas le cas, les deux fonctions $\varphi(u, w + 2\Omega)$ et $\varphi(u, w + 2\Omega')$ ne sont pas identiques. On en conclut donc que les quatre fonctions précitées,

$$\varphi(u, w), \varphi(u, w + 2\tilde{\omega}), \varphi(u, w + 2\tilde{\omega} + 2\tilde{\omega}'), \varphi(u, w + 2\tilde{\omega}'),$$

sont toutes les quatre différentes entre elles.

Mais il n'existe en tout que quatre fonctions *paires* de u, qui, mises à la place de x, satisfont à l'équation différentielle (1); ces quatre fonctions se distinguent par là, que pour $u = 0$ chacune d'elle devient égale à l'une des quatre racines de l'équation

$$R(x) = 0.$$

Ces quatre fonctions doivent donc être identiques avec les quatre fonctions paires précitées, et, en vertu des équations (9) on peut aussi les écrire de la manière suivante:

$$\varphi(u, w), \varphi(u + \tilde{\omega}, w), \varphi(u + \tilde{\omega} + \tilde{\omega}', w), \varphi(u + \tilde{\omega}', w).$$

Si nous posons

$$a = \varphi(0, w), \qquad a_1 = \varphi(\tilde{\omega}, w) = \varphi(0, w + 2\tilde{\omega}),$$

$$a_2 = \varphi(\tilde{\omega} + \tilde{\omega}', w) = \varphi(0, w + 2\tilde{\omega} + 2\tilde{\omega}'),$$

$$a_3 = \varphi(\tilde{\omega}', w) = \varphi(0, w + 2\tilde{\omega}'),$$

les quatre quantités a, a_1, a_2, a_3 représentent les 4 racines différentes de l'équation $R(x) = 0$ et l'on a

$$a = -\frac{B}{A} + \frac{1}{2\sqrt{A}} \frac{\wp''\left(\frac{w}{2}\right)}{\wp'\left(\frac{w}{2}\right)}; \qquad a_2 = -\frac{B}{A} + \frac{1}{2\sqrt{A}} \frac{\wp''\left(\frac{w}{2} + \tilde{\omega} + \tilde{\omega}'\right)}{\wp'\left(\frac{w}{2} + \tilde{\omega} + \tilde{\omega}'\right)},$$

$$a_1 = -\frac{B}{A} + \frac{1}{2\sqrt{A}} \frac{\wp''\left(\frac{w}{2} + \tilde{\omega}\right)}{\wp'\left(\frac{w}{2} + \tilde{\omega}\right)}, \qquad a_3 = -\frac{B}{A} + \frac{1}{2\sqrt{A}} \frac{\wp''\left(\frac{w}{2} + \tilde{\omega}'\right)}{\wp'\left(\frac{w}{2} + \tilde{\omega}'\right)}.$$

En posant $\wp(\tilde{\omega}) = e_1$, $\wp(\tilde{\omega} + \tilde{\omega}') = e_2$, $\wp(\tilde{\omega}') = e_3$, les relations suivantes ont lieu pour chaque valeur de l'argument u,

$$\wp\left(\frac{u}{2}\right) = -\frac{(e_2^2 - e_3^2)\sigma_1 u + (e_3^2 - e_1^2)\sigma_2 u + (e_1^2 - e_2^2)\sigma_3 u}{(e_2 - e_3)\sigma_1 u + (e_3 - e_1)\sigma_2 u + (e_1 - e_2)\sigma_3 u},$$

$$\wp'\left(\frac{u}{2}\right) = \frac{-2(e_2 - e_3)(e_3 - e_1)(e_1 - e_2)\sigma u}{(e_2 - e_3)\sigma_1 u + (e_3 - e_1)\sigma_2 u + (e_1 - e_2)\sigma_3 u},$$

$$\frac{\wp''\left(\frac{u}{2}\right)}{\wp'\left(\frac{u}{2}\right)} = -2 \cdot \frac{(e_2 - e_3)\sigma_2 u \sigma_3 u + (e_3 - e_1)\sigma_3 u \sigma_1 u + (e_1 - e_2)\sigma_1 u \sigma_2 u}{\sigma u[(e_2 - e_3)\sigma_1 u + (e_3 - e_1)\sigma_2 u + (e_1 - e_2)\sigma_3 u]}$$

$$= -2\frac{\sigma_1 u + \sigma_2 u + \sigma_3 u}{\sigma u}.$$ [1]

On a de plus

$$\frac{\sigma_1(w)}{\sigma(w)} = \sqrt{\frac{D}{A} - e_1}, \qquad \frac{\sigma_2(w)}{\sigma(w)} = \sqrt{\frac{D}{A} - e_2}, \qquad \frac{\sigma_3(w)}{\sigma(w)} = \sqrt{\frac{D}{A} - e_3}.$$

Pour chaque valeur déterminée de w, les signes correspondants des racines carrées dans ces formules sont aussi parfaitement déterminés, et l'on a

$$\sqrt{\frac{D}{A} - e_1} \cdot \sqrt{\frac{D}{A} - e_2} \cdot \sqrt{\frac{D}{A} - e_3} = -\frac{1}{2}\wp'(w) = \frac{E}{2A\sqrt{A}}.$$

En vertu de ces formules, on a donc

$$-\wp\left(\frac{w}{2}\right) = \frac{(e_2^2 - e_3^2)\sqrt{\frac{D}{A} - e_1} + (e_3^2 - e_1^2)\sqrt{\frac{D}{A} - e_2} + (e_1^2 - e_2^2)\sqrt{\frac{D}{A} - e_3}}{(e_2 - e_3)\sqrt{\frac{D}{A} - e_1} + (e_3 - e_1)\sqrt{\frac{D}{A} - e_2} + (e_1 - e_2)\sqrt{\frac{D}{A} - e_3}},$$

$$\wp'\left(\frac{w}{2}\right) = \frac{-2(e_2 - e_3)(e_3 - e_1(e_1 - e_2)}{(e_2 - e_3)\sqrt{\frac{D}{A} - e_1} + (e_3 - e_1)\sqrt{\frac{D}{A} - e_2} + (e_1 - e_2)\sqrt{\frac{D}{A} - e_3}},$$

$$\frac{1}{2\sqrt{A}}\frac{\wp''\left(\frac{w}{2}\right)}{\wp'\left(\frac{w}{2}\right)} = -\frac{1}{\sqrt{A}}\left\{\sqrt{\frac{D}{A} - e_1} + \sqrt{\frac{D}{A} - e_2} + \sqrt{\frac{D}{A} - e_3}\right\}.$$

---

[1] Voir SCHWARZ, *Formeln und Lehrsätze.*

Par conséquent la racine a de l'équation $R(x) = 0$, correspondante à la valeur particulière de w que nous avons choisie, est donnée par la formule:

$$a = -\frac{1}{\sqrt{A}} \frac{\sigma_1(w) + \sigma_2(w) + \sigma_3(w)}{\sigma(w)} - \frac{B}{A}$$

$$= -\frac{B}{A} - \frac{1}{\sqrt{A}} \left\{ \sqrt{\frac{D}{A} - e_1} + \sqrt{\frac{D}{A} - e_2} + \sqrt{\frac{D}{A} - e_3} \right\}.$$

Si l'on pose

$$h_0 = \frac{\sigma_1(w) + \sigma_2(w) + \sigma_3(w)}{\sigma(w)} = \sqrt{\frac{D}{A} - e_1} + \sqrt{\frac{D}{A} - e_2} + \sqrt{\frac{D}{A} - e_3},$$

$$h_1 = \frac{(e_2 - e_3)\sigma_1(w) + (e_3 - e_1)\sigma_2(w) + (e_1 - e_2)\sigma_3(w)}{\sigma(w)}$$

$$= (e_2 - e_3)\sqrt{\frac{D}{A} - e_1} + (e_3 - e_1)\sqrt{\frac{D}{A} - e_2} + (e_1 - e_2)\sqrt{\frac{D}{A} - e_3},$$

$$h_2 = \frac{(e_2^2 - e_3^2)\sigma_1(w) + (e_3^2 - e_1^2)\sigma_2(w) + (e_1^2 - e_2^2)\sigma_3(w)}{\sigma(w)}$$

$$= (e_2^2 - e_3^2)\sqrt{\frac{D}{A} - e_1} + (e_3^2 - e_1^2)\sqrt{\frac{D}{A} - e_2} + (e_1^2 - e_2^2)\sqrt{\frac{D}{A} - e_3},$$

on a donc

$$a = -\frac{B}{A} - \frac{h_0}{\sqrt{A}},$$

$$\varphi(u, w) = -\frac{B}{A} - \frac{h_0}{\sqrt{A}} - \frac{2}{\sqrt{A}} \frac{(e_2 - e_3)(e_3 - e_1)(e_1 - e_2)}{h_1 \wp u + h_2}.$$

En posant

$$\tilde{\omega} = \omega_1,$$

$$\tilde{\omega} + \tilde{\omega}' = \omega_2,$$

$$\tilde{\omega}' = \omega_3$$

et en désignant par λ, μ, ν les nombres 1, 2, 3 dans un ordre quelconque, on a les formules

$$\frac{\sigma_\lambda(u+2\omega_\lambda)}{\sigma(u+2\omega_\lambda)} = \frac{\sigma_\lambda u}{\sigma u},$$

$$\frac{\sigma_\lambda(u+2\omega_\mu)}{\sigma(u+2\omega_\mu)} = -\frac{\sigma_\lambda u}{\sigma u},$$

$$\frac{\sigma_\lambda(u+2\omega_\nu)}{\sigma(u+2\omega_\nu)} = -\frac{\sigma_\lambda u}{\sigma u}.$$

On obtient donc, en écrivant dans les formules précédentes, $w+2\omega_1$, $w+2\omega_2$, $w+2\omega_3$ à la place de w et en posant

$$h'_0 = \frac{\sigma_1(w) - \sigma_2(w) - \sigma_3(w)}{\sigma(w)},$$

$$h'_1 = \frac{(e_2-e_3)\sigma_1(w) - (e_3-e_1)\sigma_2(w) - (e_1-e_2)\sigma_3(w)}{\sigma(w)},$$

$$h'_2 = \frac{(e_2^2-e_3^2)\sigma_1(w) - (e_3^2-e_1^2)\sigma_2(w) - (e_1^2-e_2^2)\sigma_3(w)}{\sigma(w)},$$

$$h''_0 = \frac{-\sigma_1(w) + \sigma_2(w) - \sigma_3(w)}{\sigma(w)},$$

(14) $$h''_1 = \frac{-(e_2-e_3)\sigma_1(w) + (e_3-e_1)\sigma_2(w) - (e_1-e_2)\sigma_3(w)}{\sigma(w)},$$

$$h''_2 = \frac{-(e_2^2-e_3^2)\sigma_1(w) + (e_3^2-e_1^2)\sigma_2(w) - (e_1^2-e_2^2)\sigma_3(w)}{\sigma(w)},$$

$$h'''_0 = \frac{-\sigma_1(w) - \sigma_2(w) + \sigma_3(w)}{\sigma(w)},$$

$$h'''_1 = \frac{-(e_2-e_3)\sigma_1(w) - (e_3-e_1)\sigma_2(w) + (e_1-e_2)\sigma_3(w)}{\sigma(w)},$$

$$h'''_2 = \frac{-(e_2^2-e_3^2)\sigma_1(w) - (e_3^2-e_1^2)\sigma_2(w) + (e_1^2-e_2^2)\sigma_3(w)}{\sigma(w)},$$

les formules suivantes

$$
(15)\quad
\begin{aligned}
a &= -\frac{B}{A} - \frac{h_0}{\sqrt{A}}, \\
\varphi(u, w) &= -\frac{B}{A} - \frac{h_0}{\sqrt{A}} - \frac{2}{\sqrt{A}} \frac{(e_2 - e_3)(e_3 - e_1)(e_1 - e_2)}{h_1 \wp u + h_2}, \\
a_1 &= -\frac{B}{A} - \frac{h'_0}{\sqrt{A}}, \\
\varphi(u + \omega_1, w) &= -\frac{B}{A} - \frac{h'_0}{\sqrt{A}} - \frac{2}{\sqrt{A}} \frac{(e_2 - e_3)(e_3 - e_1)(e_1 - e_2)}{h'_1 \wp u + h'_2}, \\
a_2 &= -\frac{B}{A} - \frac{h''_0}{\sqrt{A}}, \\
\varphi(u + \omega_2, w) &= -\frac{B}{A} - \frac{h''_0}{\sqrt{A}} - \frac{2}{\sqrt{A}} \frac{(e_2 - e_3)(e_3 - e_1)(e_1 - e_2)}{h''_1 \wp u + h''_2}, \\
a_3 &= -\frac{B}{A} - \frac{h'''_0}{\sqrt{A}}, \\
\varphi(u + \omega_3, w) &= -\frac{B}{A} - \frac{h'''_0}{\sqrt{A}} - \frac{2}{\sqrt{A}} \frac{(e_2 - e_3)(e_3 - e_1)(e_1 - e_2)}{h'''_1 \wp u + h'''_2}.
\end{aligned}
$$

De cette manière on obtient les quatre fonctions *paires* de u, qui satisfont à l'équation différentielle $\frac{dx}{du} = \sqrt{R(x)}$, de même que les racines correspondantes de l'équation $R(x) = 0$, c'est à dire les valeurs que chacune de ces quatre fonctions prend pour $u = 0$, exprimées en fonctions rationelles des quantités suivantes:

$$
\frac{B}{A}, \quad \frac{1}{\sqrt{A}}, \quad \frac{\sigma_1(w)}{\sigma(w)} = \sqrt{\frac{D}{A} - e_1}, \quad \frac{\sigma_2(w)}{\sigma(w)} = \sqrt{\frac{D}{A} - e_2}, \quad \frac{\sigma_3(w)}{\sigma(w)} = \sqrt{\frac{D}{A} - e_3}.
$$

Jusqu'à présent nous n'avons assujetti les constantes A, B, C, B', A' (les coefficients de $R(x)$) à aucune condition; supposons maintenant que tous ces coefficients sont réels. — Il y a toujours dans ce cas des valeurs réelles positives de u qui satisfont à l'équation $\wp'(u) = 0$; désignons par ω la plus petite de toutes ces valeurs. De même l'équation $\bar{\wp}'(u) = 0$ peut être satisfaite par des valeurs réelles positives de u et nous désignerons par $\tilde{\omega}$ la plus petite de ces dernières.

Il faut à présent distinguer deux cas:

I. La quantité $g_2^3 - 27g_3^2$ est *positive*.

Posons dans ce cas

$$\omega_1 = \omega, \qquad \omega_2 = \omega + \tilde{\omega}i, \qquad \omega_3 = \tilde{\omega}i,$$

$$\wp(\omega_1) = e_1, \qquad \wp(\omega_2) = e_2, \qquad \wp(\omega_3) = e_3.$$

Les quantités e_1, e_2, e_3, qui représentent les racines de l'équation

$$4s^3 - g_2 s - g_3 = 0,$$

sont dans ce cas toutes trois réelles, et l'on a

$$e_1 > e_2 > e_3;$$

$(2\omega_1, 2\omega_3)$ est une paire de périodes primitives de la fonction $\wp(u)$.

II. La quantité $g_2^3 - 27g_3^2$ est *négative*.

Posons dans ce cas

$$\omega_1 = \frac{\omega - \tilde{\omega}i}{2}, \qquad \omega_2 = \omega, \qquad \omega_3 = \frac{\omega + \tilde{\omega}i}{2},$$

$$\wp(\omega_1) = e_1, \qquad \wp(\omega_2) = e_2, \qquad \wp(\omega_3) = e_3;$$

$(2\omega_1, 2\omega_3)$ représente dans ce cas aussi une paire de périodes primitives de la fonction $\wp(u)$, mais e_1 et e_3 sont dans ce cas des quantités imaginaires, conjuguées, tandis que $e_2 = -(e_1 + e_3)$ est réel. La quantité $\frac{e_1 - e_3}{i}$ est positive.

Si nous convenons de donner à $\sqrt{A}$ sa valeur positive dans le cas où A est positif, et de désigner par cette racine le produit de i par une quantité positive, dans le cas où A est négatif, on peut dans les deux cas déterminer de la manière suivante une valeur de w satisfaisant aux équations

$$\wp(w) = \frac{D}{A}, \qquad \wp'(w) = -\frac{E}{A\sqrt{A}}.$$

Les équations

$$\frac{E^2}{A^3} = \wp'(w)^2 = 4\left(\frac{D}{A} - e_1\right)\left(\frac{D}{A} - e_2\right)\left(\frac{D}{A} - e_3\right)$$

nous font voir que dans le cas I, si $A > 0$, $\frac{D}{A}$ doit être contenu soit dans l'intervalle $(e_1 \ldots \infty)$, soit dans l'intervalle $(e_3 \ldots e_2)$; si $A < 0$, $\frac{D}{A}$ appartiendra soit à l'intervalle $(-\infty \ldots e_3)$ soit à l'intervalle $(e_2 \ldots e_1)$. Dans le cas II, au contraire, $\frac{D}{A}$ doit être contenu soit dans l'intervalle $(e_2 \ldots \infty)$, soit dans l'intervalle $(-\infty \ldots e_2)$.

Désignons par $(a \ldots b)$ une quantité contenue dans l'intervalle réelle $(a \ldots b)$, par $(+)$ et $(-)$ des quantités positives ou négatives respectivement et par ε une quantité réelle satisfaisant à la condition

$$0 \leqq \varepsilon \leqq 1.$$

On a dans le cas I

$$(1) \qquad \wp(2\varepsilon\omega) = (\infty \ldots e_1), \qquad \wp'(2\varepsilon\omega) = \begin{cases} (-)\,, & \text{si } \varepsilon < \frac{1}{2}, \\ 0\,, & \text{»} \ \ \varepsilon = \frac{1}{2}, \\ (+)\,, & \text{»} \ \ \varepsilon > \frac{1}{2}. \end{cases}$$

$$(2) \qquad \wp(\omega + 2\varepsilon\tilde{\omega}i) = (e_1 \ldots e_2), \qquad \wp'(\omega + 2\varepsilon\tilde{\omega}i) = \begin{cases} (+)i, & \text{si } \varepsilon < \frac{1}{2}, \\ 0\,, & \text{»} \ \ \varepsilon = \frac{1}{2}, \\ (-)i, & \text{»} \ \ \varepsilon > \frac{1}{2}. \end{cases}$$

$$(3) \qquad \wp(2\varepsilon\omega + \tilde{\omega}i) = (e_3 \ldots e_2), \qquad \wp'(2\varepsilon\omega + \tilde{\omega}i) = \begin{cases} (+)\,, & \text{si } \varepsilon < \frac{1}{2}, \\ 0\,, & \text{»} \ \ \varepsilon = 0, \\ (-)\,, & \text{»} \ \ \varepsilon > \frac{1}{2}. \end{cases}$$

$$(4) \qquad \wp(2\varepsilon\tilde{\omega}i) = (-\infty \ldots e_3), \qquad \wp'(2\varepsilon\tilde{\omega}i) = \begin{cases} (-)i, & \text{si } \varepsilon < \frac{1}{2}, \\ 0\,, & \text{»} \ \ \varepsilon = \frac{1}{2}, \\ (+)i, & \text{»} \ \ \varepsilon > \frac{1}{2}. \end{cases}$$

Dans chacun de ces quatre cas, si l'on fait croître ε d'une manière continue de o à $\frac{1}{2}$, la fonction $\wp(u)$ parcourt tout l'intervalle indiqué, en croissant continuellement dans le 3me et dans le 4me cas, en diminuant dans le 1er et dans le 2me. Si l'on fait varier ensuite ε de $\frac{1}{2}$ jusqu'à 1, $\wp(u)$ parcourt dans chaque cas le même intervalle qu'auparavant, mais dans le sens contraire. A deux valeurs de ε, à égale distance de $\frac{1}{2}$, correspondent les mêmes valeurs de $\wp(u)$, mais des valeurs contraires de $\wp'(u)$.

Dans le cas II on a

$$(5)\qquad \wp(2\varepsilon\omega) = (+\infty \ldots e_2), \qquad \wp'(2\varepsilon\omega) = \begin{cases} (-), & \text{si } \varepsilon < \frac{1}{2}, \\ 0, & \text{» } \varepsilon = \frac{1}{2}, \\ (+), & \text{» } \varepsilon > \frac{1}{2}. \end{cases}$$

$$(6)\qquad \wp(2\varepsilon\tilde{\omega}i) = (-\infty \ldots e_2), \qquad \wp'(2\varepsilon\tilde{\omega}i) = \begin{cases} (-)i, & \text{si } \varepsilon < \frac{1}{2}, \\ 0, & \text{» } \varepsilon = \frac{1}{2}, \\ (+)i, & \text{» } \varepsilon > \frac{1}{2}. \end{cases}$$

D'après ces formules, on peut définir une quantité w, satisfaisant aux équations

$$\wp(w) = \frac{D}{A}, \qquad \wp'(w) = -\frac{E}{A\sqrt{A}}$$

de la manière suivante:

I. $g_2^3 - 27g_3^2$ *est positif.*

1) Si A est positif et $\frac{D}{A}$ appartient à l'intervalle $(+\infty \ldots e_1)$ on peut poser

$$w = 2\varepsilon\omega \quad (0 < \varepsilon < 1),$$

en remarquant que

$$\varepsilon \lesseqgtr \frac{1}{2}, \quad \text{selon que} \quad E \gtreqless 0.$$

2) Si A est négatif et $\frac{D}{A}$ se trouve dans l'intervalle $(-\infty \ldots e_3)$, on peut poser

$$w = 2\varepsilon\tilde{\omega}i \quad (0 < \varepsilon < 1),$$

et l'on a

$$\varepsilon \lesseqgtr \frac{1}{2}, \text{ selon que } E \gtreqless 0.$$

3) Si A est positif, mais $\frac{D}{A}$ se trouve dans l'intervalle $(e_3 \ldots e_2)$ on peut poser

$$w = 2\varepsilon\omega + \tilde{\omega}i \quad (0 \leqq \varepsilon < 1),$$

et l'on a

$$\varepsilon \lesseqgtr \frac{1}{2}, \text{ selon que } E \lesseqgtr 0.$$

4) Si A est négatif et $\frac{D}{A}$ appartient à l'intervalle $(e_1 \ldots e_2)$ on peut poser

$$w = \omega + 2\varepsilon\tilde{\omega}i \quad (0 \leqq \varepsilon < 1),$$

et l'on a

$$\varepsilon \lesseqgtr \frac{1}{2}, \text{ selon que } E \lesseqgtr 0.$$

II. *$g_2^3 - 27g_3^2$ est une quantité négative.*

5) Si A est positif, $\frac{D}{A}$ appartient à l'intervalle $(e_2 \ldots \infty)$ et l'on peut poser

$$w = 2\varepsilon\omega \quad (0 \leqq \varepsilon < 1),$$

$$\varepsilon \text{ étant } \lesseqgtr \frac{1}{2}, \text{ selon que } E \gtreqless 0.$$

6) Si A est négatif, $\frac{D}{A}$ se trouve dans l'intervalle $(-\infty \ldots e_2)$ et l'on peut poser

$$w = 2\varepsilon\tilde{\omega}i \quad (0 < \varepsilon < 1),$$

$$\varepsilon \text{ étant } \lesseqgtr \frac{1}{2}, \text{ selon que } E \gtreqless 0.$$

Les cas 1) et 2) se présentent lorsque toutes les quatre racines de l'équation $R(x) = 0$ sont réelles.

Les cas 3) et 4) ont lieu si tous ces racines sont imaginaires.

Enfin on a les cas 5) et 6) lorsque deux racines sont réelles et les deux autres imaginaires conjuguées.

Dans le cas 1) les quantités $\frac{D}{A} - e_1$, $\frac{D}{A} - e_2$, $\frac{D}{A} - e_3$ sont toutes trois réelles et positives; dans le cas 2) elles sont réelles et négatives; par conséquent, en vertu des équations (14) les quantités

$$\frac{h_0}{\sqrt{A}}, \frac{h_0'}{\sqrt{A}}, \frac{h_0''}{\sqrt{A}}, \frac{h_0'''}{\sqrt{A}}$$

sont dans les deux cas réelles.

Dans le cas 3) les quantités $\frac{D}{A} - e_1$, $\frac{D}{A} - e_2$ sont négatives, tandis que $\frac{D}{A} - e_3$ est positif; par conséquent

$$\frac{1}{\sqrt{A}}\sqrt{\frac{D}{A} - e_1} \quad \text{et} \quad \frac{1}{\sqrt{A}}\sqrt{\frac{D}{A} - e_2}$$

sont des quantités imaginaires, tandis que $\frac{1}{\sqrt{A}}\sqrt{\frac{D}{A} - e_3}$ est une quantité réelle.

Les quantités $\frac{h_0}{\sqrt{A}}$, $\frac{h_0'}{\sqrt{A}}$, $\frac{h_0''}{\sqrt{A}}$, $\frac{h_0'''}{\sqrt{A}}$ sont donc des quantités complexes. Les quantités a et a_3, ainsi que a_1 et a_2 sont dans ce cas-ci des quantités complexes conjuguées.

Dans le cas 4) les quantités $\frac{D}{A} - e_2$, $\frac{D}{A} - e_3$ sont positives, tandis que $\frac{D}{A} - e_1$ est négatif; par conséquent

$$\frac{\sqrt{\frac{D}{A} - e_1}}{\sqrt{A}} \text{ est réel, mais } \frac{\sqrt{\frac{D}{A} - e_2}}{\sqrt{A}}, \frac{\sqrt{\frac{D}{A} - e_3}}{\sqrt{A}} \text{ sont imaginaires.}$$

$\frac{h_0}{\sqrt{A}}$, $\frac{h_0'}{\sqrt{A}}$, $\frac{h_0''}{\sqrt{A}}$, $\frac{h_0'''}{\sqrt{A}}$ sont des quantités complexes.

Les racines a et a_1, ainsi que a_2 et a_3 sont des quantités complexes conjuguées.

Dans le cas 5) $\frac{D}{A} - e_2$ est positif, $\frac{D}{A} - e_1$, $\frac{D}{A} - e_3$ sont des quantités complexes conjuguées. Les fonctions $\frac{\sigma_1 u}{\sigma u}$, $\frac{\sigma_3 u}{\sigma u}$ prennent pour des valeurs réelles de u des valeurs complexes conjuguées, de manière que si u et u' sont des valeurs complexes conjuguées, les valeurs correspondantes

$$\frac{\sigma_1 u}{\sigma u}, \frac{\sigma_3 u'}{\sigma u'}$$

sont aussi des quantités complexes conjuguées.

Dans le cas 5), aussi bien que dans le cas 6), les quantités

$$\frac{1}{\sqrt{A}}\frac{\sigma_1 w}{\sigma w}, \frac{1}{\sqrt{A}}\frac{\sigma_3 w}{\sigma w}$$

sont des quantités complexes conjugués, tandis que

$$\frac{1}{\sqrt{A}}\frac{\sigma_2 w}{\sigma w}$$

est une quantité réelle.

Les quantités a et a_2 sont donc réelles dans les cas 5) et 6), mais a_1 et a_3 sont complexes et conjuguées.

Il résulte donc de la discussion précédente que, pour des valeurs réelles de u, les fonctions

$$\varphi(u, w), \varphi(u + \omega_1, w), \varphi(u + \omega_2, w), \varphi(u + \omega_3, w)$$

sont toutes réelles dans le cas 1) et 2) et toutes imaginaires, dans le cas 3) et 4). Il est à remarquer que dans le cas 3) la première et la troisième de même que la seconde et la quatrième de ces fonctions sont des quantités imaginaires conjuguées; dans le cas 4) ce sont au contraire la première et la seconde, de même que la 3me et la 4me de ces fonctions qui sont conjuguées.

Dans les cas 5) et 6) la première et la troisième de ces fonctions sont réelles, tandis que la seconde et la 4me sont imaginaires et conjuguées.

§ 4.

Pour pouvoir appliquer les formules du paragraphe précédent au cas qui nous occupe il faut commencer par la discussion des racines de l'équation

$$R(x) = -x^4 + 6l_1x^2 + 4c_0lx + c_0^2 - k^2 = 0.$$

Posons

$$k_0 = c_0^2 - k^2, \qquad l_0 = c_0 l.$$

En général, si

$$R(x) = Ax^4 + 4Bx^3 + 6Cx^2 + 4B'x + A'$$

et que l'on pose

$$g_2 = AA' - 4BCB' + 3C^2,$$
$$g_3 = ACA' + 2BCB' - AB'^2 - A'B^2 - C^3,$$

la condition de la réalité des racines de l'équation $R(x) = 0$ peut s'énoncer de la manière suivante.

Si $G = g_2^3 - 27g_3^2 < 0$, l'équation $R(x) = 0$ a deux racines réelles et deux racines imaginaires conjuguées.

Si $G = g_2^3 - 27g_3^2 > 0$, toutes les quatre racines sont réelles à la fois ou toutes les quatre imaginaires, conjuguées deux à deux. Le premier cas a lieu, si l'on a en plus

$$B^2 - AC > 0\,,\ 12(B^2 - AC)^2 - A^2g_2 > 0.$$

Mais si, $g_2^3 - 27g_3^2$ étant positif, l'une de ces dernières quantités est négative, toutes les quatre racines de l'équation $R(x) = 0$ sont imaginaires.

Appliquons ceci à notre cas. On aura

$$A = -1, \qquad B = 0, \qquad C = l_1, \qquad B' = lc_0 = l_0, \qquad A' = c_0^2 - k^2 = k_0,$$

$$g_2 = -k_0 + 3l_1^2,$$
$$g_3 = -l_1(k_0 + l_1^2) + l_0^2,$$

$$B^2 - AC = l_1, \qquad 12(B^2 - AC)^2 - A^2 g_2 = k_0 + 9l_1^2,$$

$$g_2^3 - 27g_3^2 = (-k_0 + 3l_1^2)^3 - 27[-l_1(k_0 + l_1^2) + l_0^2]^2$$

$$= -27\left| l_0^4 - 2l_1(k_0 + l_1^2)l_0^2 + \frac{1}{27}k_0(k_0 + 9l_1^2)^2 \right|.$$

En vertu de l'identité

$$l_1^2(k_0 + l_1^2)^2 = \frac{1}{27}k_0(k_0 + 9l_1^2)^2 + \left(\frac{-k_0 + 3l_1^2}{3}\right)^3$$

les racines de l'équation quadratique en l_0^2

$$Q(l_0^2) = l_0^4 - 2l_1(k_0 + l_1^2)l_0^2 + \frac{1}{27}k_0(k_0 + 9l_1^2)^2$$

peuvent s'écrire

$$l_0'^2 = l_1(k_0 + l_1^2) + \left(\frac{-k_0 + 3l_1^2}{3}\right)^{\frac{3}{2}},$$

$$l_0''^2 = l_1(k_0 + l_1^2) - \left(\frac{-k_0 + 3l_1^2}{3}\right)^{\frac{3}{2}},$$

formules où nous supposerons le radical $\left(\frac{-k_0 + 3l_1^2}{3}\right)^{\frac{3}{2}}$ positif, s'il est réel.

Appliquant ces formules à la discùssion des racines de l'équation $R(x) = 0$, on voit qu'il faut distinguer les cas suivants:

1°. $l_1 > 0, k_0 > 0, -k_0 + 3l_1^2 < 0$.

Dans ce cas $l_0'^2$ et $l_0''^2$ sont tous les deux imaginaires. Par conséquent $G = -27Q(l_0^2)$ est négatif pour toutes les valeurs réelles de l_0, et l'équation $R(x) = 0$ a deux racines réelles et deux racines imaginaires.

2°. $l_1 > 0, k_0 > 0, -k_0 + 3l_1^2 > 0$.

$l_0'^2$ et $l_0''^2$ sont positifs tous les deux et $l_0'^2 > l_0''^2$. Alors si

$$l_0^2 > l_0'^2 \quad \text{ou} \quad l_0''^2 > l_0^2,$$

on a $G < 0$ et l'équation $R(x) = 0$ a encore deux racines réelles et deux racines imaginaires. Si au contraire

$$l_0'^2 > l_0^2 > l_0''^2,$$

G est positif, et, vu que l'on a aussi $l_1 > 0$, $k_0 + 9l_1^2 > 0$, les quatre racines de l'équation $R(x) = 0$ sont réelles.

3°. $l_1 > 0$, $k_0 < 0$, $k_0 + 9l_1^2 > 0$.

La racine $l_0'^2$ est positive, l'autre $l_0''^2$ négative. La quantité

$$G = -27Q(l_0^2)$$

est négative, si $l_0^2 > l_0'^2$, et dans ce cas l'équation $R(x) = 0$ a quatre racines réelles. Si au contraire $l_0^2 < l_0'^2$, G est positif, et l'équation proposée n'a que deux racines réelles.

4°. $l_1 > 0$, $k_0 < 0$, $k_0 + 9l_1^2 < 0$.

Comme dans le cas précédent G est positif pour $l_0^2 < l_0'^2$, et négatif pour $l_0^2 > l_0'^2$. S'il est négatif, l'équation $R(x) = 0$ a dans ce cas quatre racines imaginaires; s'il est positif elle a deux racines réelles et deux racines imaginaires.

5°. $l_1 < 0$, $k_0 > 0$.

$l_0'^2$ et $l_0''^2$ étant négatifs ou imaginaires, la quantité $G = -27Q(l_0^2)$ reste négative pour toute valeur réelle de l_0; l'équation $R(x) = 0$ a donc deux racines réelles et deux racines imaginaires.

6°. $l_1 < 0$, $k_0 < 0$.

Ce cas est tout à fait conforme au quatrième.

Résumé:

L'équation $R(x) = 0$ à quatre racines réelles dans les deux cas suivants:

1) $l_1 > 0$, $k_0 > 0$, $-k_0 + 3l_1^2 > 0$, $l_0'^2 > l_0^2 > l_0''^2$,

où

$$l_0'^2 = l_1(k_0 + l_1^2) + \left(\frac{-k_0 + 3l_1^2}{3}\right)^{\frac{3}{2}},$$

$$l_0''^2 = l_1(k_0 + l_1^2) - \left(\frac{-k_0 + 3l_1^2}{3}\right)^{\frac{3}{2}}.$$

2) $l_1 > 0$, $k_0 < 0$, $k_0 + 9l_1^2 > 0$, $l_0^2 < l_0'^2$.

Les quatre racines de l'équation $R(x) = 0$ sont imaginaires dans les deux cas suivants:

1) $l_1 > 0$, $k_0 < 0$, $k_0 + 9l_1^2 < 0$, $l_0^2 < l_0'^2$.

2) $l_1 < 0$, $k_0 < 0$, $l_0^2 < l_0'^2$.

Enfin l'équation $R(x)=0$ a deux racines réelles dans les cas suivants:

1) $l_1>0$, $k_0>0$, $-k_0+3l_1^2<0$, pour chaque valeur de l_0.

2) $l_1>0$, $k_0>0$, $-k_0+3l_1^2>0$, $l_0^2>l_0'^2$ ou bien $l_0^2<l_0''^2$.

3) $l_1>0$, $k_0<0$, $\quad k_0+9l_1^2>0$, $l_0^2>l_0'^2$.

4) $l_1>0$, $k_0<0$, $\quad k_0+9l_1^2<0$, $l_0^2>l_0'^2$.

5) $l_1<0$, $k_0>0$, pour chaque valeur de l_0.

6) $l_1<0$, $k_1<0$, $l_0^2>l_0'^2$.

§ 5.

J'examinerai plus en détail le cas où toutes les racines de l'équation

$$R(x)=-x^4+6l_1x^2+4lc_0x+c_0^2-k^2=0$$

sont réelles.

En nous servant des mêmes dénotations que dans le § 3, on a donc dans ce cas

$$\wp(w)=\frac{D}{A}=-l_1,\qquad \wp'(w)^2=4s^3-g_2s-g_3=-l_0^2.$$

w est purement imaginaire.

e_1, e_2, e_3 sont réels et l'on a

$$e_1>e_2>e_3>-l_1.$$

Les quantités

$$\frac{\sigma_1}{\sigma}(w)=\sqrt{-(l_1+e_1)},\qquad \frac{\sigma_2}{\sigma}(w)=\sqrt{-(l_1+e_2)},\qquad \frac{\sigma_3}{\sigma}(w)=\sqrt{-(l_1+e_3)}$$

sont toutes les trois purement imaginaires, et il en est de même des quantités

$$h_0=\frac{\sigma_1}{\sigma}(w)+\frac{\sigma_2}{\sigma}(w)+\frac{\sigma_3}{\sigma}(w),$$

$$h_1=(e_2-e_3)\frac{\sigma_1}{\sigma}(w)+(e_3-e_1)\frac{\sigma_2}{\sigma}(w)+(e_1-e_2)\frac{\sigma_3}{\sigma}(w),$$

$$h_2=(e_2^2-e_3^2)\frac{\sigma_1}{\sigma}(w)+(e_3^2-e_1^2)\frac{\sigma_2}{\sigma}(w)+(e_1^2-e_2^2)\frac{\sigma_3}{\sigma}(w).$$

Le coefficient de x^4 dans $R(x)$ étant égal à -1, nous poserons donc dans ce cas, en désignant par E le produit $(e_2 - e_3)(e_3 - e_1)(e_1 - e_2)$,

$$p + qi = x_1 = \varphi(u_1) = i\left(h_0 + \frac{2E}{h_1 \wp u_1 + h_2}\right),$$

$$p - qi = x_2 = \varphi(u_2) = i\left(h_0 + \frac{2E}{h_1 \wp u_2 + h_2}\right),$$

et l'on voit que x_1 et x_2 seront des quantités imaginaires conjuguées, si u_1 et u_2 le sont. Posons

$$u_1 = \frac{u + vi}{2}, \qquad u_2 = \frac{u - vi}{2},$$

u et v étant réels.

On a alors

$$s_1 = \wp(u_1 + u_2) = \wp(u),$$

$$s_2 = \wp(u_1 - u_2) = \wp(vi).$$

Il suit de ces formules, que s_1 et s_2 sont tous les deux des quantités réelles, contenues entre les limites suivantes

$$(\infty \ldots s_1 \ldots e_1), (e_3 \ldots s_2 \ldots -\infty).$$

Pour exprimer les quantités p et q au moyen de s_1 et s_2 je me servirai des formules suivantes.

On trouve à la page 50 de l'ouvrage cité de M. SCHWARZ les formules suivantes

$$\sigma_\lambda(w)\sigma(u + v + w)\sigma(u - v)$$
$$= \sigma(u + w)\sigma(u)\sigma_\lambda(v + w)\sigma_\lambda(v) - \sigma_\lambda(u + w)\sigma_\lambda(u)\sigma(v + w)\sigma(v),$$

$$\sigma_\lambda(w)\sigma_\lambda(u + v + w)\sigma_\lambda(u - v)$$
$$= \sigma_\lambda(u+w)\sigma_\lambda(u)\sigma_\lambda(v+w)\sigma_\lambda(v) - (e_\lambda - e_\mu)(e_\lambda - e_\nu)\sigma(u+w)\sigma(u)\sigma(v+w)\sigma(v).$$

En faisant dans ces formules $w = 0$ et en donnant successivement à λ les valeurs 1, 2, 3 on trouve

$$\sigma(u + v)\,\sigma(u - v) = \sigma^2 u \sigma_1^2 v - \sigma_1^2 u \sigma^2 v$$
$$\sigma_1(u + v)\sigma_1(u - v) = \sigma_1^2 u \sigma_1^2 v - (e_1 - e_2)(e_1 - e_3)\sigma^2 u \sigma^2 v,$$
$$\sigma_2(u + v)\sigma_2(u - v) = \sigma_2^2 u \sigma_2^2 v + (e_1 - e_2)(e_2 - e_3)\sigma^2 u \sigma^2 v,$$
$$\sigma_3(u + v)\sigma_3(u - v) = \sigma_3^2 u \sigma_3^2 v - (e_1 - e_3)(e_2 - e_3)\sigma^2 u \sigma^2 v,$$

ou bien, en remarquant que

$$\sigma_2^2 u = \sigma_1^2 u + (e_1 - e_2)\sigma^2 u,$$

$$\sigma_3^2 u = \sigma_1^2 u + (e_1 - e_3)\sigma^2 u,$$

$$\sigma_2(u+v)\sigma_2(u-v) = \sigma_1^2 u \sigma_1^2 v + (e_1-e_2)(e_1-e_3)\sigma^2 u \sigma^2 v + (e_1-e_2)(\sigma^2 u \sigma_1^2 v + \sigma_1^2 u \sigma^2 v),$$

$$\sigma_3(u+v)\sigma_3(u-v) = \sigma_1^2 u \sigma_1^2 v + (e_1-e_2)(e_1-e_3)\sigma^2 u \sigma^2 v + (e_1-e_3)(\sigma^2 u \sigma_1^2 v + \sigma_1^2 u \sigma^2 v).$$

Il suit de ces formules

$$2(e_2-e_3)(e_3-e_1)(e_1-e_2)\sigma^2 u \sigma^2 v = (e_2-e_3)\sigma_1(u+v)\sigma_1(u-v)$$
$$+ (e_3-e_1)\sigma_2(u+v)\sigma_2(u-v) + (e_1-e_2)\sigma_3(u+v)\sigma_3(u-v),$$

$$2(e_2-e_3)\sigma_1^2 u \sigma_1^2 v = (e_2-e_3)\sigma_1(u+v)\sigma_1(u-v)$$
$$- (e_3-e_1)\sigma_2(u+v)\sigma_2(u-v) - (e_1-e_2)\sigma_3(u+v)\sigma_3(u-v),$$

$$2(e_2-e_3)\sigma^2 u \sigma_1^2 v$$
$$= (e_2-e_3)\sigma(u+v)\sigma(u-v) - \sigma_2(u+v)\sigma_2(u-v) + \sigma_3(u+v)\sigma_3(u-v),$$

$$2(e_2-e_3)\sigma_1^2 u \sigma^2 v$$
$$= -(e_2-e_3)\sigma(u+v)\sigma(u-v) - \sigma_2(u+v)\sigma_2(u-v) + \sigma_3(u+v)\sigma_3(u-v),$$

d'où l'on tire par division

$$\frac{\sigma_1^2 u}{\sigma^2 u} = (e_1-e_2)(e_1-e_3)\frac{(e_2-e_3)\sigma(u+v)\sigma(u-v) + \sigma_2(u+v)\sigma_2(u-v) - \sigma_3(u+v)\sigma_3(u-v)}{(e_2-e_3)\sigma_1(u+v)\sigma_1(u-v) + (e_3-e_1)\sigma_2(u+v)\sigma_2(u-v) + (e_1-e_2)\sigma_3(u+v)\sigma_3(u-v)}$$

et par conséquent, vu que $\wp u = \dfrac{\sigma_1^2 u}{\sigma^2 u} + e_1$,

$$\wp u = -\frac{E\sigma(u+v)\sigma(u-v) + (e_2^2-e_3^2)\sigma_1(u+v)\sigma_1(u-v) + (e_3^2-e_1^2)\sigma_2(u+v)\sigma_2(u-v) + (e_1^2-e_2^2)\sigma_3(u+v)\sigma_3(u-v)}{(e_2-e_3)\sigma_1(u+v)\sigma_1(u-v) + (e_3-e_1)\sigma_2(u+v)\sigma_2(u-v) + (e_1-e_2)\sigma_3(u+v)\sigma_3(u-v)},$$

$$\wp v = -\frac{-E\sigma(u+v)\sigma(u-v) + (e_2^2-e_3^2)\sigma_1(u+v)\sigma_1(u-v) + (e_3^2-e_1^2)\sigma_2(u+v)\sigma_2(u-v) + (e_1^2-e_2^2)\sigma_3(u+v)\sigma_3(u-v)}{(e_2-e_3)\sigma_1(u+v)\sigma_1(u-v) + (e_3-e_1)\sigma_2(u+v)\sigma_2(u-v) + (e_1-e_2)\sigma_3(u+v)\sigma_3(u-v)}.$$

En posant donc

$$P_a = \frac{\sigma_a(u_1+u_2)}{\sigma(u_1+u_2)}\frac{\sigma_a(u_1-u_2)}{\sigma(u_1-u_2)} = \sqrt{(s_1-e_a)(s_2-e_a)}, \qquad (a=1,2,3)$$

on a

$$\wp(u_1) + \wp(u_2) = -2\frac{(e_2^2 - e_3^2)P_1 + (e_3^2 - e_1^2)P_2 + (e_1^2 - e_2^2)P_3}{(e_2 - e_3)P_1 + (e_3 - e_1)P_2 + (e_1 - e_2)P_3},$$

$$\wp(u_1) - \wp(u_2) = \frac{-2E}{(e_2 - e_3)P_1 + (e_3 - e_1)P_2 + (e_1 - e_2)P_3},$$

$$\wp(u_1)\cdot\wp(u_2) = -\frac{(e_2 - e_3)(e_1^2 + e_2 e_3)P_1 + (e_3 - e_1)(e_2^2 + e_3 e_1)P_2 + (e_1 - e_2)(e_3^2 + e_1 e_2)P_3}{(e_2 - e_3)P_1 + (e_3 - e_1)P_2 + (e_1 - e_2)P_3}.$$

En portant ces expressions dans les formules

$$p = \frac{x_1 + x_2}{2} = i\left(h_0 + E\frac{h_1(\wp u_1 + \wp u_2) + 2h_2}{h_1^2\wp u_1\wp u_2 + h_1 h_2(\wp u_1 + \wp u_2) + h_2^2}\right.,$$

$$q = \frac{x_1 - x_2}{2i} = E\frac{h_1(\wp u_1 - \wp u_2)}{h_1^2\wp u_1\wp u_2 + h_1 h_2(\wp u_1 + \wp u_2) + h_2^2},$$

on trouve, après quelques calculs,

$$p = -i\frac{L_1 P_1 + M_1 P_2 + N_1 P_3}{LP_1 + MP_2 + NP_3},$$

$$q = \frac{E}{LP_1 + MP_2 + NP_3},$$

où j'ai posé

$$L = (e_2 - e_3)\frac{\sigma_1}{\sigma}(w) = i(e_2 - e_3)\sqrt{l_1 + e_1},$$

$$M = (e_3 - e_1)\frac{\sigma_2}{\sigma}(w) = i(e_3 - e_1)\sqrt{l_1 + e_2},$$

$$N = (e_1 - e_2)\frac{\sigma_3}{\sigma}(w) = i(e_1 - e_2)\sqrt{l_1 + e_3},$$

$$L_1 = (e_2 - e_3)\frac{\sigma_2}{\sigma}(w)\frac{\sigma_3}{\sigma}(w) = (e_2 - e_3)\sqrt{(l_1 + e_2)(l_1 + e_3)},$$

$$M_1 = (e_3 - e_1)\frac{\sigma_3}{\sigma}(w)\frac{\sigma_1}{\sigma}(w) = (e_3 - e_1)\sqrt{(l_1 + e_3)(l_1 + e_1)},$$

$$N_1 = (e_1 - e_2)\frac{\sigma_1}{\sigma}(w)\frac{\sigma_2}{\sigma}(w) = (e_1 - e_2)\sqrt{(l_1 + e_1)(l_1 + e_2)}.$$

Les quantités $l_1 + e_1$, $l_1 + e_2$, $l_1 + e_3$ sont toutes les trois positives.

Les quantités $s_1 - e_1$, $s_1 - e_2$, $s_1 - e_3$ sont aussi positives.

Les quantités $s_2 - e_1$, $s_2 - e_2$, $s_2 - e_3$ sont au contraire toutes les trois négatives.

L, M, N, P_1, P_2, P_3 sont donc imaginaires.

L_1, M_1, N_1 sont réels.

On voit que p et q ont des valeurs réelles.

Pour calculer la valeur de r je me sers de l'équation

$$2\frac{dp}{dt} = qr.$$

En général, si l'on pose

$$R(s) = R_0(s - a_0)(s - a_1)(s - a_2)(s - a_3)(s - a_4),$$

$$du_1 = \frac{s_1 ds_1}{\sqrt{R(s_1)}} + \frac{s_2 ds_2}{\sqrt{R(s_2)}}, \qquad ds_1 = \frac{\sqrt{R(s_1)}}{s_1 - s_2}(du_1 - s_2 du_2),$$

$$du_2 = \frac{ds_1}{\sqrt{R(s_1)}} + \frac{ds_2}{\sqrt{R(s_2)}}, \qquad ds_2 = \frac{\sqrt{R(s_2)}}{s_2 - s_1}(du_1 - s_1 du_2),$$

$$P_\alpha = \sqrt{c_\alpha(s_1 - a_\alpha)(s_2 - a_\alpha)}$$

$$P_{\alpha\beta} = \frac{c_{\alpha\beta} P_\alpha P_\beta}{s_1 - s_2}\left\{\frac{\sqrt{R(s_1)}}{(s_1 - a_\alpha)(s_1 - a_\beta)} - \frac{\sqrt{R(s_2)}}{(s_2 - a_\alpha)(s_2 - a_\beta)}\right\}, \qquad \begin{pmatrix}\alpha = 0, \ldots, 4\\ \beta = 0, \ldots, 4\end{pmatrix}$$

c_α et $c_{\alpha\beta}$ désignant des constantes quelconques, on trouve après quelques calculs,

$$\frac{\partial P_\alpha}{\partial u_1} = \frac{1}{2(a_\beta - a_\gamma)}\left\{\frac{P_\gamma P_{\alpha\gamma}}{c_\gamma c_{\alpha\gamma}} - \frac{P_\beta P_{\alpha\beta}}{c_\beta c_{\alpha\beta}}\right\},$$

$$\frac{\partial P_\alpha}{\partial u_2} = \frac{1}{2(a_\beta - a_\gamma)}\left\{\frac{a_\gamma P_\beta P_{\alpha\beta}}{c_\beta c_{\alpha\beta}} - \frac{a_\beta P_\gamma P_{\alpha\gamma}}{c_\gamma c_{\alpha\gamma}}\right\},$$

$$\frac{\partial P_{\alpha\beta}}{\partial u_1} = \frac{1}{2} R_0 c_{\alpha\beta} P_\alpha P_\beta,$$

$$\frac{\partial P_{\alpha\beta}}{\partial u_2} = -\frac{1}{2} R_0 a_\gamma c_{\alpha\beta} P_\alpha P_\beta - \frac{1}{2}\frac{c_{\alpha\beta} P_{\alpha\gamma} P_{\beta\gamma}}{c_\gamma c_{\alpha\gamma} c_{\beta\gamma}}.$$

Dans notre cas on a

$$R(s) = -4(s - e_1)(s - e_2)(s - e_3)(s - k_1)(s - k_2), \qquad R_0 = -4,$$

$$dt = du_1, \qquad 0 = du_2.$$

On trouve donc

$$r = \frac{2}{q}\frac{dp}{dt} = -i\frac{LP_{23} + MP_{31} + NP_{12}}{LP_1 + MP_2 + NP_3},$$

$$P_{23} = \frac{i}{s_1 - s_2}\{\sqrt{s_1 - e_1)(s_1 - k_1)(s_1 - k_2)(s_2 - e_2)(s_2 - e_3)}$$
$$- \sqrt{(s_1 - e_2)(s_1 - e_3)(s_2 - e_1)(s_2 - k_1)(s_2 - k_2}\},$$

$$P_{13} = \frac{i}{s_1 - s_2}\{\sqrt{(s_1 - e_2)(s_1 - k_1)(s_1 - k_2)(s_2 - e_1)(s_2 - e_3)}$$
$$- \sqrt{(s_1 - e_3)(s_1 - e_1)(s_2 - e_2)(s_2 - k_1)(s_2 - k_2)}\},$$

$$P_{12} = \frac{i}{s_1 - s_2}\{\sqrt{(s_1 - e_3)(s_1 - k_1)(s_1 - k_2)(s_2 - e_1)(s_2 - e_2)}$$
$$- \sqrt{(s_1 - e_1)(s_1 - e_2)(s_2 - e_3)(s_2 - k_1)(s_2 - k_2)}\}.$$

Comme nous l'avons vu plus haut, pour des valeurs réelles de t, s_1 et s_2 doivent être contenus entre les limites suivantes

$$\infty \ldots s_1 \ldots e_1, e_3 \ldots s_2 \ldots -\infty.$$

Les quantités

$$(s_1 - e_1)(s_2 - e_2)(s_2 - e_3)$$
$$(s_1 - e_2)(s_2 - e_3)(s_2 - e_1)$$
$$(s_1 - e_3)(s_2 - e_1)(s_2 - e_2)$$

sont donc positives, et les quantités

$$(s_1 - e_2)(s_1 - e_3)(s_2 - e_1)$$
$$(s_1 - e_3)(s_1 - e_1)(s_2 - e_2)$$
$$(s_1 - e_1)(s_1 - e_2)(s_2 - e_3)$$

négatives.

Pour que r soit réel pour des valeurs réelles de t, il faut que P_{23}, P_{13}, P_{12} soient réelles; pour que cela ait lieu, il faut que le produit

$$(s_1 - k_1)(s_1 - k_2)$$

soit négatif, et que le produit

$$(s_2 - k_1)(s_2 - k_2)$$

soit positif.

Les constantes d'intégration doivent donc être choisies de manière à satisfaire à l'inégalité

$$k_1 > e_1 > k_2.$$

s_1 et s_2 devront alors être contenus entre les limites

$$(k_1 \ldots s_1 \ldots e_1),\ (k_2 \ldots s_2 \ldots -\infty).$$

Il nous reste à calculer les valeurs de γ, γ', γ''.

On obtient γ'' de l'équation

$$2\frac{dq}{dt} = -rp - c_0\gamma''$$

ou

$$c_0\gamma'' = -\left(2\frac{dq}{dt} + rp\right),$$

et l'on trouve à l'aide des formules de différentiation précitées

$$c_0\gamma'' = \frac{L_1P_{23} + M_1P_{13} + N_1P_{12}}{LP_{23} + MP_{13} + NP_{12}},$$

les coefficients L, M, N, L_1, M_1, N_1 ayant les mêmes significations qu'auparavant.

La valeur de γ' peut être calculée à l'aide de l'équation

$$\frac{dr}{dt} = c_0\gamma'.$$

L'on trouve

$$c_0\gamma' = -i.\frac{(LP_1 + MP_2 + NP_3)\left(L\frac{d}{dt}P_{23} + M\frac{d}{dt}P_{13} + N\frac{d}{dt}P_{12}\right) - (LP_{23} + MP_{13} + NP_{12})\left(L\frac{d}{dt}P_1 + M\frac{d}{dt}P_2 + N\frac{d}{dt}P_3\right)}{(LP_1 + MP_2 + NP_3)^2}.$$

D'après les formules de différentiation précitées, l'on trouve

$$(LP_1 + MP_2 + NP_3)\Big(L\frac{d}{dt}P_{23} + M\frac{d}{dt}P_{13} + N\frac{d}{dt}P_{12}\Big)$$

$$-(LP_{23} + MP_{13} + NP_{12})\Big(L\frac{d}{dt}P_1 + M\frac{d}{dt}P_2 + N\frac{d}{dt}P_3\Big)$$

$$= R_0(LP_1 + MP_2 + NP_3)(LP_2P_3 + MP_3P_1 + NP_1P_2)$$

$$-(LP_{23}+MP_{31}+NP_{12})\Big(L\frac{P_3P_{13}-P_2P_{12}}{e_2-e_3} + M\frac{P_1P_{12}-P_3P_{23}}{e_3-e_1} + N\frac{P_2P_{23}-P_1P_{13}}{e_1-e_2}\Big)$$

$$= R_0(L^2 + M^2 + N^2)P_1P_2P_3 - \frac{L^2}{e_2 - e_3}P_{23}(P_3P_{13} - P_2P_{12})$$

$$- \frac{M^2}{e_3 - e_1}P_{13}(P_1P_{12} - P_3P_{23}) - \frac{N^2}{e_1 - e_2}P_{12}(P_2P_{23} - P_1P_{13})$$

$$+ MN\Big(R_0(P_2^2 + P_3^2)P_1 - P_{13}\frac{P_2P_{23} - P_1P_{13}}{e_1 - e_2} - P_{12}\frac{P_1P_{12} - P_3P_{23}}{e_3 - e_1}\Big)$$

$$+ NL\Big(R_0(P_3^2 + P_1^2)P_2 - P_{12}\frac{P_3P_{13} - P_2P_{12}}{e_2 - e_3} - P_{23}\frac{P_2P_{23} - P_1P_{13}}{e_1 - e_2}\Big)$$

$$+ LM\Big(R_0(P_1^2 + P_2^2)P_3 - P_{23}\frac{P_1P_{12} - P_3P_{23}}{e_3 - e_1} - P_{13}\frac{P_3P_{13} - P_2P_{12}}{e_2 - e_3}\Big).$$

Cette expression du nominateur de γ' peut être un peu simplifiée de la manière suivante.

En désignant par $\alpha, \beta, \gamma, \delta, \varepsilon$ les nombres 1, 2, 3, 4, 5 rangées dans un ordre quelconque et en posant, comme nous l'avons fait,

$$P_\alpha = \sqrt{(s_1 - a_\alpha)(s_2 - a_\alpha)},$$

$$P_{\alpha\beta} = \frac{\sqrt{R_0(s_1 - a_\gamma)(s_1 - a_\delta)(s_1 - a_\varepsilon)(s_2 - a_\alpha)(s_2 - a_\beta)} - \sqrt{R_0(s_1 - a_\alpha)(s_1 - a_\beta)(s_2 - a_\gamma)(s_2 - a_\delta)(s_2 - a_\varepsilon)}}{s_1 - s_2},$$

on trouve facilement les relations suivantes:

$$R_0 P_\alpha P_\beta P_\gamma - P_{\beta\gamma}\frac{P_\gamma P_{\alpha\gamma} - P_\beta P_{\alpha\beta}}{a_\beta - a_\gamma} = P_{\alpha\delta}P_{\alpha\varepsilon},$$

$$R_0 P_\alpha P_\beta P_\gamma - P_{\alpha\gamma}\frac{P_\alpha P_{\alpha\beta} - P_\gamma P_{\beta\gamma}}{a_\gamma - a_\alpha} = P_{\beta\delta}P_{\beta\varepsilon},$$

$$R_0 P_\alpha P_\beta P_\gamma - P_{\alpha\beta}\frac{P_\beta P_{\beta\gamma} - P_\alpha P_{\alpha\gamma}}{a_\alpha - a_\beta} = P_{\gamma\delta}P_{\gamma\varepsilon},$$

$$R_0(P_\beta^2 + P_\gamma^2)P_\alpha - P_{\alpha\gamma}\frac{P_\beta P_{\beta\gamma} - P_\alpha P_{\alpha\gamma}}{a_\alpha - a_\beta} - P_{\alpha\beta}\frac{P_\alpha P_{\alpha\beta} - P_\gamma P_{\beta\gamma}}{a_\gamma - a_\alpha}$$

$$= P_{\beta\delta}P_{\gamma\varepsilon} + P_{\beta\varepsilon}P_{\gamma\delta} + R_0(a_\beta - a_\gamma)^2 P_\alpha,$$

$$R_0(P_\gamma^2 + P_\alpha^2)P_\beta - P_{\alpha\beta}\frac{P_\gamma P_{\alpha\gamma} - P_\beta P_{\alpha\beta}}{a_\beta - a_\gamma} - P_{\beta\gamma}\frac{P_\beta P_{\beta\gamma} - P_\alpha P_{\alpha\gamma}}{a_\alpha - a_\beta}$$

$$= P_{\gamma\delta}P_{\alpha\varepsilon} + P_{\alpha\delta}P_{\gamma\varepsilon} + R_0(a_\gamma - a_\alpha)^2 P_\beta,$$

$$R_0(P_\alpha^2 + P_\beta^2)P_\gamma - P_{\beta\gamma}\frac{P_\alpha P_{\alpha\beta} - P_\gamma P_{\beta\gamma}}{a_\gamma - a_\alpha} - P_{\alpha\gamma}\frac{P_\gamma P_{\alpha\gamma} - P_\beta P_{\alpha\beta}}{a_\beta - a_\gamma}$$

$$= P_{\alpha\delta}P_{\beta\varepsilon} + P_{\alpha\varepsilon}P_{\beta\delta} + R_0(a_\alpha - a_\beta)^2 P_\gamma.$$

Si l'on pose donc

$$a_1 = e_1, \qquad k_1 = a_4,$$
$$a_2 = e_2, \qquad k_2 = a_5,$$
$$a_3 = e_3,$$

on trouve pour $c_0\gamma'$ l'expression suivante

$$c_0\gamma' = \frac{i}{2}\,\frac{L^2P_{14}P_{15} + M^2P_{24}P_{25} + N^2P_{34}P_{35} + MN[P_{24}P_{35} + P_{25}P_{34} + (e_2-e_3)^2P_1] + NL[P_{34}P_{15} + P_{35}P_{14} + (e_3-e_1)^2P_2] + LM[P_{14}P_{25} + P_{15}P_{24} + (e_1-e_2)^2P_3]}{(LP_1 + MP_2 + NP_3)^2}.$$

Enfin on trouve la valeur de γ, exprimée à l'aide de s_1 et de s_2, à l'aide de l'équation

$$2(p^2 + q^2) + r^2 = 2c_0\gamma + 6l_1.$$

En posant

$$L_2 = (e_2^2 - e_3^2)\frac{\sigma_1}{\sigma}(w) = i(e_2^2 - e_3^2)\sqrt{l_1 + e_1},$$

$$M_2 = (e_3^2 - e_1^2)\frac{\sigma_2}{\sigma}(w) = i(e_3^2 - e_1^2)\sqrt{l_1 + e_2},$$

$$N_2 = (e_1^2 - e_2^2)\frac{\sigma_3}{\sigma}(w) = i(e_1^2 - e_2^2)\sqrt{l_1 + e_3},$$

on trouve

$$2c_0\gamma = -4l_1 + \frac{L_2P_1 + M_2P_2 + N_2P_3}{LP_1 + MP_2 + NP_3} - \left(\frac{LP_{23} + MP_{13} + NP_{12}}{LP_1 + MP_2 + NP_3}\right)^2.$$

§ 6.

Les six quantités $p, q, r, \gamma, \gamma', \gamma''$ une fois exprimées en fonctions de s_1 et de s_2, on obtient sans peine l'expression de chaque fonction rationnelle symétrique de ces deux dernières quantités, en fonction du temps, à l'aide de formules générales.

J'emprunte les définitions suivantes, introduites dans l'analyse par M. WEIERSTRASS, au mémoire de M. KÖNIGSBERGER *Zur Transformation der Abelschen Functionen* (Journal für die reine und angewandte Mathematik, Bd. 64).

Soit

$$R(x) = A_0(x - a_0)(x - a_1)\ldots(x - a_{2\rho})$$

une fonction entière de x du degré $2\rho + 1$; supposons que $A_0, a_0, a_1, \ldots, a_{2\rho}$ soient toutes des quantités réelles et, si $A_0 > 0$,

$$a_0 > a_1 > \ldots > a_{2\rho},$$

mais si $A_0 < 0$

$$a_0 < a_1 < \ldots < a_{2\rho}.$$

Soient $u_1, \ldots, u_\rho$ ρ variables, liées aux ρ variables $x_1, \ldots, x_\rho$ par les

ρ équations suivantes, dans lesquelles $F_1(x), \ldots, F_\rho(x)$ désignent des fonctions entières de x d'un degré inférieur à ρ,

$$u_1 = \int_{a_1}^{x_1} \frac{F_1(x)dx}{\sqrt{R(x)}} + \ldots + \int_{a_{2\rho-1}}^{x_\rho} \frac{F_1(x)dx}{\sqrt{R(x)}},$$

$$\cdot \quad \cdot \quad \cdot \quad \cdot \quad \cdot \quad \cdot \quad \cdot \quad \cdot \quad \cdot \quad \cdot \quad \cdot \quad \cdot \quad \cdot \quad \cdot$$

$$u_\rho = \int_{a_1}^{x_1} \frac{F_\rho(x)dx}{\sqrt{R(x)}} + \ldots + \int_{a_{2\rho-1}}^{x_\rho} \frac{F_\rho(x)dx}{\sqrt{R(x)}}.$$

Posons

$$K_{\alpha\beta} = \int_{a_{2\beta-1}}^{a_{2\beta}} \frac{F_\alpha(x)dx}{\sqrt{R(x)}}, \qquad i\overline{K}_{\alpha\beta} = \int_{a_{2\beta-2}}^{a_{2\beta-1}} \frac{F_\alpha(x)dx}{\sqrt{R(x)}},$$

$$iK'_{\mu\nu} = i\overline{K}_{\mu1} + i\overline{K}_{\mu2} + \ldots + i\overline{K}_{\mu\nu},$$

en convenant de définir la racine carrée dans chacune de ces intégrales par la formule

$$\sqrt{R(x)} = A_0^{\rho+1}\left(\frac{x-a_0}{A_0}\right)^{\frac{1}{2}}\left(\frac{x-a_1}{A_0}\right)^{\frac{1}{2}}\cdots\left(\frac{x-a_{2\rho}}{A_0}\right)^{\frac{1}{2}}$$

Définissons ρ variables nouvelles $v_1, \ldots, v_\rho$ par les équations

$$u_1 = 2K_{11}v_1 + \ldots + 2K_{1\rho}v_\rho,$$

$$\cdot \quad \cdot \quad \cdot \quad \cdot \quad \cdot \quad \cdot \quad \cdot \quad \cdot \quad \cdot \quad \cdot \quad \cdot$$

$$u_\rho = 2K_{\rho1}v_1 + \ldots + 2K_{\rho\rho}v_\rho.$$

Supposons que ces équations, résolues par rapport à $v_1, \ldots, v_\rho$, nous donnent

$$v_1 = G_{11}u_1 + \ldots + G_{\rho1}u_\rho,$$

$$\cdot \quad \cdot \quad \cdot \quad \cdot \quad \cdot \quad \cdot \quad \cdot \quad \cdot \quad \cdot \quad \cdot$$

$$v_\rho = G_{1\rho}u_1 + \ldots + G_{\rho\rho}u_\rho.$$

Posons

$$\tau_{\alpha\beta} = 2i(G_{1\alpha}K_{1\beta} + \ldots + G_{\rho\alpha}K_{\rho\beta}),$$

d'où il suit $\tau_{\alpha\beta} = \tau_{\beta\alpha}$, et définissons la fonction $\vartheta(v_1 \ldots v_\rho)$ comme la somme de la série infinie

$$\vartheta(v_1 \ldots v_\rho) = \sum e^{\{\nu_1(2v_1+\nu_1\tau_{11}+\ldots+\nu_\rho\tau_{1\rho})+\ldots+\nu_\rho(2v_\rho+\nu_1\tau_{\rho 1}+\ldots+\nu_\rho\tau_{\rho\rho})\}\pi i};$$

la sommation, indiquée par le signe Σ, doit être effectuée de telle manière, que chacune des ρ quantités $\nu_1, \ldots, \nu_\rho$ parcourt, indépendemment des autres, toute la série des nombres entiers de $-\infty$ jusqu'à $+\infty$.

Cette fonction $\vartheta(v_1, \ldots, v_\rho)$ satisfait aux deux équations suivantes:

$$\vartheta(v_1 + p_1 \ldots v_\rho + p_\rho) = \vartheta(v_1 \ldots v_\rho),$$

$p_1 \ldots p_\rho$ désignant des nombres entiers quelconques, et

$$\vartheta(v_1 + \tau_{1\alpha} \ldots v_\rho + \tau_{\rho\alpha}) = \vartheta(v_1 \ldots v_\rho)e^{-(2v_\alpha+\tau_{\alpha\alpha})\pi i}.$$

Inversement, chaque fonction continue de $v_1 \ldots v_\rho$ satisfaisant à ces deux équations pour toutes les systèmes de valeurs $v_1 \ldots v_\rho$, doit nécessairement être égale à $\vartheta(v_1 \ldots v_\rho)$, multipliée par une constante.

Désignons par $n_1 \ldots n_\rho$ des constantes arbitraires, posons

$$\tau_\alpha = n_1\tau_{\alpha 1} + n_2\tau_{\alpha 2} + \ldots + n_\rho\tau_{\alpha\rho}$$

et définissons une nouvelle fonction $\vartheta(v_1 \ldots v_\rho \mid n_1 \ldots n_\rho)$ par l'équation

$$\vartheta(v_1 \ldots v_\rho \mid n_1 \ldots n_\rho) = \vartheta(v_1 + \tau_1 \ldots v_\rho + \tau_\rho)e^{\sum_\alpha n_\alpha(2v_\alpha+\tau_\alpha)\pi i}.$$

Les quantités $n_1 \ldots n_\rho$ s'appellent les *paramètres* de cette nouvelle fonction ϑ. Si $n'_1 \ldots n'_\rho$ désignent un autre système de ρ constantes, et que l'on pose

$$\tau'_\alpha = n'_1\tau'_{\alpha 1} + \ldots + n'_\rho\tau'_{\alpha\rho},$$

on a

$$\vartheta(v_1 + \tau'_1 \ldots v_\rho + \tau'_\rho \mid n_1 \ldots n_\rho)$$

$$= \vartheta(v_1 \ldots v_\rho \mid n_1 + n'_1 \ldots n_\rho + n'_\rho)e^{-\sum n'_\alpha(2v_\alpha+\tau'_\alpha)\pi i}$$

Si les quantités $n'_1 \dots n'_\rho$, $m_1 \dots m_\rho$ sont des nombres entiers, on a

$$\vartheta(v_1 \dots v_\rho \mid n_1 + n'_1 \dots n_\rho + n'_\rho) = \vartheta(v_1 \dots v_\rho \mid n_1 \dots n_\rho),$$

$$\vartheta(v_1 + m_1 \dots v_\rho + m_\rho \mid n_1 \dots n_\rho) = e^{\sum_a 2m_a n_a \pi i}\vartheta(v_1 \dots v_\rho \mid n_1 \dots n_\rho),$$

$$\vartheta(v_1 + \tau'_1 \dots v_\rho + \tau'_\rho \mid n_1 \dots n_\rho) = e^{-\sum n'_a(2v_a + \tau'_a)\pi i}\vartheta(v_1 \dots v_\rho \mid n_1 \dots n_\rho).$$

Posons de plus

$$\vartheta(v_1 \dots v_\rho)_\lambda = \vartheta\left(v_1 + \frac{1}{2}m_1^\lambda \dots v_\rho + \frac{1}{2}m_\rho^\lambda \,\middle|\, \frac{1}{2}n_1^\lambda \dots \frac{1}{2}n_\rho^\lambda\right) \quad (\lambda = 0, 1, 2, \dots, 2\rho)$$

où les nombres entiers $m_1^\lambda \dots m_\rho^\lambda$, $n_1^\lambda \dots n_\rho^\lambda$ sont définis par l'équation

$$\int_\infty^{a_\lambda} \frac{F_a(x)\,dx}{\sqrt{R(x)}} = m_1^\lambda K_{a1} + \dots + m_\rho^\lambda K_{a\rho} + i(n_1^\lambda K'_{a1} + n_2^\lambda K'_{a2} + \dots + n_\rho^\lambda K'_{a\rho}).$$

(On s'assure facilement que chacun des nombres entiers $m_1^\lambda \dots m_\rho^\lambda$ est égal à o ou à — 1; et chacun des nombres entiers $n_1^\lambda \dots n_\rho^\lambda$ à o ou a + 1).

En désignant ensuite par λ et μ deux nombres différents quelconques de la série $0, 1, \dots, 2\rho$, définissons les nombres entiers $m_1^\nu \dots m_\rho^\nu$, $n_1^\nu \dots n_\rho^\nu$ par les congruences

$$\left.\begin{aligned} m_a^\nu &\equiv m_a^\lambda + m_a^\mu \\ n_a^\nu &\equiv n_a^\lambda + n_a^\mu \end{aligned}\right\} \pmod 2$$

et de plus par la condition que chacun des nombres m_λ^ν doit être égal à o ou à — 1, et chacun des nombres n_λ^ν à o ou à + 1, et posons

$$\vartheta(v_1 \dots v_\rho)_{\lambda\mu} = \vartheta\left(v_1 + \frac{1}{2}m_1^\nu \dots v_\rho + \frac{1}{2}m_\rho^\nu \,\middle|\, \frac{1}{2}n_1^\nu \dots \frac{1}{2}n_\rho^\nu\right).$$

La relation entre les variables $v_1 \dots v_\rho$ et $x_1 \dots x_\rho$ peut alors être exprimée de la manière suivante.

Posons

$$\varphi(x) = (x - x_1) \dots (x - x_\rho)$$

et désignons par $\varepsilon \pm 1$, selon que $A_0 \gtrless 0$; on a alors

$$\frac{\sqrt{\varepsilon^\rho(-1)^\alpha \varphi(a_{2\alpha})}}{\sqrt[4]{R'(a_{2\alpha})}} = \frac{\vartheta(v_1 \ldots v_\rho)_{2\alpha}}{\vartheta(v_1 \ldots v_\rho)},$$

$$\frac{\sqrt{\varepsilon^\rho(-1)^{\alpha-1} \varphi(a_{2\alpha-1})}}{\sqrt[4]{R'(a_{2\alpha-1})}} = \frac{\vartheta(v_1 \ldots v_\rho)_{2\alpha-1}}{\vartheta(v_1 \ldots v_\rho)},$$

$$A_0\sqrt{\frac{\pm(a_\lambda - a_\mu)}{A_0}}\sum_1^\rho{}_\alpha \left\{\frac{\sqrt{R(x_\alpha)}}{(x_\alpha - a_\lambda)(x_\alpha - a_\mu)\varphi'(x_\alpha)}\right\} = \frac{\vartheta(v_1 \ldots v_\rho)\vartheta(v_1 \ldots v_\rho)_{\lambda\mu}}{\vartheta(v_1 \ldots v_\rho)_\lambda \vartheta(v_1 \ldots v_\rho)_\mu}.$$

Nous avons supposé que toutes les racines de l'équation $R(x) = 0$ sont réelles. M. Henoch a montré dans sa dissertation inaugurale (Berlin 1867) comment ces formules peuvent être généralisées, pour le cas où les racines de l'équation $R(x) = 0$ sont imaginaires.

§ 7.

Pour pouvoir appliquer les formules du paragraphe précédent au cas qui nous occupe, il faut ranger les cinq quantités réelles k_1, k_2, e_1, e_2, e_3 par ordre de grandeur.

Plusieurs cas peuvent se présenter ici. Le manque de temps m'empêche de les examiner tous en détail. Je me contenterai d'effectuer tous les calculs pour le cas où l'on a entre les constantes d'intégration les inégalités suivantes

$$l_1 > k > c_0 > 0\,,\; l^2 < \frac{3l_1 - k}{2}.$$

Dans ce cas, si l'on écrit

$$4s^3 - g_2 s - g_3 = 4(s + l_1)\left(s - \frac{l_1 + \sqrt{k^2 - c_0^2}}{2}\right)\left(s - \frac{l_1 - \sqrt{k^2 - c_0^2}}{2}\right) - l_0^2,$$

on s'assure facilement que les cinq quantités e_1, e_2, e_3, k_1, k_2 satisfont aux inégalités suivantes

$$\frac{l_1 + k}{2} > e_1 > e_2 > \frac{l_1 - k}{2} > e_3.$$

Si l'on désigne par a_0, a_1, a_2, a_3, a_4 les cinq racines de l'équation $R(s) = 0$, rangées par ordre de grandeur

$$a_0 < a_1 < a_2 < a_3 < a_4,$$

il faut donc poser

$$\frac{l_1 + k}{2} = a_4, \qquad \frac{l_1 - k}{2} = a_1, \qquad e_1 = a_3, \qquad e_2 = a_2, \qquad e_3 = a_0.$$

Si l'on pose alors

$$dt = du_1 = \frac{s_1 ds_1}{\sqrt{R(s_1)}} + \frac{s_2 ds_2}{\sqrt{R(s_2)}},$$

$$0 = du_2 = \frac{ds_1}{\sqrt{R(s_1)}} + \frac{ds_2}{\sqrt{R(s_2)}}$$

et que l'on détermine les quantités $K_{\alpha\beta}$, $K'_{\alpha\beta}$, $v_1 \dots v_\rho$ comme il a été montré au paragraphe précédent; v_1, v_2 seront des fonctions linéaires du temps, et l'on aura, en désignant par c_λ la valeur que prend $\vartheta(v_1, v_2)_\lambda$ pour $v_1 = 0$, $v_2 = 0$, les relations suivantes:

$$\frac{a_3 - a_2}{a_4 - a_1} = \frac{e_1 - e_2}{k} = \frac{c_2^2 c_{03}^2 c_{34}^2}{c_4^2 c_{01}^2 c_{12}^2}, \qquad \frac{l_1 + e_1}{k} = \frac{a_4 + a_1 - a_2 - a_0}{a_4 - a_1} = \frac{c_5^2 c_{23}^2}{c_4^2 c_{01}^2} - \frac{c_2^2 c_{14}^2}{c_4^2 c_{12}^2},$$

$$\frac{a_3 - a_0}{a_4 - a_1} = \frac{e_1 - e_3}{k} = \frac{c_0^2 c_{23}^2 c_{34}^2}{c_4^2 c_{01}^2 c_{12}^2}, \qquad \frac{l_1 + e_2}{k} = \frac{a_4 + a_1 - a_0 - a_3}{a_4 - a_1} = \frac{c_5^2 c_{23}^2}{c_4^2 c_{01}^2} - \frac{c_0^2 c_2^2}{c_{01}^2 c_{12}^2},$$

$$\frac{a_2 - a_0}{a_4 - a_1} = \frac{e_2 - e_3}{k} = \frac{c_5^2 c_{14}^2 c_{34}^2}{c_4^2 c_{01}^2 c_{12}^2}, \qquad \frac{l_1 + e_3}{k} = \frac{a_4 + a_1 - a_3 - a_2}{a_4 - a_2} = \frac{c_5^2 c_{03}^2}{c_4^2 c_{12}^2} - \frac{c_0^2 c_2^2}{c_{01}^2 c_{12}^2},$$

et, en posant

$$C = (a_3 - a_1)^{\frac{3}{2}} \frac{c_{01} c_{03} c_{12} c_{23} c_{14} c_{34}}{c_0^2 c_2^2 c_4^2},$$

on a

$$P_1 = \sqrt{(s_1 - e_1)(s_2 - e_1)} = \sqrt{(s_1 - a_3)(s_2 - a_3)} = -i \frac{C}{\sqrt{a_3 - a_1}} \frac{c_0 c_2 c_4}{c_{01} c_{12} c_{14}} \frac{\vartheta_3(v_1 v_2)}{\vartheta_5(v_1 v_2)},$$

$$P_2 = \sqrt{(s_1 - e_2)(s_2 - e_2)} = \sqrt{(s_1 - a_2)(s_2 - a_2)} = -i \frac{C}{\sqrt{a_3 - a_1}} \frac{c_5 c_2}{c_{12} c_{23}} \frac{\vartheta_2(v_1 v_2)}{\vartheta_5(v_1 v_2)},$$

$$P_3 = \sqrt{(s_1 - e_3)(s_2 - e_3)} = \sqrt{(s_1 - a_0)(s_2 - a_0)} = \frac{C}{\sqrt{a_3 - a_1}} \frac{c_5 c_0}{c_{01} c_{03}} \frac{\vartheta_0(v_1 v_2)}{\vartheta_5(v_1 v_2)},$$

$$P_4 = \sqrt{(s_1 - k_1)(s_2 - k_1)} = \sqrt{(s_1 - a_4)(s_2 - a_4)} = - \frac{C}{\sqrt{a_3 - a_1}} \frac{c_5 c_4}{c_{14} c_{34}} \frac{\vartheta_4(v_1 v_2)}{\vartheta_5(v_1 v_2)},$$

$$P_5 = \sqrt{(s_1 - k_2)(s_2 - k_2)} = \sqrt{(s_1 - a_1)(s_2 - a_1)} = \frac{C}{\sqrt{a_3 - a_1}} \frac{c_0 c_2 c_4}{c_{03} c_{23} c_{34}} \frac{\vartheta_1(v_1 v_2)}{\vartheta_5(v_1 v_2)},$$

$$P_{12} = \frac{\sqrt{(s_1-e_3)(s_1-k_1)(s_1-k_2)(s_2-e_1)(s_2-e_2)} - \sqrt{(s_1-e_1)(s_1-e_2)(s_2-e_3)(s_2-k_1)(s_2-k_2)}}{s_1-s_2}$$

$$= iC\frac{c_5}{c_{12}}\frac{\vartheta_{23}(v_1v_2)}{\vartheta_5(v_1v_2)},$$

$$P_{13} = \frac{\sqrt{(s_1-e_2)(s_1-k_1)(s_1-k_2)(s_2-e_1)(s_2-e_3)} - \sqrt{(s_1-e_1)(s_1-e_3)(s_2-e_2)(s_2-k_1)(s_2-k_2)}}{s_1-s_2}$$

$$= -C\frac{c_5}{c_{01}}\frac{\vartheta_{03}(v_1v_2)}{\vartheta_5(v_1v_2)},$$

$$P_{14} = \frac{\sqrt{(s_1-e_2)(s_1-e_3)(s_1-k_2)(s_2-e_1)(s_2-k_1)} - \sqrt{(s_1-e_1)(s_1-k_1)(s_2-e_2)(s_2-e_3)(s_2-k_2)}}{s_1-s_2}$$

$$= -C\frac{c_5}{c_{14}}\frac{\vartheta_{34}(v_1v_2)}{\vartheta_5(v_1v_2)},.$$

$$P_{15} = \frac{\sqrt{(s_1-e_2)(s_1-e_3)(s_1-k_1)(s_2-e_1)(s_2-k_2)} - \sqrt{(s_1-e_1)(s_1-k_2)(s_2-e_2)(s_2-e_3)(s_2-k_1)}}{s_1-s_2}$$

$$= -C\frac{\vartheta_{13}(v_1v_2)}{\vartheta_5(v_1v_2)},$$

$$P_{23} = \frac{\sqrt{(s_1-e_1)(s_1-k_1)(s_1-k_2)(s_2-e_2)(s_2-e_3)} - \sqrt{(s_1-e_2)(s_1-e_3)(s_2-e_1)(s_2-k_1)(s_2-k_2)}}{s_1-s_2}$$

$$= C\frac{c_5}{c_4}\frac{\vartheta_{02}(v_1v_2)}{\vartheta_5(v_1v_2)},$$

$$P_{24} = \frac{\sqrt{(s_1-e_1)(s_1-e_3)(s_1-k_2)(s_2-e_2)(s_2-k_1)} - \sqrt{(s_1-e_2)(s_1-k_1)(s_2-e_1)(s_2-e_3)(s_2-k_2)}}{s_2-s_2}$$

$$= -C\frac{c_5}{c_0}\frac{\vartheta_{24}(v_1v_2)}{\vartheta_5(v_1v_2)},$$

$$P_{25} = \frac{\sqrt{(s_1-e_1)(s_1-e_3)(s_1-k_1)(s_2-e_2)(s_2-k_2)} - \sqrt{(s_1-e_2)(s_1-k_2)(s_2-e_1)(s_2-e_3)(s_2-k_1)}}{s_1-s_2}$$

$$= C\frac{c_5}{c_{23}}\frac{\vartheta_{12}(v_1v_2)}{\vartheta_5(v_1v_2)},$$

$$P_{34} = \frac{\sqrt{(s_1-e_1)(s_1-e_2)(s_1-k_2)(s_2-e_3)(s_2-k_1)} - \sqrt{(s_1-e_3)(s_1-k_1)(s_2-e_1)(s_2-e_2)(s_2-k_2)}}{s_1-s_2}$$

$$= -iC\frac{c_5}{c_2}\frac{\vartheta_{04}(v_1v_2)}{\vartheta_5(v_1v_2)},$$

$$P_{35} = \frac{\sqrt{(s_1-e_1)(s_1-e_2)(s_1-k_1)(s_2-e_3)(s_2-k_2)} - \sqrt{(s_1-e_3)(s_1-k_2)(s_2-e_1)(s_2-e_1)(s_2-k_1)}}{s_1-s_2}$$

$$= -iC\frac{c_5}{c_{03}}\frac{\vartheta_{01}(v_1v_2)}{\vartheta_5(v_1v_2)},$$

$$P_{45} = \frac{\sqrt{(s_1-e_1)(s_1-e_2)(s_1-e_3)(s_2-k_1)(s_2-k_2)} - \sqrt{(s_1-k_1)(s_1-k_2)(s_2-e_1)(s_2-e_2)(s_2-e_3)}}{s_1-s_2}$$

$$= -iC\frac{c_5}{c_{34}}\frac{\vartheta_{14}(v_1v_2)}{\vartheta_5(v_1v_2)}.$$

Les six quantités $p, q, r, \gamma, \gamma', \gamma''$ une fois exprimées à l'aide des quotients $\frac{\vartheta_a(v_1, v_2)}{\vartheta(v_1, v_2)}$, dans lesquelles v_1, v_2 désignent des fonctions entières, linéaires du temps t, il s'agit de trouver les expressions des six autres cosinus $\alpha, \alpha', \alpha'', \beta, \beta', \beta''$, en fonction du temps.

Ces quantités satisfont aux équations différentielles

$$\frac{d\alpha}{dt} = \alpha' r - \alpha'' q, \qquad \frac{d\beta}{dt} = \beta' r - \beta'' q,$$

$$\frac{d\alpha'}{dt} = \alpha'' p - \alpha r, \qquad \frac{d\beta'}{dt} = \beta'' p - \beta r,$$

$$\frac{d\alpha''}{dt} = \alpha q - \alpha' p, \qquad \frac{d\beta''}{dt} = \beta q - \beta' p.$$

De ces équations il suit

$$\frac{d(\alpha + \beta i)}{dt} = (\alpha' + \beta' i)r - (\alpha'' + \beta'' i)q.$$

En divisant par $\alpha + \beta i$ et en remarquant que

$$(\alpha' + \beta' i)(\alpha - \beta i) = \alpha\alpha' + \beta\beta' + i(\alpha\beta' - \alpha'\beta) = -\gamma\gamma' + i\gamma'',$$
$$(\alpha'' + \beta'' i)(\alpha - \beta i) = \alpha\alpha'' + \beta\beta'' + i(\alpha\beta'' - \alpha''\beta) = -\gamma\gamma'' - i\gamma',$$
$$(\alpha + \beta i)(\alpha - \beta i) = \alpha^2 + \beta^2 = 1 - \gamma^2,$$

on trouve

$$\frac{d}{dt}\lg(\alpha + \beta i) = r\frac{-\gamma\gamma' + i\gamma''}{1-\gamma^2} + q\frac{\gamma\gamma'' + i\gamma'}{1-\gamma^2} = \frac{-\gamma\frac{d\gamma}{dt} + i(\gamma' q + \gamma'' r)}{1-\gamma^2}$$

$$= \frac{1}{2}\left\{\frac{\frac{d\gamma}{dt} + i(\gamma' q + \gamma'' r)}{1+\gamma} - \frac{\frac{d\gamma}{dt} - i(\gamma' q + \gamma'' r)}{1-\gamma}\right\}.$$

De la même manière on trouve

$$\frac{d}{dt}\lg(\alpha' + \beta' i) = \frac{1}{2}\left\{\frac{\frac{d\gamma'}{dt} + i(\gamma p + \gamma'' r)}{1 + \gamma'} - \frac{\frac{d\gamma'}{dt} - i(\gamma p + \gamma'' r)}{1 - \gamma'}\right\},$$

$$\frac{d}{dt}\lg(\alpha'' + \beta'' i) = \frac{1}{2}\left\{\frac{\frac{d\gamma''}{dt} + i(\gamma p + \gamma' q)}{1 + \gamma''} - \frac{\frac{d\gamma''}{dt} - i(\gamma p + \gamma' q)}{1 - \gamma''}\right\}.$$

En portant dans les seconds membres de ces équations, les valeurs de $p, q, r, \gamma, \gamma', \gamma''$, exprimées en fonctions du temps, on voit que pour obtenir les valeurs de $\alpha, \beta, \alpha', \beta', \alpha'', \beta''$ il faut intégrer des fonctions rationnelles de quotients $\frac{\vartheta_a(t + c, c_1)}{\vartheta(t + c, c_1)}$.

On peut démontrer que chacune des six quantités $\alpha, \alpha', \alpha'', \beta, \beta', \beta''$ est une fonction uniforme du temps, n'ayant que des pôles pour des valeurs finies du temps.

D'après un théorème bien connu $f(t)$ est une fonction uniforme pouvant être représentée sous forme de quotient de deux séries convergentes pour toutes les valeurs finies de t, si la condition suivante est remplie:

$$\frac{d}{dt}\lg f(t)$$

peut être développée dans l'entourage de chaque valeur finie t_0 en série convergente de la forme

$$\frac{d}{dt}\lg f(t) = m(t - t_0)^{-1} + \mathfrak{P}(t - t_0)$$

où $\mathfrak{P}(t - t_0)$ désigne une série infinie ne contenant que des termes à exposant positif, et m est égal à zéro ou à un nombre entier positif ou négatif.

Or, tel est justement le cas pour les seconds membres des équations qui définissent

$$\frac{d}{dt}\lg(\alpha + \beta i), \ \frac{d}{dt}\lg(\alpha' + \beta' i), \ \frac{d}{dt}\lg(\alpha'' + \beta'' i)$$

en fonctions du temps.

Si l'on écrit dans ces seconds membres pour $\gamma, \gamma', \gamma'', p, q, r$ leurs valeurs en fonctions du temps et si, en supposant t dans le voisinage de t_0, on les développe en séries procédant selon les puissances de $t - t_0$, on voit immédiatement que des termes à exposant négatif ne pourront entrer dans ces développements que dans les deux cas suivants:

1) Si les développements de $p, q, r, \gamma, \gamma', \gamma''$ contiennent des puissances négatives de $(t - t_0)$.

2) Si pour $t = t_0$ l'une des trois quantités $\gamma, \gamma', \gamma''$ est égale à ± 1. En ayant recours aux équations différentielles (1) et en écrivant τ au lieu de $t - t_0$, on voit que dans le 1er cas les développements de $p, q, r, \gamma, \gamma', \gamma''$ doivent avoir la forme suivante:

$$p = p_0\tau^{-1} + p_1 + p_2\tau + \ldots, \qquad \gamma = f_0\tau^{-2} + f_1\tau^{-1} + f_2 + \ldots,$$
$$q = q_0\tau^{-1} + q_1 + q_2\tau + \ldots, \qquad \gamma' = g_0\tau^{-2} + g_1\tau^{-1} + g_2 + \ldots,$$
$$r = r_0\tau^{-1} + r_1 + r_2\tau + \ldots, \qquad \gamma'' = h_0\tau^{-2} + h_1\tau^{-1} + h_2 + \ldots,$$

où les coefficients $p_0, q_0, r_0, f_0, g_0, h_0$ peuvent avoir ces deux systèmes différents de valeurs (voir § 1):

I. $$r_0 = 0, \qquad h_0 = \pm i\frac{4}{c_0},$$
$$q_0 = \pm i2, \qquad g_0 = 0,$$
$$p_0 = 0, \qquad f_0 = -\frac{4}{c_0},$$

II. $$q_0 = \pm ip_0, \qquad g_0 = \pm if_0 = -\frac{r_0}{c_0} = \mp\frac{2i}{c_0},$$
$$r_0 = \pm 2i, \qquad h_0 = 0.$$

En portant ces développements dans le second membre de l'équation

$$\frac{d\lg(\alpha'' + \beta''i)}{dt} = \frac{\gamma''\frac{d\gamma''}{dt} - i(\gamma p + \gamma' q)}{\gamma''^2 - 1}$$
$$= \frac{1}{2}\left\{\frac{\frac{d\gamma''}{dt} - i(\gamma p + \gamma' q)}{\gamma'' - 1} + \frac{\frac{d\gamma''}{dt} + i(\gamma p + \gamma' q)}{\gamma'' + 1}\right\},$$

on voit que l'on a dans le cas I

$$\frac{d}{dt}\lg(\alpha'' + \beta''i) = -2\tau^{-1} + \mathfrak{P}(\tau)$$

et dans le cas II

$$\frac{d}{dt}\lg(\alpha'' + \beta''i) = -\tau^{-1} + \mathfrak{P}(\tau).$$

Examinons maintenant le développement de $\frac{d}{dt}\lg(\alpha'' + \beta''i)$ dans le voisinage d'une valeur $t = t_0$ pour laquelle $\gamma'' = \pm 1$.

En général, si l'on pose

$$p = p_0 + p_1 t + \dots, \qquad \gamma = f_0 + f_1 t + \dots,$$
$$q = q_0 + q_1 t + \dots, \qquad \gamma' = g_0 + g_1 t + \dots,$$
$$r = r_0 + r_1 t + \dots, \qquad \gamma'' = h_0 + h_1 t + \dots,$$

les coefficients $p_1, q_1, r_1, f_1, g_1, h_1$ sont définis (en vertu des équations différentielles (1), § 2) en fonctions de $p_0, q_0, r_0, f_0, g_0, h_0$ par les équations suivantes

$$2p_1 = q_0 r_0, \qquad f_1 = r_0 g_0 - q_0 h_0,$$
$$2q_1 = -p_0 r_0 - c_0 h_0, \qquad g_1 = p_0 h_0 - r_0 f_0,$$
$$r_1 = c_0 g_0, \qquad h_1 = q_0 f_0 - p_0 g_0.$$

Les trois quantités $\gamma, \gamma', \gamma''$ sont liées par l'équation

$$\gamma^2 + \gamma'^2 + \gamma''^2 = 1.$$

Leurs valeurs initiales f_0, g_0, h_0 sont donc aussi assujetties à la condition

$$f_0^2 + g_0^2 + h_0^2 = 1.$$

Désignons par ε_1 et par ε_2 deux quantités qui satisfont à l'équation

$$\varepsilon_1^2 = 1, \qquad \varepsilon_2^2 = 1.$$

Si je pose $h_0 = \varepsilon_1$ il faut donc que l'on ait en même temps $g_0 = \varepsilon_2 i f_0$ et l'on trouve alors

$$h_1 = -\varepsilon_2 i f_0 (p_0 + \varepsilon_2 i q_0),$$
$$p_0 f_0 + q_0 g_0 = f_0(p_0 + \varepsilon_2 i q_0) = \varepsilon_2 i h_1.$$

Le développement du terme $\frac{1}{2}\frac{\frac{d\gamma''}{dt} - i\varepsilon_1(p\gamma + q\gamma')}{\gamma'' - \varepsilon_1}$ qui seul peut contenir des puissances négatives de τ, a donc la forme

$$\frac{1 + \varepsilon_1\varepsilon_2}{2}\tau^{-1} + \mathfrak{P}(\tau).$$

Si ε_1 et ε_2 ont même signe le coefficient de τ^{-1} est égal à $+ 1$. Si ε_1 et ε_2 ont des signes contraires ce coefficient est nul.

On voit donc que dans tous les cas, le développement de

$$\frac{d}{dt}\lg(\alpha'' + \beta''i)$$

a la forme voulue

$$\frac{d}{dt}\lg(\alpha'' + \beta''i) = m(t - t_0)^{-1} + \mathfrak{P}(t - t_0),$$

d'où il résulte que $\alpha'' + \beta''i$ est une fonction uniforme du temps, pouvant être mise sous la forme d'un quotient de deux séries toujours convergentes.

On arrive au même résultat concernant $\alpha + \beta i$ et $\alpha' + \beta' i$.

Je me suis aussi assurée que ces quantités peuvent être exprimées en fonctions rationnelles des quantités de la forme

$$\frac{\vartheta_\alpha(u_1 + v_1, u_2 + v_2)}{\vartheta(u_1 u_2)} e^{u_3},$$

α étant l'index d'une de 16 fonctions $\vartheta(u_1 u_2)$, u_1, u_2, u_3 désignant des fonctions linéaires, entières du temps, et v_1, v_2 désignant des constantes imaginaires.

Mais à cause de la grande complication des calculs, je ne suis pas encore arrivée à développer ces formules sous leur forme finale.

§ 8.

Il peut être intéressant de réaliser sur un modèle un cas de rotation d'un corps solide autour d'un point fixe où toutes les conditions du cas que nous venons d'étudier se trouvent remplies.

Dans les équations différentielles (1) les constantes A, B, C désignent les trois moments d'inertie principaux relativement au point fixe.

Soient maintenant A_1, B_1, C_1 les trois moments d'inertie principaux du même corps relativement à son centre de gravité et désignons par x, y, z les coordonnées d'un point dans un système d'axes de coordonnées dont l'origine se trouve au centre de gravité et dont les directions coïncident avec les trois axes d'inertie principaux passant par ce point.

On a alors, en désignant par μ la densité du corps considéré au point donc les coordonnées sont x, y, z,

$$A_1 = \iiint \mu(y^2 + z^2)\,dx\,dy\,dz, \qquad B_1 = \iiint \mu(z^2 + x^2)\,dx\,dy\,dz,$$
$$C_1 = \iiint \mu(x^2 + y^2)\,dx\,dy\,dz,$$
$$0 = \iiint \mu x\,dx\,dy\,dz, \qquad 0 = \iiint \mu y\,dx\,dy\,dz, \qquad 0 = \iiint \mu z\,dx\,dy\,dz,$$
$$0 = \iiint \mu yz\,dx\,dy\,dz, \qquad 0 = \iiint \mu zx\,dx\,dy\,dz, \qquad 0 = \iiint \mu xy\,dx\,dy\,dz.$$

Toutes ces intégrales triples doivent être étendues à tout l'intérieur du corps considéré.

Soient a, b, c les coordonnées d'un point O. Désignant par ξ, η, ζ les coordonnées d'un point de l'espace relativement à des axes parallèles aux précédents, mais dont l'origine coïncide avec le point O, on a

$$\xi = x - a, \qquad \eta = y - b, \qquad \zeta = z - c.$$

L'équation de l'ellipsoïde d'inertie dont le centre est en O est, comme on sait,

$$1 = A_1'\xi^2 + B_1'\eta^2 + C_1'\zeta^2 - 2D_1'\eta\zeta - 2E_1'\zeta\xi - 2F_1\xi\eta = \varphi(\xi, \eta, \zeta),$$

les coëfficients A_1', B_1' .... étant définis par les équations

$$A_1' = \iiint \mu(\eta^2 + \zeta^2)\,d\xi\,d\eta\,d\zeta = A_1 + M(b^2 + c^2),$$
$$B_1' = \iiint \mu(\zeta^2 + \xi^2)\,d\xi\,d\eta\,d\zeta = B_1 + M(c^2 + a^2),$$
$$C_1' = \iiint \mu(\xi^2 + \eta^2)\,d\xi\,d\eta\,d\zeta = C_1 + M(a^2 + b^2),$$
$$D_1' = \iiint \mu\eta\zeta\,d\xi\,d\eta\,d\zeta = Mbc,$$
$$E_1' = \iiint \mu\zeta\xi\,d\xi\,d\eta\,d\zeta = Mca,$$
$$F_1' = \iiint \mu\xi\eta\,d\xi\,d\eta\,d\zeta = Mab.$$

M désigne la masse totale du corps,

$$M = \iiint \mu d\xi d\eta d\zeta.$$

Soient maintenant u, v, w les coordonnées d'un point de l'espace relativement à des axes, dont l'origine est au point O, mais dont les directions coïncident avec les directions des axes principaux de l'ellipsoide d'inertie, correspondant à ce point. On a alors

$$u = \alpha\xi + \beta\eta + \gamma\zeta,$$
$$v = \alpha_1\xi + \beta_1\eta + \gamma_1\zeta,$$
$$w = \alpha_2\xi + \beta_2\eta + \gamma_2\zeta,$$
$$\xi^2 + \eta^2 + \zeta^2 = u^2 + v^2 + w^2,$$
$$\varphi(\xi, \eta, \zeta) = \lambda u^2 + \mu v^2 + \nu w^2,$$

λ, μ, ν étant des constantes positives.

De plus en désignant par u_0, v_0, w_0 les coordonnées du centre de gravité dans ce nouveau système, on a

$$-u_0 = \alpha a + \beta b + \gamma c,$$
$$-v_0 = \alpha_1 a + \beta_1 b + \gamma_1 c,$$
$$-w_0 = \alpha_2 a + \beta_2 b + \gamma_2 c.$$

Si nous imprimons maintenant au corps considéré un mouvement de rotation autour du point fixe O, les conditions du cas étudié par nous seront remplies, si l'on a

$$\lambda = \mu = 2\nu,$$
$$-w_0 = \alpha_2 a + \beta_2 b + \gamma_2 c = 0.$$

On doit donc pouvoir satisfaire aux équations suivantes

$$\varphi(\xi, \eta, \zeta) = \nu[2(u^2 + v^2) + w^2], \qquad a\alpha_2 + b\beta_2 + c\gamma_2 = 0,$$
$$\xi^2 + \eta^2 + \zeta^2 = u^2 + v^2 + w^2, \qquad \alpha_2^2 + \beta_2^2 + \gamma_2^2 = 1.$$

Voyons si l'on peut choisir le point O de manière à ce que ces équations soient remplies.

Des deux premières de ces équations, il résulte

$$\varphi(\xi, \eta, \zeta) - 2\nu(\xi^2 + \eta^2 + \zeta^2) = -\nu w^2.$$

Cette dernière équation devant être satisfaite identiquement, si l'on écrit pour w sa valeur

$$w = \alpha_2\xi + \beta_2\eta + \gamma_2\zeta,$$

on obtient, en égalant à zéro les coefficients de chaque terme dans cette identité, les équations suivantes:

$$A_1' - 2\nu = A_1 + M(b^2 + c^2) - 2\nu = -\nu\alpha_2^2,$$
$$B_1' - 2\nu = B_1 + M(c^2 + a^2) - 2\nu = -\nu\beta_2^2,$$
$$C_1' - 2\nu = C_1 + M(a^2 + b^2) - 2\nu = -\nu\gamma_2^2,$$

$$D_1' = Mbc = \nu\beta_2\gamma_2,$$
$$E_1' = Mca = \nu\gamma_2\alpha_2,$$
$$F_1' = Mab = \nu\alpha_2\beta_2.$$

Si aucune des trois constantes a, b, c (les coordonnées du point O) n'était nulle il résulterait des trois dernières équations

$$M^3a^2b^2c^2 = \nu^3\alpha_2^2\beta_2^2\gamma_2^2,$$

$$M^{\frac{1}{2}}a = \nu^{\frac{1}{2}}\alpha_2,$$

$$M^{\frac{1}{2}}b = \nu^{\frac{1}{2}}\beta_2,$$

$$M^{\frac{1}{2}}c = \nu^{\frac{1}{2}}\gamma_2,$$

ce qui est évidemment impossible par suite des équations

$$\alpha_2^2 + \beta_2^2 + \gamma_2^2 = 1,$$

$$a\alpha_2 + b\beta_2 + c\gamma_2 = 0,$$

ni M ni ν ne pouvant être $= 0$.

Si nous supposons $c = 0$, mais a et b différents de 0, nous devon poser $\gamma_2 = 0$, et par conséquent

$$Mab = \nu\alpha_2\beta_2, \qquad \alpha_2^2 + \beta_2^2 = 1, \qquad a\alpha_2 + b\beta_2 = 0,$$

d'où il suit

$$\alpha_2 = \frac{b}{\sqrt{a^2 + b^2}}, \qquad \beta_2 = \frac{-a}{\sqrt{a^2 + b^2}},$$

(le signe de $\sqrt{a^2 + b^2}$ étant déterminé arbitrairement)

$$M = \frac{-\nu}{a^2 + b^2},$$

équation impossible, vu que M et ν sont tous les deux des quantités positives.

Il faut donc supposer $b = 0$, $c = 0$. On a alors

$$A_1' = A_1, \qquad B_1' = B_1 + Ma^2, \qquad C_1' = C_1 + Ma^2,$$
$$D_1' = 0, \qquad E_1' = 0, \qquad F_1' = 0,$$
$$\varphi(\xi, \eta, \zeta) = A_1\xi^2 + (B_1 + Ma^2)\eta^2 + (C_1 + Ma^2)\zeta^2.$$

Si les trois axes d'inertie principaux A_1, B_1, C_1 relatifs au centre de gravité du corps considéré satisfont à l'équation

$$A_1 = 2(B_1 - C_1),$$

on pourra satisfaire à toutes les conditions supposées par nous, en prenant

$$a^2 = \frac{A_1 - B_1}{M},$$

car on a dans ce cas

$$B_1 + Ma^2 = A_1, \qquad C_1 + Ma^2 = C_1 + A_1 - B_1 = \frac{1}{2}A_1,$$

par conséquent

$$\varphi(\xi, \eta, \zeta) = A_1\left(\xi^2 + \eta^2 + \frac{1}{2}\zeta^2\right).$$

Remarquons seulement que pour que a soit réel, il faut et il suffit que l'on ait $B_1 > 2C_1$.

On voit donc, d'après ce calcul, qu'il est possible de réaliser mécaniquement toutes les conditions du problème que je viens d'étudier.